Current Topics in Membranes, Volume 60

Computational Modeling of Membrane Bilayers

Current Topics in Membranes, Volume 60

Series Editors

Dale J. Benos
Department of Physiology and Biophysics
University of Alabama
Birmingham, Alabama

Sidney A. Simon
Department of Neurobiology
Duke University Medical Care
Durham, North Carolina

Current Topics in Membranes, Volume 60

Computational Modeling of Membrane Bilayers

Edited by

Scott E. Feller
Department of Chemistry
Wabash College
Crawfordsville, Indiana

AMSTERDAM · BOSTON · HEIDELBERG · LONDON
NEW YORK · OXFORD · PARIS · SAN DIEGO
SAN FRANCISCO · SINGAPORE · SYDNEY · TOKYO

ELSEVIER   Academic Press is an imprint of Elsevier

Academic Press is an imprint of Elsevier
84 Theobald's Road, London WC1X 8RR, UK
Radarweg 29, PO Box 211, 1000 AE Amsterdam, The Netherlands
Linacre House, Jordan Hill, Oxford OX2 8DP, UK
30 Corporate Drive, Suite 400, Burlington, MA 01803, USA
525 B Street, Suite 1900, San Diego, CA 92101-4495, USA

First edition 2008

ISBN: 978-0-12-373893-6
ISSN: 1063-5823

For information on all Academic Press publications
visit our website at books.elsevier.com

Transferred to Digital Printing, 2011
Printed and bound in the United Kingdom

Working together to grow
libraries in developing countries

www.elsevier.com | www.bookaid.org | www.sabre.org

ELSEVIER   BOOK AID
             International   Sabre Foundation

Contents

CHAPTER   4   Molecular Dynamics Simulation of Lipid–Protein Interactions
Nicolas Sapay and D. Peter Tieleman

CHAPTER   5   Implicit Modeling of Membranes
Alan Grossfield

CHAPTER   6   Blue Matter: Scaling of N-Body Simulations to One Atom per Node
Blake G. Fitch, Aleksandr Rayshubskiy, Maria Eleftheriou, T.J. Christopher Ward, Mark Giampapa, Michael C. Pitman, Jed Pitera, William C. Swope and Robert S. Germain

Contributors

Numbers in parentheses indicate the pages on which the author's contribution begin.

Toby W. Allen (405), Department of Chemistry, University of California, One Shields Avenue, Davis, CA 95616, USA

Anton Arkhipov (313), Beckman Institute for Advanced Science and Technology, University of Illinois at Urbana-Champaign, Urbana, IL 61801, USA; Department of Physics, University of Illinois at Urbana-Champaign, Urbana, IL 61801, USA

Gary S. Ayton (181), Center for Biological Modeling and Simulation, and Department of Chemistry, University of Utah, 315 S. 1400 E. Rm 2020, Salt Lake City, UT 84112-0850, USA

Max Berkowitz (257), Department of Chemistry, University of North Carolina at Chapel Hill, Chapel Hill, NC 27599, USA

See-Wing Chiu (281), Department of Molecular and Integrative Physiology, Department of Biochemistry, UIUC programs in Biophysics, Neuroscience, and Bioengineering, National Center for Supercomputing Applications, and Beckman Institute, University of Illinois, Urbana, IL 61801, USA

T.J. Christopher Ward (159), IBM Research Division, IBM Thomas J. Watson Research Center, 1101 Kitchawan, Rd Route 134, Yorktown Heights, NY 10598, USA

Elizabeth J. Denning (385), Johns Hopkins University, School of Medicine, Departments of Physiology and of Biophysics and Biophysical Chemistry, Baltimore, Maryland, USA

Sudha Dorairaj (405), Department of Chemistry, University of California, One Shields Avenue, Davis, CA 95616, USA

Olle Edholm (91), Department of Theoretical Physics, Royal Institute of Technology, Alba Nova University Center, SE-106 91 Stockholm, Sweden

Maria Eleftheriou (159), IBM Research Division, IBM Thomas J. Watson Research Center, 1101 Kitchawan, Rd Route 134, Yorktown Heights, NY 10598, USA

Blake G. Fitch (159), IBM Research Division, IBM Thomas J. Watson Research Center, 1101 Kitchawan, Rd Route 134, Yorktown Heights, NY 10598, USA

Peter L. Freddolino (313), Center for Biophysics and Computational Biology, University of Illinois at Urbana-Champaign, Urbana, IL 61801, USA; Beckman Institute for Advanced Science and Technology, University of Illinois at Urbana-Champaign, Urbana, IL 61801, USA

Robert S. Germain (159), IBM Research Division, IBM Thomas J. Watson Research Center, 1101 Kitchawan, Rd Route 134, Yorktown Heights, NY 10598, USA

Mark Giampapa (159), IBM Research Division, IBM Thomas J. Watson Research Center, 1101 Kitchawan, Rd Route 134, Yorktown Heights, NY 10598, USA

Alan Grossfield (131), IBM T. J. Watson Research Center, 1101 Kitchawan Road, Yorktown Heights, NY 10598, USA; Present address: University of Rochester Medical Center, Department of Biochemistry and Biophysics, 601 Elmwood Ave, Box 712, Rochester, NY 14642, USA

Sergei Izvekov (181), Center for Biological Modeling and Simulation, and Department of Chemistry, University of Utah, 315 S. 1400 E. Rm 2020, Salt Lake City, UT 84112-0850, USA

Eric Jakobsson (281), Department of Molecular and Integrative Physiology, Department of Biochemistry, UIUC programs in Biophysics, Neuroscience, and Bioengineering, National Center for Supercomputing Applications, and Beckman Institute, University of Illinois, Urbana, IL 61801, USA

Mikko Karttunen (49), Department of Applied Mathematics, The University of Western Ontario, London, Ontario, Canada

Jeffery B. Klauda (1), Laboratory of Computational Biology, National Heart, Lung, and Blood Institute, National Institutes of Health, Bethesda, MD 20892, USA; Current address: Department of Chemical and Biomolecular Engineering, University of Maryland, College Park, MD 20742, USA

Libo Li (405), Department of Chemistry, University of California, One Shields Avenue, Davis, CA 95616, USA

Justin L. MacCallum (227), Dept. of Biological Sciences, University of Calgary, 2500 University Dr NW, Calgary AB, T2N 1N4, Canada

Alexander D. MacKerell Jr. (1), Department of Pharmaceutical Sciences, School of Pharmacy, University of Maryland, Baltimore, MD, 21201, USA

W.G. Noid (181), Center for Biological Modeling and Simulation, and Department of Chemistry, University of Utah, 315 S. 1400 E. Rm 2020, Salt Lake City, UT 84112-0850, USA

Y. Zenmei Ohkubo (343), Center for Biophysics and Computational Biology, Department of Biochemistry, and Beckman Institute University of Illinois at Urbana-Champaign, Urbana, IL 61801, USA

Sagar A. Pandit (281), Department of Physics, University of South Florida, Tampa, FL 47907, USA

Richard W. Pastor (1), Laboratory of Computational Biology, National Heart, Lung, and Blood Institute, National Institutes of Health, Bethesda, MD 20892, USA

Jed Pitera (159), IBM Research Division, IBM Thomas J. Watson Research Center, 1101 Kitchawan, Rd Route 134, Yorktown Heights, NY 10598, USA

Michael C. Pitman (159), IBM Research Division, IBM Thomas J. Watson Research Center, 1101 Kitchawan, Rd Route 134, Yorktown Heights, NY 10598, USA

Aleksandr Rayshubskiy (159), IBM Research Division, IBM Thomas J. Watson Research Center, 1101 Kitchawan, Rd Route 134, Yorktown Heights, NY 10598, USA

Jörg Rottler (49), Department of Physics and Astronomy, The University of British Columbia, 6224 Agricultural Road, Vancouver, BC V6T 1Z1, Canada

Benoît Roux (369), The University of Chicago, Institute of Molecular Pediatric Sciences, Gordon Center for Integrative Science, 929 East 57th Street, room W323B, Chicago, IL 60637, USA

Celeste Sagui (49), Department of Physics, North Carolina State University, Raleigh, North Carolina, USA

Nicolas Sapay (111), University of Calgary, Department of Biological Sciences, 2500 University Dr. NW, Calgary, Alberta, T2N1N4, Canada

Klaus Schulten (313), Center for Biophysics and Computational Biology, University of Illinois at Urbana-Champaign, Urbana, IL 61801, USA; Beckman Institute for Advanced Science and Technology, University of Illinois at Urbana-Champaign, Urbana, IL 61801, USA; Department of Physics, University of Illinois at Urbana-Champaign, Urbana, IL 61801, USA

H. Larry Scott (281), Department of Biological, Chemical and Physical Sciences, Illinois Institute of Technology, 3101 S. Dearborn, Chicago, IL 60616, USA

Amy Y. Shih (313), Center for Biophysics and Computational Biology, University of Illinois at Urbana-Champaign, Urbana, IL 61801, USA; Beckman Institute for Advanced Science and Technology, University of Illinois at Urbana-Champaign, Urbana, IL 61801, USA

Stephen G. Sligar (313), Center for Biophysics and Computational Biology, University of Illinois at Urbana-Champaign, Urbana, IL 61801, USA; Beckman Institute for Advanced Science and Technology, University of Illinois at Urbana-Champaign, Urbana, IL 61801, USA; Department of Biochemistry, University of Illinois at Urbana-Champaign, Urbana, IL 61801, USA

William C. Swope (159), IBM Research Division, IBM Thomas J. Watson Research Center, 1101 Kitchawan, Rd Route 134, Yorktown Heights, NY 10598, USA

Emad Tajkhorshid (343), Center for Biophysics and Computational Biology, Department of Biochemistry, and Beckman Institute University of Illinois at Urbana-Champaign, Urbana, IL 61801, USA

D. Peter Tieleman (111), University of Calgary, Department of Biological Sciences, 2500 University Dr. NW, Calgary, Alberta, T2N1N4, Canada

D. Peter Tieleman (227), Dept. of Biological Sciences, University of Calgary, 2500 University Dr NW, Calgary AB, T2N 1N4, Canada

Ilpo Vattulainen (49), MEMPHYS—Center for Biomembrane Physics, Physics Department, University of Southern Denmark, Odense, Denmark; Laboratory of Physics and Helsinki Institute of Physics, Helsinki University of Technology, Finland; Department of Physics, Tampere University of Technology, Finland; Department of Physics, North Carolina State University, Raleigh, North Carolina, USA

Richard M. Venable (1), Laboratory of Computational Biology, National Heart, Lung, and Blood Institute, National Institutes of Health, Bethesda, MD 20892, USA

Igor Vorobyov (405), Department of Chemistry, University of California, One Shields Avenue, Davis, CA 95616, USA

Gregory A. Voth (181), Center for Biological Modeling and Simulation, and Department of Chemistry, University of Utah, 315 S. 1400 E. Rm 2020, Salt Lake City, UT 84112-0850, USA

Yi Wang (343), Center for Biophysics and Computational Biology, Department of Biochemistry, and Beckman Institute University of Illinois at Urbana-Champaign, Urbana, IL 61801, USA

Thomas B. Woolf (385), Johns Hopkins University, School of Medicine, Departments of Physiology and of Biophysics and Biophysical Chemistry, Baltimore, Maryland, USA

Preface

While most would agree that laying out the future of the field of membrane simulation is a difficult thing to do, and likely a foolish exercise as well, I find writing a concise and accurate history of the field nearly equally challenging. When I began this project I intended to bring together key contributors to the study of lipid bilayer membranes using atomically detailed models. The growth of this field over the past 15 years has been spectacular and in my estimation the methods had matured to the point that the time was right to publish a book that could describe the methods so carefully developed over this period, and could highlight many of the important problems in membrane biophysics now being addressed by computational techniques. It seems that the history of these models could be written relatively neatly and cleanly. However, as a liberal arts college faculty member I actually talk to historians regularly, most of whom would argue that no history can be written in a straightforward manner and so I tread lightly in this endeavor. A reasonable timeline might begin in the 1980's with molecular dynamics (MD) simulations of water between rigid lipid headgroups (Kjellander and Marcelja, 1985) or with the early simulation of a solvent-free decanoate lipid bilayer (Vanderploeg and Berendsen, 1982). In the early 1990's simulations more resembling current models began to appear, where phospholipid molecules and water were represented with atomic detail (sometimes omitting non-polar hydrogen atoms) with complete, or nearly so, $3N$ degrees of freedom. A large number of high quality review articles have appeared covering the progression of the field (Ash *et al.*, 2004; Feller, 2000; Forrest and Sansom, 2000; Jakobsson, 1997; Kandt *et al.*, 2007; Pastor, 1994; Saiz *et al.*, 2002; Saiz and Klein, 2002; Scott, 2002; Tobias *et al.*, 1997) and the reader is urged to consult these resources for a broader picture of the history of the field than is provided by this volume.

As I began to assemble the topics and authors for this volume I found it impossible to constrain the topics only to atomistic simulation without giving an unsatisfying and incomplete picture of the field. The main cause of this difficulty is the emergence of coarser grained membrane models. The particular challenge is the multi-scale or hierarchical nature of these approaches that often uses atomistic modeling as an input to the calculations and, as described in this volume, can provide feedback to the atomic level calculations. Thus the multi-scale modeling approaches are providing both new demand for high quality atomic level modeling as well as holding promise for extending the time and length scale accessible to computational approaches. At the moment the development of molecular scale and mesoscale modeling techniques are thus intertwined, as are the research programs of many investigators, and it seemed prudent to include contributions on this topic. Providing a historical background for the coarse grained approach is especially challenging given the great diversity of methods currently being employed. In some ways this is analogous to the situation in atomistic modeling a decade or more ago when different force fields, treatments of electrostatics, and statistical mechanical ensembles were being pursued by many groups. Perhaps in a few years some brave soul can bring together a volume for this series that reviews the coarser grained models once that field has reached greater maturity in the sense of consensus on strengths and weaknesses of the various approaches.

The order of the chapters is intended to be based on the chapter authors' relative emphasis on method and application. While every chapter contains important details on methodology and every chapter addresses an important scientific problem, I have tried to organize this volume to help the reader in answering the questions "What problems can be addressed by computational models?" and "How do I go about solving my specific problem using computational models?". As my personal bias is towards understanding the fundamentals of the method, these chapters come first, but the reader will be well served to jump in to any chapter.

Chapter 1 by Pastor and co-workers focuses on the development and testing of an empirical energy function, a topic of critical importance to computational modeling efforts. Details of the energy function, both its form and its parameterization, have been continually debated by those in the membrane simulation community and their validity constantly questioned by those outside it. While sometimes distracting, this has been a positive net force, moving the various groups carrying out the generally thankless task of force field development to a point where numerous high quality force fields are available that give results in general agreement with each other and with many important experimentally measurable quantities. The careful comparison between computational model and experimental data is an focus of Chapter 1 and should serve as a guide for anyone undertaking a modeling study. Following the force field description in Chapter 1 is the contribution of Karttunen et al. who address a critically important question of implementing the force field in a molecular simulation, namely how will the electrostatic component of the force be computed. In the periodically replicated simulation cell most typically employed in molecular simulation, the computation of Coulombic interactions between atoms involves evaluating an infinite sum whose convergence is frustratingly slow. Early MD simulations of membranes typically employed truncation schemes, at least in part due to limitations in computer power, that have subsequently been shown to introduce severe artifacts into the calculations. Today the vast majority of membrane simulations compute electrostatic interactions without truncation, generally employing methods based on the Ewald summation technique. Chapter 2 reviews the strengths and weaknesses of these methods with particular emphases on computational efficiency and aspects important for membrane systems.

Chapter 3 by Edholm addresses two issues, the system size and sampling period, that are too frequently left unconsidered when designing a plan of computational research. Most often the number of molecules simulated and the length of the run are determined by the computational resources available, but this importantly reminds us that the questions that can be answered by a simulation are determined by issues of time and length scale. Thankfully, the constant advances in both hardware and algorithms has allowed the field to constantly expand the types of questions that can be addressed and Edholm describes undulatory fluctuations in a detailed membrane model that seemed impossible to study in the not distant past. In addition to rapid increases in the size and scale of bilayer simulations, recent years have seen tremendous advances in the complexity of membrane simulations. The significant challenges in carrying out studies of biologically relevant mixtures is covered in Chapter 4 by Sapay and Tieleman. Specific challenges include the accurate parameterization of amino acids that explore very different solvent environments and sampling conformations when the range of correlation times is so broad. The paucity of detailed experimental data (of the type described in Chapter 1) makes progress in this area even more difficult. But clearly this is a direction the field must move towards and this chapter should be required reading for anyone wishing to tackle a simulation of a lipid–protein complex or membrane binding peptide. While Sapay and Tieleman consider atomically detailed phospholipids to represent the membrane solvent, alternative approaches are reviewed by Grossfield in Chapter 5 where continuum representations are employed for the lipid component. The attraction of such approaches, which treat proteins and other membrane permeants in atomic detail, include the significantly improved computational efficiency and enhanced sampling rates. Drawbacks include challenges in parameterization and a presumed inability to capture all lipid–protein interactions. Nonetheless, these methods have found great success in many applications and refinements to these approaches are ongoing in numerous labs.

Chapter 6 is the contribution from the IBM Blue Matter team, the group responsible for developing a parallel molecular dynamics code for Blue Gene, currently the world's fastest supercomputer. Advances in membrane simulation have been tightly coupled to those in parallel computing due to the need to simulate large number of molecules and the desire to sample processes with long correlation times. Their chapter describes the specific challenges in simulating membrane systems, as much of the first science done on Blue Gene involved simulations of membrane proteins in explicit lipid environment, as well as providing an overview of the challenges in parallelizing atomically detailed molecular dynamics simulations. Chapter 7 is the last of the contributions with a methodological focus. Here, Voth and co-workers review the multiscale approaches to simulating membrane systems that has been so successfully developed over the past several years. These include both coarse-graining techniques, where the model maintains representations that can be identified as individual molecules, and field theoretic approaches that describe the system on longer time and length scales. An especially attractive aspect of their approach is the multiscale coupling where atomistic simulations are embedded in a mesoscopic membrane, interacting via the surface tension.

The incorporation of cholesterol and other small molecules into the lipid bilayer is described in Chapters 8–10. In Chapter 8, MacCallum and Tieleman review results for permeation by a wide variety of solutes, including models for amino acids, anesthetic molecules, ions, sugars, and carbon nanoparticles. The sampling challenges, even for these relatively small permeants, are great and the heterogeneous nature of the lipid/water interface creates an extremely complicated set of interactions that need to be understood. Chapter 9 by Berkowitz looks at the application of atomistic simulation to the study of domain formation in mixtures of phospholipid and cholesterol. While simulations have produced a substantial body of data on details of lipid–cholesterol interactions, and shed light on some important open questions raised by many years of experimental study on these systems, it is also pointed out that many details are yet to be elucidated, e.g. the nature of condensed complexes. In Chapter 10, Scott and co-workers address the issue of domain formation using a combination of atomistic molecular dynamics and mean field lattice-based simulation. Chapters 9 and 10 are highly complimentary in that the foci are on headgroup interactions and tail interactions, respectively. Additionally, in Chapter 10 the MD simulation is used as input to a mean field calculation that studies domain formation based on chain order distribution for systems of thousands of lipids on the hundred microsecond time scale.

An ambitious modeling project is described in Chapter 11 by Schulten and co-workers. They describe simulations, primarily using coarse-grained models, of protein–lipid complexes. They were able to study the formation and structure of lipoprotein assemblies using simulations on the ten microsecond time scale and to compare the resulting structures with experimental X-ray scattering data.

Chapters 11–15 explore various aspects of transport, from simulations explicitly studying movement of molecules necessary to respiration, to more general features of electrified interfaces and the features of the membrane that influence ion channel structure and function. In Chapter 12, Tajkhorshid and co-workers explore competing mechanisms for the transport of biologically relevant gas species across the membrane. They use MD simulation to relate bilayer structural features to permeation barriers and identify alternative gas conduction pathways that utilize channel proteins. Roux provides us an introduction to the field of voltage-gated potassium channels in Chapter 13, a field of tremendous activity in recent years where details of the membrane, both in terms of material properties and as the solvent for ion channel proteins, play a substantial role. He reviews the progress of the field in describing the structure and activation of this important class of proteins and shows numerous examples where computational modeling has played key roles in developing this understanding. In Chapter 14, Woolf and Denning focus their attention on computational models that have been used to understand membrane electrostatics, a key component of transport but also a fundamental aspect of membrane biophysics. This volume concludes with

Chapter 15, from Allen and co-workers, where an extensive set of simulations have been done to quantify the free energy cost of transferring a charged side chain from an aqueous environment to the lipid bilayer membrane. Their modeling approach has important implications for our understanding of protein motions during the activation of voltage gated channels and they provide a careful analysis of side chain thermodynamics that may aid in the understanding of experimentally determined hydrophobicity scales.

There are many people to acknowledge for their important roles in this book. First would have to be the authors of the individual chapters. They were extremely generous with both their wisdom and with their time and I am greatly indebted to this talented group. The staff at Elsevier have been terrific and patient. I thank Richard Pastor for introducing me to the field of membrane simulation and for showing me that there is great value in deeply understanding a complex problem. I would not have been able to continue my work in membranes without the support of the Molecular Biophysics section of the National Science Foundation and I am grateful for it. My students and colleagues at Wabash College have provided me a supportive, if somewhat unconventional, environment to pursue my research interests that has served me well. Finally, I thank my wife, Wendy, and children, Amanda and Jake. Their gift has been the most generous of all, supporting me in every way without ever being sure exactly what it is that I am doing.

References

Ash, W.L., Zlomislic, M.R., Oloo, E.O., Tieleman, D.P. (2004). Computer simulations of membrane proteins. *Biochim. Biophys. Acta Biomembranes* **1666** (1–2), 158–189.

Feller, S.E. (2000). Molecular dynamics simulations of lipid bilayers. *Curr. Opin. Colloid Interface Sci.* **5** (3–4), 217–223.

Forrest, L.R., Sansom, M.S.P. (2000). Membrane simulations: Bigger and better? *Curr. Opin. Struct. Biol.* **10** (2), 174–181.

Jakobsson, E. (1997). Computer simulation studies of biological membranes: Progress, promise and pitfalls. *Trends Biochem. Sci.* **22** (9), 339–344.

Kandt, C., Ash, W.L., Tieleman, D.P. (2007). Setting up and running molecular dynamics simulations of membrane proteins. *Methods* **41** (4), 475–488.

Kjellander, R., Marcelja, S. (1985). Polarization of water between molecular-surfaces—A molecular-dynamics study. *Chem. Scr.* **25** (1), 73–80.

Pastor, R.W. (1994). Molecular-dynamics and Monte Carlo simulations of lipid bilayers. *Curr. Opin. Struct. Biol.* **4** (4), 486–492.

Saiz, L., Klein, M.L. (2002). Computer simulation studies of model biological membranes. *Acc. Chem. Res.* **35** (6), 482–489.

Saiz, L., Bandyopadhyay, S., Klein, M.L. (2002). Towards an understanding of complex biological membranes from atomistic molecular dynamics simulations. *Biosci. Rep.* **22** (2), 151–173.

Scott, H.L. (2002). Modeling the lipid component of membranes. *Curr. Opin. Struct. Biol.* **12** (4), 495–502.

Tobias, D.J., Tu, K.C., Klein, M.L. (1997). Atomic-scale molecular dynamics simulations of lipid membranes. *Curr. Opin. Colloid Interface Sci.* **2** (1), 15–26.

Vanderploeg, P., Berendsen, H.J.C. (1982). Molecular-dynamics simulation of a bilayer-membrane. *J. Chem. Phys.* **76** (6), 3271–3276.

Previous Volumes in Series

Current Topics in Membranes and Transport

Volume 23 Genes and Membranes: Transport Proteins and Receptors* (1985)
Edited by Edward A. Adelberg and Carolyn W. Slayman

Volume 24 Membrane Protein Biosynthesis and Turnover (1985)
Edited by Philip A. Knauf and John S. Cook

Volume 25 Regulation of Calcium Transport across Muscle Membranes (1985)
Edited by Adil E. Shamoo

Volume 26 Na$^+$–H$^+$ Exchange, Intracellular pH, and Cell Function* (1986)
Edited by Peter S. Aronson and Walter F. Boron

Volume 27 The Role of Membranes in Cell Growth and Differentiation (1986)
Edited by Lazaro J. Mandel and Dale J. Benos

Volume 28 Potassium Transport: Physiology and Pathophysiology* (1987)
Edited by Gerhard Giebisch

Volume 29 Membrane Structure and Function (1987)
Edited by Richard D. Klausner, Christoph Kempf, and Jos van Renswoude

Volume 30 Cell Volume Control: Fundamental and Comparative Aspects in Animal Cells (1987)
Edited by R. Gilles, Arnost Kleinzeller, and L. Bolis

Volume 31 Molecular Neurobiology: Endocrine Approaches (1987)
Edited by Jerome F. Strauss, III, and Donald W. Pfaff

Volume 32 Membrane Fusion in Fertilization, Cellular Transport, and Viral Infection (1988)
Edited by Nejat Düzgünes and Felix Bronner

Volume 33 Molecular Biology of Ionic Channels* (1988)
Edited by William S. Agnew, Toni Claudio, and Frederick J. Sigworth

Volume 34 Cellular and Molecular Biology of Sodium Transport* (1989)
Edited by Stanley G. Schultz

---

* Part of the series from the Yale Department of Cellular and Molecular Physiology.

CHAPTER 1

Considerations for Lipid Force Field Development

Jeffery B. Klauda[*,‡], **Richard M. Venable**[*], **Alexander D. MacKerell Jr.**[†] **and Richard W. Pastor**[*]

[*]Laboratory of Computational Biology, National Heart, Lung, and Blood Institute, National Institutes of Health, Bethesda, MD 20892, USA
[†]Department of Pharmaceutical Sciences, School of Pharmacy, University of Maryland, Baltimore, MD, 21201, USA
[‡]Current address: Department of Chemical and Biomolecular Engineering, University of Maryland, College Park, MD 20742, USA

Abstract

The underlying approach to development of the CHARMM lipid force field, and the current *ab initio* and molecular dynamics methods for optimization of each term are reviewed. Results from the recent revision of the alkane force field and new results for esters illustrate the dependence of torsional surfaces on level of theory and basis set, and how changes in the surface manifest themselves in alkanes and dipalmitoylphosphatidylcholine (DPPC) bilayers. The following proper-

Current Topics in Membranes, Volume 60
Copyright © 2008, Elsevier Inc. All rights reserved

1063-5823/08 \$35.00
DOI: 10.1016/S1063-5823(08)00001-X

ties from simulation and experiment on DPPC bilayers are compared: structure factors from X-ray diffraction; deuterium order parameters; NMR spin lattice relaxation times; lipid translational diffusion constants; elastic moduli; and the dipole potential. The importance of including long-range Lennard-Jones interactions and taking finite system size into account is stressed. Theoretical and practical aspects associated with surface tensions and surface areas of lipid bilayers and monolayers are discussed.

I. INTRODUCTION

Papers describing simulations of lipid bilayers emerged in earnest in the late 1980's and early 1990's, and many included the phrase "the field is in its infancy". This fairly harmless phrase promises high potential, and requests understanding for results that are not immediately useful. Additionally, it implies the need for substantial care and feeding.

A critical component of the care and feeding has been the development of a reliable force field (FF) or potential energy function. A simulation can't proceed without it, and the availability of the appropriate FF is among the first issues that a simulator must confront when considering a new system. This chapter describes the approach taken in the development of the recent CHARMM (Chemistry at HARvard Macromolecular Mechanics) (Brooks *et al.*, 1983) all-atom lipid force fields, including the set currently under development. This approach consists of two basic steps: (1) optimization of terms in the empirical energy function using quantum mechanical (QM) calculations, molecular dynamics (MD) simulations, and experimental data on appropriate small molecules; (2) MD simulations of lipid bilayers and monolayers, and comparison with target data. This process directly links the behavior of the large assembly to the underlying physics of its components, and thereby lends confidence that agreement with experiment is not fortuitous. To provide a specific example, the recent dihedral parameters for the acyl chain of lipids in C27r (Klauda *et al.*, 2005a) were generated by high level *ab initio* calculations on butane through heptane. MD simulations were then carried out on liquid heptane, decane, tridecane, and pentadecane, to confirm that C27r yielded better agreement with experiment than the previous set, C27 (Feller and MacKerell, 2000). Then simulations on a dipalmitoylphosphatidylcholine (DPPC) bilayer were carried out, and the results were compared with experimental deuterium order parameters, X-ray diffraction density profiles, and conformational populations from infrared spectroscopy.

The core of this chapter is contained in the following two sections. Section II describes the optimization of the FF for small molecules; *i.e.*, step one of the process just outlined. The section begins with a review of *ab initio* calculations to provide a sense of the level of theory and system size accessible with presently available computers, and the specific choices used in CHARMM parameterization. Sections II.B.1 and II.B.2 detail the optimization of the electrostatic and Lennard-Jones terms, respectively. Section II.B.3 considers the torsional terms, and includes *ab initio* results for the torsional surfaces of alkanes and esters to illustrate the effects of level of theory and implicit solvation. Lastly, Section II.C reviews the assessment of C27 and C27r from MD simulations of liquid alkanes in the bulk and the interface.

Section III describes seven classes of membrane target data that are presently being used in testing of CHARMM parameter sets. These are: (A) structure factors from X-ray diffraction; (B) deuterium order parameters; (C) NMR spin lattice relaxation times; (D) translational diffusion constants; (E) elastic moduli; (F) surface areas and surface tensions of bilayers and monolayers; (G) and the dipole potential. Each topic is paired with a technical nuance (*e.g.*, finite size effects and translational diffusion), and the ordering is roughly from better to worse in terms of agreement of simulation and experiment or in ease of interpretation. As for Section II, the results of C27 and C27r are compared. The reader primarily interested in the performance of CHARMM for membranes may consider reading Section III before Section II.

Section IV summarizes the results, and considers some of the broader questions of parameter development. The remainder of this introductory section defines the terms of the CHARMM FF.

The potential energy $V(\hat{R})$ in the CHARMM FF (MacKerell, 2004) is a function of the positions of all of the atoms in the system. Like most FF used for MD simulations of macromolecules and membranes, it has the following general form:

$$
V(\hat{R}) = \sum_{\text{bonds}} K_b(b - b_0)^2 + \sum_{\text{angles}} K_\theta(\theta - \theta_0)^2
$$

$$
+ \sum_{\text{dihedrals}} \left[\sum_j K_{\varphi,j}\left(1 + \cos(n_j\varphi - \delta_j)\right) \right]
$$

$$
+ \sum_{\substack{\text{nonbonded} \\ \text{pairs } i,j}} \varepsilon_{ij}\left[\left(\frac{R_{\min,ij}}{r_{ij}}\right)^{12} - \left(\frac{R_{\min,ij}}{r_{ij}}\right)^6 \right] + \sum_{\substack{\text{nonbonded} \\ \text{pairs } i,j}} \frac{q_i q_j}{\varepsilon_D r_{ij}}.
$$

$$(1)$$

The first three terms parameterize the interactions of atoms chemically bonded to each other, and are referred to as the *intramolecular* or *internal* terms. The last two (commonly referred to as the *nonbond*, *intermolecular*, or *external* terms) describe the van der Waals and electrostatic interactions between atoms, respectively. More specifically, bond stretches and bends are harmonic, with force constants K_b and K_θ, and equilibrium values b_0 and θ_0. While more complex functional forms, such as the Urey–Bradley and improper dihedrals, are available in CHARMM, they are only used for a subset of the functionalities in lipids. Dihedral, or torsion, angles are parameterized by a cosine series where K_φ, n and δ are the force constant, multiplicity and offset, respectively. Lipid torsions are described by sets of $K_{\varphi,j}$, n_j and δ_j, where j can range from 1 to 4. Explicit coupling between neighboring torsions, which is important for proteins and peptides is described by the "CMAP" correction (Buck *et al.*, 2006; MacKerell *et al.*, 2004) has not been introduced to the lipid FF. Van der Waals interactions are treated by the well-known Lennard-Jones (LJ) "6-12" potential, where ε_{ij} is the potential energy minimum between two particles, and $R_{\mathrm{min},ij}$ is the position of this minimum. Given that the repulsive wall is multiexponential, that the attractive interaction is more accurately described as the asymptotic series $\sum_{n=3}^{\infty} r^{-2n}$, and that chemically similar (but not identical) atoms share the same LJ parameters, this part of the FF contains more uncertainty than the preceding bonded terms. It is, accordingly, the most difficult to parameterize. Lastly, q_i and q_j are the atomic partial charges, and ε_D is the dielectric constant. The membrane systems considered here include all lipids and water, so $\varepsilon_D = 1$. C27 and C27r are "additive models". That is to say, the partial charges are fixed throughout the simulation, and the electrostatic energy is the sum of 2-body terms. Polarizable models allow the partial charges to vary in response to their environment (the interactions are non-additive, or multibody), and provide a more accurate, though computationally more demanding, description of the system. There is not presently a polarizable lipid FF in CHARMM, though classical Drude based models (Rick and Stuart, 2002) for water (Lamoureux *et al.*, 2003), alkanes (Vorobyov *et al.*, 2005), and ethers (Vorobyov *et al.*, 2007) have been developed. Efforts to develop a polarizable lipid model compatible with the fluctuating charge protein model implemented in CHARMM (Patel *et al.*, 2004) are also ongoing.

The basic form of Eq. (1) is utilized for all classes of molecules in CHARMM, including proteins, DNA, carbohydrates and ethers, and these sets are all designed to be compatible (*e.g.*, a peptide to be simulated in a membrane environment does not require new parameters). Hydrogens are explicitly included to retain consistency with the other classes of molecules, and because they were found to be necessary for describing the more condensed crystal and gel phases of lipid bilayers (Venable *et al.*, 2000).

II. QUANTUM MECHANICS AND MOLECULAR DYNAMICS BASED PARAMETER OPTIMIZATION

A. *Overview of Quantum Mechanical Methods*

As emphasized in Section II, an essential component in the parameterization of the CHARMM FF is quantum mechanics. Ideally, QM calculations would be carried out on a sufficiently large cluster of molecules to include all of the essential interactions necessary for parameterizing the molecular mechanics potential en-

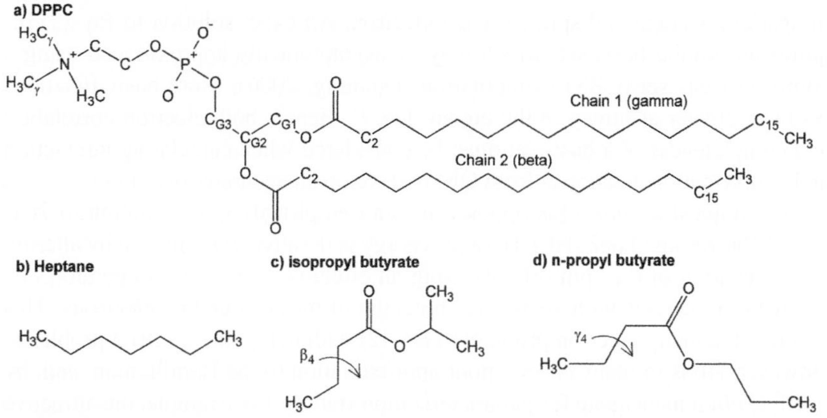

FIGURE 1 DPPC (a) and three of the compounds used to parameterize the lipid force field: (b) heptane; (c) isopropyl butyrate; (d) and *n*-propyl butyrate. Atom names for DPPC follows the *sn* IUPAC nomenclature (IUPAC, 1967). Torsions on the Chain 1 and 2 are labeled γ_i and β_i, respectively, following the Sundaralingam convention (Hauser *et al.*, 1980; Sundaralingam, 1972). Specifically, β_4 and γ_4 are the C_1–C_2–C_3–C_4 in the torsions in chain 2 and chain 1 of DPPC, respectively. The analogous torsion angles are noted in (c) and (d) for the model esters.

ergy function given by Eq. (1). For lipids in a bilayer, one might imagine that such cluster would consist of three lipids and 10 hydrating waters/lipid. DPPC (Fig. 1) contains 50 heavy atoms, so the preceding hypothetical cluster contains 180 heavy atoms. This is well outside the range of 10–12 heavy atoms that can presently be evaluated accurately with the high-level *ab initio* calculations that force field parameterizations require. In addition, limitations in low-level QM calculations with respect to London's dispersion interactions for small clusters of molecules hinder the use of QM data alone in parameter optimization (MacKerell, 2004). To understand why requires a brief review of the approximations to the exact time-independent Schrödinger equation:

$$\mathcal{H}|\Psi\rangle = \mathcal{E}|\Psi\rangle, \tag{2}$$

where $\mathcal{H}$ is the Hamiltonian operator for the system of nuclei and electrons, $|\Psi\rangle$ is the electron wave function, and $\mathcal{E}$ is the energy (Szabo and Ostlund, 1996). Even after the nuclear and electron energies are separated by applying the Born–Oppenheimer approximation, Eq. (2) can, in general, be solved analytically only for one-electron systems (this is similar to the many-body problem in classical mechanics). Numerical solutions for the exact Hamiltonian are possible for more than one electron, but for the applications considered here further simplifications are required. These are divided into simplifications to $\mathcal{H}$ (*level of theory*) and to $|\Psi\rangle$ (*basis set*). At the exact level of theory, denoted full configuration interaction (CI), all electrons interact with each other. The basis set defines the range of spatial coverage and spin for each electron. An exact solution to Eq. (2) requires an infinite basis set, which may be asymptotically approximated using a complete basis set (CBS) extrapolation (Dunning, 2000). More basis functions result in greater accuracy in the eigenvalue, $\mathcal{E}$. Hence, both electron correlation and completeness of a basis set must be considered when calculating interaction and conformational energies from QM to develop an accurate force field.

The simplest *ab initio* (as opposed to semi-empirical) approximation to $\mathcal{H}$ is that of Hartree and Fock (HF). Here the energy is iteratively minimized by altering the occupation of the spin orbitals using an effective one-electron operator, *i.e.*, an electron interacts with an average potential of the surrounding electrons. This replaces the many-electron problem in Eq. (2) with many one-electron problems. However, HF is in many cases a poor approximation to the Hamiltonian, and, by itself, is often inadequate for parameterization studies. For example, the attractive energy between two ideal gas atoms arises from electron correlation (induced dipole-induced dipole interactions and similar higher order terms). HF yields only repulsive energies because electron correlation is absent.

The electron correlation energy is defined as the additional energy beyond the HF limit (HF energy with an infinitely-sized basis set) due to explicit electron–electron interactions. Many methods have been developed to include electron correlation. Density functional theory (DFT) is the simplest and the least computationally demanding. Examples include the B3LYP (Becke, 1993) and PBE (Ernzerhof and Scuseria, 1999) functionals. However, DFT methods lack the ability to accurately describe the long-range dispersion interactions important for lipid model molecules. Second order Møller-Plesset perturbation theory (MP2) is the most efficient method that includes electron correlation and dispersion interactions (Dunning, 2000). This method incorporates a perturbation in the HF Hamiltonian, $\mathcal{H}_o$,

$$\mathcal{H}|\Psi\rangle = (\mathcal{H}_o + \mathcal{V})|\Psi\rangle = \mathcal{E}|\Psi\rangle, \tag{3}$$

where $\mathcal{V}$ is the perturbation of $\mathcal{H}_o$ from the true $\mathcal{H}$. The exact eigenfunctions and eigenvalues are expanded in a Taylor series where the second order term in the expansion is the additional energy in MP2 beyond the HF value. The coupled cluster method, CCSD(T), (Raghavachari *et al.*, 1989) is the most accurate of the commonly used correlative methods and typically approaches the full CI result (Dunning, 2000).

Just as the level of theory is important in calculating energies from QM, so too is the size of basis sets used to represent the wave function. Basis sets typically consist of Gaussian-based functions that represent orbitals, such as s, p, d and f. The correlation consistent basis sets by Dunning *et al.* (2001) have been optimized for consistent convergence of energy for correlated methods and are typically used with the abovementioned approximations of $\mathcal{H}$. In the development of the dihedral portion of the CHARMM force field, the following basis sets are used in increasing order of complexity: cc-PVnZ where n can be D (double),

T (triple), Q (quadruple), and 5 (quintuple). The computational time required for DFT, MP2, and CCSD(T) typically scales with the number of basis functions to the power of 2.5, 3.5, and 6, respectively. Consequently, these methods are especially limited by the number of basis functions or, equivalently, the number of electrons.

Computations at CCSD(T) with a large basis set such as cc-PV5Z (denoted CCSD(T)/cc-PV5Z), would result in accurate energies for most compounds. However, this is currently too computationally demanding for a system with more than three or four heavy atoms. QM hybrid methods have been developed to increase the system size based on the approximation that electron correlation and basis sets are additive. The G3 method is an example (Curtiss *et al.*, 1998; Curtiss *et al.*, 1999). It combines high-level energy calculations (*e.g.*, CCSD(T)) with small basis sets, and lower level calculations (MP2 and MP4) with larger basis sets. The G3 method results in accurate heats of formation, ionization potentials, electron affinities, and proton affinities. Klauda *et al.* (2004) developed a similar method referred to as HM-IE (the **H**ybrid **M**ethod for **I**nteraction **E**nergies) that accurately approximates the interaction energy calculated with CCSD(T) and a large basis set but uses considerably less computational time and resources. This method was developed for intermolecular energies but can also be used for intramolecular conformational energies,

$$
\begin{aligned}
E[\text{CCSD(T)/LBS}] &= E[\text{CCSD(T)/SBS}] \\
&\quad + (E[\text{CCSD(T)/LBS}] - E[\text{CCSD(T)/SBS}]) \\
&\cong E[\text{CCSD(T)/SBS}] + (E[\text{MP2/LBS}] - E[\text{MP2/SBS}]) \\
&\equiv E[\text{MP2:CC}],
\end{aligned} \tag{4}
$$

where SBS denotes the small basis set and LBS denotes the large basis set. For MP2:CC, the basis set contribution going from the CCSD(T)/SBS to CCSD(T)/LBS is approximated by the difference between the interaction energies at MP2 with the same two basis sets. Only two sets of energy calculations are required in Eq. (4), and these are calculated in the following order: (1) CCSD(T)/SBS, which includes the MP2/SBS calculations since CCSD(T) uses the MP2 result; and (2) MP2/LBS.

B. Parameterization of Force Fields

In principle, the same set of QM calculations could be used to parameterize each portion of the force field. In practice, the different terms of Eq. (1) are computed independently. There are a number of reasons for this approach. As already noted, the fragments that can presently be treated with highly accurate QM methods are relatively small, 10–12 heavy atoms, and the optimal systems for evaluating each type of interaction differ. The level of theory, basis set, and adjustments for solvent effect appropriate for these systems also differ. Finally, the approach also allows incremental, yet important improvements. For example, C27r was developed leaving all non-bonded terms and the head group torsions unchanged; *i.e.*, only the torsional potential of the acyl chains was modified.

The nonbonded terms are evaluated first for the given atom types and are described in the following two subsections. Since these are primarily parameterized for intermolecular interactions, adjustments are needed for the short-range intramolecular interactions that impact the conformational properties of a molecule via the dihedral potential. These changes are especially important for obtaining the proper secondary structure of proteins (Duan *et al.*, 2003; Feig *et al.*, 2003; Jorgensen *et al.*, 1996; MacKerell *et al.*, 1998) and conformations of dihedral states in lipids (Klauda *et al.*, 2005a). The QM methods used to parameterize the dihedral potential are discussed in subsection B.3.

1. CHARMM Electrostatic Potential Terms

Optimization of the electrostatic terms (the partial atomic charges q_i in Eq. (1)) is mostly based on QM data. Charges in most current empirical force fields are assigned either by fitting to QM electrostatic potential (ESP) maps (Bayly *et al.*, 1993; Chirlian and Francl, 1987; Henchman and Essex, 1999; Merz, 1992; Singh and Kollman, 1984), or by the supramolecule approach (MacKerell, 2004). Charge determination via ESP fitting involves adjustment of the partial atomic charges to minimize the RMS difference between the QM and empirical ESP maps. This approach may be used to rapidly obtain charges for a wide range of molecules, although the charges are representative of the gas phase and may be sensitive to conformation; these issues have been addressed in different contexts (Bush *et al.*, 1999; Jakalian *et al.*, 2000; Laio *et al.*, 2002). In the supramolecule approach, which is used in the CHARMM and the OPLS force fields (Jorgensen *et al.*, 1996), the partial atomic charges are adjusted to reproduce QM minimum interaction energies and geometries of model compounds with water or for model compound dimers; dipole moments from either QM or experiment may also be included as target data. This approach allows for local polarization associated with the interaction between the molecules to be included in the fitting, and for further optimization based on experimental data for model compound pure solvents and their free energies of solvation (Oostenbrink *et al.*, 2004; Vorobyov *et al.*, 2007). The level of theory used in determination of the intermolecular interactions in CHARMM additive FF is HF/6-31G*. While there are limitations with HF level calculations, its selection is partly based on historical reasons: its initial application to CHARMM parameterization in the 1980's was aimed at hydrogen bond interactions; these are dominated by electrostatic interactions, which are satisfactorily treated at the HF level. HF/6-31G* also tends to overestimate dipole moments, and thereby mimics the overpolarization that occurs in condensed phases. In addition, for polar, neutral compounds in CHARMM, the

QM interaction energies are scaled by 1.16 prior to their use as target data to account for limitation in the level of theory, the required overpolarization that occurs in the condensed phase, and many-body effects. These effects are also accounted for by offsetting the QM minimum interaction energy distances by 0.1 to 0.2 Å (i.e., the empirical distances should be shorter than the QM distances by that amount), which is important for reproducing the correct experimental densities. While the above approach includes a variety of assumptions and higher QM levels of theory are certainly accessible, when optimizing charges for new molecules for use with the CHARMM additive force field the same approach should be used to maintain consistency.

Inherent in the supramolecule approach to charge optimization is the water model. This is because the overall parameter set must balance solvent-solvent, solvent-solute and solute-solute interactions (the so called interaction triad). The TIP3P model (Jorgensen et al., 1983) is the standard in CHARMM. In this model the dipole moment is overestimated as required to yield the proper pure solvent properties. This overestimation leads to a water dimer interaction energy that is significantly too favorable, and this must be taken into account when adjusting the partial atomic charges of model compounds. Consequently, once chosen, a water model cannot be easily replaced. It is strongly suggested that the TIP3P model, despite its deficiencies, be used in simulations with the CHARMM lipid force field.

2. CHARMM Lennard-Jones Potential Terms

The LJ parameters are both critically important and very difficult to optimize. Their importance is based on their significant contribution to pure solvent properties (MacKerell and Karplus, 1991) and the difficulty is, in part, due to the current inability of QM methods to adequately treat dispersion interactions for large systems. For example, the use of QM data on small clusters to optimize the LJ parameters leads to poor condensed phase properties. This requires that the optimization of LJ parameters be performed by empirical fitting to reproduce thermodynamics properties from condensed phase simulations, generally of neat liquids (Jorgensen, 1986; Jorgensen et al., 1984). Properties targeted include heats of vaporization, densities, isothermal compressibilities, and heat capacities. Alternatively, heats or free energies of aqueous solvation, partial molar volumes or heats of sublimation and lattice geometries of crystals can be used as target data (MacKerell et al., 1995; Warshel and Lifson, 1970). This approach has been applied extensively in the development of the CHARMM lipid force field (Feller and MacKerell, 2000; Feller et al., 1997b; Schlenkrich et al., 1996; Vorobyov et al., 2007). However, reliance on condensed phase data alone leaves the LJ parameters underdetermined (MacKerell, 2001). This problem has been overcome by determining the relative values of the LJ parameters via high-level QM data of interactions of the model compounds with rare gases (Yin and MacKerell, 1996) while the absolute values are based on the reproduction of experimental data (Chen *et al.*, 2002; Yin and MacKerell, 1998). This approach is tedious because it requires supramolecular interactions involving rare gases; however, once satisfactory LJ parameters are optimized for atoms in a class of functional groups they typically can be directly transferred to other molecules with those functional groups without further optimization.

An extensive set of nonbond parameters for the CHARMM force fields has been produced using the preceding approaches. These parameters are correlated, and it is essential that modifications be introduced consistently, and with the water model taken into account. For example, if the partial charges for a model are changed then it is necessary to reoptimize the LJ parameters to maintain the agreement with condensed phase properties. Alternatively, alteration of either the charges or LJ parameters typically requires reoptimization of the dihedral parameters. Additional information on parameter optimization may be found elsewhere (MacKerell, 2001; MacKerell, 2004; MacKerell, 2005), including the web page of Prof. MacKerell (MacKerell, 2007).

3. CHARMM Dihedral Potential Terms

This subsection presents a description of specific QM methods used to parameterize the aliphatic and glycerol portion of the CHARMM lipid FF for DPPC. In keeping with the size limitations of accurate *ab initio* calculations, the first step of the parameterization involves choosing appropriate model compounds. Here the aliphatic portion of DPPC is modeled as heptane (Fig. 1b), and the linking region of the aliphatic chains with the glycerol is modeled by isopropyl butyrate (a branched ester, Fig. 1c) for chain 2, and *n*-propyl butyrate (a linear ester, Fig. 1d) for chain 1.

Conformational energies are evaluated from minimum energy geometries. The levels of theory and basis sets, however, need not be the same for geometry optimization and energy evaluation. The MP2 level of theory is usually sufficient for predicting the *structure* of ground state molecules. Since the computational time scales with the number of basis functions, optimization with the small yet accurate basis sets is preferred. A series of geometry optimizations with multiple basis sets is required to justify the accuracy of the methods. For pentane, the calculated MP2/cc-pVTZ energy difference from all-*trans* conformer (*tt*) and the *trans-gauche* (*tg*) conformer optimized with MP2/cc-pVDZ and MP2/6-311++G** is 0.558 and 0.553 kcal/mol, respectively (Klauda *et al.*, 2005a). The difference between the optimized structure at a double-ζ basis set (cc-pVDZ) and a triple-ζ basis set (6-311++G**) is small and geometries of all the alkanes were optimized with MP2/cc-pVDZ. Similarly, there is little basis set dependence of the β_4 torsion energy profile for isopropyl butyrate (Fig. 2). Only at $\beta_4 = 120°$ are the MP2/cc-pVTZ//MP2/cc-pVTZ energies slightly lower than MP2/cc-pVDZ//MP2/cc-pVTZ. The preceding results imply that MP2/cc-pVDZ optimizations are accurate for short alkanes and the two esters. However, this

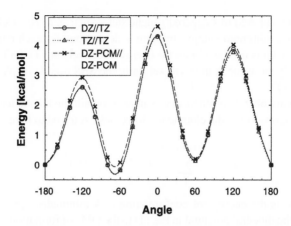

FIGURE 2 The CH_3–CH_2–CH_2–C=O (β_4 in Fig. 1) surface of isopropyl butyrate from three methods: DZ = MP2/cc-pVDZ, TZ = MP2/cc-pVTZ, DZ-PCM = MP2/cc-pVDZ with solvent correction. The notation used is as follows: (optimized level)//(single point energy). This surface is a model for the β_4 torsion in DPPC.

should not be assumed for other small molecules without similar testing procedures.

The inclusion of solvent effects on conformational energies is less straightforward than the basis set dependence on optimization. Atomistic MD simulations of hydrated lipid bilayers include the solvent effects directly with water models, such as TIP3P. However, the accuracy of these models is limited to the two-body assumption, and this lack of polarizabilty may influence conformations of the solute. The solvation effect appears to be important in obtaining the correct anomeric energy ratios for carbohydrates (Woodcock *et al.*, 2007). Therefore, more expensive QM energy calculations should be performed with solvent models, such as, polarizable continuum models (PCM) (Cances *et al.*, 1997; Cossi *et al.*, 2002) to test the validity of *in vacuo* calculations. For alkanes and esters, the solvation effects are minimal (Fig. 2). The small increase in the energy of g^- conformer with the β_4 torsion is within the uncertainty of the method. Similar results are found with solvated alkanes. These PCM calculations imply the validity of using *in vacuo* QM calculations of alkane and glycerol model compounds.

The conformational energies as a function of dihedral angle are used to parameterize the CHARMM FF. The minimum energy structure of these conformers is determined using the Berny algorithm (Schlegel, 1982) to fix a desired dihedral angle and relax the remaining degrees of freedom. An example of a torsional surface scan is shown in Fig. 2 for isopropyl butyrate. Multiple torsional surface scans are used for alkanes so that conformations are sampled from t to g and tg to g^+g^+ to g^+g^-. This is especially important because adjacent gauche states are stabilized compared to staggered gauche states such as g^+tg^+ (Klauda *et al.*, 2005b). After a minimum energy structure is determined for each conformation, Eq. (4) (MP2:CC) is used to obtain the energy effectively at CCSD(T)/cc-pVQZ, which is nearly the value at the basis set limit; the SBS used is cc-pVDZ and the LBS is cc-pVQZ.

The following objective function was used to obtain the alkane terms of C27r and the ester terms of the developmental force field, referred to here as C27r-a:

$$\chi = \sum_i^{\text{\# of QM points}} \left[U_i^{\text{QM}} - U_i^{\text{Model}}\right]^2, \tag{5}$$

where U_i and is the energy for conformation i. A minimalist approach is used when fitting the dihedral potential in Eq. (1) to the QM conformational energies in Fig. 3. The number of terms per dihedral (j) is limited to four, but initially fits with fewer terms are tested to minimize the number of parameters. Two to four terms are used for the C27r alkane force field, but fits to the ester conformation require four terms. The periodicity, n_j, in Eq. (1) is fixed as an integer and not allowed to be larger than six, which prevents overfitting to the QM energies. Similarly, the phase term δ_j is only allowed to be 0 or π. Fig. 3 includes the best fits to the QM configurations with the preceding parameter constraints.

The C27r force field improves the alkane torsional profiles over C27, which was optimized targeting lower-level QM data, both by decreasing the transitional barriers and the increasing breadth of the *gauche* potential (Fig. 3, top). The other torsions in DPPC were left unmodified in the conversion from C27 to C27r, so it is expected that comparable shifts would be observed when *ab initio* calculations at the same level are carried out on these torsions. This is indeed the case for the ester torsions (Fig. 3, middle and bottom). For example, the transitional barriers for both β_4 and γ_4 are decreased by nearly 2 kcal/mol, and the *gauche* minima are decreased by 0.7 kcal/mol. Consequently, populations of *gauche* conformers will be underpredicted with C27r. However, a simple adjustment to the torsional potential results in an excellent fit to the QM energies with only minor differences for the high energy *cis* state.

The methods described can be used to fit any torsional profile of interest. The alkane and ester torsions are simple and are parameterized independently; *i.e.*, the potential function for these torsions does not explicitly require the values of neighboring torsions even though different conformations of these neighbors may have been used for the fitting. This does not imply that the coupling between torsions is absent; it can be substantial and is manifested in the well-known "crankshaft" transitions of acyl chains in bilayers (Brown *et al.*, 1995). More complex regions, such as a glycerol moiety, may require explicit coupling between neighboring torsions. This could be corrected by the CMAP modification currently used to parameterize the ϕ/ψ linkage in peptides.

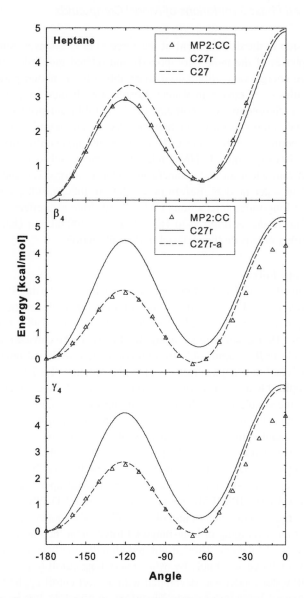

FIGURE 3 Conformational energies for the C2–C3–C4–C5 dihedral angle of heptane with all other dihedrals constrained to *trans* (top), and CH_3–CH_2–CH_2–C=O (the β_4 model) of isopropyl butyrate (middle) and n-propyl butyrate (the γ_4 model) (bottom). QM energies are at CCSD(T)/cc-pVQZ (defined MP2:CC in Eq. (4)) The lines are the surfaces of the empirical potentials, including the developmental set C27r-a.

C. Condensed Phase Simulations of Model Compounds

The next step in the development of the force field is extensive simulation of condensed phases of the model compounds and related molecules; these differ from the free energy simulations used to establish the nonbond parameters on small molecules (*e.g.*, ethane in water). Results for alkanes are presented here as illustration. Subsection C.3 compares bulk equilibrium properties (density and isothermal compressibility) for heptane, decane and tetradecane from C27, C27r and experiment; Subsection C.4 compares the nonequilibrium properties (diffusivity, viscosity, and ^{13}C NMR relaxation times). Subsection C.5 considers the alkane/air, water/air and alkane/water interfaces for C27r, and the TIP3P and TIP4P-Ew water models. Before proceeding to these results, some general comments on long-range interactions (II.C.1) and finite size effects (II.C.2) are necessary. The reader is referred to (Klauda *et al.*, 2005a) and (Klauda *et al.*, 2007) for further details on the material in this subsection.

1. Long-Range Forces

The current alkane CHARMM parameters were developed using Particle Mesh Ewald (Darden *et al.*, 1993) (PME) to include long-range electrostatics, and the analytic long range correction (LRC) (Allen and Tildesley, 1987) for the long-range LJ terms. Hence, simulations using this FF should be carried out similarly. However, while PME is applicable to multicomponent bulk and interfacial systems, the analytic long-range correction (LRC) corrections for LJ terms are only rigorously applicable to single component bulk systems. To circumvent this problem, most of the simulations to follow (including those of DPPC bilayers) have been carried out using the pressure-based long-range correction (LRC) (Lagüe *et al.*, 2004). Constant volume systems, such as alkane/vapor and lipid monolayers, cannot be directly simulated with the preceding LRC. For these the recently developed isotropic periodic sum (IPS) method (Wu and Brooks, 2005) was applied. 2-D IPS is recommended for small systems (*e.g.*, alkane/vapor and water/vapor), while the hybrid PME/IPS (PME for electrostatics and 3-D IPS for LJ) is more efficient for large systems such as monolayers. PME/IPS, though approximate because of isotropic averaging of long-range LJ forces near the interface, may also be applied to lipid bilayers. Though computationally inefficient, long-range LJ terms may be included accurately by using a very long cutoff (*e.g.*, 30 Å).

The TIP3P water model was developed with a short cutoff applied to electrostatic and Lennard Jones interactions. Therefore, in principle, it is inconsistent to simulate it with the modern methods described above. In practice, simulating with PME leads to fewer artifacts than simulating without it (Feller *et al.*, 1996), and CHARMM parameters are now developed with PME. A reparameterization of TIP4P explicitly for simulations with Ewald summation, denoted TIP4P-Ew, has recently developed (Horn *et al.*, 2004). Some results for TIP4P-Ew are included here to show the effects of water models on surface tensions, though TIP3P remains the recommended model for simulations with CHARMM parameters.

2. Finite Size Effects

It is critical to eliminate, or at least to account for, finite size effects when developing and testing parameters. These effects vary for different properties and systems and are often difficult to predict, so explicit testing is required. The system size dependence for heptane is small; *i.e.*, simulations with $N = 64, 128$ and 512 molecules yield the same averages for almost all bulk properties. An important exception to the preceding size dependencies is the self-diffusivity (or translational diffusion constant), D_s, where even $N = 512$ shows substantial finite size effects. Fortunately, a simple correction is available (Yeh and Hummer, 2004):

$$D_s = D_{PBC} + \frac{k_B T \xi}{6 \pi \eta L}, \tag{6}$$

where D_{PBC} is the diffusion constant evaluated from the simulation (in cubic periodic boundary conditions with box length L), η is the viscosity, and $\xi = 2.837297$. Equation (6) is for bulk systems, and cannot be directly applied to lipid lateral diffusion in bilayers (Section III.C). Lastly, size dependence should be rechecked when evaluating surface properties.

3. Bulk Phase Equilibrium Properties of Alkanes

An obvious first test of a potential is to verify that the system does not freeze when simulated above its melting point. This is not a trivial condition. Simulations of tetradecane with the AMBER99 (Wang *et al.*, 2000) force field led to a quasi-crystal. The configuration shown in Fig. 4 (left) formed between 0.5 and 1.5 ns, and remained stable for the total simulation time of 10 ns. In contrast, simulations with C27 and C27r correctly yield liquid densities without freezing (Fig. 4, right).

The density (ρ) and isothermal compressibility (β_T) directly probe the nonbonded contribution to the force field. Because these parameters are the same for C27 and C27r (only the torsional potential was adjusted), the results are expected to be very similar. The agreement with experiment for the density for alkanes is excellent: errors of 1.0% or less for heptane, decane, tridecane, and pentadecane (Table 1). However, neglecting the long-range LJ leads to 3% underestimate of the density.

The isothermal compressibility (or its inverse, the bulk modulus K_b) is calculated from NPT simulations from the volume V and volume fluctuations $\langle \delta V^2 \rangle$,

$$\beta_T = K_b^{-1} = -\frac{1}{V}\left(\frac{\partial V}{\partial P}\right)_T = \frac{\langle \delta V^2 \rangle}{V k_B T}, \tag{7}$$

where k_B is Boltzmann's constant, and T is the temperature. The results are not as good as the densities: approximately a 30% overestimate of experiment for

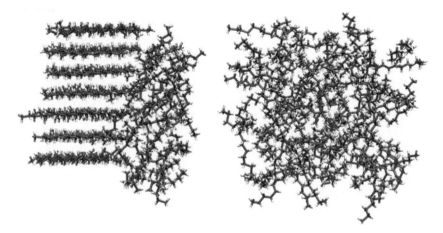

FIGURE 4 Snapshots of tetradecane simulated at 25 °C with the AMBER99 force field (left) and C27r (right) (Klauda *et al.*, 2005a).

heptane, and 13–20% for the longer alkanes (Table 1). Neglecting the long-range LJ terms increases the errors substantially; *e.g.*, β_T is approximately 50% too high for heptane. The sensitivity of β_T to the LRC suggests the LJ terms as targets for further improvements in the force field.

4. Bulk Phase Nonequilibrium Properties

Diffusion constants D_{PBC} were calculated as $1/6$ of the long-time slope of the mean squared displacement versus time,

$$\text{MSD}(t) \equiv \frac{1}{N} \left\langle \sum_{\forall i} \left(x_i(t) - x_i(0)\right)^2 + \left(y_i(t) - y_i(0)\right)^2 + \left(z_i(t) - z_i(0)\right)^2 \right\rangle,$$

(8)

where x_i, y_i and z_i are the positions of the center of mass of each particle. Self-diffusion constants D_s, with a finite size correction were then obtained from Eq. (6). Table 1 shows both the substantial correction for system size, and the near perfect agreement experiment for D_s of heptane and decane.

The shear viscosity for heptane and pentadecane was calculated using the Green–Kubo formula (Allen and Tildesley, 1987),

$$\eta = \frac{V}{kT} \int_0^\infty \left\langle P_{\alpha\beta}(t) P_{\alpha\beta}(0) \right\rangle dt,$$

(9)

where $P_{\alpha\beta}$ are the off-diagonal elements of the instantaneous pressure tensor. Since pressure is a system property, a high sampling rate and a simulation time of at least 10 ns is required to obtain accurate shear viscosities. As might be

TABLE 1

Simulation Averages and Standard Errors for Bulk Properties of Alkanes at 39 °C. Experimental Values for Density (ρ) and Viscosity (η) from (Small, 1986); Isothermal Compressibility (β_T) from (Lide, 2000); and Diffusivity (D_s) from (Douglass and McCall, 1958). D_{PBC} is the Apparent Self-Diffusivity Obtained Directly from the Mean-Squared Displacement in the Simulations; the Corrected Self-Diffusivity, D_s, is Obtained from Eq. (6)

		Alkane			
		C7	C10	C13	C15
ρ	C27r	0.661 ± 0.014	0.712 ± 0.012	0.740 ± 0.010	0.755 ± 0.009
(g/cm^3)	C27	0.660 ± 0.015	0.712 ± 0.011	0.741 ± 0.010	0.755 ± 0.009
	Exp.	0.668	0.716	0.743	0.755
β_T	C27r	18.0 ± 0.9	13.1 ± 0.3	10.7 ± 0.3	10.3 ± 0.3
(10^{-10} m^2/N)	C27	18.7 ± 0.4	12.5 ± 0.1	10.5 ± 0.3	10.0 ± 0.3
	Exp.	14.13	10.8	9.4	8.7
D_{PBC}	C27r	2.96 ± 0.03	1.39 ± 0.01	–	–
(10^{-5} cm^2/s)	C27	3.07 ± 0.04	1.28 ± 0.04	–	–
	Exp.	3.68	1.72	–	–
D_s	C27r	3.70 ± 0.03	1.72 ± 0.01	–	–
(10^{-5} cm^2/s)	C27	3.81 ± 0.04	1.61 ± 0.04	–	–
	Exp.	3.68	1.72	–	–
η	C27r	3.44 ± 0.04	–	–	20.4 ± 0.07
(10^{-4} Pa s)	C27	3.72 ± 0.03	–	–	–
	Exp.	3.46	–	–	19.4

expected from the good agreement demonstrated for the diffusion constants, the calculated viscosity from C27r for heptane matches experiment, and pentadecane overestimates experiment by only 5% (Table 1). As follows from the potential surfaces (Fig. 3 top), C27r yields somewhat more flexible chains than does C27. This plausibly explains the higher viscosity of heptane for C27.

The most dramatic difference between C27 and C27r is the NMR ^{13}C T_1 relaxation times. This is because NMR ^{13}C T_1 relaxation in liquid alkanes arises from a combination of molecular tumbling (which sensitive to the viscosity and molecular shape) and isomerization (which is sensitive to the torsional barriers) (Zhang et al., 1996). Assuming that relaxation is due to dipolar interactions between the ^{13}C nucleus and its N attached protons, the ^{13}C T_1 is

$$\frac{1}{N T_1} = \frac{1}{10} \left(\frac{\hbar \gamma_C \gamma_H}{r_{C-H}^3} \right)^2 \left[J(\omega_H - \omega_C) + 3 J(\omega_C) + 6 J(\omega_H + \omega_C) \right], \quad (10)$$

where $\hbar$ is Plank's constant divided by 2π, r_{C-H} is the effective C–H bond length, γ_H, γ_C, ω_H, and ω_C are the gyromagnetic ratios and Larmor frequencies, respec-

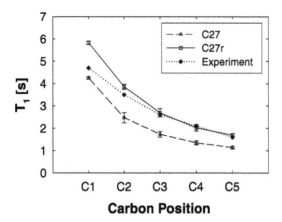

FIGURE 5 T_1 relaxation times of pentadecane from simulation (Klauda *et al.*, 2005a) and experiment (Lyerla *et al.*, 1974).

tively, of the ^{13}C and ^{1}H nuclei; $\omega_C = \gamma_C H$ and $\omega_H = \gamma_H H$, where H is the field strength. $J(\omega)$ is the spectral density of the second rank reorientational correlation function:

$$J(\omega) = \int_0^\infty \left\langle P_2(\hat{\mu}(0) \cdot \hat{\mu}(t)) \right\rangle \cos(\omega t) dt, \tag{11}$$

where P_2 is the second order Legendre polynomial and $\hat{\mu}(t)$ is the unit vector along the CH bond direction at time t. The T_1's of liquid alkanes are independent of the magnetic field strength (Brown *et al.*, 1983; Lyerla *et al.*, 1974). This occurs when all of the components of $\langle P_2(\hat{\mu}(0) \cdot \hat{\mu}(t)) \rangle$ decay rapidly compared to $1/\omega_H$, and the molecule is said to be in the motional narrowing regime. Assuming motional narrowing, and setting $r_{C\text{-}H} = 1.117\,\text{Å}$ (Ottiger and Bax, 1998), Eq. (10) reduces to

$$\frac{1}{NT_1} = \left(1.855 \times 10^{10}\ \text{s}^{-2}\right) \cdot \int_0^\infty \left\langle P_2(\hat{\mu}(0) \cdot \hat{\mu}(t)) \right\rangle dt$$
$$= \left(1.855 \times 10^{10}\ \text{s}^{-2}\right) \cdot \tau, \tag{12}$$

where τ is the rotational correlation time.

Figure 5 compares the calculated and experimental relaxation times for pentadecane. T_1's for C27r agree very well except for the end of the chain, while those from C27 are uniformly low. This is consistent with the relative *trans-to-gauche* barrier heights for the two FF (Fig. 3, top). The lower barrier for C27r leads to more isomerizations, lower relaxation times, and, from Eq. (12), higher T_1's than C27. Results for shorter alkanes do not agree as well with experiment. This likely results from overly fast rotation about the long molecular axis. However, discrepancies for short alkanes are not a large concern for development of a lipid FF because the acyl chains in a lipid are tethered to the head group and thereby cannot rotate about their long axes.

The changes to the dihedral potential also modulate the population of *trans* and *gauche* states in alkanes. There is limited experimental data available to compare the dihedral populations but, on average, C27r results in a 5% increase over C27 in the *gauche* population for all alkanes. Experimentally, Fourier transform infrared (FT-IR) spectroscopy can be used to determine the fraction of conformational states in alkanes. For tridecane, (Holler and Callis, 1989) measured 3.5 *gauche* bonds per molecule and simulations with C27 and C27r result in 2.75 and 3.04, respectively. The fraction of *trans* states with C27r for alkanes larger than decane are nearly independent of chain length with a value of 0.70, in agreement with results of Karaborni and O'Connell (1990) who obtained a fraction *trans* of 0.69 essentially independent of the chain length.

5. Interfacial Properties

The surface tension, γ, provides insight into the anisotropic forces at interfaces, and its calculation is a standard part of testing CHARMM parameters. For a planar interface whose normal is parallel to the z-axis, γ, is defined as follows:

$$\gamma = \int_{-\infty}^{+\infty} \left(P_N - P_T(z) \right) dz, \tag{13}$$

where P_N and P_T are the normal and tangential components of the pressure tensor. Because $P_N = P_T$ in the bulk, the integrand is only positive in the region of the interface. For a planar interface, P_N is independent of z and equals the bulk pressure P, so it follows from Eq. (13) that the tangential pressure is negative in the interfacial region. In fact, it is quite negative. The surface tension of the hexadecane/water interface is approximately 50 dyn/cm. Assuming the interfacial thickness is 10 Å, the average tangential pressure in the interface is therefore -500 atm.

Surface tensions in CHARMM are evaluated from the components of the stress tensor across the simulation cell (Zhang *et al.*, 1995),

$$\gamma = 0.5 \langle L_z [P_{zz} - 0.5(P_{xx} + P_{yy})] \rangle, \tag{14}$$

where L_z is the instantaneous height. The factor of 0.5 is included to take into account that the fluid is simulated as a slab and, by construction, there are two interfaces in the simulation cell. This definition works well for simple interfaces and monolayers, but is awkward for bilayers. To avoid ambiguity, the units for bilayer surface tension for bilayers are reported as dyn/cm/side.

Surface tensions of liquid/vapor interfaces, including lipid monolayers, are evaluated in the *NVT* ensemble (constant particle number, volume and temperature). No fluctuations in the cell dimensions are necessary. Liquid/liquid interfaces are most easily simulated in the *NPAT* ensemble, where P in this context indi-

TABLE 2

Surface Tensions (in dyn/cm) Calculated with a 10-Å Cutoff for Lennard-Jones Interactions, with Long-Range LJ Interactions Included by the IPS Method, and Experiment ((Small, 1986) for Alkane and Alkane/Water (Jojart and Martinek, 2007; Lemmon *et al.*, 2005) for Water). Long Range Electrostatics were Included with PME for All Simulations

System	Without long-range LJ	With long-range LJ	Expt
Heptane (25 °C)	9.0	18.7	19.8
Hexadecane (50 °C)	12.1	25.2	25.0
Water (TIP3P, 50 °C)	45.7	52.2	67.9
Water (TIP4P-Ew, 50 °C)	53.2	57.4	67.9
Hexadecane/water (TIP3P, 25 °C)	44.9	48.2	53.3
Hexadecane/water (TIP4P-Ew, 25 °C)	53.4	57.3	53.3

cates the normal pressure, and A is surface area. This allows the cell height to adjust, and the densities in the centers of the fluid slabs to relax to their bulk values.

Surface tensions converge relatively rapidly in simulations of most liquids. A precision of one dyn/cm can be obtained for simple interfaces (alkane/water and alkane/air) in several ns. As for the isothermal compressibilities, long-range LJ interactions are important and must be included for quantitative assessments of the potential. From Table 2, approximately half of the surface tension of alkane/vapor interfaces is attributable to long-range LJ terms. Once treated correctly, the surface tensions of alkanes are in excellent agreement with experiment. Both TIP3P and TIP4P-Ew underestimate the surface tension of water by 23% and 15%, respectively. The errors in surface tension of hexadecane/water are -10% for TIP3P and $+8\%$ for TIP4P-Ew. Hence, interactions of water and alkane that take place on the boundary of the acyl chains of the bilayer and waters solvating the head groups are reasonably, but not quantitatively, described, in the CHARMM lipid FF.

III. MEMBRANE TARGETS AND RELATED ISSUES

Most of the results described in this section are for DPPC at 50 °C. While unsaturated lipids are more common in biological membranes and are available in CHARMM, the wealth of experimental data available for DPPC make it a prime target for parameter testing and development. The systems generally contain 72–80 fully hydrated lipids, and were simulated for 50–100 ns at a surface area fixed to the experimental value of 64 $Å^2$/lipid. Important exceptions are noted, and the details can be found in the original references.

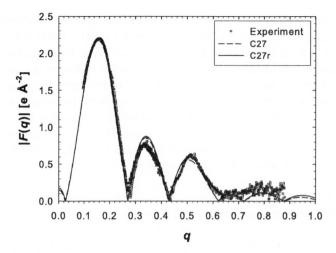

FIGURE 6 Form factors $F(q)$ for DPPC from simulation (Klauda *et al.*, 2007) and experiment (Kučerka *et al.*, 2006).

A. X-Ray Diffraction of Liquid Crystals and Structural Models

X-ray and neutron diffraction provide critical structural observables: the bilayer thickness and area, and the density distributions of its components. While comparisons with Fourier reconstructions (Levine and Wilkins, 1971), D-spacing (Lewis and Engelman, 1983; Rand and Parsegian, 1989), average atom positions (Buldt *et al.*, 1979; Zaccai *et al.*, 1979) have proven useful, the most rigorous comparison of simulation and experiment is through the structure factors, $F(q)$ (Benz *et al.*, 2005; Nagle and Tristram-Nagle, 2000). These are related to the total lipid density $\rho(z)$ by

$$F(q) = \int_{-D/2}^{D/2} \left[\rho(z) - \rho_W\right] \cos(qz) dz, \tag{15}$$

where ρ_W is the electron density of pure water, and D is the length of the unit cell perpendicular to the bilayer normal. Figure 6 compares the results of C27 and C27r with experiment. Both parameter sets yield the experimental positions of the first three lobes ($q = 0.16, 0.35$ and 0.52). The root mean squared deviation (RMSD) with experiment can provide a convenient metric for ranking. In this case, however, partly because of the relatively large experimental errors at high q and the similarity of C27 and C27r, the RMSD of the two FF are comparable.

Structure factors are very sensitive to the surface area per lipid, and can be used to estimate surface areas from simulation. For example, simulations of dimyristoylphosphatidylcholine (DMPC) bilayers were carried out at a range of surface areas that bracket the value of 60.6 Å^2/lipid obtained (with some assumptions) from the experimental data. The RMSD in $F(q)$ from simulation and experiment equaled 0.22 ± 0.007, 0.072 ± 0.007, 0.044 ± 0.003, 0.047 ± 0.002, 0.12 ± 0.001, at 55.0, 59.7, 60.7, 61.7 and 65.0 Å^2/lipid, respectively. Hence, differences of 1.0 Å^2/lipid can be distinguished in cases where the quality of the experimental data is high. Furthermore, the best agreement was obtained by the simulation with a surface area (60.7 Å^2/lipid) nearest to experiment. The success of this exercise provides further support for the CHARMM FF (Klauda *et al.*, 2006b). It also implies that simulations (with a well validated FF) can be combined with experimental diffraction data to obtain surface areas for multicomponent bilayers. This is a welcome advance because these areas are very difficult to obtain from experiment alone.

An important thrust in present day analysis is the comparison of the densities of the individual membrane components from experiment and simulations. While the component densities are easy to calculate from simulation, extracting them for experiment is not straightforward, especially for fully hydrated systems. This is because the component densities are not directly observed. Rather, they must be related to the total density by a "structural model", such as the following (Klauda *et al.*, 2006b):

$$\rho(z) = \rho_P(z) + \rho_{CH_3}(z) + \rho_{CG}(z) + \rho_{CH_2}(z) + \rho_{BC}(z), \tag{16}$$

where $\rho_P(z)$ is for the phosphate groups, $\rho_{CH_3}(z)$ for the terminal methyls, $\rho_{CG}(z)$ for the carbonyl+glycerol, $\rho_{CH_2}(z)$ for the methylenes on the hydrocarbon chains, and $\rho_{BC}(z)$ for the water + choline (BC). The functional forms are assorted Gaussian and error functions, and require a total of 11 parameters. This is reduced to 5 by introducing data (*e.g.*, the lipid molar volume) and selected assumptions. From a practical point of view, the large number of parameters and constraints makes it difficult to deduce component densities without some ambiguity.

In spite of these limitations, a component analysis provides a very useful test of the simulation. Figure 7 shows the results for DPPC. It is clear that the major features of the density, the phosphate peak and the methyl trough, are largely reproduced, though a careful analysis of the latter reveals that experiment and simulation are statistically different. Is the present 7% disagreement of simulation and experiment for the total density in the bilayer midplane acceptable, or is there reason to try to do better? The component analysis indicates that the agreement is partly the result of cancellation of errors: the methyl density is 21% low and the methylene is 25% high. Such difference is of greater concern, and will be monitored in future parameter development. Other discrepancies include the water distribution, which partially reflects the approximations inherent in the TIP3P water model.

While C27r yields nearly quantitative agreement with experiment for fluid phase DPPC and DMPC at the experimental surface area per lipid, the situation with gels is less clear. The gel phase is tightly packed and motion is restricted,

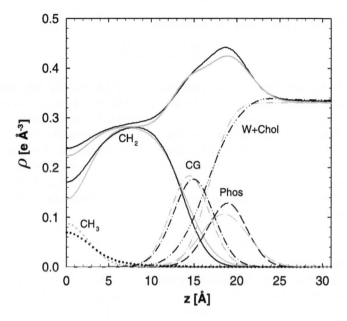

FIGURE 7 The electron density of the DPPC bilayer from simulation (black) (Klauda *et al.*, 2007) and experiment (Kučerka *et al.*, 2006) (grey). The individual component densities are also shown: CH_3 = methyl, CH_2 = methylene, CG = carbonyl-glycerol, P = phosphate, and W + Chol = water and choline.

making equilibration difficult. The utility of the gel phase as a target for parameterization is limited until these issues are resolved.

B. Deuterium Order Parameters and the Signature at C2

The deuterium order parameter, S_{CD}, provides an indispensable measure of both disorder and fine-structure in lipid bilayers. However, its interpretation is not entirely straightforward, and values calculated in simulation do not necessarily correspond to quantities that can be measured. The subsection attempts to clarify some potential misconceptions and highlights a target that deserves more attention.

Experimentally, a fraction of the hydrogens at selected carbons are replaced by deuteriums, and the residual quadrupolar coupling of the CD bond is measured. If the CD bond is axially averaged about a director, the spectrum will appear as a symmetric powder pattern (Seelig, 1977). S_{CD} is proportional to the distance between peaks (or splitting, $\Delta \nu$) in the powder pattern and is related to angle θ

that the CD bond makes with the director by

$$S_{CD} = \left| \left\langle \frac{3 \cos^2 \theta - 1}{2} \right\rangle \right|, \tag{17}$$

where the brackets signify an average over all orientations. The coincidence of the director and normal has been established for fluid phase bilayers using measurements on oriented samples, so θ can be assumed to be the instantaneous angle of the CH vector and the bilayer normal in most applications. The absolute value sign, which is not always included in definitions, is included here as a reminder that S_{CD}, like $\Delta \nu$, is always positive. Residual dipolar couplings of CH and HH can be negative (Gross *et al.*, 1997), and therefore contain additional information.

To a reasonable approximation, S_{CD} can be written as the product of three order parameters:

$$S_{CD} = S_{int} \times S_{wobble} \times S_{collective}, \tag{18}$$

where S_{int} is associated with internal motions such as *gauche-trans* isomerization, S_{wobble} arises from the rigid-body diffusive rotation motion of single lipids sometimes denoted as "wobble in a cone", and $S_{collective}$ comes from collective rotations of groups of lipids. However, S_{CD} only reflects motions that are averaged on the time scale of deuterium quadrupolar coupling constant, 170 KHz, or approximately 10^{-5} s. Hence, very slow collective motions (*e.g.*, those on the second time scale) do not contribute to the average in Eq. (17).

The preceding conditions place some restrictions on simulators. For example, DPPC gels are known *not* to be axially averaged (Davis, 1983), so reporting S_{CD} from a simulation of a gel is inappropriate. In principle, axial averaging (or its absence) can be demonstrated. If the z-axis is taken to be the axis of averaging, axial averaging implies that the averages $\langle \hat{x}^2 \rangle = \langle \hat{y}^2 \rangle$, and $\langle \hat{x} \rangle = \langle \hat{x}\hat{y} \rangle = \langle \hat{x}\hat{z} \rangle = \langle \hat{y} \rangle = \langle \hat{y}\hat{z} \rangle = 0$, where the "hat" signifies the appropriate unit vector projections. In effect, the lipid rotationally diffuses about the z-axis and averages out the x and y projections within 10^{-5} s. Clearly it is not practical to demonstrate that a gel is not axially averaged at present, though it is easy to demonstrate axial averaging for the fluid state.

Present simulations of lipid bilayers are usually on the 50–100 Å length scale, and 10–100 ns time scale. Consequently, motions on larger length scales (*e.g.*, 1000 Å) whose time scales *are* well averaged on the 10^{-5} s time scale would not be observed in simulations of a small system because of constraints imposed by periodic boundary conditions. If the contribution of such motions to S_{CD} is not negligible, the agreement with experiment obtained for small systems is fortuitous.

The experimental and simulated order parameters for DPPC are plotted in Fig. 8. There are two very characteristic features of the order parameter profile. The first is a plateau region for the chain carbons extending from carbonyl region to around carbon 10, followed by a drop-off to lower values. The value of

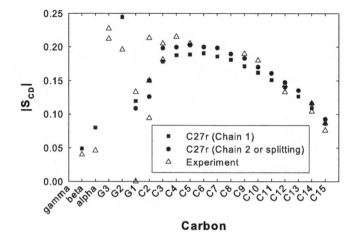

Carbon

FIGURE 8   Deuterium order parameters for DPPC from simulation ((Klauda *et al.*, 2005a) for chains; glycerol and headgroup unpublished) and experiment (Gally *et al.*, 1981; Seelig and Seelig, 1974; Seelig and Seelig, 1975; Seelig *et al.*, 1977).

the plateau region, approximately 0.2, indicates intermediate disorder, and is a basic target for any force field. While the decomposition of S_{CD} into its components is approximate, a range of 0.6–0.7 for S_{wobble} appears reasonable (Feller *et al.*, 1997a). Assuming that the contribution of collective motions is small, $S_{int} = 0.29$–0.33 for the plateau, and it drops further down the chain. S_{int} for these carbons is largely determined by the quality of the alkane potential, which highlights the utility of parameterizing the FF from the molecular fragments. Like $F(q)$, S_{CD} is sensitive to the surface area per lipid (Feller *et al.*, 1997a). This dependence appears more related to S_{wobble} than S_{int} for small deviations about the experimental area, and has a simple interpretation. A medium length alkane such as hexadecane rapidly isomerizes and tumbles in all directions; *i.e.*, there is no residual anisotropy. In a bilayer, the acyl chains isomerize and have similar conformational distributions as alkanes (Fig. 9). The difference in the two environments (from the perspective of the hydrocarbons) is that the chains of lipids are tethered at one end to the bilayer/water interface, while those in the alkane/water are not constrained in this manner. Expansions in the bilayer area are compensated by contractions in the bilayer thickness, leaving the bilayer interior substantially unchanged. However, wobbling is restricted. An expansion in the area increases the effective wobble angle, and thus decreases S_{wobble}.

The other characteristic feature is the splitting at carbon 2 on chain 2 (values of 0.09 and 0.15). Carbon 2 of chain 1 ($S_{CD} = 0.205$), in contrast, groups with the other carbons in the plateau region. This inequivalence of the chains has been observed in numerous deuterium studies of both saturated and unsaturated lipids (Davis, 1983; Seelig, 1977), and appears to be critical structural

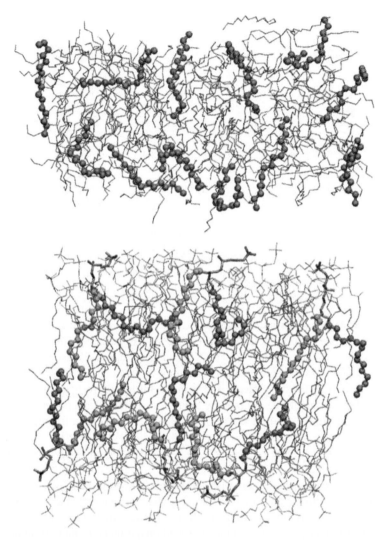

FIGURE 9 Snapshots from simulations of hexadecane/water (top) and a DPPC bilayer (bottom). 12 chains are highlighted in each graphic (chains 1 and 2 of DPPC are colored green and red, respectively), and the waters are omitted for clarity.

feature of bilayers in their liquid-crystal phase. Given the striking "chair-like" structure in the glycerol region of lipids in their crystal forms (Pascher *et al.*, 1992), some residual inequivalence of the chains might be expected. Therefore, it is disconcerting that C27r does *not* reproduce this feature of the order parameter profile. Rather, $S_{CD} \approx 0.16$ for all 4 CH vectors at carbon 2. A substantial spitting at C1 of the glycerol (labeled G1 in Fig. 8) is also not reproduced by the parameter set. The order parameters of the other glycerol carbons and the head group are reasonably obtained. It is possible that hydrogen bonds between the carbonyl oxygens and hydrogens on the chain and glycerol promote the observed pattern.

C. NMR Spin-Lattice Relaxation Times and Possible Collective Effects

Just as the deuterium order parameter probes the conformational preference of specific carbons of lipids in membranes, the spin lattice (or T_1) relaxation time yields information on the underlying dynamics. Values even at a single frequency are useful. Measurements taken at multiple fields enable a decomposition of the dynamics into fast and slow motions, and offers further targets for parameterization. Conversely, motional models are typically required to relate experimental decay constants to specific motions. Simulations are very useful in this regard, as the applicability of a motional model can be evaluated and alternative models examined. ^{13}C T_1 are the focus here, though other NMR relaxation data such as deuterium (Brown et al., 2002) and ^{31}P T_1's (Roberts and Redfield, 2004) are also being evaluated (Klauda et al., 2008b).

As already noted in Section II.C.4, T_1's of liquid alkanes are independent of the magnetic field strength. In contrast, $\langle P_2(\hat{\mu}(0) \cdot \hat{\mu}(t)) \rangle$ for lipids in bilayers contains multiple decays, including fast "alkane-like" ones. Hence, an important step in the validation of a lipid FF is a demonstration that experimental T_1's for liquid alkanes comparable in length to the acyl chains of lipids are reproduced. This was described in Section II.C.4 for C27r.

Figure 10 shows the overall very good agreement of simulated (calculated from Eq. (10) and experimental T_1's for DPPC at 500 MHz, including the alkane-like region near the bilayer midplane, and the more restricted head group/water interface (Klauda et al., 2008a). For carbons where both vesicle and multilayer data are available, simulations with C27r tend to be closer to the latter. This is encouraging, in that the multilayers are flat on a large length scale and thereby correspond more closely with the simulation geometry (where periodicity enforces flatness). The discrepancy of simulation and experiment at C3 is interesting. The simulated T_1 is too low, implying that motion is too slow. From Fig. 1, the dynamics of C3 is directly impacted by the β_4 and γ_4 torsional surfaces, which, from Fig. 3 (middle and bottom), contains a *trans-to-gauche* barrier that is approximately 2 kcal/mol too high for C27r. It is reasonable to anticipate that these changes will yield lead to faster dynamics and thereby better agreement with experimental T_1's at C3.

The vesicle data of (Brown et al., 1983) was obtained at a very wide range of frequencies. While its interpretation has been somewhat controversial, it has proven very valuable to the development of the field. As shown Fig. 11, the data for the average of carbons 4-13 (taken from 15 to 125 MHz) are remarkably linear

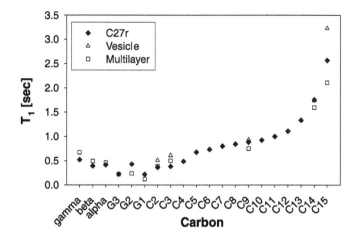

FIGURE 10   Spin lattice relaxation times at 500 MHz (hydrogen) for C27r and experiments on vesicles (Brown *et al.*, 1983) and multilayers (Klauda *et al.*, 2008a). The experimental points at C9 are the average of C4–C13. T_1's for the acyl chain carbons are averaged.

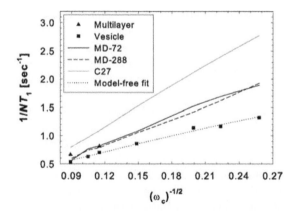

FIGURE 11 $1/T_1$ vs. $\omega_C^{-1/2}$ of carbons 4–13 (averaged) for experiments on vesicles (Brown *et al.*, 1983), multilayers (Klauda *et al.*, 2008a), simulations with C27r at two system sizes, and C27 (Pastor *et al.*, 2002), and a model free fit to experiment (Eq. (19)) with $A_j^2 = 0.035$, $\tau_j = 24.5$ ps and $\tau_s = 2.1$ ns. The model-free parameters published earlier (Szabo, 1986), $A_j^2 = 0.035$, $\tau_j = 20.4$ ps and $\tau_s = 1.8$ ns, were obtained with $r_{C–H} = 1.1$ Å and thereby differ somewhat from those obtained here ($r_{C–H} = 1.117$ Å).

when plotted versus $\omega_C^{-1/2}$; $1/T_1$ for the other carbons show similar behavior. This dependence was interpreted by Brown to indicate collective motions. Szabo (1986) analyzed the same data with a "model-free" formalism (Lipari and Szabo,

1982), where the spectral densities (Eq. (10)) are written:

$$J(\omega) = \left(1 - A_j^2\right)\tau_j + \frac{A_j^2 \tau_S}{1 + (\tau\omega)^2}, \tag{19}$$

where τ_j is a fast relaxation time for each carbon and A_j^2 is the generalized order parameter (not to be confused with deuterium order parameter), and τ_S is a slow relaxation common to all of the carbons. A single slow relaxation time is more consistent with single lipid diffusive reorientation than with collective motion. (Restricted diffusion leads to multiexponential decays (Szabo, 1984), but one tends to dominate.) The excellent fit of Eq. (19) to the experimental data (Fig. 1, dotted line) indicates that a non-collective model for the frequency dependent T_1's is at least plausible in this frequency range. The results of Brownian and MD simulations (Pastor *et al.*, 2002) support this proposal, and have associated τ_S with a pendulum-like diffusive orientation (or "wobble") of individual lipids.

The fast relaxation times obtained by Szabo's analysis have been quantitatively reproduced in MD simulations (Pastor *et al.*, 2002), which has further validated the alkane parameters and confirmed the proposal of Brown that the fast dynamics corresponds to torsional dynamics of the acyl chains. However, as evident in Fig. 11, C27 does not lead to the observed frequency dependence. This discrepancy can be explained in three ways: (1) C27 is inadequate; (2) the simulation system (72 lipids) is too small; (3) the data contain artifacts because of the high curvature of the vesicles, which are only about 250 Å in diameter. Recent results suggest that the answer is largely a force field issue (Klauda *et al.*, 2008a). Specifically, T_1's from the multilayers are not substantially differently from those from the vesicles (Fig. 10). Simulations with C27r with 72 and 288 lipids yield almost identical T_1's (Fig. 11), and they are in substantially better agreement with both multilayer and the vesicle data. While this does not rule out collective motions on longer length/lower frequency scales, it points to errors in the C27 FF. The argument is analogous to that provided for pentadecane in Section II.C.4: the *trans-to-gauche* barrier of C27 is 0.5 kcal/mol too high, which would decrease the fast component of T_1, and lead to the overestimate of $1/T_1$ shown in Fig. 10. The barrier is reduced for C27r, and the agreement with experiment at high field is substantially improved. The frequency dependence for C27r is also in closer agreement with experiment than for C27. This can partially be explained by the broadening of the torsional potential influenced by the revision of C27. This broadening and the increased isomerization rate amplify the contribution of fast motions to the spectral density (A_j^2 is reduced in Eq. (19)), and the apparent slope in a plot of $1/T_1$ vs. $\omega_C^{-1/2}$ is decreased. A final resolution of the discrepancy of simulations must await further improvements in the FF, simulations of larger systems including vesicles, and NMR measurements of DPPC multilayers at lower fields. Nevertheless, the calculation of NMR T_1's provides an important assessment of a parameter set.

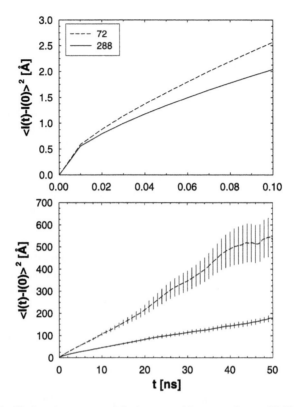

FIGURE 12 The lateral mean squared displacement of the center of mass of lipids for systems of 72 and 288 lipids over the intervals 0–100 ps (top) and 0–50 ns (bottom). Standard errors are denoted with vertical bars (Klauda *et al.*, 2006a).

D. Translational Diffusion Constants and a Finite Size Effect

Experimental values of the lateral translational diffusion constant, D_ℓ, for lipids in bilayers have been available for many years, but it is only until recently that simulations could be carried out for sufficient lengths to calculate them. The slope of D_ℓ is obtained from the mean squared displacement in the xy plane vs. time, which equals $4 \times D_\ell$, analogous to translational diffusion in a bulk phase (Eq. (8)). Adjustment for net translation of each monolayer is neither necessary nor recommended. Figure 12 plots the MSD from simulations of 72 and 288 lipids. Three features are evident: (1) The short-time slopes of the two systems are similar; (2) they differ from the long-time slopes; (3) the long-time slopes are very different from each other.

The first two features are easy to understand from Fig. 13, which plots the CM trajectory of one of the lipids from the system containing 288 lipids. The CM

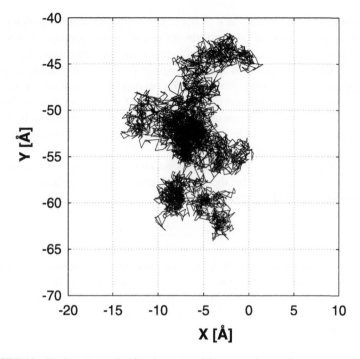

FIGURE 13 The lateral trace (in 10-ps intervals) of the center of mass of a lipid over 50 ns in the simulation of 288 lipids (Klauda *et al.*, 2006a).

fluctuates in a local region and then makes two jumps to other regions in the 50-ns trajectory. Others lipids make different numbers of jumps, as expected from a Poisson process, but the qualitative behavior is similar. This is the underlying reason for the two slopes in MSD: the local motion yields the rapid increase at short time; the infrequent (though substantial) jumps produce the smaller slope at long times. Strictly speaking, there is only a single translational diffusion constant for the system, and this is related to the long-time slope. It can be measured by NMR or photobleaching. The experimental values of 1.25×10^{-7} cm^2/s at 50 °C (Vaz *et al.*, 1985) and 1.52×10^{-7} cm^2/s at 51 °C (Scheidt *et al.*, 2005) are fairly close to 0.95×10^{-7} cm^2/s obtained from simulation for the 288 system, but differ greatly from the 2.92×10^{-7} cm^2/s from the 72 lipid system (Klauda *et al.*, 2006a). The short-time diffusion constant, sometimes denoted, D_ℓ^{cage}, is obtained from neutron scattering, and equals 12.0×10^{-7} cm^2/s at 55 °C (Tabony and Perly, 1991). The simulated values, 13.9 and 14.7×10^{-7} cm^2/s for the large and the small systems, respectively, agree well with this value.

The difference in D_ℓ for the two systems is a striking example of a finite size artifact. This is analyzed in detail using a Poisson analysis based on clusters calculated from the CM trajectory (Klauda *et al.*, 2006a), but a qualitative explanation gets to the core of the effect. Each leaflet in the 72 lipid system is a periodic 6×6 array of lipids. A jump of one perturbs its neighbors and propagates across the periodic boundaries leading to bursts of highly correlated transitions. In contrast, the disturbance associated with a jump dissipates before the periodicity of the 288 lipid system comes into play. Removing the net displacement of each leaflet substantially lowers the apparent diffusion constant of the smaller system, but only hides the artifact and is therefore not correct. A more general lesson is: if an effect seems especially interesting, confirm it's not an artifact. In this case, quadrupling the size of the system was well worth it.

As a final point, a jump model affords estimates that set the time scale of the diffusion process and, thereby, the statistical errors in the simulation. The diffusion constant is related to the jump rate λ by

$$D_\ell = \frac{\lambda b^2}{4}, \tag{20}$$

where b is the jump length (7.5 Å from the analysis in (Klauda *et al.*, 2006a)). Rounding to one significant figure, $\lambda = 4 \times (1 \times 10^{-7})/(8 \times 10^{-8})^2 = 0.06 \times 10^9$ s^{-1}. This yields 3 transitions per lipid in the 50 ns simulation (one jump every 15 ns), and approximately $N = 900$ total transitions for 288 lipid system. From Poisson statistics, the statistical error in N, and therefore D_ℓ, is $N^{-1/2} = 1/30$, or 3%.

E. Elastic Moduli and Slow Averaging

The malleability of surface area of lipids is vital to their function in cells, so the elastic moduli of bilayers are an important target for parameterization. The bulk elastic modulus K_b is the inverse of the isothermal compressibility (Section II.C.3). The surface area elastic modulus, K_a, can be evaluated from derivatives or from fluctuations, as follows

$$K_a = A \left(\frac{d\gamma}{dA} \right)_T = \frac{k_B T \langle A_{tot} \rangle}{\langle \delta A_{tot}^2 \rangle}, \tag{21}$$

where $\langle V \rangle$, $\langle A_{tot} \rangle$, $\langle \delta V^2 \rangle$ and $\langle \delta A_{tot}^2 \rangle$ are the average (total) volumes and areas, and their fluctuations. K_a has been measured for a range of phosphatidylcholines (Rawicz *et al.*, 2000). After correcting surface fluctuations associated with undulations, the experimental values of K_a are all about 240 ± 15 dyn/cm. Hence, while K_a does not provide a sensitive measure for distinguishing phosphatidylcholines, it does provide a robust value for parameterization. The surface area compressibility is very sensitive to other components, including alcohols (Ly and Longo, 2004) and cholesterol (Evans and Rawicz, 1990). There are not many measurements of the bulk compressibility. The value for DPPC (Mitaku *et al.*, 1978),

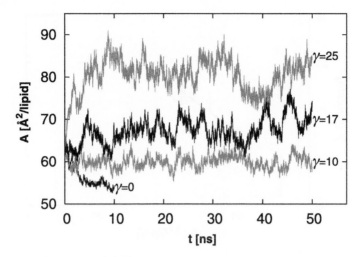

FIGURE 14 $NP\gamma T$ trajectories of DPPC carried out at 4 different applied surface tensions, including $\gamma = 0$ (Venable *et al.*, 2006).

2.1×10^{10} dyn/cm^2, is comparable to incompressible fluids; *e.g.*, K_b for water (20 °C) and pentadecane (39 °C) are 2.2 and 1.1×10^{10} dyn/cm^2, respectively.

It is useful to discuss two technical points before proceeding to the results of the calculations. First, the statistical errors in compressibilities of bilayers are large. Figure 14 plots the time series of a 50-ns trajectory of DPPC carried out at constant surface tension $\gamma = 17$ dyn/cm, PME for electrostatics, and a cutoff of 10 Å for the LJ interactions. (These simulations were carried out before the introduction of the IPS method for LJ terms in CHARMM.) In spite of large fluctuations, the average surface area can be obtained with a 1% statistical error for a trajectory of this length. However, the statistical error in the *fluctuations* of a quantity is always higher than that of the quantity. Consequently, the error in K_a is dominated by errors in $\langle \delta A_{tot}^2 \rangle$, leading to $K_a = 92 \pm 25$ dyn/cm. Similar considerations explain the large statistical error in the volume compressibility, $K_b = 1.5 \pm 0.4 \times 10^{10}$ dyn/cm^2.

The second technical point is that K_a can also be evaluated from the derivative relation in Eq. (21). Because the *NPAT* and $NP\gamma T$ ensembles yield statistically equivalent $\gamma - A$ isotherms (Fig. 15), this allows some pooling of simulations at different surface area and surface tensions, and interpolation to areas not explicitly simulated. However, care must be exercised to only use the linear regime of the $\gamma - A$ isotherm. Figure 15 shows the $\gamma - A$ isotherm for a combination of *NPAT* and $NP\gamma T$ simulations of DPPC. Fitting the points between 59 and 68 Å^2/lipid (a total of 175 ns of data) yields $K_a = 138 \pm 26$ dyn/cm for $A = 64$ Å^2/lipid.

The simulated results for K_b and K_a underestimate experiment by over 30%. It is likely that a substantial portion of this error arises from the neglect of

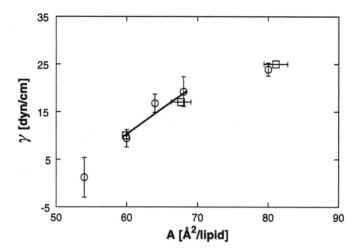

FIGURE 15 Surface tension–surface area isotherm for DPPC simulated at *NPAT* (circles) and *NPγT* (squares). The solid line between 60 and shows the surface area range used to calculate the compressibility from $d\gamma/dA$ at $A = 64$ Å^2/lipid (Venable *et al.*, 2006).

long-range LJ terms. Two results suggest this possibility. As already noted in Section II.C.3, addition of a long-range correction improves agreement of simulation and experiment for neat alkanes. While K_b for bilayers is most appropriately evaluated from $NP\gamma T$ simulations, volume fluctuations from recent from *NPAT* simulations with a pressure-based LRC and with IPS yield 1.8 and 1.7×10^{10} dyn/cm^2, respectively; *i.e.*, approximately halfway between the values of 1.5 and 2.1×10^{10} dyn/cm^2 obtained by simulations with no long-range LJ and experiment. In summary, long-range Lennard-Jones terms should not be ignored when evaluating elastic constants.

F. Surface Area in Bilayers, and the Value of the Surface Tension

A particularly thorny aspect of force field development is the surface area per lipid. This is partly because the quantity is difficult to obtain experimentally, and accurate values are only available for only a few single component bilayers (Nagle and Tristram-Nagle, 2000; Wiener *et al.*, 1991). Hence there are few targets for calibration of parameters. To proceed further requires a discussion of the surface tension of a bilayer and a very delicate balance of forces.

C27r yields a surface tension of approximately 20 dyn/cm/side for a DPPC bilayer consisting of 72 lipids when simulated at the experimental surface area. The values for other phosphatidylcholines are comparable ((Klauda *et al.*, 2006b) and unpublished data). In contrast, the experimental surface tension of pure black

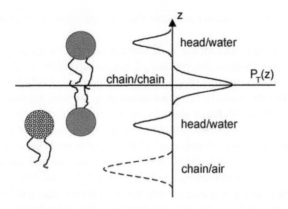

FIGURE 16 A schematic of the tangential profiles for a lipid bilayer (heads represented with grey circles) and a monolayer (stippled head).

lipid films is only 1 dyn/cm (Tien and Diana, 1968). Furthermore, based on arguments regarding the balance of forces in self assembled systems (Israelachvili *et al.*, 1977; Jahnig, 1996), the surface tension of a large flaccid vesicle is zero.

There are two different physical arguments that lead, in principle, to small or zero γ for a macroscopic sized bilayer. The first is sketched in Fig. 16. The tangential pressure in the head group/water interface is negative, as would be expected for an interface of water and an organic liquid. This negative pressure is balanced by a *positive* tangential pressure in the bilayer midplane arising from repulsive interactions of the chains. Integrating across the entire bilayer then leads to a net surface tension of zero at all length scales, even a microscopic patch. If this argument is correct, C27r is clearly flawed.

The other explanation invokes the large undulations over thousands of angstroms that are known to occur in flaccid bilayers (Brochard *et al.*, 1976). These entropic fluctuations balance the positive surface tension in small patches, and, on a *macroscopic* length scale, lead to a surface tension of near zero. In this view, the tangential pressure in the bilayer midplane could be positive, but not sufficiently positive to balance the negative tangential pressure in the head group/water interface. Estimates of the finite size effect for simulation-sized patches vary. Initial estimates (Feller and Pastor, 1996) yielded the range 20–50 dyn/cm, which is comparable to the values obtained from simulation with C27r. A more rigorous analytical approach (Marsh, 1997) led to 4–6 dyn/cm. Pipette aspiration experiments (Rawicz *et al.*, 2000) indicate that the tension required to smooth out a flaccid bilayer is less than a dyn/cm. Recent simulations based on the CHARMM potential on up to 288 (Castro-Roman *et al.*, 2006) and 256 (Herce and Garcia, 2006) lipids have *not* shown evidence of system-size dependent surface tension. While much larger systems will be required to conclusively disprove the finite-size hypothesis, most evidence indicates that surface tensions obtained with C27r are too large.

One's view of the appropriate local surface tension, necessarily zero or potentially positive, impacts the process of parameterization. In the former, the requirement for zero surface tension can be (and perhaps should be) imposed. As an example, Berger *et al.* (1997), when developing an early version of a lipid FF used in many GROMOS based simulations, found that their original parameters yielded a positive surface tension, and then reduced the charges of head groups to yield zero surface tension. Sonne *et al.* (2007) have recently reparameterized the partial charges of the CHARMM C27r FF to obtain zero surface tension of DPPC near the experimental surface area. We have adopted the view that the surface tensions *could* be positive in a microscopic patch of membrane, and have therefore not imposed zero surface tension as a constraint. This is based on the notion that the realistic surface tension should arise from proper treatment of fundamental terms in the FF, and the realization that more than one variation of the parameters can yield zero surface tension. Nevertheless, surface tensions are routinely evaluated in the development process even though they are not used as a specific target.

Theoretical considerations aside, simulations of most bilayers using C27r must either be carried out at the experimental surface area in the *NPAT* ensemble or at $NP\gamma T$ with an applied surface tension of approximately 20 dyn/cm to obtain sensible results. Most of our simulations related to parameterization are carried out at *NPAT*. Although the surface tension–surface area isotherms for *NPAT* and $NP\gamma T$ are equivalent (Fig. 15), it is difficult to pick the precise value of the surface tension that yields the experimental area. However, simulating at *NPT* (equivalent to $\gamma = 0$) leads to disaster: the surface contracts from 64 Å^2/lipid to 52 Å^2/lipid (a gel-like state) in approximately 10 ns (Fig. 14). This because the "pulling in" of the surface tension of the system is not opposed by an applied surface tension. The same considerations apply to the GROMOS96 45A3 (Chandrasekhar *et al.*, 2003) and General Amber Force Field (GAFF) (Jojart and Martinek, 2007). Both show positive surface tensions at the experimental lipid surface area.

While there is arguably some uncertainty in the surface tension of a microscopic patch of a bilayer, the situation with monolayers is quite clear. Therefore, monolayers are an excellent target for parameterization. Figure 17 compares the $\gamma-A$ isotherms for DPPC monolayers from simulation with C27r and experiment. The agreement is remarkably good, and indicates that the CHARMM FF captures the many interactions of lipid/water and alkane/air interfaces. Other parameter sets, including those explicitly tuned to yield zero surface tension for bilayers should be tested against monolayers.

The large positive surface tensions of monolayers, approximately 40 dyn/cm for DPPC at 64 Å^2/lipid are consistent with both zero or positive surface tension for bilayers. In monolayers the chain/chain region is replaced by the chain/air

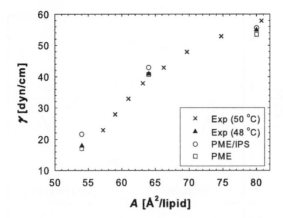

FIGURE 17 The simulated DPPC lipid monolayer surface tension at three surface areas for PME/IPS and PME with an $r_c = 10$ Å (Klauda *et al.*, 2007). The experimental surface tensions, shown for 50 °C (Crane *et al.*, 1999) and 48 °C (Somerharju *et al.*, 1985) were obtained from Π / A isotherms using the using the relation $\Pi = \gamma_0 - \gamma$, where γ_0 is the surface tension of pure water at the experimental temperature.

interface. The tangential pressure profile thus contains two regions of negative pressure, as sketched in Fig. 16. Because the monolayer is macroscopically flat, there is no cancellation by undulations. If the surface tension of the chain/air interface is assumed to be equal to that of hexadecane/air (25 dyn/cm), the head group/water surface tension of a monolayer is about 15 dyn/cm. This value is close to the bilayer surface tensions presently obtained with C27r.

As a final technical point, the *NPT* simulation described above was carried out in tetragonal boundary conditions (the surface area and height can respond independently). A bilayer simulated in cubic conditions, where the shape of the cell is preserved, will not shrink substantially when simulated at *NPT*. This is because the pressure increases upon even a small volume contraction prevent further decrease in either height or surface area.

G. Dipole Potentials and Polarizable Force Fields

The dipole potential in a neutral membrane arises from the nonuniform distribution charges normal to interface. Given the importance of electrostatic signaling by molecules including phosphoinositides such as PIP2 and PIP3, the dipole potential is a natural target for a FF. Unfortunately, the absolute dipole potential is not possible to obtain for simple interfaces such as water/air (Paluch, 2000) or alkane/air, and values do not seem to be available for alkane/water interfaces. Hence, unlike the surface tension, simple systems do not provide undisputable target data. Furthermore, the dipole potential of bilayers is difficult to measure,

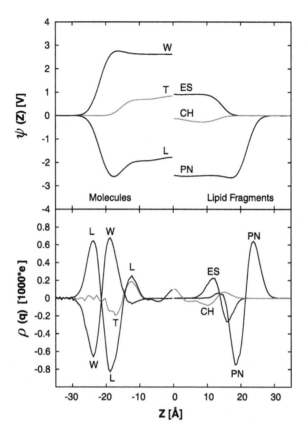

FIGURE 18 The simulated dipole potential of DPPC for the lipids (L), water (W) and total (T) (top left), and the acyl chains (CH), ester (ES) and PN groups (top right). CH includes carbons 2–16 of each chain; ES contains the glycerol C_{G1} and C_{G2}, C_1 of each chain and all atoms between; PN includes C_{G3} and phosphocholine group. The fragments are electrically neutral, and include bound hydrogens. The bottom plots the charge distribution for each of the preceding entities.

and many results are considered controversial because of localization of voltage sensitive dyes. The range 225–250 mV is considered reasonable (Clarke, 2001), although recent experiments based on freeze-fracture raise the possibility that the true value is higher (Wang *et al.*, 2006).

The dipole potential is calculated from a simulation by double integration of Poisson's equation,

$$\psi(z) - \psi(0) = -\frac{4\pi}{\varepsilon_0} \int_0^\infty \int_{z'}^\infty \rho_c(z'') dz'' dz', \qquad (22)$$

where ρ_c is the time averaged charge density. The results are shown in the top panel of Fig. 18. Starting with the left side, the total potential drop proceeds from the center of the bilayer ($z = 0$) to the center of the water ($z = -33$). The value is 0.850 V, well over 3 times that of experiment. This impressive level of disagreement can be seen as the sum of two much larger terms: a 2.7 V drop for the water, and a 1.9 V rise for the lipid. The right hand side of the plot shows the contributions of the chains, the ester and PN groups. The PN group dominates, while the chains contribute relatively little, as expected from the small partial charges on the carbons and hydrogens. The potential drop associated with the ester group approximately equals that of the entire system. Further insight is obtained from the component charge distributions. While the large (and mostly compensating) values in the head group/water interface are not surprising, the excess positive charge in the center the bilayer is a possible cause for concern and will be monitored in future development.

It is clear from Fig. 18 that better agreement with experiment could be obtained in a variety of ways; *e.g.*, increasing the magnitude of the PN contributions or decreasing that of the water. However, as noted in Section II, the charges on these components were developed in part by fitting to solvation data. Studies of polarizable models have shown that the charges and dipoles of additive modes tend to be too large, a consequence of correctly obtaining the free energies of solvation. Charges in polarizable models are smaller, and dipole potentials of simple systems are substantially smaller (Patel and Brooks, 2006; Vorobyov *et al.*, 2007). It is possible that the correct description of the dipole potential drop in membranes will require an explicit treatment of polarizabilty.

IV. SUMMARY AND PERSPECTIVES

This chapter contains two broad themes. The first is the underlying approach to development of the CHARMM force field. This begins with *ab initio* calculations of small molecules, adjustments as necessary to reproduce target data of condensed phases of these and closely related molecules, and special attention to the balance of solute-solute, solvent-solvent, and solute-solvent interactions. After the small molecule results are satisfactory, the potential function suitable for simulation of the large assembly (in this case a lipid bilayer) is formulated, and simulations are carried out to compare with appropriate target data. It is not likely that the basic approach will change in the near future, because it provides a fundamental connection to the underlying physics of the system.

The second theme of this chapter is the particular *ab initio* and molecular dynamics methods, and the targets for validation. These change continually, and sometimes rapidly. Advances in methodology and increases in computer speed and memory will permit *ab initio* calculations at higher levels of theory, more complete basis sets for larger systems, and explicit treatment of condensed phases. Molecular dynamics trajectories will be longer and on larger assemblies. Additional high-quality experimental data on lipid systems, including mixtures, will become available. Even the form of the potential energy function, Eq. (1), will change as more complex interactions (*e.g.*, polarizabilty) can be treated efficiently. In this sense, the chapter describes a "snapshot" of the force field, and its development and validation.

So, how good is the CHARMM lipid force field? This is actually two questions. The first regards the comparison with the experimental target data. The second pertains what can be accomplished by a simulation using the force field. The following paragraphs consider both.

A simulation focusing on the bilayer interior can be carried out with high confidence. This is because the densities, isothermal compressibilities, diffusivities, viscosities (Table 1), surface tensions (Table 2), and NMR T_1's of medium length bulk alkanes (Fig. 5) are in virtually quantitative agreement with experiment. As discussed in Section III.B and illustrated in Fig. 9, the conformational distributions of alkanes and the acyl chains of bilayers are nearly identical for chains of the same length. The excellent agreement of simulation and experiment for the acyl chain deuterium order parameters (Fig. 8) follows, in part, from the careful parameterization of alkanes.

Density profiles for bilayers of DMPC and DPPC (Fig. 7) also agree well with X-ray diffraction data, and many features of the head group region are within 15–30% of experiment. These include the NMR T_1's (Fig. 10), lateral diffusion constants (Fig. 12), the surface area elastic modulus (Fig. 15), and the surface pressure-surface area monolayer isotherm (Fig. 17). Hence, the FF provides the essential bilayer environment (hydrophobic interior/hydrophilic surface) necessary for the stability of membrane proteins. Less satisfactory are the deuterium order parameters of the glycerol-carbonyl region (Fig. 8), and the dipole potential drop (Fig. 18). These deficiencies potentially limit studies concerned with detailed lipid protein interactions, and the interactions of ionic species with the head group/water interface.

CHARMM parameters for singly and polyunsaturated chains have been validated in a manner comparable to DPPC. Parameters for phosphatidylethanolamine and cholesterol are available, but have not been exhaustively tested. Sphingomylin and glycolipids are under study. Hence, some biologically more relevant membranes can be simulated, though with somewhat less confidence than pure phosphatidylcholine bilayers. Because the surface areas of multicomponent bilayers are not well known, it is advisable to simulate at a range of surface areas in order to bracket the correct value. If the effect under consideration is very sensitive to the surface area, results of the simulation must be interpreted with extra caution.

Naturally, simulations must be carried out appropriately to obtain the mostly good agreement with experiment described in this chapter. In this chapter, the effects of long-range Lennard-Jones interactions, finite system size, and surface tension were highlighted. Long-range LJ interactions make substantial contributions to the densities, compressibilities (Section II.C.3), and surface tensions (Section II.C.5 and Table 2) of liquid alkanes. Bilayers are more than half alkane, and should be simulated with the isotopic periodic sum method, the pressure-based long-range correction, or a very long cutoff. Translational diffusion is sensitive to finite size effects in both neat fluids and bilayers, though the origins of the effects appear to be different. Equation (6) indicated that the alkane diffusion constants calculated directly with the simulation were increased 25% by the finite size correction (Table 1). In contrast, the lateral diffusion constant of DPPC decreased by more than a factor or three when the system size increased from 72 to 288 lipids (Fig. 12). With this notable exception, systems of 72 lipids are adequate for evaluating the targets described in Section III. This does not imply that larger systems are not important for parameter testing. The simulation of large systems will probe collective motions accessible by NMR relaxation times obtained at lower frequencies (<15 MHz), and lend insight into role of undulations. The most controversial aspect of the CHARMM lipid FF is the presence of a positive surface tension for DPPC at its experimental surface area. From a purely practical point of view, simulations must be carried out at constant surface area (NVT or $NPAT$) or constant surface tension ($NP\gamma T$); the two ensembles are equivalent (Fig. 15). The latter is recommended when it is necessary to include area fluctuations, such as when simulating peptide insertion. Simulating at $NP\gamma T$ is no more difficult than simulating NPT, and the surface area will be maintained (Fig. 14). Simulating complex systems at several different applied surface tensions is recommended.

Present force field development efforts for DPPC are focused on the glycerol-carbonyl region. Resolving the discrepancy of calculated and experimental deuterium order parameters (Fig. 8) at carbon 2 of the chains and carbon 1 of the glycerol is the highest priority. A change in structure of this region would modulate the dipole potential, which is the sum of large negative and positive contributions (Fig. 16). Correctly describing the inequivalence of the chains near the head group should also shift the methyl and methylene distributions in the bilayer center (Fig. 7). Specifically, offsetting chain 2 slightly closer to the lipid/water interface than chain 1 would reduce the methylene density and increase the methyl density and thereby improve agreement with experiment. The effect on the surface tension is less clear, given that surface tensions of monolayers agree well with experiment (Fig. 17). Nevertheless, the bilayer surface tension includes contributions from both the surface and the interior (Fig. 16), so some changes are expected.

A polarizable lipid force field is a longer term goal. The results of simulations with such a FF will provide insight into the some of the discrepancies with experiment discussed here, notably the dipole potential, and may prove essential for simulating membranes with charged lipids and ions. Proceeding to more detail, combined quantum mechanics/molecular mechanics (QM/MM) will be required to simulate chemical reactions in membranes. At the other extreme, coarse grained models allow simulations of huge patches of membranes and thereby can be used to study fusion, mixed phases, and raft formation. Reliable potentials will be required in each case. The basic approach outline in this chapter will allow their development.

Acknowledgements

This work has profited from many valuable interactions with both experimentalists and theorists. We especially thank Bernard Brooks, Michael Brown, Scott Feller, Klaus Gawrisch, Adrian Parsegian, John Nagle, Attila Szabo, Douglas Tobias and Steven White. This research was supported in part by the Intramural Research Program of the NIH, National Heart, Lung and Blood Institute. Some of the simulation results presented here utilized the high-performance computational capabilities of the Biowulf PC/Linux cluster at the National Institutes of Health, Bethesda, MD. (http://biowulf.nih.gov) Financial support from the NIH (GM51501 and GM072558) and computational support from the DOD ASC Major Shared Resource Computing and PSC Pittsburgh Supercomputing Center to ADM Jr. are acknowledged.

References

Allen, M.P., Tildesley, D.J. (1987). "Computer Simulations of Liquids". Clarendon Press, Oxford.

Bayly, C.I., Cieplak, P., Cornell, W.D., Kollman, P.A. (1993). A well-behaved electrostatic potential based method using charge restraints for deriving atomic charges—the RESP model. *J. Phys. Chem.* **97**, 10269–10280.

Becke, A.D. (1993). Density-functional thermochemistry. 3. The role of exact exchange. *J. Chem. Phys.* **98**, 5648–5652.

Benz, R.W., Castro-Roman, F., Tobias, D.J., White, S.H. (2005). Experimental validation of molecular dynamics simulations of lipid bilayers: A new approach. *Biophys. J.* **88**, 805–817.

Berger, O., Edholm, O., Jahnig, F. (1997). Molecular dynamics simulations of a fluid bilayer of dipalmitoylphosphatidylcholine at full hydration, constant pressure, and constant temperature. *Biophys. J.* **72**, 2002–2013.

Brochard, F., de Gennes, P.G., Pfeuty, P. (1976). Surface-tension and deformations of membrane structures—Relation to 2-dimensional phase-transitions. *J. Phys.* **37**, 1099–1104.

Brooks, B.R., Bruccoleri, R.E., Olafson, B.D., States, D.J., Swaminathan, S., Karplus, M. (1983). CHARMM—A Program for Macromolecular Energy, Minimization, and Dynamics Calculations. *J. Comput. Chem.* **4**, 187–217.

Brown, M.F., Ribeiro, A.A., Williams, G.D. (1983). New view of lipid bilayer dynamics from ^{2}H and ^{13}C NMR relaxation-time measurements. *Proc. Nat. Acad. Sci. USA Biol. Sci.* **80**, 4325–4329.

Brown, M.F., Thurmond, R.L., Dodd, S.W., Otten, D., Beyer, K. (2002). Elastic deformation of membrane bilayers probed by deuterium NMR relaxation. *J. Am. Chem. Soc.* **124**, 8471–8484.

Brown, M.L., Venable, R.M., Pastor, R.W. (1995). A method for characterizing transition concertedness from polymer dynamics computer-simulations. *Biopolymers* **35**, 31–46.

Buck, M., Bouguet-Bonnet, S., Pastor, R.W., MacKerell, A.D. (2006). Importance of the CMAP correction to the CHARMM22 protein force field: Dynamics of hen lysozyme. *Biophys. J.* **90**, L36–L38.

Buldt, G., Gally, H.U., Seelig, J., Zaccai, G. (1979). Neutron-diffraction studies on phosphatidyl-choline model membranes. 1. Head group conformation. *J. Mol. Biol.* **134**, 673–691.

Bush, B.L., Bayly, C.I., Halgren, T.A. (1999). Consensus bond-charge increments fitted to electrostatic potential or field of many compounds: Application to MMFF94 training set. *J. Comput. Chem.* **20**, 1495–1516.

Cances, E., Mennucci, B., Tomasi, J. (1997). A new integral equation formalism for the polarizable continuum model: Theoretical background and applications to isotropic and anisotropic dielectrics. *J. Chem. Phys.* **107**, 3032–3041.

Castro-Roman, F., Benz, R.W., White, S.H., Tobias, D.J. (2006). Investigation of finite system size effects in molecular dynamics simulations of lipid bilayers. *J. Phys. Chem. B* **110**, 24157–24164.

Chandrasekhar, I., Kastenholz, M., Lins, R.D., Oostenbrink, C., Schuler, L.D., Tieleman, D.P., van Gunsteren, W.F. (2003). A consistent potential energy parameter set for lipids: dipalmitoylphosphatidylcholine as a benchmark of the GROMOS96 45A3 force field. *Eur. Bioph. J. Biophys. Lett.* **32**, 67–77.

Chen, I.J., Yin, D.X., MacKerell Jr., A.D. (2002). Combined ab initio/empirical approach for optimization of Lennard-Jones parameters for polar-neutral compounds. *J. Comput. Chem.* **23**, 199–213.

Chirlian, L.E., Francl, M.M. (1987). Atomic charges derived from electrostatic potentials—A detailed study. *J. Comput. Chem.* **8**, 894–905.

Clarke, R.J. (2001). The dipole potential of phospholipid membranes and methods for its detection. *Adv. Colloid Interface Sci.* **89**, 263–281.

Cossi, M., Scalmani, G., Rega, N., Barone, V. (2002). New developments in the polarizable continuum model for quantum mechanical and classical calculations on molecules in solution. *J. Chem. Phys.* **117**, 43–54.

Crane, J.M., Putz, G., Hall, S.B. (1999). Persistence of phase coexistence in disaturated phosphatidylcholine monolayers at high surface pressures. *Biophys. J.* **77**, 3134–3143.

Curtiss, L.A., Raghavachari, K., Redfern, P.C., Rassolov, V., Pople, J.A. (1998). Gaussian-3 (G3) theory for molecules containing first and second-row atoms. *J. Chem. Phys.* **109**, 7764–7776.

Curtiss, L.A., Redfern, P.C., Raghavachari, K., Rassolov, V., Pople, J.A. (1999). Gaussian-3 theory using reduced Moller–Plesset order. *J. Chem. Phys.* **110**, 4703–4709.

Darden, T., York, D., Pedersen, L. (1993). Particle mesh Ewald—An NLog(N) method for Ewald sums in large systems. *J. Chem. Phys.* **98**, 10089–10092.

Davis, J.H. (1983). The description of membrane lipid conformation, order and dynamics by H-2-NMR. *Biochim. Biophys. Acta* **737**, 117–171.

Douglass, D.C., McCall, D.W. (1958). Diffusion in paraffin hydrocarbons. *J. Phys. Chem.* **62**, 1102–1107.

Duan, Y., Wu, C., Chowdhury, S., Lee, M.C., Xiong, G.M., Zhang, W., Yang, R., Cieplak, P., Luo, R., Lee, T., Caldwell, J., Wang, J.M., Kollman, P. (2003). A point-charge force field for molecular mechanics simulations of proteins based on condensed-phase quantum mechanical calculations. *J. Comput. Chem.* **24**, 1999–2012.

Dunning, T.H. (2000). A road map for the calculation of molecular binding energies. *J. Phys. Chem. A* **104**, 9062–9080.

Dunning, T.H., Peterson, K.A., Wilson, A.K. (2001). Gaussian basis sets for use in correlated molecular calculations. X. The atoms aluminum through argon revisited. *J. Chem. Phys.* **114**, 9244–9253.

Ernzerhof, M., Scuseria, G.E. (1999). Assessment of the Perdew–Burke–Ernzerhof exchange-correlation functional. *J. Chem. Phys.* **110**, 5029–5036.

Evans, E., Rawicz, W. (1990). Entropy-driven tension and bending elasticity in condensed-fluid membranes. *Phys. Rev. Lett.* **64**, 2094–2097.

Feig, M., MacKerell Jr., A.D., Brooks, C.L. (2003). Force field influence on the observation of π-helical protein structures in molecular dynamics simulations. *J. Phys. Chem. B* **107**, 2831–2836.

Feller, S.E., MacKerell Jr., A.D. (2000). An improved empirical potential energy function for molecular simulations of phospholipids. *J. Phys. Chem. B* **104**, 7510–7515.

Feller, S.E., Pastor, R.W. (1996). On simulating lipid bilayers with an applied surface tension: Periodic boundary conditions and undulations. *Biophys. J.* **71**, 1350–1355.

Feller, S.E., Pastor, R.W., Rojnuckarin, A., Bogusz, S., Brooks, B.R. (1996). Effect of electrostatic force truncation on interfacial and transport properties of water. *J. Phys. Chem.* **100**, 17011–17020.

Feller, S.E., Venable, R.M., Pastor, R.W. (1997a). Computer simulation of a DPPC phospholipid bilayer: Structural changes as a function of molecular surface area. *Langmuir* **13**, 6555–6561.

Feller, S.E., Yin, D.X., Pastor, R.W., MacKerell Jr., A.D. (1997b). Molecular dynamics simulation of unsaturated lipid bilayers at low hydration: Parameterization and comparison with diffraction studies. *Biophys. J.* **73**, 2269–2279.

Gally, H.U., Pluschke, G., Overath, P., Seelig, J. (1981). Structure of Escherichia-Coli membranes—Glycerol auxotrophs as a tool for the analysis of the phospholipid headgroup region by deuterium magnetic-resonance. *Biochemistry* **20**, 1826–1831.

Gross, J.D., Warschawski, D.E., Griffin, R.G. (1997). Dipolar recoupling in MAS NMR: A probe for segmental order in lipid bilayers. *J. Am. Chem. Soc.* **119**, 796–802.

Hauser, H., Pascher, I., Sundell, S. (1980). Conformation of phospholipids—Crystal-structure of a lysophosphatidylcholine analog. *J. Mol. Biol.* **137**, 249–264.

Henchman, R.H., Essex, J.W. (1999). Generation of OPLS-like charges from molecular electrostatic potential using restraints. *J. Comput. Chem.* **20**, 483–498.

Herce, H.D., Garcia, A.E. (2006). Correction of apparent finite size effects in the area per lipid of lipid membranes simulations. *J. Chem. Phys.* **125**.

Holler, F., Callis, J.B. (1989). Conformation of the hydrocarbon chains of sodium dodecyl sulfate molecules in micelles: An FTIR Study. *J. Phys. Chem.* **93**, 2053–2058.

Horn, H.W., Swope, W.C., Pitera, J.W., Madura, J.D., Dick, T.J., Hura, G.L., Head-Gordon, T. (2004). Development of an improved four-site water model for biomolecular simulations: TIP4P-Ew. *J. Chem. Phys.* **120**, 9665–9678.

Israelachvili, J.N., Mitchell, D.J., Ninham, B.W. (1977). Theory of self-assembly of lipid bilayers and vesicles. *Biochim. Biophys. Acta* **470**, 185–201.

IUPAC (1967). The nomenclature of lipids. *J. Lipid Res.* **8**, 523–528.

Jahnig, F. (1996). What is the surface pension of a lipid bilayer membrane? *Biophys. J.* **71**, 1348–1349.

Jakalian, A., Bush, B.L., Jack, D.B., Bayly, C.I. (2000). Fast, efficient generation of high-quality atomic charges. AM1-BCC model: I. Method. *J. Comput. Chem.* **21**, 132–146.

Jojart, B., Martinek, T.A. (2007). Performance of the general Amber force field in modeling aqueous POPC membrane bilayers. *J. Comput. Chem.* **28**, 2051–2058.

Jorgensen, W.L. (1986). Optimized intermolecular potential functions for liquid alcohols. *J. Phys. Chem.* **90**, 1276–1284.

Jorgensen, W.L., Chandrasekhar, J., Madura, J.D., Impey, R.W., Klein, M.L. (1983). Comparison of simple potential functions for simulating liquid water. *J. Chem. Phys.* **79**, 926–935.

Jorgensen, W.L., Madura, J.D., Swenson, C.J. (1984). Optimized intermolecular potential functions for liquid hydrocarbons. *J. Am. Chem. Soc.* **106**, 6638–6646.

Jorgensen, W.L., Maxwell, D.S., Tiradorives, J. (1996). Development and testing of the OPLS all-atom force field on conformational energetics and properties of organic liquids. *J. Am. Chem. Soc.* **118**, 11225–11236.

Karaborni, S., O'Connell, J.P. (1990). Molecular-dynamics simulations of hydrocarbon chains. *J. Chem. Phys.* **92**, 6190–6194.

Klauda, J.B., Garrison, S.L., Jiang, J., Arora, G., Sandler, S.I. (2004). HM-IE: quantum chemical hybrid methods for calculating interaction energies. *J. Phys. Chem. A* **108**, 107–112.

Klauda, J.B., Brooks, B.R., MacKerell Jr., A.D., Venable, R.M., Pastor, R.W. (2005a). An ab initio study on the torsional surface of alkanes and its effect on molecular simulations of alkanes and a DPPC bilayer. *J. Phys. Chem. B* **109**, 5300–5311.

Klauda, J.B., Pastor, R.W., Brooks, B.R. (2005b). Adjacent gauche stabilization in linear alkanes: Implications for polymer models and conformational analysis. *J. Phys. Chem. B* **109**, 15684–15686.

Klauda, J.B., Brooks, B.R., Pastor, R.W. (2006a). Dynamical motions of lipids and a finite size effect in simulations of bilayers. *J. Chem. Phys.* **125**, 144710.

Klauda, J.B., Kučerka, N., Brooks, B.R., Pastor, R.W., Nagle, J.F. (2006b). Simulation-based methods for interpreting X-ray data from lipid bilayers. *Biophys. J.* **90**, 2796–2807.

Klauda, J.B., Wu, X.W., Pastor, R.W., Brooks, B.R. (2007). Long-range Lennard-Jones and electrostatic interactions in interfaces: Application of the isotropic periodic sum method. *J. Phys. Chem. B* **111**, 4393–4400.

Klauda, J.B., Eldho, N.V., Gawrisch, K., Brooks, B.R., Pastor, R.W. (2008a). Collective and noncollective models of NMR relaxation in lipid vesicles and multilayers. *J. Phys. Chem. A.* In press.

Klauda, J.B., Roberts, M.F., Redfield, A.G., Brooks, B.R., Pastor, R.W. (2008b). Rotation of lipids in membranes: MD simulation, ^{31}P spin-lattice relaxation, and rigid-body dynamics. *Biophys. J.* **94**, 3074–3083.

Kučerka, N., Tristram-Nagle, S., Nagle, J.F. (2006). Closer look at structure of fully hydrated fluid phase DPPC bilayers. *Biophys. J.* **90**, L83–L85.

Lagüe, P., Pastor, R.W., Brooks, B.R. (2004). Pressure-based long-range correction for Lennard-Jones interactions in molecular dynamics simulations: Application to alkanes and interfaces. *J. Phys. Chem. B* **108**, 363–368.

Laio, A., VandeVondele, J., Rothlisberger, U. (2002). D-RESP: Dynamically generated electrostatic potential derived charges from quantum mechanics/molecular mechanics simulations. *J. Phys. Chem. B* **106**, 7300–7307.

Lamoureux, G., MacKerell Jr., A.D., Roux, B. (2003). A simple polarizable model of water based on classical Drude oscillators. *J. Chem. Phys.* **119**, 5185–5197.

Lemmon, E.W., McLinden, M.O., Friend, D.G. (2005). Thermophysical properties of fluid systems. In: Linstrom, P.J., Mallard, W.G. (Eds.), NIST Chemistry WebBook, NIST Standard Reference Database Number 69. National Institute of Standards and Technology, Gaithersburg, MD.

Levine, Y.K., Wilkins, M.H.F. (1971). Structure of oriented lipid bilayers. *Nature New Biology* **230**, 69–72.

Lewis, B.A., Engelman, D.M. (1983). Lipid bilayer thickness varies linearly with acyl chain-length in fluid phosphatidylcholine vesicles. *J. Mol. Biol.* **166**, 211–217.

Lide, D.R. (2000). "CRC Handbook". CRC Press, Boca Raton, FL.

Lipari, G., Szabo, A. (1982). Model-free approach to the interpretation of nuclear magnetic-resonance relaxation in macromolecules. 1. Theory and range of validity. *J. Am. Chem. Soc.* **104**, 4546–4559.

Ly, H.V., Longo, M.L. (2004). The influence of short-chain alcohols on interfacial tension, mechanical properties, area/molecule, and permeability of fluid lipid bilayers. *Biophys. J.* **87**, 1013–1033.

Lyerla, J.R., McIntyre Jr., H.M., Torchia, D.A. (1974). A ^{13}C nuclear magnetic resonance study of alkane motion. *Macromolecules* **7**, 11–14.

MacKerell Jr., A.D. (2001). Atomistic models and force fields. In: Becker, O.M., MacKerell, A.D., Roux, B., Watanabe, M. (Eds.), Computational Biochemistry and Biophysics. Marcel Dekker, New York, pp. 7–38.

MacKerell Jr., A.D. (2004). Empirical force fields for biological macromolecules: Overview and issues. *J. Comput. Chem.* **25**, 1584–1604.

MacKerell Jr., A.D. (2005). Interatomic potentials: Molecules. In: Yip, S. (Ed.), Handbooks of Material Modeling. Springer, Netherlands, pp. 509–525.

MacKerell Jr., A.D. (2007). http://www.pharmacy.umaryland.edu/faculty/amackere/param/force_field_dev.htm.

MacKerell Jr., A.D., Karplus, M. (1991). Importance of attractive van der Waals contribution in empirical energy function models for the heat of vaporization of polar liquids. *J. Phys. Chem.* **95**, 10559–10560.

MacKerell Jr., A.D., Wiorkiewiczkuczera, J., Karplus, M. (1995). An all-atom empirical energy function for the simulation of nucleic-acids. *J. Am. Chem. Soc.* **117**, 11946–11975.

MacKerell Jr., A.D., Bashford, D., Bellott, M., Dunbrack, R.L., Evanseck, J.D., Field, M.J., Fischer, S., Gao, J., Guo, H., Ha, S., Joseph-McCarthy, D., Kuchnir, L., Kuczera, K., Lau, F.T.K., Mattos, C., Michnick, S., Ngo, T., Nguyen, D.T., Prodhom, B., Reiher, W.E., Roux, B., Schlenkrich, M., Smith, J.C., Stote, R., Straub, J., Watanabe, M., Wiorkiewicz-Kuczera, J., Yin, D., Karplus, M. (1998). All-atom empirical potential for molecular modeling and dynamics studies of proteins. *J. Phys. Chem. B* **102**, 3586–3616.

MacKerell Jr., A.D., Feig, M., Brooks, C.L. (2004). Extending the treatment of backbone energetics in protein force fields: Limitations of gas-phase quantum mechanics in reproducing protein conformational distributions in molecular dynamics simulations. *J. Comput. Chem.* **25**, 1400–1415.

Marsh, D. (1997). Renormalization of the tension and area expansion modulus in fluid membranes. *Biophys. J.* **73**, 865–869.

Merz, K.M. (1992). Analysis of a large data-base of electrostatic potential derived atomic charges. *J. Comput. Chem.* **13**, 749–767.

Mitaku, S., Ikegami, A., Sakanishi, A. (1978). Ultrasonic studies of lipid bilayer—Phase-transition in synthetic phosphatidylcholine liposomes. *Biophys. Chem.* **8**, 295–304.

Nagle, J.F., Tristram-Nagle, S. (2000). Structure of lipid bilayers. *Biochim. Biophys. Acta Rev. Biomembr.* **1469**, 159–195.

Oostenbrink, C., Villa, A., Mark, A.E., Van Gunsteren, W.F. (2004). A biomolecular force field based on the free enthalpy of hydration and solvation: The GROMOS force-field parameter sets 53A5 and 53A6. *J. Comput. Chem.* **25**, 1656–1676.

Ottiger, M., Bax, A. (1998). Determination of relative N–H^N N–C′, C$^\alpha$–C′, and C$^\alpha$–H$^\alpha$ effective bond lengths in a protein by NMR in a dilute liquid crystalline phase. *J. Am. Chem. Soc.* **120**, 12334–12341.

Paluch, M. (2000). Electrical properties of free surface of water and aqueous solutions. *Adv. Colloid Interface Sci.* **84**, 27–45.

Pascher, I., Lundmark, M., Nyholm, P.G., Sundell, S. (1992). Crystal-structures of membrane-lipids. *Biochim. Biophys. Acta* **1113**, 339–373.

Pastor, R.W., Venable, R.M., Feller, S.E. (2002). Lipid bilayers, NMR relaxation, and computer simulations. *Acc. Chem. Res.* **35**, 438–446.

Patel, S., Mackerell, A.D., Brooks, C.L. (2004). CHARMM fluctuating charge force field for proteins: II. Protein/solvent properties from molecular dynamics simulations using a nonadditive electrostatic model. *J. Comput. Chem.* **25**, 1504–1514.

Patel, S.A., Brooks, C.L. (2006). Revisiting the hexane–water interface via molecular dynamics simulations using nonadditive alkane–water potentials. *J. Chem. Phys.* **124**.

Raghavachari, K., Trucks, G.W., Pople, J.A., Head-Gordon, M. (1989). A 5th-order perturbation comparison of electron correlation theories. *Chem. Phys. Lett.* **157**, 479–483.

Rand, R.P., Parsegian, V.A. (1989). Hydration forces between phospholipid—Bilayers. *Biochim. Biophys. Acta* **988**, 351–376.

Rawicz, W., Olbrich, K.C., McIntosh, T., Needham, D., Evans, E. (2000). Effect of chain length and unsaturation on elasticity of lipid bilayers. *Biophys. J.* **79**, 328–339.

Rick, S.W., Stuart, S.J. (2002). Potentials and algorithms for incorporating polarizability in computer simulations. In: Reviews in Computational Chemistry, Vol 18, pp. 89–146.

Roberts, M.F., Redfield, A.G. (2004). High-resolution ^{31}P field cycling NMR as a probe of phospholipid dynamics. *J. Am. Chem. Soc.* **126**, 13765–13777.

Scheidt, H.A., Huster, D., Gawrisch, K. (2005). Diffusion of cholesterol and its precursors in lipid membranes studied by H-1 pulsed field gradient magic angle spinning NMR. *Biophys. J.* **89**, 2504–2512.

Schlegel, H.B. (1982). Optimization of equilibrium geometries and transition structures. *J. Comput. Chem.* **3**, 214–218.

Schlenkrich, M., Brinkman, J., MacKerell Jr., A.D., Karplus, M. (1996). An empirical potential energy function for phospholipids: Criteria for Parameter optimization and applications. In: Merz, K.M., Roux, B. (Eds.), Membrane Structure and Dynamics. Birkhauser, Boston, pp. 31–81.

Seelig, J. (1977). Deuterium magnetic-resonance—Theory and application to lipid-membranes. *Quart. Rev. Biophys.* **10**, 353–418.

Seelig, A., Seelig, J. (1974). The dynamic structure of fatty acyl chains in a phospholipid bilayer measured by deuterium magnetic resonance. *Biochemistry* **13**, 4839–4845.

Seelig, A., Seelig, J. (1975). Bilayers of dipalmitoyl-3-*sn*-phosphatidylcholine conformational differences between the fatty acyl chains. *Biochim. Biophys. Acta* **406**, 1–5.

Seelig, J., Gally, H.U., Wohlgemuth, R. (1977). Orientation and flexibility of choline head group in phosphatidylcholine bilayers. *Biochim. Biophys. Acta* **467**, 109–119.

Singh, U.C., Kollman, P.A. (1984). An approach to computing electrostatic charges for molecules. *J. Comput. Chem.* **5**, 129–145.

Small, D.M. (1986). Handbook of lipid research. In: Hanahan, D.J. (Ed.), The Physical Chemistry of Lipids: From Alkanes to Phospholipids. Plenum Press, New York.

Somerharju, P.J., Virtanen, J.A., Eklund, K.K., Vainio, P., Kinnunen, P.K.J. (1985). 1-palmitoyl-2-pyrenedecanoyl glycerophospholipids as membrane probes—Evidence for regular distribution in liquid-crystalline phosphatidylcholine bilayers. *Biochemistry* **24**, 2773–2781.

Sonne, J., Jensen, M.O., Hansen, F.Y., Hemmingsen, L., Peters, G.H. (2007). Reparameterization of all-atom dipalmitoylphosphatidylcholine lipid parameters enables simulation of fluid bilayers at zero surface tension. *Biophys. J.* **92**, 4157–4167.

Sundaralingam, M. (1972). *Ann. N.Y. Acad. Sci. USA* **195**, 324–355.

Szabo, A. (1984). Theory of fluorescence depolarization in macromolecules and membranes. *J. Chem. Phys.* **81**, 150–167.

Szabo, A. (1986). Nuclear magnetic resonance relaxation and the dynamics of proteins and membranes. *Ann. N.Y. Acad. Sci.* **482**, 44–50.

Szabo, A., Ostlund, N.S. (1996). "Modern Quantum Chemistry: Introduction to Advanced Electronic Structure Theory". Dover Publication, Inc., Mineola.

Tabony, J., Perly, B. (1991). Quasi-elastic neutron-scattering measurements of fast local translational diffusion of lipid molecules in phospholipid-bilayers. *Biochim. Biophys. Acta* **1063**, 67–72.

Tien, H.T., Diana, A.L. (1968). Bimolecular lipid membranes—A review and a summary of some recent studies. *Chem. Phys. Lipids* **2**, 55–101.

Vaz, W.L.C., Clegg, R.M., Hallmann, D. (1985). Translational diffusion of lipids in liquid-crystalline phase phosphatidylcholine multibilayers—A comparison of experiment with theory. *Biochemistry* **24**, 781–786.

Venable, R.M., Brooks, B.R., Pastor, R.W. (2000). Molecular dynamics simulations of gel (L-beta I) phase lipid bilayers in constant pressure and constant surface area ensembles. *J. Chem. Phys.* **112**, 4822–4832.

Venable, R.M., Skibinsky, A., Pastor, R.W. (2006). Constant surface tension molecular dynamics simulations of lipid bilayers with trehalose. *Mol. Simul.* **32**, 849–855.

Vorobyov, I., Anisimov, V.M., Green, S., Venable, R.M., Moser, A., Pastor, R.W., MacKerell, A.D. (2007). Additive and classical drude polarizable force fields for linear and cyclic ethers. *J. Chem. Theory Comput.* **3**, 1120–1133.

Vorobyov, I.V., Anisimov, V.M., MacKerell, A.D. (2005). Polarizable empirical force field for alkanes based on the classical drude oscillator model. *J. Phys. Chem. B* **109**, 18988–18999.

Wang, J.M., Cieplak, P., Kollman, P.A. (2000). How well does a restrained electrostatic potential (RESP) model perform in calculating conformational energies of organic and biological molecules? *J. Comput. Chem.* **21**, 1049–1074.

Wang, L.G., Bose, P.S., Sigworth, F.J. (2006). Using cryo-EM to measure the dipole potential of a lipid membrane. *Proc. Nat. Acad. Sci. USA* **103**, 18528–18533.

Warshel, A., Lifson, S. (1970). Consistent force field calculations. 2. Crystal structures, sublimation energies, molecular and lattice vibrations, molecular conformations, and enthalpies of alkanes. *J. Chem. Phys.* **53**, 582–594.

Wiener, M.C., King, G.I., White, S.H. (1991). Structure of a fluid dioleoylphosphatidylcholine bilayer determined by joint refinement of X-ray and neutron-diffraction data. 1. Scaling of neutron data and the distributions of double-bonds and water. *Biophys. J.* **60**, 568–576.

Woodcock, H., Moran, D., Pastor, R.W., MacKerell, A.D., Brooks, B.R. (2007). Ab initio modeling of glycosyl torsions and anomeric effects in a model carbohydrate: 2-Ethoxy tetrahydropyran. *Biophys. J.* **93**, 1–10.

Wu, X.W., Brooks, B.R. (2005). Isotropic periodic sum: A method for the calculation of long-range interactions. *J. Chem. Phys.* **122**, 044107.

Yeh, I.C., Hummer, G. (2004). System-size dependence of diffusion coefficients and viscosities from molecular dynamics simulations with periodic boundary conditions. *J. Phys. Chem. B* **108**, 15873–15879.

Yin, D.X., MacKerell Jr., A.D. (1996). Ab initio calculations on the use of helium and neon as probes of the van der Waals surfaces of molecules. *J. Phys. Chem.* **100**, 2588–2596.

Yin, D.X., MacKerell Jr., A.D. (1998). Combined ab initio empirical approach for optimization of Lennard-Jones parameters. *J. Comput. Chem.* **19**, 334–348.

Zaccai, G., Buldt, G., Seelig, A., Seelig, J. (1979). Neutron-diffraction studies on phosphatidylcholine model membranes. 2. Chain conformation and segmental disorder. *J. Mol. Biol.* **134**, 693–706.

Zhang, Y.H., Feller, S.E., Brooks, B.R., Pastor, R.W. (1995). Computer-simulation of liquid/liquid interfaces. I. Theory and application to octane/water. *J. Chem. Phys.* **103**, 10252–10266.

Zhang, Y.H., Venable, R.M., Pastor, R.W. (1996). Molecular dynamics simulations of neat alkanes: The viscosity dependence of rotational relaxation. *J. Phys. Chem.* **100**, 2652–2660.

CHAPTER 2

Electrostatics in Biomolecular Simulations: Where Are We Now and Where Are We Heading?

Mikko Karttunen*, Jörg Rottler[†], Ilpo Vattulainen[‡,§,¶] and Celeste Sagui[‖]

*Department of Applied Mathematics, The University of Western Ontario, London, Ontario, Canada
[†]Department of Physics and Astronomy, The University of British Columbia,
6224 Agricultural Road, Vancouver, BC V6T 1Z1, Canada
[‡]MEMPHYS—Center for Biomembrane Physics, Physics Department,
University of Southern Denmark, Odense, Denmark
[§]Laboratory of Physics and Helsinki Institute of Physics, Helsinki University of Technology, Finland
[¶]Department of Physics, Tampere University of Technology, Finland
[‖]Department of Physics, North Carolina State University, Raleigh, North Carolina, USA

Abstract

In this review, we discuss current methods and developments in the treatment of electrostatic interactions in biomolecular and soft matter simulations. We re-

Current Topics in Membranes, Volume 60
1063-5823/08 $35.00
DOI: 10.1016/S1063-5823(08)00002-1

view the current 'work horses', namely, Ewald summation based methods such the Particle-Mesh Ewald, and others, and also newer real-space methods such as multigrid methods, and local algorithms for Coulomb's law. We also pay attention to boundary conditions. Although periodic boundary conditions are used most commonly, it is often desirable to have systems that are confined or have boundaries. Finally, we briefly describe some current and available software for the computation of electrostatics in biomolecular and soft matter simulations.

I. INTRODUCTION

Over the past decade, computer simulations have truly become an invaluable tool and they have provided insight into such diverse problems as Alzheimer's disease (Tsigelny *et al.*, 2007), predicting properties of novel quantum devices (Sun *et al.*, 2003), and revealing some of the fundamental mechanisms in friction (Mosey *et al.*, 2005). Numerous reviews of computational modeling exist covering various fields. For biomolecular simulations, the topic of this paper, see *e.g.*, Refs. (Karplus and McCammon, 2002; Vattulainen and Karttunen, 2006).

In this review, we focus on biological and soft matter systems. The emphasis will be on the computation of electrostatics interactions. Biological systems are characterized by the presence of charges, as most biological molecules such as DNA, proteins, glycomolecules, water, and lipids are either charged or polar (Alberts *et al.*, 1994; Boroudjerdi *et al.*, 2005). Since long-range charge-charge interactions are ubiquitous in classical atomistic representations, and multipole and non-additive polarization effects are very important, electrostatics clearly plays a major role in these systems. Furthermore, biological molecules are embedded in water—everything in molecular biology occurs in an aqueous environment—and each water molecule is represented by at least three charged centers. In a typical biomolecular simulation more than 95% of the simulation time is spent computing interactions involving water molecules.

The importance of electrostatics is further highlighted by numerous cellular functions not explicitly involving any of the above biomolecules but rather one of the products we use on a daily basis, salt. Its role in a multitude of biological processes such as signal transduction, osmosis and functions of molecular motors, and ion channels is well known. Salt also stabilizes cellular membranes and is involved in, *e.g.*, energy storage processes in terms of asymmetric transmembrane salt distributions.

Developments at the algorithmic and methodological level are the key to progress. Although steady, the increase in processor speed will not necessarily lead to significant improvements in systems sizes or simulation times unless algorithmic developments are carefully coupled to the new computing architectures. Two state-of-the-art examples are illustrated below.

In biomolecular simulations, one of the most exciting developments is the CHARMM++ interface of Kale and Kirshnan (1996). It is based on smart load balancing which enables vastly better parallelization than before. Recently, Sanbonmatsu and Tung (2007) used NAMD (Kalé *et al.*, 1999; Phillips *et al.*, 2005) with CHARM++ to simulate a ribosome system of about 2.64 million atoms with full Particle Mesh Ewald (PME) electrostatics parallelized over 1024 CPUs. Both the number of atoms and processors are current records in this type of simulations.

In a different type of simulation, the current record exceeds billion particles. Vashishta *et al.* (2006) have reported 18.9 billion particle multi-resolution MD simulations and 1.4 million atom density functional simulations. In their case, electrostatic interactions were computed using the Fast Multipole Method (FMM) combined with conjugate gradient-based minimization. In summary, we have come a long way from the pioneering works of Rahman (1964) and Alder and Wainwright (1957) over 40 years ago.

The rest of this chapter is organized as follows. In the next section, we review the importance of long-range interactions in biological systems. In Section III, we focus on the following methods: simple truncation, reaction-field which is a clear improvement upon the former, Ewald summation based methods, and two promising real-space methods, multigrid algorithms and local algorithms for Coulomb's law. In Section V we discuss some of the latest developments and applications related to both particle-mesh and real space methods. Finally, we finish with some notes on software, accuracy, and speed in Section VI.

II. IMPORTANCE OF LONG-RANGE INTERACTIONS IN SIMULATIONS OF BIOLOGICAL SYSTEMS

Computationally, electrostatics poses a major challenge. Like gravitation, it is long-ranged and decays as $1/r$. In general, we define a long-range interaction as one for which $V(r) \sim 1/r^a$, where $a < d$, and d is the dimension of space.

Starting from the more fundamental level, a significant body of work has been devoted to bridging quantum mechanical and classical regimes. These so-called *QM/MM methods* allow for the description of, say, active sites in enzymes on a quantum mechanical level, while the rest of the system is treated classically. One of the major issues is the treatment of electrostatics in the combined QM/MM simulation (Schaefer *et al.*, 2005). Second, the development of realistic *polarizable force fields* for molecules such as water is highly relevant to describe biological systems more accurately.

Polarizability is important even in the case of seemingly simple systems such as xenon. Its highly polarizable nature is not well understood, which also reflects the fact that the mechanism of *xenon as an anesthetic* is far from being well understood (Stimson *et al.*, 2005). Water, another polarizable molecule, is one of the

most difficult molecules to model and a massive amount of work has been devoted to the development and testing of force fields for water, see *e.g.* (Mark and
Nilsson, 2002; Guillot, 2002; Hess and van der Vegt, 2006). It still seems, however, that we have a long way to go to develop a true understanding of *water and
hydrogen bonding* under varying thermodynamic conditions. Importantly, water
and, in particular, protons are fundamental in developing methods for *constant
pH* molecular dynamics simulations (Adcock and McCammon, 2006).

A related problem concerns the *description of ions*. Even the most popular force
fields used in biomolecular simulations differ substantially in terms of their ion
parameters. For example, the van der Waals size of Na^+ ions in current GRO-
MACS and CHARMM force fields differs by a factor of about two from one
another (Patra and Karttunen, 2004; Gurtovenko and Vattulainen, 2007). It is thus
easy to realize why different force fields for ions may give rise to significantly different behavior in terms of structure and dynamics (Patra and Karttunen, 2004).

Finally, while all the above examples highlight the relevance of long-range
interactions in atom-scale molecular simulations, there is reason to emphasize
the importance of coarse grained modeling as well (Karttunen *et al.*, 2004;
Nielaba *et al.*, 2002). Coarse grained models facilitate considerations over large
scales and allow studies of processes that are not feasible by atomistic approaches.
Yet, the role of electrostatic interactions in coarse grained models is as important
as in atomistic ones, thus efforts have been directed to clarify related methodological issues (Groot, 2003a, 2003b).

Electrostatic interactions are thus fundamental in most complex soft-matter and
biological processes. The folding of proteins and polypeptides, fusion of vesicles,
the dynamics of polyelectrolytes such as DNA, and the dynamics in tightly packed
cell membrane structures such as lipid rafts are just some of the examples. Is it
possible to reduce the computational cost related to electrostatic interactions as
they take a lion's share of the CPU time? One is tempted to consider ways to
speed up the dynamics at the expense of accuracy in calculation of long-range
interactions. Is this justified? In some cases perhaps yes, if conducting the study
is not feasible otherwise. Yet there is always reason for concern, and great care
should be taken to make sure that the treatment of electrostatics chosen for the
simulations is not a source of significant error. Doing this in practice is not so
difficult, since it is rather straight-forward to carry out a set of complementary
simulations with different schemes for electrostatics, and then compare the results
for quantities that are most sensitive to the choice of description for electrostatic
interactions. Examples of such sensitive quantities are presented below.

III. OVERVIEW OF METHODS AND THEIR APPLICABILITY

One should employ schemes that treat electrostatic interactions as accurately
as it is feasible, yet, at the same time, one should also worry about the associated

computational cost. The efficiency is an issue already in studies of small molecules such as lipids, since they readily form large molecular aggregates such as vesicles, whose dynamics takes place over large time scales. Furthermore, such systems are always embedded in water which adds a very large number of molecules in the simulation. The efficiency is an even more important issue in the case of larger molecules such as proteins, not to mention large protein complexes such as the ribosome comprised of more than 50 individual proteins.

The above two requirements may be problematic and partly conflicting for a number of reasons. On one hand, if we employed schemes such as those based on the Ewald summation (Sagui and Darden, 1999a; Sagui *et al.*, 2004), we could rely on the fact that the electrostatic interactions are treated accurately, thus avoiding significant artifacts due to the scheme chosen for electrostatics. The price we have to pay when we use highly accurate schemes, however, is the computational cost that can be major (although that does not necessarily have to be the case Patra *et al.*, 2007). On the other hand, if we tried to reduce the cost by using schemes such as the truncation method, where electrostatic interactions are treated up to some cut-off distance only, then the results may be influenced by the truncation scheme: it has been observed in numerous studies that the truncation scheme can indeed lead to serious artifacts, including shortcomings in the description of structure as well as dynamics of systems composed of water (Alper and Levy, 1989; Alper et al., 1993a, 1993b; Feller *et al.*, 1996; Mark and Nilsson, 2002; Hess, 2002; Yonetani, 2006), peptides (Smith and Pettitt, 1991; Schreiber and Steinhauser, 1992), proteins (York *et al.*, 1993; Saito, 1994; Ibragimova and Wade, 1998; Zuegg and Greedy, 1999; Faraldo-Gómez *et al.*, 2002; Kessel *et al.*, 2004; Tieleman *et al.*, 2002), DNA (York *et al.*, 1995; Norberg and Nilsson, 2000), and lipid membranes (Venable *et al.*, 2000; Patra et al., 2003, 2004; Anézo *et al.*, 2003; Khelashvili and Scott, 2004). Rather generally one can conclude that the truncation of electrostatics can have a significant influence on the properties of a given system, altering *e.g.* the stability of DNA and proteins, and leading to artificial ordering in systems such as water.

A. Truncation Schemes for Electrostatics

Of the many techniques available for treating long-range electrostatic interactions, the truncation method is in a certain way very appealing, since it can provide a considerable reduction in the computational load compared with schemes such as Ewald or PME (Patra *et al.*, 2007), if the truncation distance is not very large (Patra *et al.*, 2007). Consequently, it is frequently used in atom-scale simulations when computational requirements are substantial due to large system sizes or long time scales, which may be the case, *e.g.*, in studies of lipid–protein systems (Zangi *et al.*, 2002), self-assembly of lipids (Marrink *et al.*, 2001), and membrane fusion (Marrink and Tieleman, 2002). Additionally, truncation together with short-ranged interactions is used commonly in coarse grained models such as dissipative particle dynamics and other explicit solvent approaches. The use of truncation in all such large-scale systems that require excessive amounts of computing time is likely acceptable, since otherwise studies of those systems and related complex processes would not be feasible at all. There is reason to emphasize, however, that the accuracy of electrostatics should not be sacrificed at the expense of speed, meaning that the validity of the simulations using truncation should first be confirmed through careful testing.

What happens if the long-ranged electrostatic interaction is truncated abruptly at some cut-off distance, r_{cut}. Perhaps not surprisingly, it has turned out that the truncation of electrostatic interactions may affect structural properties of a variety of atomistic molecular systems like water (Alper and Levy, 1989; Alper et al., 1993a, 1993b; Feller *et al.*, 1996; Mark and Nilsson, 2002; Yonetani, 2006), peptides (Smith and Pettitt, 1991; Schreiber and Steinhauser, 1992), proteins (York *et al.*, 1993), DNA (York *et al.*, 1995; Norberg and Nilsson, 2000), and lipids (Patra et al., 2003, 2004, 2007; Anézo *et al.*, 2003; Khelashvili and Scott, 2004). In coarse grained models, the effects of truncation are considerably less characterized. Here, we demonstrate this issue in the context of atomistic models for lipid membranes.

The radial distribution function (RDF), $g(r)$, describes the probability of finding a pair of atoms a distance r apart, relative to the probability expected for a completely random distribution at the same density (Allen and Tildesley, 1993). For the ideal gas characterized by the absence of conservative interactions, the RDF is therefore equal to one for all distances r. Interacting systems in turn are characterized by RDFs that express peaks and valleys describing regions of high and low density, respectively. Hence, the RDF is an excellent means to gauge the ordering of any system. Here we use this quantity to consider truncation induced effects in a fully hydrated lipid membrane comprised of saturated DPPC (dipalmitoylphosphatidylcholine) phospholipids. For details of the model, see (Patra et al., 2003, 2004, 2007).

Figure 1 shows the intermolecular RDFs between the nitrogen (N) (top) and phosphorous (P) atoms in the DPPC head groups in a lipid bilayer. The results for the PME technique turn out to be as expected, since in that case there is a strong main peak at about 0.85 nm, characterizing short-scale fluid-like order, and essentially no structure beyond 1.5 nm. This is consistent with the fluid phase characterized by the lack of long-range ordering in the plane of the membrane. The truncation techniques, however, give rise to *pronounced peaks exactly at the truncation distance*. For the truncation distance of 1.8 nm, we find that the RDF has a prominent peak exactly at 1.8 nm. When the truncation distance is increased, the peak moves accordingly.

The above finding gives rise to major concern, since it suggests that the truncation of electrostatic interactions induces artificial structure at the scale of the truncation distance. Is this only of local nature or does it also change the phase

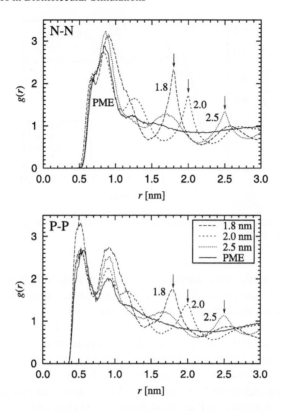

FIGURE 1   Radial distribution function $g(r)$ between the two central atoms in the head group of a DPPC molecule: RDFs for N–N and P–P pairs (Patra *et al.*, 2003). Cutoff distances are indicated by arrows.

behavior of the system? Unfortunately, it turns out that the worst case scenario really comes about. Figure 2 depicts the RDF between center of mass positions of the lipids in the bilayer plane. We find that the artificial ordering due to truncation leads to additional peaks that are entirely unphysical, thus changing the phase behavior of the membrane from a fluid to some semi-fluid-like phase characterized by artificial ordering at the length scale of the truncation distance. At the same time, the PME method does not yield any artificial structure of the same kind.

Without any doubt, the above artifacts in structural quantities are due to the fact that the truncation method does not account for the long-range component of electrostatic interactions. Apparently, the truncation of electrostatic interactions gives rise to artificial ordering in the bilayer plane, thus changing the phase behavior of the system. Further, since the RDF for center of mass positions of lipids changes, a variety of thermodynamic response functions such as the bilayer compressibility also change. What is also alarming is the observation that the height of the artifi-

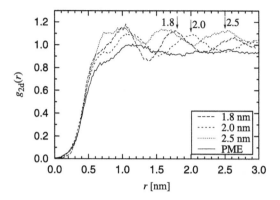

FIGURE 2 Radial distribution function $g_{2d}(r)$ for the center of mass positions of the DPPC molecules (Patra *et al.*, 2003).

cial peaks does not decrease considerably with increasing truncation distance. In a related study for liquid water, Yonetani (2006) has found that increasing the cutoff distance may even make things worse: the artifacts were enhanced by increasing the cutoff distance. Van der Spoel and van Maaren considered this matter in more detail and concluded that there is a subtle interplay between truncation distance, *group-based* cutoff, and the quadrupole moment of water (van der Spoel and van Maaren, 2006). Summarizing, we may conclude that the truncation of long-range interactions may induce artifacts not only in local (short-scale) properties but also in *large-scale properties* such as phase behavior and elasticity.

While related artifacts have been observed in a number of different soft matter systems, it would seem intriguing to find that the artifacts are more pronounced in lipid membranes. The explanation of why this happens becomes rather simple when one considers the dielectric constant. Unlike in many other biomolecular simulations, in lipid membranes the effective dielectric constant of the system is not that of water, about 80. Instead, the dielectric constant inside the membrane is about 2–4, and in the vicinity of the water–membrane interface it is about 20–40. Screening of long-range interactions in a membrane is therefore considerably weaker compared to pure water, thus enhancing the magnitude of artifacts compared with solvated water-like systems. It is also worth mentioning that a recent membrane simulation study by Edholm et al. (Wohlert and Edholm, 2004) indicated how the influence of truncation is rather subtle since that seems to depend on certain practical implementation details such as the choice of charge groups. Clearly, overall, this is a major issue and warrants particular attention in large-scale simulations.

While the present example has centered on lipid membranes, it is likely that similar kinds of findings would be faced in many other soft matter systems. To be on the safe side, we conclude that the abrupt truncation of electrostatic interactions should be avoided, if at all possible. One method that improves over the simple truncation schemes is the reaction field method, discussed below.

B. Switching or Shifting Functions

The easiest way to avoid some the problems arising from direct truncation is to use a switching function—often also called as shifting function. The latter term is somewhat misleading as the potential is not shifted as in the case of the truncated and shifted Lennard-Jones potential.

What a switching function does is the following: at some distance r_{switch}, the potential is modified for $r > r_{switch}$ in such a way that the potential is reduced to zero at the cutoff distance r_{cut}. This has the advantage of avoiding discontinuities present in abrupt truncation as now both the resulting force and its derivative are continuous for all r. In practice, this can be done by using a third order polynomial for $r > r_{switch}$. It is worth noticing that the switching distance is arbitrary. A common choice is to choose $r_{switch} = r_{short-range}$, i.e., the same as the cutoff for short-range interactions. Switching functions are also commonly use for Lennard-Jones interactions. Recent studies by van der Spoel and van Maaren also support the use of shifting functions instead of plain truncation (van der Spoel and van Maaren, 2006).

C. Reaction Field for Electrostatics

The next improvement is to consider the far field using a mean-field approximation. The basic idea is to treat all electrostatic interactions explicitly within the cutoff distance r_{cut}. For larger distance, the system is described by a mean-field approximation, typically based on the Poisson–Boltzmann equation. This approach is based on Onsager's reaction-field idea dating back to 1936 (Onsager, 1936). In the reaction-field approximation, the continuum part is described by a dielectric constant ϵ, and one has to treat the boundary between the explicit and continuum descriptions with some care. This is particularly so in inhomogeneous systems and systems having surfaces. That is easier to understand if one considers the following example: the dielectric constant of water is about 78 whereas in the lipid bilayer core it is about 2–5 (Koehorst et al., 2004). This change occurs over a very short distance, and in practice there is a dielectric discontinuity at the water–bilayer interface.

For $r > r_{cut}$ the system is treated on a mean-field level and is thus completely described by its dielectric constant ϵ. The potential is

$$\mathcal{V}(r) = \frac{q_i q_j}{4\pi \epsilon_0 r} \left[1 + \frac{\epsilon - 1}{2\epsilon + 1} \left(\frac{r}{r_{cut}} \right)^3 \right] - \frac{q_i q_j}{4\pi \epsilon_0 r_{cut}} \frac{3\epsilon}{2\epsilon + 1},$$

for $r \leqslant r_{cut}$. \hfill (1)

The second term makes the potential vanish at $r = r_{cut}$.

In practice, the reaction field approach works reasonably well. Studies by Anézo *et al.* (2003) have indicated that the reaction field works considerably better than abrupt truncation. Similar conclusions have been drawn by Patra *et al.* (2007). As an example of practical model systems where the reaction field technique has been used recently, let us mention simulations of lipid raft membranes (Niemela *et al.*, 2007), in which case extensive testing prior to actual production simulations confirmed that the approach did not compromise the results.

D. Ewald-Based Methods

Ewald, a crystallographer and physicist, first tackled the problem of long-range interactions between particles and their infinite periodic images in 1921 (Ewald, 1921). His goal was to determine the electrostatic energy of ionic crystals (*i.e.*, their Madelung constant). The problem presented by ionic crystals is that the Coulomb series that describes them is only conditionally convergent, *i.e.*, its limit depends on how the series is summed. Using a Jacobi theta transform, Ewald converted the slowly, conditionally convergent sum for the Coulomb potential into two sums that converge rapidly and absolutely: the *direct* and the *reciprocal* sums. In the process, the conditionally convergent character of the series has to be hidden somewhere. The Ewald sum is augmented with a (computationally inexpensive) surface term (DeLeeuw *et al.*, 1980) that depends on the dipole moment of the unit cell U, the dielectric constant of the surrounding medium, and the order of summation; that is, on the asymptotic shape of the array of copies of U. If the surrounding medium has infinite dielectric ("tin-foil" boundary conditions), the surface term vanishes, leaving the absolutely convergent sums for energies and forces. Although the surface term is only rarely used, its use has been recommended for calculations of the dielectric constant of the sample (Boresch and Steinhauser, 1997). Ewald-based methods use the error function erf and the complementary error function erfc to write the Coulomb interaction as $1/r = \mathrm{erfc}(\beta r)/r + \mathrm{erf}(\beta r)/r$ (plus the surface term). The first term is short-range and gives rise to the direct sum, while the second term is long-range, and is handled in the reciprocal sum via Fourier transforms. Such splitting of the Coulombic interactions involves a smooth function of r, and therefore avoids any form of discontinuity in the forces or higher-order derivatives of $1/r$. Ewald-based methods include the original Ewald summation (Ewald, 1921), the particle–particle particle-mesh method (Hockney and Eastwood, 1981; Pollock and Glosli, 1996), the particle-mesh Ewald algorithm (Darden *et al.*, 1993; Essmann *et al.*, 1995), and the Fast Fourier–Poisson (York and Yang, 1994) method.

Possible artifacts of the Ewald summation methods have been explored by many authors in the past (see for instance, Smith and Pettitt (1996, 1997) and

DeLeeuw *et al.* (1986)). The general conclusion is that the artifacts are negligible for high dielectric solvents, with typical simulation cell sizes. More recently, Villarreal and Montich (2005) analyzed the octaalanine peptide with charged termini in explicit solvent, for which severe effects due to the use of Ewald sums were predicted using continuum electrostatics. They carried out very extensive simulations for different cell sizes and showed that there were no significant periodicity-induced artifacts, and that incomplete sampling is likely to affect the results to a larger extent than any artifacts induced by the use of Ewald sums. These conclusions are supported by the work of Monticelli et al. (2004, 2006) for the formation of various secondary structure forming peptides during very long time simulations.

1. Ewald Summations

Under periodic boundary conditions (PBC), the system to be simulated comprises a unit simulation cell U, whose edges are given by the vectors $\mathbf{a}_1$, $\mathbf{a}_2$ and $\mathbf{a}_3$ (not necessarily orthogonal). The volume of U is given by $V = \mathbf{a}_1 \cdot \mathbf{a}_2 \times \mathbf{a}_3$. A lattice vector is defined as $\mathbf{n} = n_1\mathbf{a}_1 + n_2\mathbf{a}_2 + n_3\mathbf{a}_3$ for all integer triples (n_1, n_2, n_3), with n_1, n_2, n_3 not all zero. A standard way (DeLeeuw *et al.*, 1980; Smith, 1994; Caillol, 1994) of treating Coulombic interactions under PBC is to model the system as a large but finite ensemble of copies of the unit cell U, which is then immersed in a dielectric medium. Each particle at position $\mathbf{r}_i$ within the cell interacts with all the other particles in U, with their and its own periodic images, as well as with the reaction field induced in the surrounding dielectric medium. Infinite lattice sums are obtained in the limit of an infinite ensemble. The traditional Coulomb (monopolar) potential $\phi_0(\mathbf{r}_i)$ acting on charge i is expressed as:

$$\phi_0(\mathbf{r}_i) = \sum_{\mathbf{n}}{}' \sum_{j=1}^{N} \frac{q_j}{|\mathbf{r}_i - \mathbf{r}_j + \mathbf{n}|}, \tag{2}$$

where the prime indicates that terms with $i = j$ and $\mathbf{n} = 0$ are omitted.

The Coulomb infinite series converges slowly to a finite limit only if U is electrically neutral, and the limit of the series depends on the order of summation, *i.e.*, the series is *conditionally convergent*. This is better understood if one considers the simple lattice sum $\sum_{\mathbf{n} \neq 0} |\mathbf{n}|^{-p}$ which converges for $p > 3$ and diverges for $p \leqslant 3$. If one carries out a multipolar expansion of the electronic density, octupolar ($p = 4$) and higher-order terms in the Coulomb potential will therefore converge *absolutely*, but the monopolar and dipolar terms will not (in the quadrupolar case, the electrostatic potential will be conditionally convergent, but the electric field and force will converge absolutely). For a neutral unit cell, monopolar and dipolar terms converge but only conditionally, which in the past gave rise to polemic discussions in the literature.

The Ewald method has also been extended to higher-order multipoles (Smith, 1982; Toukmaji *et al.*, 2000; Aguado and Madden, 2003; Sagui *et al.*, 2004). To illustrate the application of the Ewald method for these more general cases, we consider a system where electrostatics is represented simply by partial charges and dipoles: N point charges $q_1, q_2, \ldots, q_N$ and N point dipoles $\mathbf{p}_1, \mathbf{p}_2, \ldots, \mathbf{p}_N$ at positions $\mathbf{r}_1, \mathbf{r}_2, \ldots, \mathbf{r}_N$ within the unit cell, assuming $q_1 + q_2 + \cdots + q_N = 0$ (otherwise, a simple neutralizing "plasma" needs to be added). As before, the edges of the unit cell are denoted by vectors $\mathbf{a}_\alpha$, $\alpha = 1, 2, 3$, and the conjugate reciprocal vectors $\mathbf{a}_\alpha^*$ are defined by the relations $\mathbf{a}_\alpha^* \cdot \mathbf{a}_\beta = \delta_{\alpha\beta}$ (the Kronecker delta), for $\alpha, \beta = 1, 2, 3$. The point charge q_i at position $\mathbf{r}_i$ has fractional coordinates $s_{\alpha i}$, $\alpha = 1, 2, 3$, defined by $s_{\alpha i} = \mathbf{a}_\alpha^* \cdot \mathbf{r}_i$. The charges and dipoles interact with each other, and with their periodic images. The electrostatic energy of the unit cell is then written as:

$$
E(\mathbf{r}_1, \ldots, \mathbf{r}_N) = \frac{1}{2} \sum_{\mathbf{n}}{}' \sum_{i,j} (q_i + \mathbf{p}_i \cdot \nabla_i)(q_j - \mathbf{p}_j \cdot \nabla_i)\left(\frac{1}{|\mathbf{r}_i - \mathbf{r}_j + \mathbf{n}|}\right),
$$
(3)

where the outer sum is over the vectors $\mathbf{n} = n_1 \mathbf{a}_1 + n_2 \mathbf{a}_2 + n_3 \mathbf{a}_3$, the prime indicating that terms with $i = j$ and $\mathbf{n} = 0$ are omitted.

For molecular systems, corrections are introduced to account for the "masked pairs", which are atom pairs $(i, j) \in M$, M the masked list, whose nonbond interactions should not be calculated, since they are accounted for by other terms in the potential. These masked pairs are easily omitted from the direct sum. However, since all pair interactions are unavoidably included in the Fourier treatment of the reciprocal sum, the reciprocal part of the masked pair contributions must be separately subtracted. Computationally, this is conveniently achieved by skipping these pairs in the direct sum and separately adding the sum over $(i, j) \in M$ of the "adjusted" interaction energy, which is the difference between the direct sum interaction and the standard Coulomb interaction between the charge, dipole pairs. Similarly, the self-energy, which is the reciprocal part of the interactions of the charge, dipole pairs with themselves, must be removed. This is achieved by calculating the adjusted interaction energy of the charge, dipole pair with an identical pair at distance r and taking the limit as $r \to 0$ (the result is independent of the path taken). The electrostatic energy can then be written as a sum of four terms (Toukmaji *et al.*, 2000), the direct, the reciprocal, the adjusted and the self terms: $E = E_{\text{dir}} + E_{\text{rec}} + E_{\text{adj}} + E_{\text{self}}$.

The direct sum energy is then given by

$$
E_{\text{dir}} = \frac{1}{2} \sum_{\mathbf{n}}{}^* \sum_{i,j=1}^{N} (q_i + \mathbf{p}_i \cdot \nabla_i)(q_j - \mathbf{p}_j \cdot \nabla_i)\frac{\text{erfc}(\beta|\mathbf{r}_j - \mathbf{r}_i + \mathbf{n}|)}{|\mathbf{r}_j - \mathbf{r}_i + \mathbf{n}|}, \quad (4)
$$

where the asterisk over the sum denotes that the terms with $\mathbf{n} = 0$ and either $j = i$ or $(i, j) \in M$ are omitted. The adjusted energy has a similar form to Eq. (4), with erfc replaced by $-$erf:

$$E_{\text{adj}} = -\frac{1}{2} \sum_{(i,j)\in M} (q_i + \mathbf{p}_i \cdot \nabla_i)(q_j - \mathbf{p}_j \cdot \nabla_i)\frac{\text{erf}(\beta|\mathbf{r}_j - \mathbf{r}_i + \mathbf{n}|)}{|\mathbf{r}_j - \mathbf{r}_i + \mathbf{n}|}. \tag{5}$$

The reciprocal energy E_{rec} has the same form as the traditional monopolar case, but it requires a generalization of the structure factor $S(\mathbf{m})$ to include the dipolar interactions. With the reciprocal lattice vectors $\mathbf{m}$ defined by $\mathbf{m} = m_1\mathbf{a}_1^* + m_2\mathbf{a}_2^* + m_3\mathbf{a}_3^*$ with m_1, m_2, m_3 integers not all zero, the reciprocal electrostatic energy is given by

$$E_{\text{rec}} = \frac{1}{2\pi V} \sum_{\mathbf{m}\neq 0} \frac{\exp(-\frac{\pi^2 \mathbf{m}^2}{\beta^2})}{\mathbf{m}^2} S(\mathbf{m})S(-\mathbf{m}). \tag{6}$$

Here, $S(\mathbf{m})$ is the structure factor

$$S(\mathbf{m}) = \sum_{j=1}^{N} [q_j + 2\pi i \mathbf{p}_j \cdot \mathbf{m}] \exp(2\pi i \mathbf{m} \cdot \mathbf{r}_j), \tag{7}$$

where $\exp(2\pi i \mathbf{m} \cdot \mathbf{r}_j) = \exp(2\pi i(m_1 s_{1j} + m_2 s_{2j} + m_3 s_{3j}))$ and $s_{\alpha j}, \alpha = 1, 2, 3$, are the fractional coordinates of site j, defined above.

Finally, the self-energy term is given by:

$$E_{\text{self}} = -\frac{\beta}{\sqrt{\pi}} \sum_{i}^{N} \left(q_i^2 + \frac{2\beta^2}{3}|\mathbf{p}_i|^2\right). \tag{8}$$

The electrostatic field and force on atom i at position $\mathbf{r}_i$ are computed as the negative gradient of the electrostatic potential $\phi(\mathbf{r}_i)$ and electrostatic energy $E(\mathbf{r}_i)$, respectively:

$$\mathbf{E}(\mathbf{r}_i) = -\nabla_i \phi(\mathbf{r}_i),$$
$$\mathbf{F}(\mathbf{r}_i) = -\nabla_i E(\mathbf{r}_i) = -(q_i + \mathbf{p}_i \cdot \nabla_i)\nabla_i \phi(\mathbf{r}_i),$$
$$\boldsymbol{\tau}(\mathbf{r}_i) = \mathbf{p}_i \times \mathbf{E}(\mathbf{r}_i). \tag{9}$$

An important difference from the monopolar case is that for non-zero dipoles the force is no longer parallel to the field, and a torque $\boldsymbol{\tau}(\mathbf{r}_i)$ needs to be calculated. This torque, in turn, leads to extra terms in the atomic forces (Toukmaji et al., 2000).

Clearly all the equations above recover the traditional Ewald expression when the dipoles are set to zero. The erfc function in the Ewald direct sum originates in Gaussian screening functions centered at each point particle. The reciprocal sum is given by the Fourier series representation of the solution to Poisson's equation in PBC, using the sum of the compensating Gaussians, again centered at each point particle, as a source distribution. The relative contributions and computational cost of the direct and reciprocal sum can be tuned by varying the width (the β parameter in the above expressions) of the Gaussian distributions. Narrow Gaussians (large βs) make the erfc in the direct sum decay fast, and a shorter cutoff is needed for the direct sum, while more terms are needed in the reciprocal space. Wide Gaussians have the opposite effect: the exponential in the reciprocal sum decays fast and fewer terms are needed, while the direct sum needs more terms. In particular, if the Gaussians are chosen to vanish (within a prescribed tolerance) at a standard cutoff distance independent of system size N, conventionally taken to coincide with the cutoff of the Lennard-Jones interactions, the direct sum is $\mathcal{O}(N)$, but the reciprocal sum becomes $\mathcal{O}(N^2)$. If, on the other hand, the Gaussians are chosen to vanish at a cutoff of half the cell size, the direct sum is over all minimum image pairs, and thus is $\mathcal{O}(N^2)$, while the reciprocal sum is $\mathcal{O}(N)$. By varying the cutoff with the square root of the cell size, one can obtain an optimal scaling of $\mathcal{O}(N^{3/2})$ for the direct and reciprocal sums (Perram *et al.*, 1988). The total force, energy, etc. are independent of β and this fact is important when programming modifications to the Ewald method: if the results vary with β, then something is wrong.

2. Particle-Mesh Methods

The Particle Mesh (PM) based approaches use Fast Fourier Transform (FFT) to accelerate the solution of Poisson's equation in systems with PBC. When a fixed cutoff is applied to the direct Ewald sum, say 8 or 9 Å (angstroms), the number of terms needed in the reciprocal sum is $\mathcal{O}(N)$. Since each such term is nominally of order N to calculate, the cost of calculating the reciprocal sum is $\mathcal{O}(N^2)$. A discrete Fourier transform of N coefficients is also $\mathcal{O}(N^2)$, but the FFT performs this task in $\mathcal{O}(N \log N)$ operations. All the PM methods transform the reciprocal sum into a sum over coefficients of the discrete Fourier transform of a gridded charge distribution, whose calculation is accelerated to $\mathcal{O}(N \log N)$ via three-dimensional FFTs. The methods differ in how they transform the continuous charge density due to the sum of compensating Gaussians onto a regular three-dimensional grid, and how they compensate for the loss of accuracy introduced in this process. These methods have been carefully studied and compared in the literature (Pollock and Glosli, 1996; Toukmaji and Board, 1996; Darden *et al.*, 1997; Deserno and Holm, 1998; Sagui and Darden, 1999a, 1999b), and therefore our review will be brief.

Hockney and Eastwood (1981) introduced first the Particle-Mesh method, wherein the Coulombic potential was obtained by solving the discretized Poisson's equation on a regular grid under PBC via FFTs. Since the interactions between nearby particles were poorly represented, they later developed the Particle–Particle–Particle-Mesh approach (PPPM or P3M), wherein the interaction is split into short- and long-range contributions using switching functions, and the long-range potential is obtained with the aid of FFTs as before. Although switching functions led to a smoother charge density on the grid, the method was still not accurate enough for MD of molecular systems (Hansen, 1986; Greengard, 1988), and the connection with the Ewald summation was not made, probably because the authors did not apply Gaussian screening functions. As a result, the PM methods were ignored by the community, that favored the multipole expansion approaches such as the Fast Multipole Method (FMM) (Greengard, 1988). As was later shown, the use of Gaussian screening functions and an optimal Green's function (Ferrell and Bertschinger, 1994) greatly improved the accuracy of the method.

The PME algorithm of Darden et al. (1993, 1995) took the Ewald sum as the starting line. Instead of considering charge interpolation and optimization of the Green's function, the method approximated the complex exponentials in the reciprocal sum by local polynomial interpolation: Lagrange interpolation in the original implementation (Darden *et al.*, 1993), and Euler spline interpolation, based on cardinal B-splines in the smooth PME (Essmann *et al.*, 1995). B-splines are factored into three one-dimensional B-splines which can be differentiated analytically via recursion relations. The forces are obtained analytically from the gradient of the energy in the smooth PME method at very low cost, requiring only 2 FFTs (versus 4 FFTs for equivalent accuracy in the P3M method). The actual steps in the algorithms for either P3M or PME are well-known and described in the literature, and so are the differences and similarities between the two methods (Pollock and Glosli, 1996; Darden *et al.*, 1997; Deserno and Holm, 1998; Sagui and Darden, 1999a, 1999b). The main results of very careful comparisons is that both methods are equally accurate (Darden *et al.*, 1997; Deserno and Holm, 1998; Sagui and Darden, 1999b), but P3M is less efficient. (Originally, the PME method used B-spline interpolation, but *least-squares B-spline approximation* gives superior accuracy (Pollock and Glosli, 1996), and this form has been in use since 2000 (Sagui and Darden, 1999b).) PME has become the most widely used method for computing electrostatic interactions in molecular simulations with explicit solvent. The method is implemented in most of the major biomolecular codes: AMBER, CHARMM, NAMD, GROMACS, MOIL, DL_POLY, etc.

Another variant of the Particle Mesh approach is the Fast Fourier Poisson (FFP) method, proposed by York and Yang (1994). This method directly samples the compensating Gaussians onto the grid, and avoids loss of accuracy from interpolation by use of a clever identity, later used in multigrid methods. Due to the cost of sampling the Gaussians, this method is not competitive for the modest accuracies appropriate for present force fields, but it appears to be more efficient than the above methods for high accuracy requirements.

E. Real-Space Methods

1. Multigrid Methods

The introduction of multigrid methods for biomolecular simulations was motivated by the need to optimize the calculation of long-range electrostatics interactions to parallelization. Given the recent breakthroughs in the treatment of FFT, the PME method in codes such as NAMD or PMEMD in AMBER are highly parallelizable—at present more so than multigrid methods—and it is not clear whether the purported parallelization advantages of multigrid will actually prevail, but this is an area of intense research and new developments are expected. The multigrid methods, however, have other advantages: ease of handling any sort of boundary conditions, including highly mixed boundary conditions; great functionality in implicit solvent methods, like Poisson–Boltzmann approaches; possible advantages in the implementation of highly accurate, continuous electrostatics (Cisneros *et al.*, 2005); possible adaptability in efficient multiple-time-step integrators that separate fast and slow interactions (Qian and Schlick, 2002); etc. In particular, multigrid or the related multilevel methods (Holst and Saied, 1993; Holst *et al.*, 1994; Holst and Saied, 1995) have been very successful in the *static* calculation of electrostatic potentials for large, complex molecules in implicit solvent, specially with the use of finite elements. In this review we will not address implicit methods, and will restrict ourselves to multigrid as applied to molecular dynamics, generally with explicit solvent.

First introduced in the 1970s by Brandt (1977, 1994, 2001), multigrid methods, which can solve linear and non-linear elliptic partial differential equations and integro-differential equations on grids with K points in $\mathcal{O}(K)$ operations, have reached maturity (Hackbush and Trottenberg, 1982; Jespersen, 1984; Hackbush, 1985; Wesseling, 1991; Briggs *et al.*, 2000; McCormick, 1989; Adams, 1989; Press *et al.*, 1992; Beck, 2000; Trottenberg *et al.*, 2001). The algorithms are flexible, so that, in principle, one can adapt a general multigrid technique to a specific problem. Here, we only briefly describe the method. Essentially, the multigrid approach is based on the observation that the individual frequency components in Fourier space of the error are most effectively reduced on grids whose spacing is similar in magnitude to the wavelength of the error component (Brandt, 1977, 1994, 2001). Thus, when one considers iterations on a fine grid, they readily reduce the high-frequency components of the error (*i.e.*, those with wavelengths comparable to the grid spacing), but are rather slow in eliminating the corresponding low-frequency components. Multigrid algorithms aim to take advantage of these observations by working with different grids. By introducing successive auxiliary grids with larger mesh spacings, it is possible to eliminate the lower-frequency components. The resulting accelerated convergence on all length scales allows for the solution of elliptical partial differential equations on grids with K points in $\mathcal{O}(K)$ operations (Brandt, 1977). For condensed-matter systems with a density near that of water, the total number of grid points K is linearly proportional to the number of atoms N; *i.e.*, the multigrid is truly an $\mathcal{O}(N)$ method. Although other well-known relaxation methods such as the Gauss–Seidel, the alternating direction implicit method and the iterative SOR methods share many important advantages of the multigrid methods (no FFTs, ease of parallelization, no need for periodic boundary conditions), their convergence rate is always slower than that of multigrid methods. For instance, in Ref. (Press *et al.*, 1992) it is shown that in an *optimal* SOR method the number of iterations necessary to reduce the initial error by a factor of 10^{-P} is proportional to $pK^{1/3}$, *i.e.*, it increases as the system size increases. The total work for convergence is given by the product of the number of iterations ($\propto pK^{1/3}$) times the cost per iteration ($\propto K$), so an *optimal* SOR method scales as $\mathcal{O}(K^{4/3})$ or $\mathcal{O}(N^{4/3})$. Therefore, the total work increases in a non-linear fashion, since the cost per iteration also increases as the size of the system increases. In multigrid methods, on the other hand, the number of iterations is constant, independent of system size, so that the work is truly $\mathcal{O}(N)$.

Multigrid methods have been successfully applied in calculations of the electronic structure (White *et al.*, 1989; Briggs *et al.*, 1996; Beck *et al.*, 1997; Beck, 1997, 2000; Fattebert and Bernholc, 2000; Lee *et al.*, 2000). Here, we are interested in the application of multigrid methods to *classical* biomolecular simulations. The first attempt in this direction used the Fast Multipole Method (FMM) (Greengard and Rokhlin, 1987). Originally, the FMM, which has a complexity of $\mathcal{O}(N)$, was considered especially suited for parallelization (Greengard and Rokhlin, 1987; Board *et al.*, 1992; Pollock and Glosli, 1996). However, optimized codes for the FMM method are quite elaborate, especially for non-cubic cells. The FMM methods are significantly slower than the PM methods on single processors (Pollock and Glosli, 1996) (and these days in parallel computers too) and, more importantly, the FMM do not conserve energy during MD simulations unless a higher than usual accuracy is enforced (*e.g.* 12th order multipoles) (Zhou and Berne, 1995; Bishop *et al.*, 1997; Sagui and Darden, 1999a, 2001). A first attempt to use multigrid in the context of biomolecular simulations consisted of an adaptive multigrid technique implemented into the original FMM (Zaslavsky and Schlick, 1998). This technique is competitive with the FMM only for low accuracy (comparable to FMM with 4th–6th order multipoles). Molecular dynamics was not implemented, but given what is known about lack of energy conservation for FMM with lower-order multipoles, one would expect that the method did not represent a solution for MD.

Sagui and Darden (2001) introduced a multigrid method appropriate for molecular dynamics of biomolecules in explicit solvent. Along general lines, the method consists of: (i) a charge assignment scheme, which both interpolates and smoothly assigns the charges onto a grid; (ii) the solution of Poisson's equation on the grid via multigrid methods; (iii) the back-interpolation of the forces from the grid to the particle space. The method can be also considered an Ewald-based method since it uses a similar decomposition in direct and reciprocal sums, where the real-space multigrid method (instead of FFTs) is used to solve for the "reciprocal" sum. For classical calculations, the charge assignment scheme (*i.e.*, the spreading of the charges on the mesh) was found to be crucial for the performance of the method. To obtain reasonable accuracies, the charge density on the grid must be as smooth as possible. This means that the charges must be spread over a fairly large number of grid points, that can quickly become an important time consuming part of the algorithm. The authors introduced two methods for interpolating the charge: a lattice diffusion multigrid scheme (LDM) and a lattice Gaussian multigrid scheme (LGM). In the LDM method, charges are interpolated onto the grid first via B-splines, and then a diffusion scheme is applied to smoothly spread the charge. The method has an adjustable parameter, the diffusion constant, and gives relative force errors of approximately 0.5×10^{-3}. Not satisfied with the accuracy of the method and the presence of the adjustable parameter, the authors introduced the LGM, which uses Gaussians as interpolating functions and gave relative force errors of 10^{-4} or less, and relative energy errors of approximately 10^{-5}—at considerably more cost than the LDM method. Both methods showed excellent energy conservation in the nanosecond scale. The high accuracy of the LGM method was achieved by using, instead of the traditional partition in the Ewald method, the partition proposed in the FFP (York and Yang, 1994).

In addition, since standard discretization schemes proved not accurate enough, a deferred defect correction scheme (Brandt, 1977; Hackbush, 1985; Briggs *et al.*, 1996) for the LGM was applied. The deferred defect correction scheme uses a double discretization. A Hermitian representation of Poisson's equation is used on the finest grid, while on the coarser grids the seven-point central finite-difference operator is used. The gain in accuracy in the Hermitian methods is obtained not by including more points in the Laplacian representation, but by considering that the entire differential equation is satisfied at several grid points. The resulting formulas, therefore, involve the value of the function *and the derivatives* at several grid points. The big advantage of the Hermitian representation versus a higher order representation of the Laplacian operator is that for a given accuracy, the former requires less grid points, and therefore, less communication between processors in a parallel implementation.

Since then, several other variations of multigrid have attempted to speed up the method (Skeel *et al.*, 2002; Goedecker, 2003; Izaguirre *et al.*, 2005). The challenge has been to obtain a method that is efficient, while preserving accuracy, since relative force errors for reliable biomolecular MD trajectories need to be of the order of 10^{-4}. Goedecker (2003) proposed a modified multigrid V cycle scheme that employs wavelets and is more efficient than the standard V cycle. Instead of transferring the residue to the coarser grid levels, this scheme works with copies of the charge density on the different grid levels that were obtained from the underlying charge density on the finest grid by wavelet transformations. The authors show that halfway V cycles with the restriction and prolongation steps based on wavelet theory are the most efficient approach for the solution of the three-dimensional Poisson's equation on the grid. Currently, it is not clear whether wavelets will improve efficiency when applied to charge and force interpolation.

Two recent studies (Banerjee and Board, 2005; Shan *et al.*, 2005) have applied convolutions to accelerate the charge assignment and force interpolation steps in the LGM method (which simply used separability of three-dimensional Gaussians to pre-tabulate one-dimensional Gaussians). Both schemes effectively reduce the number of mesh points at which the charge spreading function is nonzero by proceeding in two stages. In the first stage, charges are spread onto the mesh, and in the second stage a convolution with a Gaussian is performed to further spread the charges. Shan *et al.* (2005) used another Gaussian (with a smaller support) for the first stage of charge spreading, and this same Gaussian was used for the final interpolation to calculate the forces. Banerjee and Board (2005), on the other hand, used a set of weights on the grid that were further convolved with the second Gaussian. These same weights are used for the back interpolation to compute the forces, but in a different manner. On a single-processor, both convolution-based methods reduce the computational burden by a factor of about 2 when compared to the earlier LGM. Parallelization studies were not reported in Ref. (Shan *et al.*, 2005). The convolutions implemented in Ref. (Banerjee and Board, 2005) perform well in parallel environment although the multigrid implementation itself performs poorly in communication constrained parallel environments and needs further improvement of parallel performance.

Among other novel applications of the multigrid method there has been its implementation in dissipative particle dynamics (DPD) by Groot (2003a), and its application to higher-order multipoles by Sagui *et al.* (2004). Groot incorporated electrostatics into DPD by solving for the electrostatic field locally on a grid with a multigrid method. The method captures the essential features of electrostatic interactions and satisfies screening of the potential near a charged surface and the Stillinger–Lovett moment conditions. The method is specially applicable to (coarse-grained) simulations of mixed surfactants near oil–water interfaces, or to simulations of Coulombic polymer–surfactant interactions.

F. Local Algorithms for Coulomb's Law

The interactions between charged particles and electromagnetic fields are fundamentally described by Maxwell's equations (Jackson, 1999). The fast electrostatic methods discussed so far, however, are all based on the simplification of potential theory, which is equivalent to the static limit of Maxwellian electrodynamics: the magnetic field $\mathbf{B}$ and all currents $\mathbf{J}$ have vanished, $\dot{\mathbf{B}} = -c\,\mathrm{curl}\,\mathbf{E} = 0$, and the electric field $\mathbf{E}$ obeys Gauss' law, $\mathrm{div}\,\mathbf{E} = \rho$. In this limit, $\mathbf{E} = -\nabla\phi$ and the electrostatic potential is the solution of Poisson's equation $\nabla^2\phi = -\rho$ (we work in units of $\epsilon_0 = \mu_0 = 1$). The Coulomb interaction is instantaneous and must be obtained by solving an elliptic partial differential equation everywhere in space. This need for finding a *global* solution is the fundamental origin of the high computational cost of electrostatics.

In full electrodynamics, however, electromagnetic waves propagate with finite speed of light c as described by the laws of Ampere and Faraday:

$$\dot{\mathbf{B}} = -c \operatorname{curl} \mathbf{E},$$

$$\dot{\mathbf{E}} = c \operatorname{curl} \mathbf{B} - \mathbf{J}, \tag{10}$$

where $\mathbf{J}$ denotes the current due to moving charges. In contrast to the Poisson equation, solutions to these hyperbolic wave equations (10) require only *local* operations. Mawxell's equations furthermore conserve Gauss' law if it is obeyed as an initial condition. One might therefore expect that the Coulomb interaction will also be generated when coupling charges directly to the time varying electromagnetic fields and solving for their dynamics; this is routinely done in plasma physics.

For such an approach to be meaningful in a condensed matter context it is of course not possible to use the true physical value of c to propagate the interactions. In 2002, however, Maggs and Rossetto (2002) showed rigorously that the Coulomb interaction is obtained even in the thermodynamic limit *independent* of the value of c. They studied a constrained (partial) partition function of the form

$$\mathcal{Z}(\{\mathbf{r}_i\}) = \int \mathcal{D}\mathbf{E} \prod_{\mathbf{r}} \delta(\operatorname{div}\mathbf{E}(\mathbf{r}) - \rho(\{\mathbf{r}_i\})) e^{-\mathcal{U}/k_B T}, \tag{11}$$

where the energy $\mathcal{U} = \int \mathbf{E}^2/2\, d^3\mathbf{r}$ and the charge density $\rho(\mathbf{r}) = \sum_i e_i \delta(\mathbf{r} - \mathbf{r}_i)$, and showed that a Gibbs distribution characterized by $1/r$ interactions is generated when the electric field $\mathbf{E} = -\operatorname{grad}\phi + \mathbf{E}_{tr}$ is allowed to fluctuate under the constraint of Gauss' law. The partition function factorizes and the Coulombic part completely decouples from the unconstrained or transverse (free) electromagnetic field $\mathbf{E}_{tr}$ (Alastuey and Appel, 2000), which is not present in conventional electrostatics. This result is nontrivial and shows that the Coulomb interaction is more than the $c = \infty$ limit of Maxwell's equations; it is fundamentally a consequence of Gauss' law and does not require curl $\mathbf{E} = 0$. Rather than quenching the curl degrees of freedom in the electric field to zero, one can perform an integration over them and obtain the Coulomb interaction even though the electrostatic energy $U \geqslant \int (\operatorname{grad}\phi)^2/2\, d^3\mathbf{r}$. As discussed below, performing this integration while maintaining Gauss' law requires only local operations and obviates the need for computing long ranged interactions at every timestep. Based on this observation, linear scaling $\mathcal{O}(N)$ methods for molecular dynamics and Monte Carlo have been constructed, see Section V.B

IV. NON-PERIODIC SYSTEMS AND DIFFERENT BOUNDARY CONDITIONS

The Ewald summation as presented above is applicable to fully periodic systems. It can be modified, with a price though, to systems of slab geometry, *i.e.*, system having two of the coordinates periodic and one of them non-periodic. As confined systems and systems having surfaces are common and of great interest, there has been a lot of effort to develop methods for simulations of such systems. Most of the methods are, one way or another, based on the Ewald formulas. The main problem is that they scale very badly, typically as $\mathcal{O}(N^2)$. One significant advantage of the real-space methods, multigrid, fast multipoles and the local electrostatics algorithm, is that periodicity is not required. In other words, they can accommodate different boundary conditions naturally without a loss in the scaling properties of the algorithm. At the moment, however, these methods are not implemented in major simulation packages. The hope is, however, that the situation will change.

In addition to the above, there are two other methods worth mentioning, namely Lekner summation (Lekner, 1989, 1991, 1998) and the so-called MMM2D of Arnold and Holm (2002b, 2002a). Unlike Ewald-based methods which use the potential energy as a starting point, they use forces instead. J. Lekner was the first to show how different boundary conditions may be naturally accommodated by such an approach. The Lekner summation scales essentially as $\mathcal{O}(N^2)$ but avoids some of the pitfalls of the Ewald-based methods. MMM2D has better scaling properties $\mathcal{O}(N^{5/3})$ and is clearly better for large systems.

In the following, we will first discuss the Ewald-based methods as they are, despite their problems, often used. Then, we finish this section with a short discussion of the Lekner and MMM2D methods. Quantitative comparisons between different methods have been done by Widmann and Adolf (1997), Mazars (2005), and to smaller extent Arnold and Holm (2002b, 2002a). In addition, the Fast Multipole Method and Ewald methods have been compared under different conditions in Refs. (Sølvason *et al.*, 1995; Pollock and Glosli, 1996)

A. Ewald Based

The Ewald summation can be modified for of slab geometry. In the following, let us assume that z-direction is the non-periodic one.

There are many variants based on the Ewald summation, namely the Heyes–Barber–Clarke (Heyes *et al.*, 1977), or HBC approach, the Hautman–Klein approach (Hautman and Klein, 1992), de Leeuw–Perram (de Leeuw and Perram, 1979), and the Nijboer–de Wette (Nijboer and de Wette, 1957) approaches being the most well-known ones. Since the system is non-periodic in the z-direction, we use the modified Ewald summation containing a correction term that takes into account the 2D periodic nature of the system (Yeh and Berkowitz, 1999). It is

possible to modify the Ewald sum directly which retains its scaling properties, as
pointed out by Yeh and Berkowitz (1999). They showed that the potential energy
for the 2-d periodic system can written using the Ewald summation as follows,

$$U_{2\text{dEW}}(r) = \frac{1}{2} \sum_{i=1}^{N} \sum_{j=1}^{N} \sideset{}{'}\sum_{|\mathbf{n}|=0}^{\infty} q_i q_j \frac{\text{erfc}(\alpha |\mathbf{r}_{ij} + \mathbf{n}|)}{|\mathbf{r}_{ij} + \mathbf{n}|} \tag{12}$$

$$+ \frac{1}{2\pi V} \sum_{i=1}^{N} \sum_{j=1}^{N} \sum_{\mathbf{k} \neq 0} q_i q_j \frac{4\pi^2}{k^2} \exp\left(-\frac{k^2}{4\alpha}\right) \tag{13}$$

$$\times \cos(\mathbf{k} \cdot \mathbf{r}_{ij}) - \frac{\alpha}{\sqrt{\pi}} \sum_{i=1}^{N} q_i^2 + J(\mathbf{M}, P). \tag{14}$$

Here, q_i and q_j are the two charges, α is Ewald parameter, $\mathbf{k}$ is the reciprocal
lattice vector. $V = L_x \times L_y \times L_z$ is the volume of the system. The system was
periodic in the x–y-plane and $L_x = L_y$ is chosen for simplicity. The last term
in Eq. (14) is the correction term that accounts for the non-periodicity in the z-
direction. It depends on the geometry (cubic, rhombic, etc.) used for the Ewald
summation (P) and the total dipole moment ($\mathbf{M}$) of the system.

Next, we have to determine the term $J(\mathbf{M}, P)$ in the above equation. The slab
geometry corresponds to having the system surrounded (in z-direction) by a vacuum. For a spherical summation geometry we can write (Yeh and Berkowitz,
1999)

$$J(\mathbf{M}, P) = \frac{2\pi}{(2\varepsilon + 1)V} |\mathbf{M}|^2. \tag{15}$$

In vacuum $\varepsilon = 1$ and using a rectangular geometry the above reduces to

$$J(\mathbf{M}, P) = \frac{2\pi}{V} M_z^2, \tag{16}$$

where M_z is the z-component if the total dipole moment. Equation (16) can be
used directly simulations and it is trivial to implement. This method can be easily
applied in conjunction with the PME or P3M methods giving $\mathcal{O}(N \log N)$ scaling.
Small practical issues are how to choose the Ewald parameter α, the number of
k-vectors, and the amount of empty space in the non-periodic direction.

B. Lekner Summation

As mentioned above, Lekner's starting point was to use the force exerted by
particle j on particle i instead of the potential, *i.e.*,

$$\vec{F}_i = q_i q_j \sum \frac{\vec{r}_i - \vec{r}_j}{|\vec{r}_i - \vec{r}_j|^3}.$$

After that, Lekner moved to dimensionless displacements for the charges (*i.e.*, defining $r_{i;x/y/z} - r_{j;x/y/z} = \eta_{x/y/z} L_{x/y/z}$, for each component, respectively). Then, the resulting dimensionless formulas are first subjected to an Euler transformation. That is followed by the application of the Poisson–Jacobi identity, and finally the integral representations of Bessel functions are used. After the transformations one is left with formulas for forces (as well as potentials) which can be used for periodic and non-periodic systems alike. Instead of the Fourier transformation required in PM methods, here, one is left with Bessel functions of the zeroth and first kind together with trigonometric functions. In addition to being used to compute electrostatic forces, this method has also been employed in simulations of vortices which interact through logarithmic potentials in two dimensions (Grønbech-Jensen, 1997).

C. MMM2D and ELC

It is also worth mentioning that the Lekner method has been later extended and modified by Sperb and Strebel (Sperb, 1994, 1998, 1999; Strebel and Sperb, 2001) such that the modified method scales as $\mathcal{O}(N \log N)$ in fully periodic three-dimensional systems. In 2D periodic cases the scaling is worse, $\mathcal{O}(N^{7/5})$. Using the work of Sperb and Strebel as a starting point, Arnold and Holm (2002b, 2002a) have developed and improved the method further in 2D periodic systems. They coined the name MMM2D and essentially using the Poisson summation formula were able to improve the scaling to $\mathcal{O}(N^{5/3})$, which makes the method better than the Lekner summation or the 2D modification of the Ewald summation. In addition to the scaling properties, MMM2D has the advantage that it is possible to perform rigorous error analysis on it.

In addition to MMM2D, Holm *et al.*, have combined their approach with the method of Yeh and Berkowitz as described above. The method is called Electrostatic Layer Correction (ELC). The basic idea is to use the fully periodic Ewald (or any variant of it) and then apply layer corrections. The advantage is that the method scales as $\mathcal{O}(N \log N)$ if PME or P3M is used for the full 3D-summations.

V. APPLICATIONS AND LATEST METHODOLOGICAL DEVELOPMENTS

A. Particle-Mesh Methods

The latest developments in PM methods have been centered on both accuracy and efficiency via improvements in parallelization. Improvements in accuracy have been obtained by the inclusion of higher-order multipoles and continuous functions for the representation of electrostatics, as well as new descriptions for polarization.

Electrostatic interactions have traditionally been modeled using an atom-centered point charge ('partial charge') representation of the molecular charge density (Sagui and Darden, 1999a). The most popular methods for extracting charges from molecular wavefunctions are based on fitting atomic charges to the molecular electrostatic potential (MEP), computed with *ab initio*, density functional theory or semiempirical wavefunctions. The charge fitting procedure consists of minimizing the squared deviation between the Coulombic potential produced by the atomic charges and the MEP. These non-bond potentials are then expressed as a sum of spherically isotropic atom–atom potentials. Such representations are believed to be an important source of error in current force fields (Stone, 1996).

The fit to the MEP can be improved either by adding more charge sites (Dixon and Kollman, 1997) or by including higher order multipoles at the atomic sites or bonds. Even with these improvements the fit to the MEP remains poor in regions near the atomic nuclei, where the charge densities overlap. As a consequence, the electrostatic interaction energy must be corrected for "penetration" effects at close range (Wheatley and Mitchell, 1994) (usually this error is absorbed into the exchange repulsion term); and the optimal values of the point multipoles may be poorly determined (Chipot *et al.*, 1998; Bayly *et al.*, 1993; Francl and Chirlian, 1999). Nevertheless the use of off-center charges and/or higher order atomic point multipoles can significantly improve the treatment of electrostatics.

The *distributed* multipole analysis first introduced by Stone assigns distributed multipole moments to several sites in the molecule (*i.e.*, atoms and bond midpoints), and gives a more accurate representation of the electrostatic potential than one-center, molecular multipole expansion. The generalization of Ewald summation to atomic multipoles up to quadrupoles was given by Smith (1982). Since then, a few groups (Toukmaji *et al.*, 2000; Aguado and Madden, 2003; Ren and Ponder, 2003; Sagui *et al.*, 2004) have extended the Ewald method to take into account multipoles at the atomic and other point sites. However, the multipoles greatly increase the cost of calculations within the Ewald framework. For instance, an electrostatic representation including charges, dipoles and quadrupoles costs approximately 100 times more than a representation with only charges, using the Ewald formalism, thus rendering multipolar representations in biomolecular simulations prohibitively expensive. In order to surmount this difficulty, PME-based methods have been introduced. A first approach was introduced by Toukmaji *et al.* (2000), who developed—in addition to the classical Ewald treatment—a PME based treatment of fixed and induced point dipoles. Both methods have been implemented into the sander molecular dynamics module of AMBER (versions 6 to 9), along with several variants of an iterative scheme for solving the induced dipoles, as well as a Car–Parrinello extended Lagrangian scheme, wherein the induced dipoles have an associated fictitious mass and evolve with the rest of the system. The latter, when combined with the PME implementation, is quite efficient; for a 1 fs timestep it is only approximately 1.33 times more expensive than a calculation including only charges. However, it requires a gentle temperature control and cannot use larger time steps. DNA simulations with the AMBER ff02 force field using the PME-based approach with the extended Lagrangian formalism lead to stable trajectories (Baucom et al., 2004; Babin et al., 2006a, 2006b), with some improvements over previous force fields.

Sagui et al. (2004, 2005) recently developed an efficient implementation of higher order multipoles in a cartesian tensor formalism, adopting the maximum rank expansion. The long-range electrostatic interactions are divided in two sums according to the usual Ewald scheme: the *direct* sum, which evaluates the fast-varying, particle–particle interactions, considered up to a given cutoff in real space; and the *"reciprocal"* sum, which evaluates the smoothly varying, long-range part of the interaction. Equations (3) to (9) give a general idea on how the Ewald formalism can be extended to higher-order multipoles. When implementing multipoles, however, one has to take care of additional physics that is not present in the usual treatment of charges. First, the higher-order multipoles produce additional contributions to the reciprocal virial, that arise from the dependence of the structure factor on the reciprocal lattice vector. Second, all the multipolar components that appear in the expressions of energy, forces, etc. are given in a global coordinate system. It is necessary therefore to transform the local multipole moments—generally defined in reference to the molecule—to a global framework before any calculation starts. This is achieved by defining "frames" (local orthogonal coordinate systems). Third, to carry out molecular dynamics, the torques produced by every multipole need to be converted into atomic forces.

In order to accelerate the evaluation of the Ewald-like sums, the authors used for the direct sum a McMurchie–Davidson formalism (McMurchie and Davidson, 1978) originally developed for the evaluation of quantum molecular integrals over Cartesian Gaussians, with extensions due to Challacombe et al. (1995). The reciprocal part was implemented in three different ways: using an Ewald scheme, a PME-based approach and a multigrid-based approach (reviewed in the next section). The standard matrix implementation of multipole interactions up to hexadecapole–hexadecapole costs three orders of magnitude more than charge–charge interactions. Due to the factorizability of B-splines, the smooth PME is very efficient. For instance, for the same grid density and spline order, calculating reciprocal sum interactions up to hexadecapole–hexadecapole is only twice as expensive as calculating charge–charge interactions. Therefore, by transferring more of the computation of the interactions to reciprocal space and using a small cutoff for the direct sum, it is possible to preserve the accuracy of the calculations at a moderate cost. In fact, a considerably accurate calculation of interactions up to hexadecapole–hexadecapole costs only a factor of 8.5 more (for relative force errors of $\sim 5 \times 10^{-4}$) than a regular AMBER implementation with only charge–charge interactions. Furthermore, a 'regular' cutoff of 8 Angstroms for the Coulomb summation (with the acceleration provided by the McMurchie–Davidson scheme) is approximately six times more expensive and has a force error two orders of magnitude larger than the complete calculation for which most of the interaction is computed in reciprocal space via the PME method. This code has been adapted to achieve a fast implementation of the AMOEBA (Atomic Multipole Optimized Energetics for Biomolecular Applications) force field of Ren and Ponder (2002, 2003, 2004) and Ponder and Case (2003). The force field includes fixed atomic multipoles up to quadrupole level as well as inducible atomic dipoles using a Thole damping model. A PME-based implementation of the multipolar code for AMOEBA has been released through AMBER 9.

More recently, Cisneros et al. (2005, 2006) and Piquemal *et al.* (2006) have striven to achieve accurate electrostatics using continuous functions. The group introduced a force field based on density fitting termed the Gaussian electrostatic model (Piquemal *et al.*, 2006) (GEM) that improves the description of the non-bonded interactions. The method relies on density fitting methodology to reproduce each contribution of the constrained space orbital variation (CSOV) energy decomposition scheme, by expanding the electronic density of the molecule in Gaussian functions centered at specific sites. Originally they used s-type functions, and later they extended the formalism to express the Coulomb and exchange components of the force field in terms of auxiliary basis sets of arbitrary angular momentum. Since the basis functions with higher angular momentum have directionality, a reference molecular frame formalism similar to that defined in Ref. (Sagui *et al.*, 2004) was employed for the rotation of the fitted expansion coefficients. McMurchie–Davidson recursion relations were also used. The use of Hermite Gaussian functions has the additional advantage of facilitating a point multipole decomposition determination at each expansion site. Computational speed was investigated by reciprocal space based formalisms, mainly the PME and the FFP methods. Frozen-core (Coulomb and exchange-repulsion) intermolecular interaction evaluations with the formalism were tested on several homo- and hetero dimers, and water boxes (up to 1024 molecules) showing excellent agreement with the CSOV calculated reference contributions (with errors below 0.1 kcal for electrostatics for the dimers and RMS accuracies of 10^{-3} and 10^{-4} for the water boxes). All calculations showed a significant computational speed improvement with respect to the reference CSOV calculation.

B. Local Electrostatics Algorithm

1. Application to Molecular Dynamics

In molecular dynamics (MD), it is natural to sample Eq. (11) through propagative field dynamics as given by Eq. (10). The equations of motion of the particles have the usual form $m\ddot{\mathbf{r}}_i = F_{el}$, where the electrostatic force on a charge e_i at position $\mathbf{r}_i$ is given by $\mathbf{F}_{el} = e_i \mathbf{E}(\mathbf{r}_i)$. The Lorentz force, $e_i(\mathbf{v} \times \mathbf{B})/c$, has been dropped since we are only interested in electrostatics, and it can be shown that the resulting equations of motion obey a generalized Liouville theorem (Rottler and

Maggs, 2004b). In addition, the algorithm (Rottler and Maggs, 2004b) imposes Langevin thermostats on both particle and field degrees of freedom,

$$m\dot{\mathbf{v}}_i = e_i\mathbf{E}(\mathbf{r}_i) - \gamma_1\mathbf{v}_i + \vec{\xi}_1,$$
$$\dot{\mathbf{B}} = -c\,\text{curl}\,\mathbf{E} - \gamma_2\mathbf{B} + \vec{\xi}_2, \tag{17}$$

where the damping γ_j and the noise $\vec{\xi}_j$ are related by the fluctuation dissipation theorem. This procedure ensures ergodicity and in particular removes any unwanted constraints, in particular div $\mathbf{B} = 0$. Note however that although the noise on $\mathbf{B}$ is completely general it does not destroy Gauss' law for $\mathbf{E}$.

The algorithm can now be operated in two modes: a first possibility is to maintain both particle and field degrees of freedom at the same temperature T. In this mode, the electric forces also contain noise that acts on the particles like an additional thermostat. This mode is most natural in the spirit of Eq. (11) and also the most efficient, as it allows the simultaneous integration of the equations of motion with the same timestep (see below). Alternatively, one may remove the noise on the magnetic field and quench the field to zero temperature using damping only. This limit is very close in spirit to the Car–Parrinello method in *ab initio* molecular dynamics (Car and Parrinello, 1985), where a constrained Lagrangian is used to generate fictitious dynamics of electronic wavefunctions that follow the ground state solutions adiabatically. Here, it is the electric field that will approach the (unique) solution of Poisson's equation.

Two implementations are currently available that demonstrate the merit of the approach. In Ref. (Rottler and Maggs, 2004b), Rottler and Maggs discretized electric and magnetic fields on a grid and integrated them with the Yee algorithm (Yee, 1966). A Verlet scheme was used to jointly advance fields and particles, and the presence of an effective Coulombic interaction in both modes was verified numerically. The implementation was recently improved to obtain higher accuracy, and its parallelizability was explored on a contemporary computer cluster (Rottler, 2007). An independent implementation by Pasichnyk and Duenweg (2004) based on a vector potential formulation reached similar conclusions and also presented an alternative derivation of the method using a Lagrangian formulation and the variational principle.

The principal technical challenge lies in efficient coupling of charges and fields. As in all mesh based methods, the charges need to be interpolated from the continuum onto the electrostatic grid to generate the current $\mathbf{J}$, and the electric field must be extrapolated to compute the electrostatic force $\mathbf{F}_{el}$. Similar to the multigrid method, high accuracy requires smooth interpolation over a large volume; this is the dominant computational cost in the algorithm. The initial implementations (Rottler and Maggs, 2004b; Pasichnyk and Duenweg, 2004) used only low order spline interpolation schemes, but more recent schemes using Gaussian charge spreading (Maggs, 2005) eliminated lattice artefacts and reached much higher accuracy comparable to the schemes used in the multigrid methods. The displacement of a charge e_i generates a local charge fluctuation in the interpolating volume. The resulting local current flows through this volume in such a way that the electric field $\mathbf{E}$ on the mesh obeys Gauss' law after the particle displacement and hence at all times. The forces on the charges are computed from the principle of virtual work: $F_\alpha = -\delta U / \delta r_{i\alpha}$. The simplest possible discretization of the expression $\mathcal{U} = \int \mathbf{E}^2 / 2 \, d^3 \mathbf{r}$ results in truncation errors equivalent to the usual 7-point discretization of the Laplacian, but higher order schemes are also available (Maggs, 2005).

In the local electrostatics method, the speed of light appears as an explicit optimization parameter. If one is only interested in thermodynamic averages, Eq. (11) guarantees that the trajectories visit states with the correct Coulombic weights independent of the choice of c. One expects, however, that dynamical correlations become modified when c becomes of order the particle velocities. Reference (Rottler and Maggs, 2004b) examined this effect in velocity autocorrelations and showed that unmodified correlations can be obtained upon increasing c but without lowering the integration timestep. The method then yields results equivalent to those of conventional fast Coulomb solvers in periodic boundary conditions.

2. Application to Monte Carlo

Monte Carlo (MC) simulations have in the past rarely been employed for the simulation of charged systems, because, unlike in MD, the electrostatic energy must be recomputed after every particle move (an exception are simulations based on pretabulation of the Coulomb interaction). However, MC is often the preferred choice for equilibrium problems due to superior stability and larger timesteps. The present reformulation of the electrostatic problem as a local algorithm provides a remedy and for the first time, permits efficient MC simulation with electrostatics.

In their initial proposal of a MC algorithm for a charged lattice gas, Maggs and Rossetto (2002) suggested to sample Eq. (11) with the Metropolis algorithm and two types of MC moves. First, a particle is displaced along the link connecting two nodes of the lattice, and the electric field on the traversed link is modified to obey Gauss law as illustrated in Fig. 3. The resulting energy change is purely local and is used to determine the acceptance probability. Second, an integration over the curl degrees of freedom of the electric field is performed by locally modifying the field by a pure rotation; this can be done efficiently by grouping the field into plaquettes (see Fig. 4). The two MC moves can be randomly mixed and generate configurations with Coulombic weights. In MC, the particle and field degrees of freedom are kept at the same temperature; minimizing the energy by accepting only those field changes that decrease the energy would instead converge the field configuration to the solution of Poisson's equation. It can be shown (Maggs, 2002) that this procedure generates a field dynamics described by the modified Maxwell

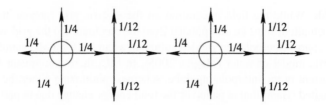

FIGURE 3   Displacement of a charge $e = +1$ from the left to the right node on a lattice in the MC algorithm. The numbers denote the value of the electric field **E** which lives on the links of the lattice (here in 2D with lattice constant a). The field always satisfies Gauss' law, but is not in the minimum energy configuration. Upon moving the charge, a local current $J = -e/a^2$ flows along the traversed link so that the Gauss constraint is upheld in the new configuration.

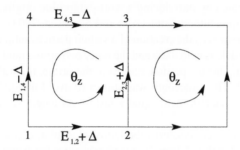

FIGURE 4   Integration over transverse (curl) degrees of freedom in the local MC algorithm. Four link fields are grouped into a plaquette and are modified by a random change Δ in such a way that **E** is modified only by a pure rotation.

equations

$$\mathbf{B} = -c \operatorname{curl} \mathbf{E} + \gamma_1,$$
$$\dot{\mathbf{E}} = c \operatorname{curl} \mathbf{B} - \mathbf{J}, \tag{18}$$

or equivalently

$$\dot{\mathbf{E}} = c^2 \nabla^2 \mathbf{E} - \mathbf{J} - \operatorname{curl} \gamma_1. \tag{19}$$

In the absence of current (*i.e.* stationary charges), the electric field obeys a diffusion equation and the magnetic field no longer appears explicitly. Diffusive dynamics is naturally generated in MC and a valid choice as long as the transition rates obey detailed balance. The diffusion constant c^2 of the electric field is a freely tuneable parameter and now depends on the ratio of particle to field updates (Maggs, 2002, 2004).

In order to simulate continuous systems such as polyelectrolyte and membrane models, Rottler and Maggs extended the lattice gas algorithm to an offlattice MC scheme (Rottler and Maggs, 2004a) and applied it to simple electrolytes and polar fluids. While the field integration on the electric grid remains the same, charge fluctuations must be interpolated from the continuum to the grid as in MD. Gaussian charge interpolation yielded sufficient accuracy for a MC simulation of an atomistic model of water (Maggs, 2005). In MC, there is a greater freedom for the current in the interpolating cube; schemes minimizing either the number of sites visited (Hamiltonian paths) or the total energy change due to particle displacement have been successfully employed. As in MD, the charge interpolation is the biggest contributor to the computational effort.

Since its inception, the local MC algorithm has attracted considerable attention and hence experienced a series of improvements. In order to simulate efficiently low density systems, the electric field must be rapidly equilibrated in large, mostly empty regions. Local update schemes permit only diffusive spreading. This problem can be eliminated by introducing a worm algorithm originally proposed for the simulation of quantum systems (Alet and Sorensen, 2003). The worm algorithm can be understood as the creation of a virtual particle/antiparticle pair on the same site. One particle is held fixed and the other performs a random walk through the simulation cell. One the particle returns to the origin, the walk is globally accepted or rejected. This procedure was shown to avoid the diffusive dispersion and equilibrates the electric field equally fast on all length scales (Levrel *et al.*, 2005).

A second improvement concerns the lattice algorithm at low temperatures. It was soon realized that the Metropolis acceptance rates for discrete particle moves on the lattice rapidly deteriorate when the electrostatic energy barrier is much larger than the thermal energy. Levrel and Maggs presented a solution that prevents the slowdown in mobility, which consists in temporary charge spreading (Levrel and Maggs, 2005). Duncan and collaborators addressed the same issue and proposed a related scheme that combines the particle move with an extended field change (Duncan *et al.*, 2005).

One of the important advantages of the local electrostatic algorithm is that it naturally generalizes to electrostatics in inhomogeneous dielectric environments that are described by a spatially varying dielectric function $\epsilon(\mathbf{r})$ (Maggs, 2004; Duncan *et al.*, 2006). The electrostatics of polarizable media is conveniently described by introducing the electric displacement field $\mathbf{D} = \epsilon(\mathbf{r})\mathbf{E}$, which then satisfies div $\mathbf{D} = \rho(r)$. The electrostatic energy generalizes to $\mathcal{U} = \int \mathbf{D}^2/2\epsilon(\mathbf{r}) \, d^3\mathbf{r}$. This situation can be easily treated in the local algorithms with virtually no overhead: the electric field on the links of the lattice is replaced by $\mathbf{D}$ and $\epsilon(\mathbf{r})$ is discretized onto the links to compute U. Mobile dielectric inclusions can be treated on the same footing.

When treating situations with dielectric contrast with the local algorithms, it is important to realize that the algorithm generates an additional apparent dipole–dipole interaction that is absent in traditional electrostatics (Maggs, 2004). The finite temperature fluctuations in the electric field give rise to a Keesom/Lifshitz interaction (Israelachvili, 1992). For spherical dielectric inclusions separated by a distance r, this interaction is short ranged, $\sim k_B T / r^6$. It may be viewed as occurring naturally in thermal systems, or may be removed by a multiboson method as described in (Duncan et al., 2006; Duncan and Sedgewick, 2006).

3. General Applicability

The local algorithms for MD and MC are by construction true linear scaling, $\mathcal{O}(N)$ methods, which has also been verified numerically. The prefactor to scaling is mainly controlled by the required degree of accuracy, which determines the volume over which the charges are interpolated. The integration of the field degrees of freedom is very simple and adds only a small additional overhead. Among all methods discussed in this chapter, the local electrostatic formulations are the newest and most experimental algorithms. Current implementations offer medium range accuracy sufficient for coarse grained simulations. Further development should be aimed at optimized charge spreading schemes that minimize interpolation volume while maintaining maximum accuracy.

For small simulations on a single processor, the local MD algorithm does not offer any speed advantages over the Ewald-based methods, since the charges must be spread over larger volumes. The algorithm, however, be more efficient on large ($N > 10^5$) systems in highly parallelized environments. If the problem is suitable for MC dynamics, however, the method offers a great potential for speedup at any system size. Being a real-space method, the local algorithms share with the multigrid methods the following additional advantages:

- Good parallelizability on commodity supercomputer cluster due to reduced communication requirements. For a simulation cell of volume L^3, the communication cost scales only as $L^{2/3}$ as opposed to L^3 in FFT.
- No need for periodic boundary conditions as in the reciprocal space methods. Non-periodic boundary conditions and slab geometries can be realized without difficulty; boundary effects in finite systems can for instance be treated with the boundary element method. The lattice version of the MC algorithm has recently been used to study ions between charged plates (Duncan et al., 2005). An extension to constant potential boundary conditions has also been proposed (Levrel, 2006).
- Ability to treat static and mobile dielectric contrast. Note that the Fourier transformation does not diagonalize the Poisson equation $\nabla \cdot (\epsilon(\mathbf{r}) \nabla \phi) = -\rho$ in the presence of a spatially varying dielectric function $\epsilon(\mathbf{r})$. The large dielectric contrast between hydrocarbons ($\epsilon \approx 2$) and water ($\epsilon \approx 78$) is routinely ignored in many implicit solvent models for polyelectrolytes and charged complex fluids; including such effects will require real-space schemes.

The latter two points are also fundamental to ion transport through membrane channels and DNA translocation through nanopores driven by electric fields. The importance of the large dielectric contrast between water and membrane molecules for the function of ion channels is well known and must be accounted for in any quantitative simulation. The present algorithms hold the promise of improved modeling of such systems on the mesoscale.

VI. FINAL NOTES ON SOFTWARE, ACCURACY, AND SPEED

Let us finish this review with a brief account on software. The discussion related to GROMACS and NAMD below refer to the published versions, GROMACS 3.3 and NAMD 2.6. Both are being actively developed and progress is generally rapid.

All other issues aside, but in terms of speed and scalability in atomistic (bio-molecular) simulations it is fair say that currently there are two software packages that are above the rest, namely GROMACS and NAMD. GROMACS is generally accepted as the fastest one on a single processor CPU due to its use of inner loops written in assembler. GROMACS is also well-known to be the speediest if only a small number of processors (less than 32) is needed. After about 32 (older version 16), GROMACS (v. 3.3) does not scale very well, but for less than 32 CPUs, it is the uncrowned king. For simulations using a very large number of processors, there is currently only one choice: NAMD. It has superior scaling compared to any other package and holds the current 'world record' of being able to simulate a biomolecular system over 2.6 million atoms using 1024 processors.

As for the very latest, tentative results for the new GROMACS 4.0 beta version indicate that its scaling is very good up to (at least) hundreds of computing nodes (Lindahl, 2007; Hess *et al.*, 2007). Obviously, this is another good news for the biomolecular simulation community. Amber (Pearlman *et al.*, 1995) has also undergone dramatic improvements. With the new PMEMD (Particle mesh Ewald molecular dynamics) (Duke, 2006), parallelization to at least few hundreds of processors is very efficient. Additionally, there is reason to follow the development of DESMOND (Bowers *et al.*, 2006), which seems to be a particularly promising and efficient simulation package for (parallel) biomolecular simulations.

All Ewald summation based methods depend on Fourier transforms. The FFTW library (Frigo and Johnson, 2005) is by far the most common choice, and almost all packages such as GROMACS, Amber (Case and Cheatham, 2005), NAMD (Kalé *et al.*, 1999; Phillips *et al.*, 2005), LAMMPS (Plimpton, 2005), and Espresso (Limbach *et al.*, 2006) use FFTW or offer it as an option. Despite the dependence on Fourier transform, it is worth noticing that the common misconception that using truncation or reaction-field is always faster than an Ewald based method is simply not true even for parallel computation (Patra *et al.*, 2007). In a typical simulation, an Ewald based method, such as PME, offers often (at least) roughly equal performance, or even better, depending on the chosen cutoff distance. Thus there is no reason to avoid Ewald based method as they offer far better accuracy than any truncation scheme. Of the new methods for non-periodic systems, both MMM2D and ELC are implemented in the Espresso simulation suite. The former is also given with an extension to 1-d periodic systems (only one dimension is periodic). The MMM2D code is also available through the CPC program library (at cpc.cs.qub.ac.uk/summaries/ADQM).

The hope is that the real space methods, the Fast Multipole Method (Greengard and Rokhlin, 1997; Hrycak and Rokhlin, 1998; Ying *et al.*, 2004), multigrid methods (Sagui and Darden, 2001; Izaguirre *et al.*, 2005), and the approach based on the local electrostatic algorithms (Maggs and Rossetto, 2002; Rottler and Maggs, 2004b), may offer yet more speed-up and better scaling. That is particularly important for multiscale modeling in biological and soft matter systems that typically contain a lot of water, and thus charges.

In summary, fast computation of electrostatic interactions remains to be one of the bottlenecks in biomolecular simulations. As discussed, progress in algorithms and methods has been rapid and promising, and as the very recent massive simulations show, we have very exciting times ahead of us.

Acknowledgements

We thank A.C. Maggs for a critical reading of parts of the manuscript. This work has been supported by Academy of Finland (I.V., M.K.), Emil Aaltonen foundation (M.K.), NSF under Grant Nos. ITR-0121361 and CAREER DMR-0348039 (C.S.), the Natural Sciences and Engineering Research Council of Canada (J.R., M.K.). We also thank The Finnish IT Center for Science—CSC, the HorseShoe supercomputing facility at the University of Southern Denmark, and the Southern Ontario SharcNet (www.sharcnet.ca) grid computing facility for computer resources.

References

Adams, J. (1989). Mudpack: Multigrid Fortran software for the efficient solution of linear elliptic partial differential equations. *Appl. Math. Comput.* **34**, 113–146.

Adcock, S.A., McCammon, J.A. (2006). Molecular dynamics: Survey of methods for simulating the activity of proteins. *Chem. Rev.* **106**, 1589–1615.

Aguado, A., Madden, P.A. (2003). Ewald summation of electrostatic multipole interactions up to the quadrupolar level. *J. Chem. Phys.* **119**, 7471–7483.

Alastuey, A., Appel, W. (2000). On the decoupling between classical coulomb matter and radiation. *Physica A* **276**, 508.

Alberts, B., Bray, D., Lewis, J., Raff, M., Roberts, K., Watson, J.D. (1994). "Molecular Biology of the Cell", 3rd ed. Garland Publishing.

Alder, B.J., Wainwright, T.E. (1957). Phase transition for a hard sphere system. *J. Chem. Phys.* **27**, 1208.

Alet, F., Sorensen, E. (2003). Cluster Monte Carlo algorithm for the quantum rotor model. *Phys. Rev. E* **67**, 015701(R).

Allen, M.P., Tildesley, D.J. (1993). "Computer Simulation of Liquids". Clarendon Press, Oxford.

Alper, H.E., Levy, R.M. (1989). Computer-simulations of the dielectric-properties of water—studies of the simple point-charge and transferable intermolecular potential models. *J. Chem. Phys.* **91**, 1242–1251.

Alper, H.E., Bassolino, D., Stouch, T.R. (1993a). Computer-simulation of a phospholipid monolayer-water system—the influence of long-range forces on water-structure and dynamics. *J. Chem. Phys.* **98**, 9798–9807.

Alper, H.E., Bassolino-Klimas, D., Stouch, T.R. (1993b). The limiting behavior of water hydrating a phospholipid monolayer—a computer simulation study. *J. Chem. Phys.* **99**, 5547–5559.

Anézo, C., de Vries, A.H., Höltje, H.-D., Tieleman, D.P. (2003). Methodological issues in lipid bilayer simulations. *J. Phys. Chem. B* **107**, 9424–9433.

Arnold, A., Holm, C. (2002a). MMM2D: A fast and accurate summation method for electrostatic interactions in 2d slab geometries. *Comput. Physics Commun.* **148**, 327–348.

Arnold, A., Holm, C. (2002b). A novel method for calculating electrostatic interactions in 2d periodic slab geometries. *Chem. Phys. Lett.* **354**, 324–330.

Babin, V., Baucom, J., Darden, T.A., Sagui, C. (2006a). Molecular dynamics simulations of DNA with polarizable force fields: Convergence of an ideal B-DNA structure to the crystallographic structure. *J. Phys. Chem. B* **110**, 11571–11581.

Babin, V., Baucom, J., Darden, T.A., Sagui, C. (2006b). Molecular dynamics simulations of polarizable DNA in crystal environment. *Int. J. Quantum Chem.* **106**, 3260–3269.

Banerjee, S., Board Jr., J.A. (2005). Efficient charge assignment and back interpolation in multigrid methods for molecular dynamics. *J. Comput. Chem.* **26**, 957–967.

Baucom, J., Transue, T., Fuentes-Cabrera, M.A., Krahn, J.M., Darden, T., Sagui, C. (2004). Molecular dynamics simulations of the d(CCAACGTTGG)$_2$ decamer in crystal environment: Comparison of atomic point-charge, extra-point and polarizable force fields. *J. Chem. Phys.* **121**, 6998–7008.

Bayly, C.I., Cieplak, P., Cornell, W.D., Kollman, P.A. (1993). A well-behaved electrostatic potential based method using charge restraints for deriving atomic charges—the RESP model. *J. Phys. Chem.* **97**, 10269–10280.

Beck, T.L. (1997). Real space multigrid solution of electrostatics problems and the Kohn–Sham equations. *Int. J. Quantum Chem.* **65**, 447–486.

Beck, T.L. (2000). Real-space mesh techniques in Density-Functional Theory. *Rev. Mod. Phys.* **72**, 1041–1080.

Beck, T.L., Iyer, K.A., Merrick, M.P. (1997). Multigrid methods in Density Functional Theory. *Int. J. Quantum Chem.* **61**, 341–348.

Bishop, T., Skeel, R., Schulten, K. (1997). Difficulties with multiple stepping and fast multipole algorithm in molecular dynamics. *J. Comput. Chem.* **18**, 1785–1791.

Board, J.A., Causey, J.W., Leathrum, J.F., Windemuth, A., Schulten, K. (1992). Accelerated molecular dynamics simulation with the parallel fast multipole algorithm. *Chem. Phys. Lett.* **198**, 89–94.

Boresch, S., Steinhauser, O. (1997). Presumed versus real artifacts of the Ewald summation technique: The importance of dielectric boundary conditions. *Ber. Bunseges. Phys. Chem.* **101**, 1019–1029.

Boroudjerdi, H., Kim, Y.-W., Naji, A., Netz, R.R., Schlagberger, X., Serr, A. (2005). Statics and dynamics of strongly charged soft matter. *Phys. Rep.* **416**, 129–199.

Bowers, K.J., Chow, E., Xu, H., Dror, R.O., Eastwood, M.P., Gregersen, B.A., Klepeis, J.L., Kolossvary, I., Moraes, M.A., Sacerdoti, F.D., Salmon, J.K., Shan, Y., Shaw, D.E. (2006). Scalable algorithms for molecular dynamics simulations on commodity clusters. *SC2006*, pp. 1–13.

Brandt, A. (1977). Multi-level adaptive solutions to boundary value problems. *Math. Comput.* **31**, 333–390.

Brandt, A. (1994). Rigorous quantitative analysis of multigrid I: Constant coefficients two-level cycle with L2-norm. *SIAM J. Num. Anal.* **31**, 1695.

Brandt, A. (2001). Multiscale scientific computation: Review 2001. In: Barth, T.J., Chan, T.F., Haimes, R. (Eds.), "Multiscale and Multiresolution Methods: Theory and Applications." Springer-Verlag, Heidelberg.

Briggs, E.L., Sullivan, D.J., Bernholc, J. (1996). A real-space multigrid-based approach to large-scale electronic structure calculations. *Phys. Rev. B* **54**, 14362–14375.

Briggs, W.L., Henson, V.E., McCormick, S.F. (2000). "A Multigrid Tutorial", 2nd ed. SIAM, Philadelphia, PA.

Caillol, J.P. (1994). Comments on the numerical simulations of electrolytes in periodic boundary conditions. *J. Chem. Phys.* **101**, 6080–6090.

Car, R., Parrinello, M. (1985). Unified approach for molecular dynamics and Density Functional Theory. *Phys. Rev. Lett.* **55**, 2471.

Case, D.A., Cheatham III, T.E., Darden, T., Gohlke, H., Luo, R., Merz Jr., K.M., Onufriev, A., Simmerling, C., Wang, B., Woods, R. (2005). The Amber biomolecular simulation programs. *J. Comput. Chem.* **26**, 1668–1688.

Challacombe, M., Schwegler, E., Almlöf, J. (1995). Recurrence relations for calculation of the cartesian multipole tensor. *Chem. Phys. Lett.* **241**, 67–72.

Chipot, C., Angyan, J.G., Millot, C. (1998). Statistical analysis of distributed multipoles derived from molecular electrostatic potentials. *Mol. Phys.* **94**, 881–895.

Cisneros, G.A., Piquemal, J.P., Darden, T.A. (2005). Intermolecular electrostatic energies using density fitting. *J. Chem. Phys.* **123**, 044109.

Cisneros, G.A., Piquemal, J.P., Darden, T.A. (2006). Generalization of the Gaussian electrostatic model: Extension to arbitrary angular momentum, distributed multipoles, and speedup with reciprocal space methods. *J. Chem. Phys.* **125**, 184101.

Darden, T.A., York, D.M., Pedersen, L.G. (1993). Particle mesh Ewald: An N log(N) method for Ewald sums in large systems. *J. Chem. Phys.* **98**, 10089–10092.

Darden, T., Toukmaji, A., Pedersen, L. (1997). Long-range electrostatic effects in biomolecular simulations. *J. Chim. Phys.* **94**, 1346–1364.

de Leeuw, S.W., Perram, J.W. (1979). Electrostatic lattice sums for semi-infinite lattices. *Mol. Simul.* **37**, 1313–1322.

DeLeeuw, S.W., Perram, J.W., Smith, E.R. (1980). Simulation of electrostatic systems in periodic boundary conditions I: Lattice sums and dielectric constants. *Proc. R. Soc. London* **A373**, 27–56.

DeLeeuw, S.W., Perram, J.W., Smith, E.R. (1986). Computer simulation of the static dielectric constant of systems with permanent dipole moments. *Annu. Rev. Phys. Chem.* **37**, 245–270.

Deserno, M., Holm, C. (1998). How to mesh up Ewald sums I: A theoretical and numerical comparison of various particle mesh routines. *J. Chem. Phys.* **109**, 7678–7693.

Dixon, R.W., Kollman, P.A. (1997). Advancing beyond the atom-centered model in additive and non-additive molecular mechanics. *J. Comput. Chem.* **18**, 1632–1646.

Duke, R.E. (2006). PMEMD. In Amber 9 manual: amber.scripps.edu/doc9/amber9.pdf. AMBER 9, D.A. Case et al., University of California, San Francisco, pp. 225–230.

Duncan, A., Sedgewick, R.D. (2006). Accelerated multiboson algorithm for Coulomb gases with dynamical dielectric effects. *Phys. Rev. E* **73**, 066711.

Duncan, A., Sedgewick, R.D., Coalson, R.D. (2005). Improved local lattice approach for Coulombic simulations. *Phys. Rev. E* **71**, 046702.

Duncan, A., Sedgewick, R.D., Coalson, R.D. (2006). Local simulation algorithms for Coulomb gases with dynamical dielectric effects. *Phys. Rev. E* **73**, 016705.

Essmann, U., Perera, L., Berkowitz, M.L., Darden, T., Lee, H., Pedersen, L.G. (1995). A smooth particle mesh Ewald method. *J. Chem. Phys.* **103**, 8577–8593.

Ewald, P. (1921). Die Berechnung optischer und elektrostatischer Gitterpotentiale. *Ann. Phys. (Leipzig)* **64**, 253–287.

Faraldo-Gómez, J.D., Smith, G.R., Sansom, M.S.P. (2002). Setting up and optimization of membrane protein simulations. *Eur. Biophys. J.* **31**, 217–227.

Fattebert, J.L., Bernholc, J. (2000). Towards grid-based O(N) DFT methods: Optimized non-orthogonal orbitals and multigrid acceleration. *Phys. Rev. B* **62**, 1713.

Feller, S.E., Pastor, R.W., Rojnuckarin, A., Bogusz, A., Brooks, B.R. (1996). Effect of electrostatic force truncation on interfacial and transport properties of water. *J. Phys. Chem.* **100**, 17011–17020.

Ferrell, R., Bertschinger, E. (1994). Particle-mesh methods on the connection machine. *J. Mod. Phys.* **5**, 933–956.

Francl, M.M., Chirlian, L.A. (1999). The pluses and minuses of mapping atomic charges to electrostatic potentials. In: Lipkowitz, K., Boyd, D.B. (Eds.), "Reviews in Computational Chemistry", vol. 14. VCH Publishers, New York, NY.

Frigo, M., Johnson, S.G. (2005). The design and implementation of FFTW3. *Proc. IEEE* **93**, 216–231.

Goedecker, S. (2003). Combining multigrid and wavelet ideas to construct more efficient multiscale algorithms. *J. Theor. Comput. Chem.* **2**.

Greengard, L.F. (1988). "The Rapid Evaluation of Potential Fields in Particle Systems". The MIT Press, Cambridge, MA.

Greengard, L., Rokhlin, V. (1987). A fast algorithm for particle simulations. *J. Comput. Phys.* **73**, 325–348.

Greengard, L., Rokhlin, V. (1997). A fast algorithm for particle simulation. *J. Comput. Phys.* **135**, 280–292.

Grønbech-Jensen, N. (1997). Lekner summation of long range interactions in periodic systems. *Int. J. Mod. Phys. C* **8** (6), 1287.

Groot, R.D. (2003a). Electrostatic interactions in dissipative particle dynamics—simulation of poly-electrolytes and anionic surfactants. *J. Chem. Phys.* **118**, 11265–11277.

Groot, R.D. (2003b). Erratum: Electrostatic interactions in dissipative particle dynamics—simulation of polyelectrolytes and anionic surfactants. *J. Chem. Phys.* **119**, 10454.

Guillot, B. (2002). A reappraisal of what we have learn during three decades of computer simulations on water. *J. Mol. Liq.* **101**, 219–260.

Gurtovenko, A.A., Vattulainen, I. (2007). Ion leakage through transient water pores in protein-free lipid membranes driven by transmembrane ionic charge imbalance. *Biophys. J.* **92**, 1878–1890.

Hackbush, W. (1985). "Multigrid Methods and Applications". Springer-Verlag, Berlin.

Hackbush, W., Trottenberg, U. (1982). "Multigrid Methods". Springer-Verlag, Berlin.

Hansen, J.-P. (1986). Molecular-dynamics simulations of coulomb systems in two and three dimensions. In: Ciccotti, G., Hoover, W.G. (Eds.), "Molecular Dynamics Simulation of Statistical–Mechanical Systems". North-Holland, Amsterdam.

Hautman, J., Klein, M.L. (1992). An Ewald summation method for planar surfaces and interfaces. *Mol. Phys.* **75** (2), 379–395.

Hess, B. (2002). Determining the shear viscosity of model liquids from molecular dynamics simulations. *J. Chem. Phys.* **116**, 209–217.

Hess, B., van der Vegt, N.F.A. (2006). Hydration thermodynamic properties of amino acid analogues: A systematic comparison of biomolecular force fields and water models. *J. Phys. Chem. B* **110**, 17616–17626.

Hess, B., Kutzner, C., van der Spoel, D., Lindahl, E. (2007). GROMACS 4.0: Efficient, load-balanced molecular dynamics. Submitted for publication.

Heyes, D.M., Barber, M., Clarke, J.H.R. (1977). Molecular dynamics computer simulation of surface properties of crystalline potassium chloride. *J. Chem. Soc., Faraday Trans.* **2**, 1485–1496.

Hockney, R.W., Eastwood, J.W. (1981). "Computer Simulation Using Particles". McGraw-Hill, New York.

Holst, M., Saied, F. (1993). Multigrid solution of the Poisson–Boltzmann equation. *J. Comput. Chem.* **14**, 105–113.

Holst, M., Kozack, R.E., Saied, F., Subramaniam, S. (1994). Treatment of electrostatic effects in proteins: Multigrid-based Newton iterative method for solution of the full Poisson–Boltzmann equation. *Proteins Struct. Fun. Gen.* **18**, 231–241.

Holst, M., Saied, F. (1995). Numerical solution of the nonlinear Poisson–Boltzmann equation: Developing more robust and efficient methods. *J. Comput. Chem.* **16**, 337–364.

Hrycak, T., Rokhlin, V. (1998). An improved fast multipole algorithm for potential fields. *SIAM J. Sci. Comput.* **19**, 1804–1826.

Ibragimova, G.T., Wade, R.C. (1998). Importance of explicit salt ions for protein stability in molecular dynamics simulation. *Biophys. J.* **74**, 2906–2911.

Israelachvili, J.N. (1992). "Intermolecular and Surface Force". Academic Press, New York.

Izaguirre, J.A., Hampton, S.S., Matthey, T. (2005). Parallel multigrid summation for the N-body problem. *J. Parallel Distrib. Comput.* **65**.

Jackson, J.D. (1999). "Classical Electrodynamics". Wiley, New York.

Jespersen, D. (1984). Multigrid methods for partial differential equations. In: "Studies in Numerical Analysis", vol. 24. Mathematical Association of America, Washington, DC.

Kale, L.V., Kirshnan, S. (1996). Charm++: Parallel programming with message-driven objects. In: Wilson, G.V., Lu, P. (Eds.), "Parallel Programming using C++". MIT Press, Boston, pp. 175–213.

Kalé, L., Skeel, R., Bhandarkar, M., Brunner, R., Gursoy, A., Krawetz, N., Phillips, J., Shinozaki, A., Varadarajan, K., Schulten, K. (1999). NAMD2: Greater scalability for parallel molecular dynamics. *J. Comput. Phys.* **151**, 283–312.

Karplus, M., McCammon, J.A. (2002). Molecular dynamics simulations of biomolecules. *Nature Struct. Biol.* **9**, 646–652.

Karttunen, M., Vattulainen, I., Lukkarinen, A. (2004). "Novel Methods in Soft Matter Simulations". Springer-Verlag, New York.

Kessel, A., Tieleman, D.P., Ben-Tal, N. (2004). Implicit solvent model estimates of the stability of model structures of the alamethicin channel. *Eur. Biophys. J.* **33**, 16–28.

Khelashvili, G.A., Scott, H.L. (2004). Combined Monte Carlo and molecular dynamics simulation of hydrated 18:0 sphingomyelin-cholesterol lipid bilayers. *J. Chem. Phys.* **120**, 9841–9847.

Koehorst, R.B.M., Spruijt, R.B., Vergeldt, F.J., Hemminga, M.A. (2004). Lipid bilayer topology of the transmembrane-α–helix of M13 major coat protein and bilayer polarity profile by site-directed fluorescence spectroscopy. *Biophys. J.* **87**, 1445–1455.

Lee, I.H., Kim, Y.H., Martin, R.M. (2000). One-way multigrid method in electronic structure calculations. *Phys. Rev. B* **61**, 4597.

Lekner, J. (1989). Summation of dipolar fields in simulated liquid–vapor interfaces. *Physica A* **157**, 826–838.

Lekner, J. (1991). Summation of Coulomb fields in computer-simulated disordered systems. *Physica A* **176**, 485–498.

Lekner, J. (1998). Coulomb forces and potentials in systems with an orthorhombic unit cell. *Mol. Simul.* **20** (6), 357.

Levrel, L. (2006). Physique statistique de l'interaction coulombienne. Ph.D. thesis, Université Paris 6.

Levrel, L., Maggs, A.C. (2005). Monte Carlo algorithms for charged lattice gases. *Phys. Rev. E* **72**, 016715.

Levrel, L., Alet, F., Rottler, J., Maggs, A.C. (2005). Local simulation algorithms for Coulombic interactions. *Pramana (Proceedings Statphys 22)* **64**, 1001.

Limbach, H.J., Arnold, A., Mann, B.A., Holm, C. (2006). ESPResSo—an extensible simulation package for research on soft matter systems. *Comput. Phys. Commun.* **174**, 704–727.

Lindahl, E. (2007).

Maggs, A.C. (2002). Paper 2. *J. Chem. Phys.* **117**, 1975.

Maggs, A.C. (2004). Paper 3. *J. Chem. Phys.* **120**, 3108.

Maggs, A.C. (2005). Monte Carlo simulation of a model of water. *Phys. Rev. E* **72**, 040201.

Maggs, A.C., Rossetto, V. (2002). Local simulation algorithms for Coulomb interactions. Paper 1. *Phys. Rev. Lett.* **88**, 196402.

Mark, P., Nilsson, L. (2002). Structure and dynamics of liquid water with different long-range interaction truncation and temperature control methods in molecular dynamics simulations. *J. Comput. Chem.* **23**, 1211–1219.

Marrink, S.J., Lindahl, E., Edholm, O., Mark, A.E. (2001). Simulation of the spontaneous aggregation of phospholipids into bilayers. *J. Am. Chem. Soc.* **123**, 8638–8639.

Marrink, S.J., Tieleman, D.P. (2002). Molecular dynamics simulation of spontaneous membrane fusion during a cubic-hexagonal phase transition. *Biophys. J.* **83**, 2386–2392.

Mazars, M. (2005). Lekner summations and Ewald summations for quasi-two-dimensional systems. *Mol. Phys.* **103**, 1241–1260.

McCormick, S.F. (1989). "Multilevel Adaptive Methods for Partial Differential Equations". SIAM, Philadelphia, PA.

McMurchie, L.E., Davidson, E.R. (1978). One- and two-electron integrals over cartesian Gaussian functions. *J. Comput. Phys.* **26**, 218–231.

Monticelli, L., Colombo, G. (2004). The influence of simulation conditions in molecular dynamics investigations of model beta-sheet peptides. *Theor. Chem. Acc.* **112**, 145–157.

Monticelli, L., Simões, C., Belvisi, L., Colombo, G. (2006). Assessing the influence of electrostatic schemes on molecular dynamics simulations of secondary structure forming peptides. *J. Phys. Condens. Matter* **18** (14), S329–S345.

Mosey, N.J., Müser, M.H., Woo, T.K. (2005). Molecular mechanisms for the functionality of lubricant additives. *Science* **307**, 1612–1615.

Nielaba, P., Mareschal, M., Ciccotti, G. (Eds.) (2002). Bridging Time Scales: Molecular Simulations for the Next Decade. Springer-Verlag, New York.

Niemela, P., Ollila, S., Hyvonen, M.T., Karttunen, M., Vattulainen, I. (2007). Assessing the nature of lipid raft membranes. *PLoS Comput. Biol.* **3**, 304–312.

Nijboer, B.R.A., de Wette, F.W. (1957). On the calculation of lattice sums. *Physica* **23**, 309–321.

Norberg, J., Nilsson, L. (2000). On the truncation of long-range electrostatic interactions in DNA. *Biophys. J.* **79**, 1537–1553.

Onsager, L. (1936). Electric moments of molecules in liquids. *J. Am. Chem. Soc.* **58**, 1486–1493.

Pasichnyk, I., Duenweg, B. (2004). Coulomb interactions via local dynamics: A molecular-dynamics algorithm. *J. Phys. Condens. Matter* **16**, S3999.

Patra, M., Karttunen, M. (2004). Systematic comparison of force fields for microscopic simulations of NaCl in aqueous solutions: Diffusion, free energy of hydration, and structural properties. *J. Comput. Chem.* **25**, 678–689.

Patra, M., Karttunen, M., Hyvonen, M.T., Falck, E., Lindqvist, P., Vattulainen, I. (2003). Molecular dynamics simulations of lipid bilayers: Major artifacts due to truncating electrostatic interactions. *Biophys. J.* **84**, 3636–3645.

Patra, M., Karttunen, M., Hyvonen, M.T., Falck, E., Vattulainen, I. (2004). Lipid bilayers driven to a wrong lane in molecular dynamics simulations by subtle changes in long-range electrostatic interactions. *J. Phys. Chem. B* **108**, 4485–4494.

Patra, M., Hyvonen, M.T., Falck, E., Sabouri-Ghomi, M., Vattulainen, I., Karttunen, M. (2007). Long-range interactions and parallel scalability in molecular simulations. *Comput. Phys. Commun.* **176**, 14–22.

Pearlman, D.A., Case, D.A., Caldwell, J.W., Ross, W.S., Cheatham III, T.E., DeBolt, S., Ferguson, D., Seibel, G., Kollman, P. (1995). Amber, a package of computer programs for applying molecular mechanics, normal mode analysis, molecular dynamics and free energy calculations to simulate the structural and energetic properties of molecules. *Comput. Phys. Commun.* **91**, 1–41.

Perram, J.W., Petersen, H.G., DeLeeuw, S.W. (1988). An algorithm for the simulation of condensed matter that grows as the 3/2 power of the number of particles. *Mol. Phys.* **65**, 875–893.

Phillips, J.C., Braun, R., Wang, W., Gumbart, J., Tajkhorshid, E., Villa, E., Chipot, C., Skeel, R.D., Kale, L., Schulten, K. (2005). Scalable molecular dynamics with NAMD. *J. Comput. Chem.* **26**, 1781–1802.

Piquemal, J.P., Cisneros, G.A., Reinhardt, P., Gresh, N., Darden, T.A. (2006). Towards a force field based on density fitting. *J. Chem. Phys.* **124**, 104101.

Plimpton, S.J. (2005). Fast parallel algorithms for short-range molecular dynamics. *J. Comput. Phys.* **117**, 1–19.

Pollock, E., Glosli, J. (1996). Comments on PPPM, FMM, and the Ewald method for large periodic Coulombic systems. *Comput. Phys. Commun.* **95**, 93–110.

Ponder, J.W., Case, D.A. (2003). Force fields for protein simulation. *Adv. Protein Chem.* **66**, 27.

Press, W.H., Teukolsky, S.A., Vettering, W.T., Flannery, B.P. (1992). "Numerical Recipes in FORTRAN: The Art of Scientific Computing", 2nd ed. Cambridge University Press, Cambridge.

Qian, X., Schlick, T. (2002). Efficient multiple-time-step integrators with distance-based force splitting for Particle-Mesh-Ewald molecular dynamics simulations. *J. Chem. Phys.* **116**, 5971–5983.

Rahman, A. (1964). Correlations in the motion of atoms in liquid argon. *Phys. Rev.* **136**, A405.

Ren, P., Ponder, J.W. (2002). A consistent treatment of inter- and intramolecular polarization in molecular mechanics calculations. *J. Comput. Chem.* **23**, 1497–1506.

Ren, P., Ponder, J.W. (2003). Polarizable atomic multipole water model for molecular mechanics simulation. *J. Phys. Chem. B* **107**, 5933–5947.

Ren, P., Ponder, J.W. (2004). Temperature and pressure dependence of the AMOEBA water model. *J. Phys. Chem. B* **108**, 13427–13437.

Rottler, J. (2007). Local electrostatics algorithm for classical molecular dynamics simulations. *J. Chem. Phys.* **127**, 134104.

Rottler, J., Maggs, A.C. (2004a). A continuum, $\mathcal{O}(N)$ Monte Carlo algorithm for charged particles. *J. Chem. Phys.* **120**, 3119.

Rottler, J., Maggs, A.C. (2004b). Local molecular dynamics with coulombic interactions. *Phys. Rev. Lett.* **93**, 170201.

Sagui, C., Darden, T.A. (1999a). Molecular dynamics simulations of biomolecules: Long-range electrostatic effects. *Annu. Rev. Biophys. Biomol. Struct.* **28**, 155–179.

Sagui, C., Darden, T.A. (1999b). P3M and PME: A comparison of the two methods. In: Pratt, L.R., Hummer, G. (Eds.), Simulation and Theory of Electrostatic Interactions in Solution, vol. 492. AIP, Melville, NY.

Sagui, C., Darden, T.A. (2001). Multigrid methods for classical molecular dynamics simulations of biomolecules. *J. Chem. Phys.* **114**, 6578–6591.

Sagui, C., Pedersen, L.G., Darden, T.A. (2004). Towards an accurate representation of electrostatics in classical force fields: Efficient implementation of multipolar interactions in biomolecular simulations. *J. Chem. Phys.* **120**, 73–87.

Sagui, C., Roland, C., Pedersen, L.G., Darden, T.A. (2005). New distributed multipole methods for accurate electrostatics in large-scale biomolecular simulations. In: Leimkuhler, B., Chipot, C., Elber, R., Laaksonen, A., Mark, A., Schlick, T., Schuette, C., Skeel, R. (Eds.), New Algorithms for Macromolecular Simulations, vol. 49. Springer-Verlag, Berlin.

Saito, M. (1994). Molecular-dynamics simulations of proteins in solution—artifacts caused by the cutoff approximation. *J. Chem. Phys.* **101**, 4055–4061.

Sanbonmatsu, K.Y., Tung, C.-S. (2007). High performance computing in biology: Multimillion atom simulations of nanoscale systems. *J. Struct. Biol.* **157**, 470–480.

Schaefer, P., Riccardi, D., Cui, Q. (2005). Reliable treatment of electrostatics in combined QM/MM simulation of macromolecules. *J. Chem. Phys.* **123**, 014905.

Schreiber, H., Steinhauser, O. (1992). Cutoff size does strongly influence molecular-dynamics results on solvated polypeptides. *Biochemistry* **31**, 5856–5860.

Shan, Y., Klepeis, J.L., Eastwood, M.P., Dror, R.O., Shaw, D.E. (2005). Gaussian split Ewald: A fast Ewald mesh method for molecular simulation. *J. Chem. Phys.* **122**, 054101.

Skeel, R.D., Tezcan, I., Hardy, D.J. (2002). Multiple grid methods for classical molecular dynamics. *J. Comput. Chem.* **23**.

Smith, W. (1982). Point multipoles in the Ewald summation. *CCP5 Information Quarterly* **4**, 13–25.

Smith, E.R. (1994). Calculating the pressure in simulations using periodic boundary conditions. *J. Stat. Phys.* **77**, 449–472.

Smith, P.E., Pettitt, B.M. (1991). Peptides in ionic solutions: A comparison of the Ewald and switching function techniques. *J. Chem. Phys.* **95**, 8430–8441.

Smith, P.E., Pettitt, B.M. (1996). Ewald artifacts in liquid state molecular dynamics simulations. *J. Chem. Phys.* **105**, 4289–4293.

Smith, P.E., Pettitt, B.M. (1997). On the presence of rotational Ewald artifacts in the equilibrium and dynamical properties of a zwitterionic tetrapeptide in aqueous solution. *J. Phys. Chem. B* **101**, 3886–3890.

Sølvason, D., Kolafa, J., Petersen, H.G., Perram, J.W. (1995). A rigorous comparison of the Ewald method and the fast multipole method in two dimensions. *Comput. Phys. Commun.* **87**, 307–318.

Sperb, R. (1994). Extension and simple proof of Lekner's summation formula for Coulomb forces. *Mol. Simul.* **13**, 189–193.

Sperb, R. (1998). An alternative to Ewald sums, part 1: Identities for sums. *Mol. Simul.* **20**, 179–200.

Sperb, R. (1999). An alternative to Ewald sums, part 2: The Coulomb potential in a periodic system. *Mol. Simul.* **22**, 199–212.

Stimson, L.M., Vattulainen, I., Róg, T., Karttunen, M. (2005). Exploring the effect of xenon on biomembranes. *Cell. Mol. Biol. Lett.* **10**, 563–569.

Stone, A.J. (1996). "The Theory of Intermolecular Forces". Clarendon Press, Oxford.

Strebel, R., Sperb, R. (2001). An alternative to Ewald Sums. Part 3: Implementation and results. *Mol. Simul.* **27**, 61–74.

Sun, Q.-f., Guo, H., Wang, J. (2003). A spin cell for spin current. *Phys. Rev. Lett.* **90**, 258301.

Tieleman, D.P., Hess, B., Sansom, M.S.P. (2002). Analysis and evaluation of channel models: Simulations of alamethicin. *Biophys. J.* **83**, 2393–2407.

Toukmaji, A., Board, J.A. (1996). Ewald sum techniques in perspective: A survey. *Comput. Phys. Commun.* **95**, 78–92.

Toukmaji, A., Sagui, C., Board, J.A., Darden, T. (2000). Efficient PME-based approach to fixed and induced dipolar interactions. *J. Chem. Phys.* **113**, 10913–10927.

Trottenberg, U., Oosterlee, C., Schüller, A. (2001). "Multigrid". Academic Press, San Diego, CA.

Tsigelny, I.F., Bar-On, P., Sharikov, Y., Crews, L., Hashimoto, M., Miller, M.A., Keller, S.H., Platoshyn, O., Yuan, J.X.-J., Masliah, E. (2007). Dynamics of α-synuclein aggregation and inhibition of pore-like oligomer development by β-synuclein. *FEBS J.* **274**, 1862–1877.

van der Spoel, D., van Maaren, P.J. (2006). The origin of layer structure artifacts in simulations of liquid water. *J. Chem. Theor. Comput.* **2**, 1–11.

Vashishta, P., Kalia, R.K., Nakano, A. (2006). Multimillion atom simulations of dynamics of oxidation of an aluminum nanoparticle and nanoindentation on ceramics. *J. Phys. Chem. B* **110**, 3727–3733.

Vattulainen, I., Karttunen, M. (2006). Modeling of biologically motivated soft matter systems. In: Rieth, M., Schommers, W. (Eds.), "Computational Nanotechnology". American Scientific Press.

Venable, R.M., Brooks, B.R., Pastor, R.W. (2000). Molecular dynamics simulations of gel (L_β I) phase lipid bilayers in constant pressure and constant surface area ensembles. *J. Chem. Phys.* **112**, 4822–4832.

Villarreal, M.A., Montich, G.G. (2005). On the Ewald artifacts in computer simulations. The test-case of the octaalanine peptide with charged termini. *J. Biomol. Struct. Dyn.* **23**, 135–142.

Wesseling, P. (1991). "An Introduction to Multigrid Methods". Wiley, New York.

Wheatley, R.J., Mitchell, J.B.O. (1994). Gaussian multipoles in practice: Electrostatic energies for intermolecular potentials. *J. Comput. Chem.* **15**, 1187–1198.

White, S.R., Wilkins, J.W., Teter, M.P. (1989). Finite-element method for electronic-structure. *Phys. Rev. B* **39**, 5819–5833.

Widmann, A.H., Adolf, D.B. (1997). A comparison of Ewald summation techniques for planar surfaces. *Comput. Phys. Commun.* **107**, 167–186.

Wohlert, J., Edholm, O. (2004). The range and shielding of dipole–dipole interactions in phospholipid bilayers. *Biophys. J.* **87**, 2433–2445.

Yee, K.S. (1966). Numerical solution of initial boundary value problems. *IEEE Trans. Antennas Propag.* **14**, 302–307.

Yeh, I.-C., Berkowitz, M.L. (1999). Ewald summation for systems with slab geometry. *J. Chem. Phys.* **111** (7), 3155–3162.

Ying, L., Biros, G., Zorin, D. (2004). A kernel-independent adaptive fast multipole algorithm in two and three dimensions. *J. Comput. Phys.* **196**, 591–626.

Yonetani, Y. (2006). Liquid water simulation: A critical examination of cutoff length. *J. Chem. Phys.* **124**, 204501.

York, D., Yang, W. (1994). The Fast Fourier Poisson (FFP) method for calculation Ewald sums. *J. Chem. Phys.* **101**, 3298–3300.

York, D.M., Darden, T.A., Pedersen, L.G. (1993). The effect of long-range electrostatic interactions in simulations of macromolecular crystals—a comparison of the Ewald and truncated list methods. *J. Chem. Phys.* **99**, 8345–8348.

York, D.M., Yang, W.T., Lee, H., Darden, T., Pedersen, L.G. (1995). Toward the accurate modeling of DNA—the importance of long-range electrostatics. *J. Am. Chem. Soc.* **117**, 5001–5002.

Zangi, R., de Vocht, M.L., Robillard, G.T., Mark, A.E. (2002). Molecular dynamics study of the folding of hydrophobin SC3 at a hydrophilic/hydrophobic interface. *Biophys. J.* **83**, 112–124.

Zaslavsky, L.Y., Schlick, T. (1998). An adaptive multigrid technique for evaluating long-range forces in biomolecular simulations. *Appl. Math. Comput.* **97**, 237–250.

Zhou, R., Berne, B. (1995). A new molecular dynamics method combining the reference system propagator algorithm with a fast multipole method for simulating proteins and other complex systems. *J. Chem. Phys.* **103**, 9444–9459.

Zuegg, J., Greedy, J.E. (1999). Molecular dynamics simulations of human prion protein: Importance of correct treatment of electrostatic interactions. *Biochemistry* **38**, 13862–13876.

CHAPTER 3

Time and Length Scales in Lipid Bilayer Simulations

Olle Edholm

Department of Theoretical Physics, Royal Institute of Technology,
Alba Nova University Center, SE-106 91 Stockholm, Sweden

Abstract

Time and length scales for different kinds of local and collective motions in lipid bilayers that may be probed in molecular dynamics simulations are discussed. The possibility to determine parameters in continuum models using simulation techniques is discussed. Agreements and discrepancies between simulations and experimental data are presented and discussed. The chapter covers large scale undulations and peristaltic thickness fluctuations, hydrocarbon chain reorientation motion, lipid lateral diffusion and motion of the entire monolayers with respect to each other and with respect to the water.

I. INTRODUCTION

The equilibrium motions in lipid bilayers range from rattling of individual lipid hydrocarbon tails on picosecond time scales to the collective motion of all lipids in μm sized patches extending over many milliseconds. Experimentally, the former

Current Topics in Membranes, Volume 60
1063-5823/08 $35.00
DOI: 10.1016/S1063-5823(08)00003-3

region is accessible through spectroscopic methods (König and Sackmann, 1996), while the latter can be studied by microscopy (Evans, 1991). In addition, there are non-equilibrium processes like for instance cell division and various sorts of phase transitions occurring in simpler model systems, that include large temporal and spatial scale motions. Processes of this type span about ten orders of magnitude in time and about five orders of magnitude in length.

Atomistic simulations like molecular dynamics have since about 40 years proved to be a powerful tool to study of the bridging between atomic scale motions and collective macroscopic behavior. Early simulations of simple liquids like argon (Rahman, 1964) or even water (Rahman and Stillinger, 1971) included at those times typically less than 1000 atoms and could with the computers of those times be run for times corresponding to 10 or maybe 100 ps. This was still large and long enough to deduce information about the macroscopic and long time properties of those systems. The reason for this is that in these system the essential dynamics occurs at such short time and length scales. There are no long time correlation functions extending far beyond picoseconds and no spatial correlations extending for many nanometers in simple liquids or water.

At the other end of complexity we may compare to protein simulations. There one would need much longer time scales to solve the protein folding problem. However, protein simulations have been quite successful by limiting the studies to the simulation of small fluctuations around a well known X-ray structure.

The first molecular dynamics simulations of a lipid bilayer van der Ploeg and Berendsen (1982) used a simplified (coarse grained) model for the lipid and did not include explicit water. Later, explicit water was included and the lipids was modeled in a more realistic way (Egberts and Berendsen, 1988). During the following decade, simulation methods for lipid bilayers made substantial progress (Pastor, 1994; Tu et al., 1996; Tieleman et al., 1997). For a long time, researchers were however still restricted to small systems with the order of 100 lipids, and sub-nanosecond time scales. Now, in the first decade of the present millennium, simulations can routinely be performed for several tenths of nanoseconds with several hundred lipids, corresponding to 10 nm length scales. When necessary, a few simulations with thousands of lipids (20–50 nm length scales) can be done, and smaller systems can in a few cases be simulated for up to the order of a microsecond.

Still, the atomistic simulation methods, obviously suffer from that we cannot cover all interesting properties of lipid membranes at large time and length scales. There are several reasons for why lipid bilayer simulations have problems compared to for instance simulations of simple liquids. First of all the range of time and length scales by which different processes occur in membranes is vast, which calls for long time simulations of large systems. Secondly, when simulating a bulk system, a certain minimum number of the smallest repeated unit of the system is need. For simple liquids, this is about 1000 atoms in three dimensions and 100 atoms in two dimensions. Membranes are two dimensional systems but each elementary building block consists in the third dimension of 2 lipids and maybe 50 waters, which is 250–500 atoms. Thus, the minimum number of included atoms will be 25–50 000. Finally, strong electrostatic interactions calls for a small time step and a longer cutoff or the use of Ewald summation methods. This adds another couple of orders of magnitude to the computer time needed for a simulation of given length.

Lipids can be modeled at different levels of detail. This includes all atom models or united atoms models in which non-polar hydrogens have been included into the carbons into united CH, CH_2 and CH_3 atoms. These models seem to be accurate enough to describe most properties of the bilayers quite well. The united atom models give faster simulations since they reduce the number of atoms. The factor is, however not a factor nine, as one would expect in a pure hydrocarbon system but smaller since a large part of the system consists of water that remain unaltered by the handling of the lipid hydro-carbons. One may have some concern about whether the lack of anisotropy of the hydrocarbon groups in the united atom model will affect the hydrocarbon packing and structure in dense gel phases.

There have been several efforts to construct coarse grained particle models in order to reduce computational times. Typically new atom-like particle are then constructed out of 3–4 lipid united atoms. Similarly 3–4 water molecules are grouped into a new molecule. The smaller number of atoms reduces computational times, but even more important for the performance of these models is that the interaction forces can be made weaker and more short ranged. This means that the integration time step can be increased and cutoffs can be shortened. This requires, however, a careful parametrization that properly balances of the forces between different atoms. Most efficient are the DPD (dissipative particle dynamics) models (Venturoli and Smit, 1999), which usually have no attraction between atoms but just different soft core repulsions. The next level of complication is a Lennard-Jones model (Goetz *et al.*, 1999), in which the character of the particles is represented by different Lennard-Jones parameters. Finally there are models that include short range electrostatics as well as Lennard-Jones interactions (Marrink *et al.*, 2004). These are more realistic at the price of computational performance.

It is also possible to go in the opposite direction and move into models at the electronic structure level or classical polarizable models. We have not, yet, seen much of that, but that will certainly be something for the future.

It is important to give careful attention the level of detail of the lipid model. What level needed depends on the purpose of the intended simulation. For some studies coarse grained models are the only alternatives since the desired results would never be produced in reasonable computing time with more detailed models. For other purposes, it is clear that the coarse grained models are too coarse and more detail is needed. It will here be shown that atomistic models can properly describe the equilibrium properties and dynamics of lipid bilayers up to fairly large length and times scales. It will also be shown where the limitations are and where the coarse grained models may help us.

This chapter is organized as follows. In Section II (possible) finite size effects will be discussed. Section III will discuss large scale undulations and the interface between molecular dynamics simulations and continuum field theories for these motions. These motions occur at large spatial scales and are slow. In Section IV area and thickness fluctuations occurring and constant volume are treated. They occur at intermediate length scales. In Section V the motion of the hydrocarbon tails is discussed. These motions are fairly local in space but contain all time scales from picoseconds to hundreds of nanoseconds. Finally the translational diffusion of single lipids, entire monolayers and bilayers is discussed in Section VI.

II. AREA PER LIPID AND FINITE SIZE EFFECTS

In the early simulations (van der Ploeg and Berendsen, 1982) a cooperative tilt extending over the entire periodic simulation box was observed in a bilayer consisting of 2×16 decanoate molecules (size of the box about 2 nm) while a system with double linear dimensions (2×64 decanoate molecules) did not show this artifact. In later simulations, including typically 64–72 double chain lipids several authors had problems reproducing the experimental area per phospholipid which should be 0.64 ± 0.015 nm^2 for dipalmitoylphosphatidylcholine (DPPC) according to Nagle and Tristram-Nagle (2000). In general many simulation set-ups seemed to result in areas per lipid that were up to 15% too small at surface tension zero. For this reason (Chiu et al., 1995) did setup simulations at a negative lateral pressure of 100 bar corresponding to a positive applied surface tension of 50 mN/m (25 mN/m for each monolayer surface) to get the correct area per lipid. A similar approach was also advocated by Feller et al. (1995). Still other researchers seemed to obtain reasonable areas per lipid with slightly different force fields (Tu et al., 1996; Berger et al., 1997). This lead to a discussion in which Jähnig (1996) argued, based on basic thermodynamics, that a system which is free to change its area should find equilibrium at a free energy minimum with respect to area. Since surface tension is defined as the derivative of free energy with respect to area this means per definition that surface tension should be zero. Thus, he claimed that bilayer membranes of vesicles and cells should have zero surface tension. Against this, Feller and Pastor (1996) put forward the idea that there could be a finite size effect. The 15% too small area per lipid observed in many simulated systems would then be an effect of too small system sizes together with periodic boundary conditions. In addition, it was early noted that cutoff schemes and Ewald summation tended to result in different inter-facial properties for water/air and water/octane surfaces (Feller et al., 1996). Some support for the idea about finite size effects was given in a study of Lindahl and Edholm (2000a), which indicated an increase in surface area per lipid from 0.611 to 0.630 and 0.633 nm^2 upon increasing the number of lipids from 64 to 256 and 1024. Extrapolation versus inverse system size gave 0.635 ± 0.005 nm^2 for an infinite system, which means that the difference in area per lipid was not more than 4% when going from the smallest systems up to infinite system size. This was far to little to explain the observed differences of 15–20% in area per lipid. A similar study performed without using cutoffs of the electrostatic interactions (Wohlert and Edholm, 2004) but a particle mesh Ewald technique (Darden *et al.*, 1993) resulted in the same area per lipid when extrapolated to infinite system size, but a finite size effect that was even smaller, more like 1% instead of 4%. This seems to indicate that systems consisting of a few hundred lipids should be large enough to avoid dramatic finite size effects. Perhaps more important than the size of the system is to use cutoff schemes for electrostatic interactions with care, if at all. Patra *et al.* (2003) showed that cutoff schemes could result in areas per lipid that were far different from those obtained with PME methods. Wohlert and Edholm (2004) showed that these results were a consequence of how the cutoff was performed (the use of so called charge groups and how these were constructed) not of the cutoffs in themselves.

In judging these results, especially the early ones, one should also keep in mind that the area per lipid does fluctuate a lot in small systems. The free energy, F, can as a function of the area, A, of the system close to a minimum be written in an expansion up to second order

$$F(A) = \frac{K_A}{2A_0}(A - A_0)^2, \tag{1}$$

with A_0 being the area at the minimum and K_A being the area compressibility modulus

$$K_A = A\left(\frac{\partial \gamma}{\partial A}\right) \approx 250 \,[\text{dyn/cm} = \text{mN/m}]. \tag{2}$$

Here, γ is the surface tension as a function of area (zero for $A = A_0$). The value 250 for K_A is an experimental value for DPPC taken from the review of Nagle and Tristram-Nagle (2000). Putting the average free energy equal to $k_B T/2$ (k_B being Boltzmann's constant and T the absolute temperature) and $A_0 = Na_0/2$, with N being the number of lipids, we may solve for the relative area fluctuations

$$\sqrt{\left\langle \left(\frac{A - A_0}{A_0}\right)^2 \right\rangle} = \sqrt{\frac{2k_B T}{K_A N a_0}} = \frac{0.23}{\sqrt{N}}. \tag{3}$$

This means that we should expect area fluctuations that may be up to about 3% in the smallest systems. Since, these at least in part may have very long periods in time, one needs long equilibration's and long sampling times to get accurate areas per lipid. This should not be much of a problem now, but was certainly one with many of the early simulations.

Finally, it should be noted that Castro-Román *et al.* (2006) in a recent study compares experimental X-ray and neutron scattering data directly with the same data calculated from simulations. They compare different force fields and different system sizes (72–288 lipids). The conclusions are that finite size effects are negligible, but that there are differences in the detailed electron densities that calls for further work and refinement of the force fields.

The area compressibility modulus can be obtained from simulations in different ways. Either Eq. (3) may be used to calculate it from the relative area fluctuations or better by fitting relative area fluctuations for different system sizes against system size. Alternatively simulations at different fixed areas gives the area compressibility as the derivative of surface tension with respect to area from Eq. (2). Feller and Pastor (1999) found reasonable agreement with experiment with the derivative method but found too large area fluctuations to account for the experimental area compressibility's with the first method. With the fluctuation method Lindahl and Edholm (2000a) obtained a value of 250–300 mN/m for DPPC in fair agreement with the experimental value 250 mN/m reported for DPPC in the review of Nagle and Tristram-Nagle (2000). For dimyristoylpalmitoylphosphatidylcholin (DMPC) (Wohlert and Edholm, 2006) reports 264 ± 20 mN/m from simulations at 303 K while the experimental value at the same temperature is 234 mN/m (Rawicz *et al.*, 2000). The derivative method gave 300 mN/m for DPPC (Lindahl and Edholm, 2000b). Some studies do, however, indicate that there is a system size dependence for calculated area compressibility modulus (den Otter, 2005).

III. UNDULATIONS

Undulations are large scale elastic deformations, which may be observed in model membranes as well as in some cells. There is a class of continuum models, which go back to Helfrich (1973) and are discussed in some detail in the book of Safran (1994) that describes these deformations successfully. The basic assumption is that the membrane is modeled as an infinitely thin mathematical surface and that there is an energetic cost for having a curvature. We first review the formalism of this continuum theory and do then show that atomistic simulations can reach large enough time and length scales to bridge the gap to these models and predict the parameters of these models.

In its most general form the curvature free energy per unit area is

$$f_c = 2k_c(H - c_0)^2 + \bar{k}K, \tag{4}$$

where H is the mean curvature, and K the Gaussian curvature. These curvatures can be expressed in terms of the two principal radii of curvature, R_1 and R_2 as

$$H = \frac{1}{2}\left[\frac{1}{R_1} + \frac{1}{R_2}\right] \quad \text{and} \quad K = \frac{1}{R_1 R_2}. \tag{5}$$

The three parameters in Eq. (4) are, the bending modulus (k_c), the spontaneous curvature (c_0) and the saddle splay modulus ($\bar{k}$). For a bilayer consisting of two symmetric leaflets, the spontaneous curvature will by symmetry be zero. This is usually the case in simulations and in artificial model systems but certainly not always in real biological systems. The saddle splay modulus measures the energetic cost for saddle like deformations of the bilayer. This term can often be neglected since it usually is small and/or weakly varying. It is therefore neglected in most cases. In a periodic system this term gives a net contribution that is exactly zero due to the theorem of Gauss–Bonnet. We now study small deformations of a flat symmetric membrane and use a coordinate system with the z coordinate perpendicular to the membrane and x and y in the flat membrane surface. If we then introduce the membrane surface as $z = u(x, y)$, we may write the bending free energy as a functional of this field

$$F_{bend}[u] = 0.5k_c \iint |\nabla^2 u(x, y)|^2 \, dx \, dy. \tag{6}$$

If there is a nonzero surface tension, γ, in the system, we have to add a term to take into account that a nonzero first derivative of u will increase the surface area and therefore cost free energy

$$F_{bend}[u] = 0.5 \iint \left[k_c |\nabla^2 u(x, y)|^2 + \gamma |\nabla u(x, y)|^2 \right] dx \, dy. \tag{7}$$

We note that for symmetry reason, no linear (or other odd) terms in u and its derivatives occur. The goal is then to find the probability distribution for $u(x, y)$ and various averages of it. This is done by making a Fourier expansion

$$u(x, y) = \sum_q \hat{u}(\mathbf{q}) e^{i\mathbf{q} \cdot \mathbf{r}}. \tag{8}$$

If this is inserted in Eq. (7) one obtains

$$F_{bend}[u] = \frac{A}{2} \sum_q \left[k_c q^4 |\hat{u}(\mathbf{q})|^2 + \gamma q^2 |\hat{u}(\mathbf{q})|^2 \right], \tag{9}$$

where A is the total (projected) area of the membrane. The different Fourier modes are decoupled and the energy is quadratic in the amplitudes. This means that there is an exact analytical solution to the statistical mechanical equilibrium problem. The mean square average amplitudes of the different Fourier modes are obtained by invocation of the equipartition theorem as

$$\langle \hat{u}^2(\mathbf{q}) \rangle = \frac{k_B T}{A(k_c q^4 + \gamma q^2)}. \tag{10}$$

There is no limit for the system sizes and length scales that can be handled by this model. In a more general cases where the energy also contains non-quadratic terms in the amplitudes one may have to use approximate perturbation methods around the exact solutions for the quadratic Hamiltonian or when this is a poor approximation Monte Carlo methods. The static properties of these models are well understood. The total mean square fluctuations in real space is obtained by summing up all Fourier modes, usually in an integral approximation

$$\langle u^2 \rangle = \frac{k_B T}{4\pi\gamma} \ln\left[\frac{1 + \frac{A\gamma}{\pi^2 k_c}}{1 + \frac{a^2\gamma}{\pi^2 k_c}}\right], \tag{11}$$

where a is a short wave length cutoff (that typically should reflect the molecular dimensions) at which the continuum model breaks down. The most interesting limiting case, that of zero surface tension, $\gamma = 0$ gives

$$\langle u^2 \rangle = \frac{A k_B T}{4\pi^3 k_c}, \tag{12}$$

while the limit $k_c = 0$ gives

$$\langle u^2 \rangle = \frac{k_B T}{4\pi\gamma} \ln\left[\frac{A}{a^2}\right]. \tag{13}$$

Equation (12) suffers from that the integral approximation is not that good for small q and it can be improved by doing the summation numerically (Lindahl and Edholm, 2000a) which results in the equation

$$\langle u^2 \rangle = \frac{A k_B T}{8.3\pi^3 k_c}. \tag{14}$$

In both cases the undulations grow with system size and will diverge for infinite systems. It turns, however out that the coefficients in front of the diverging expressions are small enough so that these expressions and this model may hold even for μm or mm sized systems. The surface tension, γ, is usually a parameter that may be controlled in simulations and given a specific value that often is zero. In contrast, the bending modulus k_c, is material constant that is to be determined. This has been done experimentally, see for instance (Evans and Rawicz, 1990; Duwe and Sackmann, 1990). From simulations, the model may be tested by plotting the spectral density, $\hat{u}^2(q)$, against the wave number in a double logarithmic plot. One should then find a straight line with slope -4, from the position of which the bending modulus can be deduced. This has been done for a coarse grained model (Goetz et al., 1999) and for atomistic models of DPPC (Lindahl and Edholm, 2000a) and glycerolmonoolein (GMO) (Marrink and Mark, 2001). For a tensionless system and wave numbers below about 1 nm^{-1} a q^{-4} variation of the spectral density is obtained in agreement with the continuum model. A bending modulus of 4×10^{-20} J is obtained for DPPC (Lindahl and Edholm, 2000a) in fair agreement with the experimental value 5×10^{-20} J (Evans and Rawicz, 1990). For GMO (Marrink and Mark, 2001) a similar value is obtained. It is worth noting that due to the limited size of the simulated systems (about 20×20 nm) the analysis is restricted to a wave number interval 0.2–1 nm^{-1} which contains only a handful of Fourier modes. In contrast experiments are mainly sensitive to a wide region of much smaller wave numbers.

At smaller spatial scales the continuum models break down, but before that, in a wave number interval 1–5 nm^{-1} it is possible to identify a region with a different type of disorder, for which the term protrusions has been coined (Lipowsky and Grotehans, 1993). Here, the spectral density can be written as

$$\langle \hat{u}^2(q) \rangle = \frac{k_B T}{A \gamma_p q^2}, \tag{15}$$

with γ_p being a constant of dimension surface tension, but having another value than the (macroscopic) surface tension. It takes on values of 50 mN/m for DPPC (Lindahl and Edholm, 2000a) and the smaller value 20 mN/m for GMO (Marrink and Mark, 2001). The protrusions involve a few or single lipids and take place independently for the two halves to the bilayer. There will be a restoring force similar to a surface tension, since the modes tend to increase the local surface area. These motions are only present at small length-scales and do not interfere with the undulations like a macroscopic surface tension would do.

Bilayers under applied stress have also been simulated and analyzed in (Marrink and Mark, 2001). For such systems, the surface tension will dominate and suppress long wave length undulations with wave numbers less than $\sqrt{\gamma/k_c}$. The second term in the denominator of Eq. (10) will then dominate. With a surface tension of 16 mN/m and $k_c = 5 \times 10^{-20}$ J this starts occurring at $q = 0.6$ nm^{-1}. Since protrusions dominate in this region, undulations will never be observed. For GMO at this surface tension that was also the result observed by Marrink and Mark (2001). The system passed from a low wave number region in which undulations was suppressed by the surface tension directly into a high wave number region dominated by protrusions. For smaller surface tensions, there should be possible to identify an intermediate region that is dominated by undulations having the q^{-4} variation of the spectral density surrounded by regions with q^{-2} variation with different constants corresponding to the macroscopic surface tension at low wave numbers and the protrusions at high wave numbers.

Instead of using equilibrium fluctuations to determine the free energy in Eq. (6), this may also be probed directly by forcing the system out of equilibrium. Using techniques such as umbrella sampling and measuring potentials of mean constraint force, coarse grained lipid models have been probed by den Otter and Briels (2003). The results indicate good agreement with the equilibrium methods close to the free energy minimum. If the system is pushed far enough out of equilibrium, clear deviations from the simple harmonic model of Eq. (6) may, however, be observed.

A. Dynamics of the Undulating Modes

From the potential energy in Eq. (7) it is straightforward to construct a wave equation for the dynamics of undulating membrane modes

$$\frac{\partial^2 u(x, y, t)}{\partial t^2} + \frac{k_c}{\rho}\left[\frac{\partial^4 u(x, y, t)}{\partial x^4} + \frac{\partial^4 u(x, y, t)}{\partial y^4}\right] = 0 \qquad (16)$$

with ρ being the mass density (per area) of the membrane. The solution would be a sum of non-damped uncoupled oscillations

$$u(\mathbf{r}, t) = \sum_{\mathbf{q}} c(\mathbf{q})e^{i\omega(q)t}e^{i\mathbf{q}\cdot\mathbf{r}} \qquad (17)$$

with angular frequencies given by the dispersion relation $\omega(q) = q^2\sqrt{k_c/\rho}$. This would result in oscillation periods ranging from a couple of ps for the highest wave numbers up to a few hundred ps for the longest wave lengths accessible in present simulations. This is clearly not what is observed in simulations, where one rather sees an exponential damping on ns time scales. Several efforts have been undertaken to include hydrodynamic damping from the water as well as internal bilayer friction into the treatment. A small but far from complete bibliography consists of (Kramer, 1971; Milner and Safran, 1987; Seifert and Langer, 1994; Evans and Yeung, 1994; Yeung and Seifert, 1995; Granek, 1997; Lin and Brown, 2004; Lin and Brown, 2005).

The hydrodynamic flow of water around the bilayer was included in the Stokes' approximation for low Reynolds numbers by Kramer (1971). This yielded in the appropriate limit exponential relaxation in time with a rate

$$\Gamma = \frac{1}{\tau} = \frac{k_c}{4\eta}q^3. \qquad (18)$$

Here η is the viscosity of water. This would for wave lengths around 10–20 nm give time constants ranging from half a ns up to couple of ns. The simulations of Lindahl and Edholm (2000a) indicated agreement with this model. One should, however, keep in mind that the short length of the simulations and the limited size of the system necessarily resulted in fairly poor statistics over a small wave number interval. Recently Shkulipa et al. (2006a, 2006b), have tested a more complete model of Seifert and Langer (1994) using coarse grained lipids in simulations. The main difference to the older models is that here it is assumed that the friction between the two monolayers also contribute to the damping of the undulations. The result of Eq. (18) is regained at low wave numbers. At wave numbers much larger than

$$q_c = \frac{\eta K_A}{b_{\ell\ell}(k_c + d^2 K_A)} \qquad (19)$$

with $b_{\ell\ell}$ being the friction coefficient between the monolayers, K_A the area compressibility modulus and d the distance between the middle of the bilayer and the monolayer neutral surfaces, the behavior switches and the relaxation rate of the bending modes is given by

$$\Gamma = \frac{1}{\tau} = \frac{k_c + d^2 K_A}{4\eta}q^3. \qquad (20)$$

Here $k_c + d^2 K_A$ is an effective bending modulus that typically could be an order of magnitude larger than k_c. Thus relaxation rates will increase dramatically at large wave numbers. Realistic values for the parameters taken from simulations and experiments indicate q_c:s in the interval 0.001–1 nm^{-1} which means that we for the system sizes reached in simulations almost always should be in the limit were undulations do relax according to Eq. (20) not Eq. (18).

IV. PERISTALTIC FLUCTUATIONS

For a membrane of finite thickness, we may consider volume, area and thickness fluctuations. The volume compressibility modulus of membranes, $K_V = -V(\partial P/\partial V)$, is known to be about 2×10^4 bar from simulations Berger et al. (1997) and experiments Braganza and Worcester (1986). This value is close to that of liquid hydrocarbons. The area compressibility modulus $K_A = A(\partial \gamma/\partial A)$ is 250 mN/m for DPPC in the liquid crystalline phase (see e.g. the review of Nagle and Tristram-Nagle (2000), and considerably larger in the gel phase. From these figures, it is obvious that it is much easier to change the area per lipid in the bilayer than to change the volume. A pressure change ΔP will result in a relative volume change of

$$\frac{\Delta V}{V} = -\frac{\Delta P}{K_V} \approx -5 \times 10^{-5} \Delta P. \qquad (21)$$

On the other hand, a change of the lateral pressure by ΔP_L over a membrane of thickness $h = 4$ nm will result in a relative area change of

$$\frac{\Delta A}{A} = -\frac{h \Delta P_L}{K_A} \approx -1.6 \times 10^{-3} \Delta P_L. \qquad (22)$$

It is thus about 30 times easier to change the area compared to the volume. Therefore, we may in a lowest order approximation disregard the small volume density fluctuations in membranes and concentrate on area fluctuations that necessarily have to be anti-correlated with thickness fluctuations to occur at constant volume. This means that what we really consider are shape fluctuations of the membrane lipids. Since these occur at constant volume, the term peristaltic fluctuations has been coined. If we chose to describe these fluctuations by the membrane thickness we may introduce a thickness field $h(x, y)$ that plays the same role in describing he peristaltic fluctuations as $u(x, y)$ in the preceding section played for the undulations. Since the thickness fluctuations are anti-correlated to fluctuations in the area per lipid field $a(x, y)$, the size of these fluctuations is determined by the area compressibility modulus. In contrast to the undulations that grow indefinitely with wave length, the peristaltic fluctuations are limited in size since the membrane otherwise would break. In a continuum approximation, the peristaltic thickness fluctuations may thus be described by an expression of the following type for the (free) energy functional (Lindahl and Edholm, 2000a)

$$F_{\text{per}}[h] = \frac{1}{2} \iint \left[k_e |h - h_0|^2 + \gamma |\nabla h|^2 + k_d |\nabla^2 h|^2 \right] dx \, dy. \qquad (23)$$

Here $h(x, y)$ is the bilayer thickness at the point (x, y) and h_0 the preferred equilibrium thickness. The first term, describes the cost of energy to deviate from this preferred thickness of the bilayer. The second term is a surface tension term, which describes the cost of energy to change the surface area of the system. The coefficient γ is the surface tension (coefficient). We will here consider a system which is free to adjust its area without applied surface tension that is with $\gamma = 0$. The third term describes the cost of energy to have a curvature in the bilayer. The constants k_e and k_d found from a simulation of a DPPC bilayer (Lindahl and Edholm, 2000a) are $k_e = 1 \times 10^{-21}$ J/nm^4, $k_d = 5 \times 10^{-21}$ J. The constants have here been divided by 4 compared to original article since $h = 2u_{\text{per}}$ is the bilayer thickness while the variable u_{per} used in that article is half that thickness. The function $h(x, y)$ can be decomposed into Fourier modes

$$h(x, y) = h_0 + \sum_{\mathbf{q}} \hat{h}(\mathbf{q}) e^{i \mathbf{q} \cdot \mathbf{r}}. \qquad (24)$$

Due to the symmetry of the energy function, the mean square of the Fourier coefficients will for a large enough system only be functions of the size of the wave vector not of its direction. The total number of Fourier components, N (plus the constant zero mode), is determined from a high wave number cut off. It does not make sense to introduce wave lengths into the undulating surface that are much smaller than the size of lipid molecules. Therefore, it is reasonable to chose N equal to the number of lipids in a monolayer, which means that the highest wave numbers is $\pi \sqrt{N/A} \approx 4$ nm^{-1}. In addition there will be a cutoff at low wave number due to the finite size of the system. This will be at $\pi \sqrt{1/A} = \pi \sqrt{N/A} \sqrt{1/N} \approx 4/\sqrt{N}$ nm^{-1}. For a tension-free system, we get the free energy

$$F_{\text{per}}[h] = \frac{A}{2} \sum_{\mathbf{q}} \left[k_d q^4 + k_e \right] |\hat{h}(q)|^2. \qquad (25)$$

Since the energy is quadratic in the amplitude and its derivatives, we have equipartition of the energy between the Fourier modes and do get the mean square amplitudes of modes with nonzero wave number as

$$\langle |\hat{h}(q)|^2 \rangle = \frac{k_B T}{A} \frac{1}{k_d q^4 + k_e}. \qquad (26)$$

The root mean square fluctuations in local bilayer thickness can now be calculated using Parseval's theorem

$$\sigma_h^2 \equiv \langle (h - h_0)^2 \rangle = \frac{1}{A} \iint \left(h(x, y) - h_0 \right)^2 dA$$

$$= \sum_q \hat{h}^2(q) = \sum_{n,m} \hat{h}^2 \left(\frac{2\pi \sqrt{n^2 + m^2}}{\sqrt{A}} \right), \tag{27}$$

where N is the number of lipids in a monolayer and A the area of the system. The sums in Eq. (27) can be performed using an integral approximation. This gives the thickness fluctuations

$$\sigma_h^2 = \frac{k_B T}{4\pi \sqrt{k_d k_e}} \left[\arctan\left(\frac{N\pi^2}{A} \sqrt{\frac{k_d}{k_e}} \right) - \arctan\left(\frac{\pi^2}{A} \sqrt{\frac{k_d}{k_e}} \right) \right]. \tag{28}$$

Here we have used a circular low-q cut off at $\pi/\sqrt{A}$ and a similar high-q cutoff at $\pi\sqrt{N/A}$. The argument of the first arctan function will with the values of k_d and k_e from simulations be around 35 which is large enough to make it a good approximation to put this term equal to $\pi/2$. This term comes from the high-q cutoff and this shows that results are insensitive to the precise upper-q limit and that this can be put equal to infinity. The argument of the second arctan function is inversely proportional to system size and will thus go to zero for a large enough system. The parameters can be combined into a correlation length

$$\lambda = \left(\frac{k_d}{k_e} \right)^{1/4} \approx 1.5 \text{ nm}, \tag{29}$$

which combines with the area per lipid into the dimensionless parameter

$$N_0 = \frac{N}{A} \pi^2 \lambda^2 = \frac{N}{A} \pi^2 \sqrt{\frac{k_d}{k_e}} \approx 35. \tag{30}$$

The parameter N_0 is a measure of the typical size (in number of lipids per monolayer) of a region of correlated thickness. Finally we have a parameter with dimension surface tension (area compressibility)

$$\sqrt{k_e k_d} \approx 2.2 \text{ mN/m}. \tag{31}$$

In terms of these parameters we can write

$$\sigma_h^2 = \frac{k_B T}{8\sqrt{k_e k_d}} \left[1 - \frac{2}{\pi} \arctan\left(\frac{N_0}{N} \right) \right]. \tag{32}$$

The maximum value of σ_h obtained in the limit of N going to infinity will then be

$$\sigma_h^{\max} = \left[\frac{(k_B T)^{1/2}}{2\sqrt{2}(k_e k_d)^{1/4}} \right] \approx 0.5 \text{ nm}. \tag{33}$$

The maximum root mean square thickness fluctuations ± 0.5 nm in a bilayer of thickness typically 3.5 nm, are reasonable and in agreement with results from a direct calculation of this quantity from a simulation. The second term in Eq. (32) cannot be neglected for very small systems but already for N about 100 it will be 10% of the first term resulting in root mean square fluctuations that reach about 90% of their maximum value.

V. HYDROCARBON CHAIN DYNAMICS

The hydrocarbon chain dynamics can be experimentally probed by deuterium and carbon-13 nuclear magnetic resonance (NMR). These techniques measure NMR relaxation rates that are proportional to the Fourier transform at the Larmor frequency of orientation correlation functions of the C–H(D) bonds. Experiments done at different frequencies (Brown et al., 1983) indicate a T_1 relaxation rate of the form

$$R_1 = \frac{1}{T_1} = A + \frac{B}{\sqrt{\omega}} \tag{34}$$

for spectrometer frequencies ranging form 2.5 MHz (deuterium) up to hundreds of MHz (carbon-13). The low frequencies are specially interesting since they reflect the very slowly decaying long time tail of the correlation functions. The relevant correlation functions for NMR can be calculated directly in the time domain from molecular dynamics simulations (Feller et al., 1999; Lindahl and Edholm, 2001; Pastor et al., 2002; Wohlert and Edholm, 2006). To reach the lowest frequencies one needs, however, very long simulations reaching the order of a μs. Simulations show that the appropriate time correlation functions decay slowly and a fit to an algebraic decay with and exponent around $-1/2$ is good although a stretched exponential or the sum of four exponential functions also would work reasonably. The $t^{-1/2}$ decay gives the experimentally observed $\omega^{-1/2}$ after Fourier transformation. The numerical agreement between simulation and experiment is fair but there are also some differences. The algebraic decay of the correlation functions is observed over about 4 orders of magnitude in time which range from 10 ps up to 100 ns. This is observed in fairly small systems consisting of about 100 lipids. This means that the correlation function decay reflects a process that is a very extended in time but still is quite local in space. This contrasts to collective processes like undulations and peristaltic fluctuations for which time scales and spatial scales usually are coupled and the dynamics gets very slow but only for motions involving many lipids. The isomerisations of the hydrocarbon chain dihedral angles, that are responsible for the main part of the decay of the orientation correlation functions are strongly correlated within the chains and with motions in neighboring chains. These correlations, which still are quite local, make the last part of the decay of the correlation functions towards zero very slow.

VI. LATERAL DIFFUSION OF LIPIDS

The lateral diffusion coefficient of the lipids in the membrane plane can be defined as

$$D_{\text{lat}} = \lim_{t \to \infty} \frac{1}{4} \frac{d}{dt} \langle [\mathbf{r}_i(t + t_0) - \mathbf{r}_i(t_0)]^2 \rangle_{i,t_0}, \tag{35}$$

where $\mathbf{r}_i$ the center of mass position in two dimensions of a lipid and the average is performed over all lipids, ($i = 1, N$) of the system and over all initial times t_0. The question is just what an infinite time is. The simulations have to be performed over long enough time to get a mean square displacement that is linear in time. Early simulations tended to get diffusion coefficients that were one the large side compare to at least some experiments. It seems now clear (Böckmann et al., 2003; Klauda et al., 2006; Wohlert and Edholm, 2006) that the root mean square displacement versus time is strongly curved for short times (about 5 ns) during which the slope continuously decreases to finally attain a constant slope that gives a diffusion constant of 0.95, 1.2 or 1.3×10^{-7} cm^2/s for DPPC at 323 K with slightly different setups (Klauda et al., 2006; Lindahl and Edholm, 2001; Patra et al., 2004) and 0.79×10^{-7} cm^2/s for DMPC 303 K (Wohlert and Edholm, 2006). For palmitoyloleylphosphatidylcholine (POPC) (Böckmann et al., 2003) observe a diffusion coefficient that is a factor 2–4 smaller, 2.6–3.9×10^{-8} cm^2/s depending on sodium ion concentration. With these diffusion constants one needs 15–60 ns for the root mean square displacement to reach typical inter-lipid distances. One would expect that times of the order 10 times this value are necessary for reaching time scales at which a macroscopic continuum model like the diffusion equation with certainty is valid. This is, however, not the main problem. Since the lipids are quite large flexible molecule, the center of mass of the molecule may also move due to conformational changes of the lipid molecule. From the mean square displacement of the individual atoms one can see that atoms far down in the tail can move almost a nm in a few ns at which time scale the lipid heads move very little. This is nicely illustrated by the pictures in (Pastor et al., 2002). Thus a mean square displacement of the center of mass is obtained by wagging of the lipid tail. This will also be diffusive, but on a faster timescale than the diffusion due to that lipids actually exchange place. In (Wohlert and Edholm, 2006) a good fit of the mean square displacement of their data for DMPC was obtained to the expression

$$\langle [\mathbf{r}_i(t + t_0) - \mathbf{r}_i(t_0)]^2 \rangle_{i,t_0} = \frac{4D_1 r_0^2 t}{r_0^2 + 4D_1 t} + 4D_2 t. \tag{36}$$

Here, the first term describes the fast diffusion at short time scales with a diffusion constant D_1. This term will, however, continuously decrease its slope versus time until it at long timescales reaches the constant maximum value r_0^2. A fit of this expression to the data of Böckmann et al. (2003); Klauda et al. (2006); Wohlert and Edholm (2006) gives in all cases a fast diffusion coefficient, D_1 that is about one order of magnitude larger than the slower long time diffusion constant, D_2 and maximum displacements due to the rattling of the tails in the range 0.3–0.7 nm. This gives a consistent interpretation of the lateral diffusion data from simulations. It does also give an atomic scale interpretation of the explanation given by (Vaz and Almeida, 1991) for the differences in lateral diffusion coefficients measured by techniques sensitive at different time scales. This is also in agreement with neutron scattering experiments (König et al., 1992) that measure different mean displacements of protons in tails and lipid heads which are consistent with the r_0-numbers obtained from simulation data.

An additional problem that lacks a conclusive explanation is the finite size effect upon the diffusion constant observed in simulations. Comparison of earlier studies often suffer from that the larger and smaller systems have not been simulated for the same time or that the setups are not absolutely identical. In (Klauda et al., 2006), this problem is not present and a reduction of the lateral diffusion constant with a factor 2.5 is still seen upon increasing the system size from 72 to 288 lipids.

A. The Relative Motion of the Entire Monolayers

Diffusive motions of the entire monolayers with respect to each other and with respect to the water layer have been reported in a number of simulations (Lindahl and Edholm, 2001; Anezo et al., 2003; Patra et al., 2004; Klauda et al., 2006; Wohlert and Edholm, 2006). Such motions occur quite easily in simulations with periodic boundary conditions, especially when the systems are small, since there are no resistance at a lateral boundary. They may be considered as artifacts and have to be subtracted from the mean square displacements before these are used to calculate lateral diffusion coefficients. This motion does, however, have an interest in itself since it allows the calculation of friction coefficients between the monolayers and between monolayers and water. The friction coefficient per unit area $(b_{\ell\ell})$ between the monolayers obtained in this way from equilibrium simulations is of the order 10^6 Ns/m³. Similar values are also obtained by non-equilibrium methods for coarse grained lipid models (Shkulipa et al., 2005). This is a bit disturbing since experimental measurements indicate values that are about two orders of magnitude larger (Evans and Yeung, 1994; Raphael and Waugh, 1996). The friction coefficient between water and a monolayer, $b_{\ell w} = 4 \times 10^6$ Ns/m³ (Wohlert and Edholm, 2006), is on the other hand reasonable. If one assumes, stick boundary conditions for the water at the monolayer interface, one may attribute this friction to a continuously distributed viscous resistance of the water. Using hydrodynamics, see e.g. (Landau and Lifshitz, 1959) one may derive the relation $\eta = b_{\ell w} h/3$, were h is the thickness of the water layer and η the viscosity of water. This gives with the value of $b_{\ell w} = 4 \times 10^6$

and a thickness of 2 nm for the water layer, a viscosity of 2.7 cP which is a bit larger than the viscosity of bulk water (1 cP) but still has the correct order of magnitude.

B. The Dynamic Structure Factor

Inelastic X-ray scattering can measure the dynamic structure factor, $S(q, \omega)$, directly (Chen *et al.*, 2001; Weiss *et al.*, 2003). The main part of the loss of energy for the X-rays is due to the scattering against the carbons of the tails. The results are usually interpreted in the framework of the theory of molecular hydrodynamics. A review of the theory and proper references are given in (Weiss *et al.*, 2003). The q and ω values accessible by this technique are in the range 2–30 nm^{-1} and 2–25 meV respectively. This means in space wave lengths up to about 3 nm and in time motions that occur on ps time scales. These are space and time regions which are easily accessible in molecular dynamics simulations and these kind of simulations have since more than 30 years been used to interpret corresponding inelastic neutron scattering data in liquids. (Tarek *et al.*, 2001) have performed an exploratory study of the possibilities of an analysis of this type of simulation data from membrane simulations. The main problem with simulations as well as experiments seems to be that molecular hydrodynamics predicts that the dynamic structure factor as a function of ω should have a central (Raleigh) peak surrounded by two symmetric (Brillouin) peaks. The peaks are, however, located very close to each other so that both experiments and simulations need extremely extensive sampling to resolve the different peaks and determine the parameters of the molecular hydrodynamics theory. The fast short time lateral diffusion coefficient can be obtained from the width of the central peak in fair agreement experiment and other computational methods. To resolve the peaks accurately, densely sampled data from a 50–100 ns simulation of a fairly small bilayer would probably be sufficient. The problem is not the length of the simulation but that extensive amounts of densely sampled data is needed.

References

Anezo, C., de Vries, A.H., Höltje, H.-D., Tieleman, D.P., Marrink, S.-J. (2003). Methodological issues in lipid bilayer simulations. *J. Phys. Chem. B* **107**, 9424–9433.

Berger, O., Edholm, O., Jähnig, F. (1997). Molecular dynamics simulation of a fluid bilayer of dipalmitoylphosphatidylcholine at full hydration, constant pressure and constant temperature. *Biophys. J.* **72**, 2002–2013.

Braganza, L.F., Worcester, D.L. (1986). Structural changes in lipid bilayers and biological membranes caused by hydrostatic pressure. *Biochemistry* **25**, 7484–7488.

Brown, M.F., Ribeiro, A.A., Williams, G.D. (1983). New view of lipid bilayer dynamics from ^{2}H and ^{13}C NMR relaxation time measurements. *Proc. Natl. Acad. Sci. USA* **80**, 4325–4329.

Böckmann, R.A., Hac, A., Heimburg, T., Grubmüller, H. (2003). Effect of sodium chloride on a lipid bilayer. *Biophys. J.* **85**, 1647–1655.

Castro-Román, F., Benz, R.W., White, S.H., Tobias, D.J. (2006). Investigation of finite size effects in molecular dynamics simulations of lipid bilayers. *J. Phys. Chem. B* **110**, 24157–24164.

Chen, S.H., Liao, C.Y., Huang, H.W., Weiss, T.M., Bellisent-Funel, M.C., Sette, F. (2001). Collective dynamics in fully hydrated phospholipid bilayers studied by inelastic X-ray scattering. *Phys. Rev. Lett.* **86**, 740–743.

Chiu, S., Clark, M., Balaji, V., Subramaniam, S., Scott, H., Jakobsson, E. (1995). Incorporation of surface tension into molecular dynamics simulation of an interface: A fluid phase lipid bilayer membrane. *Biophys. J.* **69**, 1230–1245.

Darden, T., York, D., Pedersen, L. (1993). Particle mesh Ewald: An $N - \log(N)$ method for Ewald sums in large systems. *J. Chem. Phys.* **98**, 10089–10092.

Duwe, H.P., Sackmann, E. (1990). Bending elasticity and thermal excitations of lipid bilayer vesicles: Modulation by solutes. *Physica A* **163**, 410–428.

Egberts, E., Berendsen, H.J.C. (1988). Molecular dynamics simulation of a smectic liquid crystal with atomic detail. *J. Chem. Phys.* **98**, 1712–1720.

Evans, E., Rawicz, W. (1990). Entropy driven tension and bending elasticity in condensed-fluid membranes. *Phys. Rev. Lett.* **64**, 2094–2097.

Evans, E. (1991). Entropy-driven tension in vesicle membranes and unbinding of adherent vesicles. *Langmuir* **7**, 1900–1908.

Evans, E., Yeung, A. (1994). Hidden dynamics in rapid changes of bilayer shape. *Chem. Phys. Lipids* **73**, 39–56.

Feller, S.E., Pastor, R.W. (1996). On simulating lipid bilayers with and applied surface tension: Periodic boundary conditions and undulations. *Biophys. J.* **71**, 1350–1355.

Feller, S.E., Pastor, R.W. (1999). Constant surface tension simulations of lipid bilayers: The sensitivity of surface areas and compressibilities. *J. Chem. Phys.* **111**, 1281–1287.

Feller, S.E., Zhang, Y., Pastor, R.W. (1995). Computer simulation of liquid/liquid interfaces. II Surface tension–area dependence of a bilayer and a monolayer. *J. Chem. Phys.* **103**, 10267–10272.

Feller, S.E., Pastor, R.W., Rojnuckarin, A., Bogusz, S., Brooks, B.R. (1996). The effect of electrostatic force truncation on interfacial and transport properties of water. *J. Phys. Chem.* **100**, 17011–17020.

Feller, S.E., Huster, D., Gawrisch, K. (1999). Interpretation of NOSEY cross-relaxation rates from molecular dynamics simulation of a lipid bilayer. *J. Am. Chem. Soc.* **121**, 8963–8964.

Goetz, R., Gompper, G., Lipowsky, R. (1999). Mobility and elasticity of self-assembled membranes. *Phys. Rev. Lett.* **82**, 221–224.

Granek, R. (1997). From semi-flexible polymers to membranes: Anomalous diffusion and reptation. *J. Phys. II (France)* **7**, 1761–1788.

Helfrich, W. (1973). Elastic theory of lipid bilayers: Theory and possible experiments. *Z. Naturforsch. C* **28**, 693–703.

Jähnig, F. (1996). What is the surface tension of a lipid membrane? *Biophys. J.* **71**, 1348–1349.

Klauda, J.B., Brooks, B.R., Pastor, R.W. (2006). Dynamical motions of lipids and finite size effect on simulations of bilayers. *J. Chem. Phys.* **125**, 144710.

König, S., Pfeiffer, W., Bayerl, T., Richter, D., Sackmann, E.E. (1992). Molecular dynamics of lipid bilayers studied by incoherent quasi-elastic neutron scattering. *J. Phys. II (France)* **2**, 1589–1615.

König, S., Sackmann, E. (1996). Molecular and collective dynamics of lipid bilayers. *Curr. Opin. Colloid Interface Sci.* **1**, 78–82.

Kramer, L. (1971). Theory of light scattering from fluctuations of membranes and monolayers. *J. Chem. Phys.* **55**, 2097–2105.

Landau, L.D., Lifshitz, E.M. (1959). "Fluid Mechanics". Pergamon Press, London, Great Britain.

Lin, L.C.-L., Brown, F.L.H. (2004). Brownian dynamics in Fourier space: Membrane simulations over long length and time scales. *Phys. Rev. Lett.* **93**, 256001.

Lin, L.C.-L., Brown, F.L.H. (2005). Dynamic simulations of membranes with cytoskeletal interactions. *Phys. Rev. E* **72**, 011910.

Lindahl, E., Edholm, O. (2000a). Mesoscopic undulations and thickness fluctuations in lipid bilayers from molecular dynamics simulations. *Biophys. J.* **79**, 426–433.

Lindahl, E., Edholm, O. (2000b). Spatial and energetic/entropic decomposition of surface tension in lipid bilayers from molecular dynamics simulations. *J. Chem. Phys.* **113**, 3882–3893.

Lindahl, E., Edholm, O. (2001). Molecular dynamics simulations of NMR relaxation rates and slow dynamics in lipid bilayers. *J. Chem. Phys.* **115**, 4938–4950.

Lipowsky, R., Grotehans, S. (1993). Hydration vs. protrusion forces between lipid bilayers. *Europhys. Lett.* **23**, 599–604.

Marrink, S.-J., Mark, A.E. (2001). Effect of undulations on surface tension in simulated bilayers. *J. Phys. Chem. B* **105**, 6122–6127.

Marrink, S.-J., de Vries, A., Mark, A.E. (2004). Coarse grained model for semi-quantitative lipid simulations. *J. Phys. Chem. B* **108**, 750–760.

Milner, S.T., Safran, S.A. (1987). Dynamical fluctuations of droplet microemulsions and vesicles. *Phys. Rev. A* **36**, 4371–4379.

Nagle, J.F., Tristram-Nagle, S. (2000). Structure lipid bilayers. *Biochem. Biophys. Acta* **1469**, 159–195.

den Otter, W.K., Briels, W.J. (2003). The bending rigidity of an amphiphilic bilayer from equilibrium and nonequilibrium molecular dynamics. *J. Chem. Phys.* **118**, 4712–4720.

den Otter, W.K. (2005). Area compressibility and buckling of amphiphilic bilayers in molecular dynamics simulations. *J. Chem. Phys.* **123**, 214906.

Pastor, R.W. (1994). Computer simulations of lipid bilayers. *Curr. Opin. Struct. Biol.* **4**, 486–492.

Pastor, R.W., Venable, R.M., Feller, S.E. (2002). Lipid bilayers, NMR relaxation and computer simulation. *Acc. Chem. Res.* **35**, 438–446.

Patra, M., Karttunen, M., Hyvönen, T., Falck, E., Lindqvist, P., Vattulainen, I. (2003). Molecular dynamics simulations of lipid bilayers: Major artifacts due to truncating electrostatic interactions. *Biophys. J.* **84**, 3636–3645.

Patra, M., Karttunen, M., Hyvönen, T., Falck, E., Vattulainen, I. (2004). Lipid bilayers driven to a wrong lane in molecular dynamics simulations by subtle changes in long-range electrostatic interactions. *J. Phys. Chem. B* **108**, 4485–4494.

Rahman, A. (1964). Correlations of the motions of atoms in liquid argon. *Phys. Rev. A* **136**, 405–411.

Rahman, A., Stillinger, F.H. (1971). Molecular dynamics study of liquid water. *J. Chem. Phys.* **55**, 3336–3359.

Raphael, R.M., Waugh, R.E. (1996). Accelerated interleaflet transport of phosphatidylcholine molecules in membranes under deformation. *Biophys. J.* **71**, 1374–1388.

Rawicz, W., Olbrich, K.C., McIntosh, T., Needham, D., Evans, E. (2000). Effect of chain length and unsaturation on elasticity of lipid bilayers. *Biophys. J.* **79**, 328–339.

Safran, S.A. (1994). "Statistical Thermodynamics of Surfaces, Interfaces, and Membranes". Addison–Wesley, Reading, Massachusetts.

Seifert, U., Langer, S.A. (1994). Hydrodynamics of membranes: The bilayer aspect and adhesion. *Biophys. Chem.* **49**, 13–22.

Shkulipa, S.A., den Otter, W.K., Briels, W.J. (2005). Surface viscosity, diffusion and intermonolayer friction: Simulating sheared amphiphilic bilayers. *Biophys. J.* **89**, 823–829.

Shkulipa, S.A., den Otter, W.K., Briels, W.J. (2006a). Thermal undulations of lipid bilayers relax by intermonolayer friction at submicrometer length scales. *Phys. Rev. Lett.* **96**, 178302.

Shkulipa, S.A., den Otter, W.K., Briels, W.J. (2006b). Simulations of the dynamics of thermal undulations in lipid bilayer in the tensionless state and under stress. *J. Chem. Phys.* **125**, 234905.

Tarek, M., Tobias, D.J., Chen, S.-H., Klein, M.L. (2001). Short wave length collective dynamics in phospholipid bilayers: A molecular dynamics study. *Phys. Rev. Lett.* **87**, 238101.

Tieleman, D.P., Marrink, S.-J., Berendsen, H.J.C. (1997). A computer perspective of membranes: Molecular dynamics studies of lipid bilayer systems. *Biochim. Biophys. Acta* **1331**, 235–270.

Tu, K., Tobias, D.J., Blasie, J.K., Klein, M.L. (1996). Molecular dynamics investigation of the structure of a fully hydrated gel-phase dipalmitoylphosphatidylcholine bilayer. *Biophys. J.* **70**, 595–608.

van der Ploeg, P., Berendsen, H.J.C. (1982). Molecular dynamics simulation of a bilayer membrane. *J. Chem. Phys.* **76**, 3271–3276.

Vaz, W.L., Almeida, P.F. (1991). Microscopic versus macroscopic diffusion in one-component fluid phase lipid bilayer membranes. *Biophys. J.* **60**, 1553–1554.

Venturoli, M., Smit, B. (1999). Simulating the self-assembly of model membranes. *Phys. Chem. Commun.* **2**, 10.

Weiss, T.M., Chen, P.-J., Sinn, H., Alp, E.A., Chen, S.-H., Huang, H.W. (2003). Collective chain dynamics in lipid bilayers by inelastic X-ray scattering. *Biophys. J.* **79**, 426–433.

Wohlert, J., Edholm, O. (2004). The range and shielding of dipole/dipole interactions in phospholipid bilayers with cholesterol. *Biophys. J.* **87**, 2433–2445.

Wohlert, J., Edholm, O. (2006). Dynamics in atomistic simulations of phospholipid membranes. Nuclear magnetic resonance relaxation rates and lateral diffusion. *J. Chem. Phys.* **125**, 20473.

Yeung, A., Seifert, U. (1995). Unexpected dynamics in shape fluctuations of bilayer vesicles. *J. Phys. II (France)* **5**, 1501–1523.

CHAPTER 4

Molecular Dynamics Simulation of Lipid–Protein Interactions

Nicolas Sapay and D. Peter Tieleman

University of Calgary, Department of Biological Sciences, 2500 University Dr. NW, Calgary, Alberta, T2N1N4, Canada

Abstract

Membrane proteins and membrane-active peptides are attractive targets for computer simulations, both because of their intrinsic importance and the difficulties associated with experiments on such systems. Such simulations require parameter sets for both lipids and proteins and sufficient sampling to average over the motions of lipids and proteins, which may occur on a time scale of hundreds of nanoseconds. We briefly review available parameters and discuss technical issues

Current Topics in Membranes, Volume 60
1063-5823/08 $35.00
DOI: 10.1016/S1063-5823(08)00004-5

in combining separate parameter sets for lipids and for proteins in a single simulation. We show that different combinations of non-bonded parameters lead to significantly different results. We discuss relevant time scales as related to obtaining sufficient sampling to draw valid conclusions based on a simulation, and emphasize the urgent need for experimental data that allows a critical validation of simulation approaches.

I. INTRODUCTION

A. Simulations of Membrane Proteins

As soon as lipid models became available, simulations of membrane-bound peptides and proteins have been attempted. In many ways these studies jumped the gun, as we know now that the lipid models themselves are difficult to test due to the long time scales involved in equilibration of lipids (tens of nanoseconds, versus picoseconds—1 nanosecond in the first simulations), and because of the lack of accurate experimental data that provide critical tests of the lipid models. However, these initial studies made clear that there is tremendous interest in simulations of membrane-bound peptides and membrane proteins. Key biological processes including signaling, membrane fusion, bioenergetics, proteins secretion, host-defense mechanisms and transport in general take place at or inside the membrane. There have been several reviews of membrane protein simulations (Ash et al., 2004; Forrest and Sansom, 2000; Gumbart et al., 2005; Jakobsson, 1997; Mátyus et al., 2007; Tieleman et al., 1997). In this review we focus on simulation concerns that are particularly important for mixed lipid/protein systems with an emphasis on accurately describing lipid–protein interaction parameters. There are other issues in simulating membrane proteins besides the choice of force field and sampling concerns. We have recently described what we consider the best approach for setting up such simulations (Kandt et al., 2007), and we have previously explored the effect of different algorithmic choices in pure DPPC simulations (Anézo et al., 2003) and of different electrostatic treatments in model ion channels (Tieleman et al., 2002).

B. Simulation Issues

In general, molecular dynamics (MD) simulations rely on (1) the ability to accurately describe interactions between atoms and (2) on sufficient sampling to explore enough of phase space to draw meaningful conclusions from simulations. These are requirements in any form of molecular simulation, and they are particularly problematic in the characterization of lipid–protein interactions.

First, it is difficult to accurately describe interactions between atoms in a system that consists of lipids, proteins/peptides, water, and possibly ions and other biomolecules. Indeed, simulations of nucleic acids interacting with lipids have been described in the context of DNA transfection (*e.g.* (Bandyopadhyay *et al.*, 1999) and simulations of carbohydrates have been described in several contexts, including the possible role of carbohydrates in cryoprotection of cells (reviewed in this issue, MacCallum and Tieleman (2003)).

We review the current state of the art in lipid force fields below. In summary, significant progress has been made and some lipids, in particular phosphatidylcholine (PC) lipids are quite accurate, but in our opinion there is currently no comprehensive lipid force field that allows confident simulations of all types of phospholipids, which themselves are only a fraction of all biologically relevant lipids. A second challenge is that it is generally not trivial to combine force field parameters from a good lipid force field with parameters from a good protein force field. Even if they have been derived using the same principles, molecular dynamics parameters are empirical, and there is no guarantee that a combination of a good protein force field with a good lipid force field will give a reliable combination. In fact, recent force field tests that focused on the properties of protein parameters in solvents of different polarities showed significant errors, although the general behavior of all side chains is well represented (MacCallum and Tieleman, 2003; Villa and Mark, 2002; Shirts *et al.*, 2003). Because membrane processes involve partitioning side chains and parts of proteins between very different environments within the membrane, there is little reason to believe the accuracy of such simulations is better than the accuracy of these detailed tests of force fields for soluble proteins.

Third, sampling of proteins and peptides in lipids and of lipids by themselves remains a major limiting factor. Ideally, a simulation of a mixture of lipids should give the same results regardless of the starting structure. In practice, phospholipids diffuse on a time scale of hundreds of nanoseconds, at the upper end of what is possible by 'normal' computational efforts. Peptides in solution fold and unfold on time scales of hundreds of nanoseconds to microseconds. There is little guarantee in simulations of peptide binding to a membrane that we are able to sample the relevant conformational space of those peptides, even in microseconds. For simulations of large membrane proteins, the details of interactions with lipids may not be so important. However, even in this case it is very important how simulations are set up, as it is quite possible to create high energy systems (due to unfavorable contacts, tension or strain, *etc.*) that over the course of the 'production' simulation simply evolve towards a more favorable arrangement; in this case, the ensemble of structures sampled is not representative of an equilibrium, and therefore scarcely relevant.

In a sense, these are standard simulation problems, but there is an additional issue that is more particular to lipid–protein interactions. How does one judge the quality of a simulation of a lipid–protein system? In general, available structural data on lipids is limited, with significant error margins, and often from experiments that are not straightforward to interpret. Accurate measurements on membrane proteins and peptides that can be used to validate and improve models are even rarer. Some of the most powerful techniques, including X-ray and neutron diffraction, solid state Nuclear Magnetic Resonance (NMR) and Infra Red (IR) spectroscopy, are difficult to interpret. Even if we can address the sampling problem, we will also need accurate experimental data to validate simulations.

In this chapter we briefly review available force fields for simulations of lipids and proteins, discuss some problems with combining different force fields in a single simulation, review recent simulations that give an indication of the scope of the sampling problem for lipid–protein simulations, and discuss the importance and lack of experimental data that is a critical test for simulations.

II. FORCE FIELDS

A. General Form of Force Fields

In molecular dynamics, the force field consists of the equations chosen to model the potential energy and their associated parameters. The general form of the potential energy is a sum of terms similar to:

$$
V(r) = \sum_{\text{bonds}} k_b(b - b_0)^2 + \sum_{\text{angles}} k_\theta(\theta - \theta_0)^2
$$
$$
+ \sum_{\text{dihedrals}} k_\phi\big(1 + \cos(n\phi - \phi_0)\big) + \sum_{\text{impropers}} k_\psi(\psi - \psi_0)^2
$$
$$
+ \sum_{\substack{\text{non-bonded} \\ \text{pairs}(i,j)}} 4\varepsilon_{ij}\left[\left(\frac{\sigma_{ij}}{r_{ij}}\right)^{12} - \left(\frac{\sigma_{ij}}{r_{ij}}\right)^{6}\right] + \sum_{\substack{\text{non-bonded} \\ \text{pairs}(i,j)}} \frac{q_i q_j}{\varepsilon_D r_{ij}}. \quad (1)
$$

The potential function V depends on r, the position vector of particles in the system, expressed as in terms of distances and angles between atoms. Despite occasional extra terms in some force fields, the functional form of the potential function is essentially the same in all common force fields, so that they are not fundamentally different. Indeed, the potential function embodies the most serious assumptions, and will provide an upper limit to the accuracy of any parameter set to describe the interactions between atoms.

In practice, the parameters in Eq. (1) are generally stored in data files with lists corresponding to each term in the sum. The force field includes a list of particle types corresponding to the different types of atoms that occur in a system. For example, the carbon atom in a methyl group $-CH_3$ and in a carbonyl group $C=O$ can be described by two different particle types with two different charges (q in Eq. (1)) and van der Waals parameters (ε, σ), as well as specific bonds, angles, and dihedral parameters. The force field includes a list of parameters to describe the bonds (k_b, b_0), angles (k_θ, θ_0), and dihedrals $(k_\phi, \phi_0, n, k_\psi, \psi_0)$ connecting all combinations of different particle types necessary to model a particular molecule or class of molecules, *e.g.* proteins, nucleic acids, or lipids.

Major force fields commonly used in MD simulations of biomolecules are AMBER (Case *et al.*, 2005; Ponder and Case, 2003), CHARMM (MacKerell *et al.*, 1998), GROMOS (Scott *et al.*, 1999; van Gunsteren *et al.*, 1996) and OPLS (Jorgensen *et al.*, 1996). Parameters in each force field are supposed to be internally consistent but this is not necessarily true between different force fields. This is in part because each force field has been originally developed in conjunction with a particular model of water: the TIP3P model with some modifications (Jorgensen *et al.*, 1983) for CHARMM and AMBER, the SPC model (Berendsen *et al.*, 1981) for GROMOS96, *etc.*, and in part because different force fields are based on their own parameterization philosophy. Interestingly, recent and ongoing reparameterization of common force fields appears to lead to some degree of convergence in parameters and in accuracy. Hess and van der Vegt compared hydration free energies of OPLS-AA, AMBER99 and GROMOS 53A6, and found that their accuracy was comparable. However, the choice of water model has a significant effect on hydration free energies for a given force field (Hess and van der Vegt, 2006), emphasizing the importance of considering a water model together with other force field parameters.

Force fields also have important subversions, as the parameterization procedure has been changed over time as more elaborate calculations became possible. In particular, recent versions of several common force fields use high-level quantum mechanics and full corrections to long-range interactions, whereas many initial versions were parameterized with short cut-offs of non-bonded interactions (of the order of 1.0 nm and sometimes less) and used rather approximate quantum mechanics by today's standards. Long-range corrections to electrostatics in particular strongly affect the properties of water models, to the point that in many ways combining recent versions of a force field with older parameters gives the same problems as combining two different force fields.

All major force fields have been extensively optimized over the past twenty years with particular emphasis on the treatment of proteins. Originally, they were developed with a united-atom scheme, *i.e.* only non-hydrogen atoms were explicitly described. For example, a chemical group like $-CH_3$ is represented only by a single 'atom', with a modified mass of 15 a.m.u. and non-bonded parameters to implicitly take into account the hydrogen atoms. This reduces the number of atoms and pair interactions in a simulation, although the extent of reduction depends on the exact simulations. For proteins in water it makes little difference, for lipids and liquid alkanes it makes a very large difference. For boxes with the same number of cyclohexane molecules, the all-atom version was 15 times slower to simulate than the united atom version, primarily because of the much higher particle density, in addition to the obvious but smaller effect of having a larger number of atoms in the all-atom case (Tieleman *et al.*, 2006). United atom representations also reduce the number of parameters, which may be appropriate if there is not enough experimental or higher-level theoretical data to fit these parameters. The improvement of computer hardware and software has allowed the development of more complex representations. Today, CHARMM, AMBER and OPLS use an all-atom scheme where hydrogens are explicitly represented. Only GROMOS is still developed with a united atom representation.

B. Lipid Force Fields

Most major force fields have developed a specific set of parameters for apolar solvents, common organic molecules, as well as for common phospholipids (typically at least DPPC, POPC, DOPC). However, only two phospholipid force fields are in common use today: an all-atom force field that part of the official CHARMM distribution and a united-atom force field by Berger *et al.* (1997), created by combining parameters taken from united-atom versions of OPLS and AMBER with some modifications to charges and lipid chain parameters. Both CHARMM based lipids and Berger lipids reproduce the available experimental information on the structure and dynamics of phospholipid bilayers reasonably well, particularly for the experimentally well-studied phosphatidylcholine lipids, and in our opinion there is no compelling experimental information that indicates either force field is substantially better, although their errors differ (Benz *et al.*, 2005). In addition to these two force fields, there has been recent progress in developing GROMOS96-based lipids (Chandrasekhar *et al.*, 2003), some use of AMBER parameters to describe lipids (*e.g.* Pertsin *et al.*, 2007), and there are several modifications of older force fields (*e.g.* the chapter by Erik Jakobsson in this book). Only CHARMM currently provides a full set of parameters for all types of biomolecules, whereas recent GROMOS96 developments are moving in that direction. Other choices of lipid parameters require combining these parameters with a second set of parameters unless one is interested in pure lipids.

The parameterization of all lipid force fields is typically based on *ab initio* and other calculations on molecules mimicking lipid fragments, such as dimethylphosphate and liquid alkanes, for several reasons. Lipids are complex molecules but share a limited number of functional groups, so that a building block approach similar to proteins seems sensible. Detailed structural and thermodynamic information on complete phospholipids is scarce, but more readily available for well-chosen model compounds. Finally, QM calculations have a high computational cost and are limited in size for high levels of theory. This general approach assumes that bond, angle and dihedral parameters as well as charge distributions in model compounds like dimethylphosphate or butane are reasonably close to those of phospholipids.

In practice, lipid force fields continue to have issues when parameters that work well for the model compounds are combined for use in full bilayers. The chapter by MacKerell *et al.* in this issue describes in detail the current procedure for CHARMM. Ad-hoc adjustments can improve the results for a single lipid or class of lipids (phosphatidylcholine), but this often runs counter to chemical intuition and in practice often has turned out to be not transferable to other lipids, such as the closely related phosphatidylethanolamine or phosphatidylserine lipids. For example, Chiu et al. have reparameterized GROMOS-based lipids to give the right area for PC (Chiu *et al.*, 1999), and Sonne and coworkers have shown that is possible to partially solve the area per lipid problem of CHARMM27 by treating DPPC molecules as a whole and not as sum of building blocks (Sonne *et al.*, 2007a). A key problem appears to be that lipid bilayers are extremely sensitive to an accurate balance between water–water, water–lipid, and lipid–lipid interactions, at different levels along the bilayer normal.

Finally, one should remember that the force field parameters themselves are not the only determinants in lipid bilayer simulations. The macroscopic boundary conditions as well as the method used to compute electrostatic and even Lennard-Jones interactions have also an important role as they effectively change the balance between water–water, water–lipid, and lipid–lipid interactions (Anézo *et al.*, 2003).

III. COMBINING FORCE FIELDS

A. Why Combining Force Fields?

Currently CHARMM is the only the force field with a consistent parameter set for all types of biomolecules, but several other parameter sets are only dedicated to one or few types of molecule. Using CHARMM for lipid–protein simulations is a reasonable option, but there are several issues to consider. First, CHARMM currently requires simulations at fixed area or a significant surface tension to obtain reasonable areas per lipids (Sonne *et al.*, 2007a). A fixed area is not desirable for membrane protein simulations because this area is not known in advance, and conformational changes in proteins may require a change in area. A protein like the mechanosensitive channel MscL undergoes a very large change in its cross-sectional area in the membrane during gating (Betanzos *et al.*, 2002). Similarly, antimicrobial peptides inserting into a bilayer could cause a large change in area; fixing the area of stretching the bilayer by a surface tension will change the energetic of insertion and might prevent insertion completely. Second, CHARMM is an all-atom force field. Although this is quite reasonable and consistent, in many cases the all-atom nature of the lipids is not likely to play a significant role in membrane protein behavior, while the cost of simulating all-atom lipids is much higher than the cost of simulating united-atom lipids. Third, there may be reasons of consistency with other work to prefer different force fields. We have experimented with several combinations of force fields to simulate membrane proteins, primarily for reasons 1 and 2 above.

As described above, the potential energy functions in common force fields are not fundamentally different (Eq. (1)), though several variations exist, as well as several inconvenient alternative implementations of the same physics that require conversions of parameters that introduce modest changes in the force field beyond simple unit conversions. An example of a variation of the potential function is the CMAP backbone potential of CHARMM (MacKerell *et al.*, 2004), which cannot currently be used in GROMACS (www.gromacs.org) without some changes to the code. Examples of conversion issues are described in detailed below: the treatment of proper dihedrals and the treatment of non-bonded (van der Waals and electrostatics) parameters.

B. Combining Dihedrals Terms

In general, force fields represent dihedral angles ϕ between atoms i, j, k and l with a periodic function of the following form:

$$V_{\text{proper}}(\phi) = k_\phi \big(1 + \cos(n\phi - \phi_0)\big). \tag{2}$$

Here k_ϕ is the amplitude of the potential, ϕ_0 the phase angle, and n the multiplicity of ϕ (*i.e.* the number of local minima in the function). Because this function has the same depth for each potential well, it is combined with Lennard-Jones and Coulomb interactions to represent, *e.g.*, the difference in minima for trans and gauche angles in hydrocarbons. The distance between i and l atoms is so small that regular strength non-bonded interactions are normally too strong, and the general approach is to scale them. Unfortunately, the scaling factors used are not the same in different force fields (see below).

Some force fields allow the description of one dihedral by more than one set of parameters (*i.e.* more than one triplet of (k_ϕ, ϕ_0, n), like in CHARMM). This is a way to manage complex dihedral potentials, *e.g.* those involving the phosphorus atom in phospholipids or dihedrals in sugar rings.

An alternative approach is to use a more complex description of the dihedral potential, *e.g.* the cosine expansion of the Ryckaert–Bellemand's potential (Ryckaert and Bellemans, 1978):

$$V_{\text{Ryckaert–Bellemans}}(\phi) = \sum_{i=0}^{n} C_i \big(\cos(\phi - 180)\big)^n. \tag{3}$$

OPLS uses yet a different representation, which can be rigorously converted to the RB potential (www.gromacs.org). In this case, Lennard-Jones and Coulomb interactions between atoms i and l are normally excluded, because this function itself has enough detail to represent different values for different minima.

C. Combining Non-Bonded Parameters

Non-bonded interactions of force fields consist of van der Waals and electrostatics interactions. Their calculation is based on atom pairs. Most of the time, force fields use the Lennard-Jones potential (Lennard-Jones, 1924) to describe the van der Waals interactions between a pair (i, j) of atoms:

$$V_{\text{Lennard-Jones}}(r_{ij}) = 4\varepsilon_{ij} \left[\frac{\sigma_{ij}^{12}}{r_{ij}^{12}} - \frac{\sigma_{ij}^{6}}{r_{ij}^{6}} \right]. \tag{4}$$

With r_{ij} the distance between i and j and ε_{ij} and σ_{ij}, two constants based on Lennard-Jones parameters of atoms i and j. Problems arise because different force fields do not necessarily use the same rule to determine ε_{ij} and σ_{ij}. Some use a geometric average to define σ_{ij}:

$$\varepsilon_{ij} = \sqrt{\varepsilon_i \varepsilon_j},$$
$$\sigma_{ij} = \sqrt{\sigma_i \sigma_j}. \tag{5}$$

While others use an arithmetic average:

$$\varepsilon_{ij} = \sqrt{\varepsilon_i \varepsilon_j},$$
$$\sigma_{ij} = \frac{1}{2}(\sigma_i + \sigma_j). \tag{6}$$

Additionally, some force fields, like CHARMM, may prefer the use of the r_{min} parameter instead of σ_{ij}:

$$r_{\text{min}} = 2^{1/6} \sigma_{ij}. \tag{7}$$

With σ_{ij} representing the distance at which the energy between i and j is null and r_{min}, the distance at which the energy between i and j is minimal. The conversion between r_{min} and σ_{ij} is simple. However, it is generally not possible to systematically mix arithmetic and geometric averages to compute σ_{ij}. In addition, some force fields specify explicitly for each pair of atom types how the combination is treated. How does one combine such a force field with one that relies on automatic rules to determine these combinations (see below)? The user must therefore make a choice, and validate this choice.

A different issue comes from the 1–4 interactions inside molecules. 1–4 interactions are non-bonded interactions (van der Waals and electrostatics) between two atoms separated by 3 bonds, and are tightly coupled to the energy surface for dihedral rotations. As such they have a strong influence on the conformation of biomolecules. Force fields typically include them in different ways:

1. 1–4 interactions are ignored, but taken into account implicitly through dihedral parameters. An example is the Ryckaert–Bellemans potential for butane.

2. 1–4 interactions used the same Lennard-Jones and Coulomb parameters are regular non-bonded interactions, but are uniformly scaled by a factor, potentially different for Lennard-Jones and Coulomb. The OPLS-AA force field scales down both interactions by a factor 2.

3. Specific 1–4 Lennard-Jones parameters are specified for each atom type of the system and electrostatic interactions remain unchanged. This is the case with CHARMM and GROMOS. In most cases, 1–4 Lennard-Jones parameters are just regular parameters scaled down to compensate for the close proximity of 1–4 neighbors.

There is no fundamental problem with combining, *e.g.*, a protein and a lipid force field with a different treatment of 1–4 interactions, since 1–4 interactions only occur inside molecules, not between. However, there could be technical difficulties. Most major MD engines apply a 1–4 scaling factor universally on all 1–4 pairs of the simulated system. It is generally possible to define separately 1–4 Lennard-Jones parameters, but this is not the case for charges. In this case, the 1–4 interactions can be completely described by proper dihedral potentials, by adding a supplemental potential for each concerned dihedral. This re-parameterization of dihedrals gives good results with the combination of OPLS-AA and Berger force fields (Tieleman *et al.*, 2006), or see Section III.E below for details). It is also possible to "cheat" with GROMACS by defining two times each pair of 1–4 interactions in the topology file. Therefore, their contribution to the electrostatics interactions will be calculated twice. If a scaling factor of 0.5 is applied, the electrostatics will remain unchanged.

D. Testing the Combined Force Field

The first obvious test of simulations with combined force fields is to check if the properties of molecules described solely by the original force fields have not changed. If the only change is a conversion of units or implementation in a different program, the exact energies should be reproduced. If, *e.g.*, the proper dihedrals parameters have been modified to integrate 1–4 interactions, the new dihedral parameters must reproduce the original total potential energy as a function of the dihedral angle, although the result will not be exactly the same because fitting is not exact, and because the coupling between dihedrals and angles is not the same in the two different treatments (Tieleman *et al.*, 2006). However, testing beyond this trivial comparison is harder, because it requires data to validate the simulations. Although this is the same problem one encounters in developing any force field, in this case the problem is harder because there is little critical lipid–protein specific experimental data. Two currently important and feasible test systems are thermodynamic data on small molecules and experimental studies of model transmembrane helices. Below we describe examples of both.

E. A Case Study: WALP23 and Polyleucine in a DOPC Bilayer and in Hydrophobic Solvent

We recently tested a number of different ways to combine a protein force field with a different lipid force field (Tieleman *et al.*, 2006). The protein force fields were (1) ffgmx as implemented in GROMACS. This is a united atom force field based on GROMOS87 with a number of improvements in water–carbon interactions and aromatic residues. (2) The all-atom OPLS-AA force field. The lipid force field was based on an older OPLS united-atom force field and AMBER united-atom, as described by Berger *et al.* (1997). To combine this force field with ffgmx no special adjustments are required, but there are different ways to treat the Lennard-Jones interactions between the lipid atoms and the protein atoms, which are described by different force fields. In much of the literature using this combination, following early simulations of alamethicin (Tieleman *et al.*, 1999a; Tieleman *et al.*, 1999b), the interaction parameters between protein and lipid were taken from a special table that treated interactions of the protein with the lipid as if the protein 'saw' atom types from the protein force field. Thus a CH_3 group in the protein, described by ffgmx, interacted with a CH_3 group in the lipid by ffgmx parameters, while CH_3 groups in the lipid interacted with other CH_3 groups in the lipid by Berger parameters.

A modification of the lipids themselves was required for the combination with the OPLS-AA force field, since OPLS-AA scales all 1–4 non bonded interactions by 0.5 (no scaling is applied in GROMOS). To avoid the scaling issue in a general way we refitted the dihedral potential terms for the lipid to include 1–4 non-bonded interactions of the lipids in the proper dihedral potential. This means 1–4 interactions are not included explicitly anymore, and are no longer dependent on the scaling factors used for 1–4 interactions by the force field that describes proteins (or nucleic acids, carbohydrates). The non-bonded interactions between the lipids and the protein in this case where based on combination rules, with no adjustments. The results are shown in Table I.

TABLE I

Area per Lipid in Simulations of a 72 Lipid DOPC Bilayer in the Presence of Hydrophobic Transmembrane Helical Peptides (Tieleman *et al.*, 2006). Areas Are in nm^2 with a Standard Error Based on Block Averaging on the Last 40 ns. GROMOS87 Simulations Used the SPC Water Model, OPLS-AA Simulations Used TIP4

System	GROMOS87 + Berger	OPLS-AA + Berger (with refitted dihedrals)
Pure DOPC	0.645 (0.005)	0.621 (0.009)
Leu22 (1×)	0.578 (0.002)	0.631 (0.003)
WALP23 monomer (1×)	0.568 (0.003)	0.635 (0.007)
WALP23 dimer (1×)	0.595 (0.003)	0.641 (0.004)

TABLE II

Partial Volume of WALP23 Peptide in Water and Octane, with Different Force Field Combinations (Tieleman *et al.*, 2006). Simulations Have Been Run at 298 K. Octane Topologies Parameters Have Been Created from the Atom Types Defining the –CH_2– and –CH_3 Groups in GROMOS87 Amino Acids and Berger's Lipids

Solvent parameter	Peptide parameter	$V_{solvent}$ (Å^3/molecule)	Density (g/L)[a]	$V_{peptide}$ (nm^3) (std. error)
SPC	GROMOS87	30.23	989.7	7.81 (0.02)
TIP3P	OPLS-AA	29.84	1002.5	9.15 (0.02)
TIP4P	OPLS-AA	29.74	1004.9	7.44 (0.03)
GROMOS87	GROMOS87	248.83	762.3	4.14 (0.03)
Berger	GROMOS87	294.44	644.3	3.01 (0.02)
Berger	OPLS-AA	294.44	644.3	3.71 (0.02)

[a]Density of water at 298 K = 997 g/L; density of the octane at 293 K = 702 g/L.

Table I shows that the area per lipid with the refitted dihedrals is smaller. We believe this is because the refitted dihedrals are almost exactly the same as the original dihedrals, but only for the equilibrium geometry of the lipids. As soon as angles fluctuate, the response of 1–4 interactions and a pure dihedral potential are different. This effect is surprisingly large. A second significant result is that the presence of a transmembrane helix causes the area of the box to shrink when the GROMOS87 force field is used, even although a hydrophobic helix was added to the same number of lipids. The OPLS-AA combination does not show this effect. Although we do not know the correct answer, an increase in area appears more reasonable.

This is confirmed by calculations of the partial volume of the peptide in water and octane (Table II). The WALP23 peptide should interact more strongly with octane than water because of its hydrophobicity. A reasonable decrease of its partial volume should be therefore observed in octane.

When GROMOS87 is combined with its own –CH_2 and –CH_3 atom types for the octane, the WALP23 peptide becomes completely unfolded after a few nanoseconds. This shows that the interaction of GROMOS87 amino acid parameters with themselves is strong, as expected. When GROMOS87 is combined with Berger's parameters using combination rules, WALP23 remains folded but its partial volume is significantly smaller than for the OPLS-AA + Berger combination. The peptide–lipid interaction in the GROMOS + Berger combination are thus not well balanced. Such results may have been partially expected since Berger's lipid contain a large number of parameters from a united-atom version of OPLS (Berger *et al.*, 1997).

IV. SAMPLING CONCERNS

A. Time Scales and Sampling in Simulations

If MD simulations are run sufficiently long, the system will explore all its relevant states with the appropriate weight. In practice, an exhaustive exploration of all possible states is not possible since the calculation time is limited by the computer hardware and straightforward MD is unable to cross high energetic barriers. Practical limits dictate an upper boundary of hundreds of ns for a typical membrane system at the moment. This is of the same order of magnitude as the time scale on which lipids switch place in a bilayer, *i.e.* short compared to large-scale equilibration of the bilayer, lipid mixtures, or lipid–peptide interactions. Although it is not possible to state how long a given simulation has to be to obtain good averaging, there are several studies in the literature that help provide a reasonable estimate for common cases. We outline a number of scenarios.

In an extensive set of comparative simulations of the same DPPC bilayer, Anézo et al. have estimated ca. 10–20 ns is required for small well-tested DPPC bilayers to equilibrate properly (Anézo *et al.*, 2003), while slow fluctuations in area or, for constant area simulations, pressure, require sampling during several times that period to obtain reasonable averages. If lateral diffusion is of interest, longer simulations are required (Wohlert and Edholm, 2006). Interestingly, the time scale required for a simulation depends on the size of the system too. Larger bilayers develop slow undulatory motions that are a feature of real bilayers as well, but are very difficult to sample. Ironically, the 10–20 ns simulations that give good averages for small DPPC bilayers do so by virtue of simulating a system that suppresses larger scale slow fluctuactions, essentially an artefact of a small system.

Several studies have examined the interaction of ions with lipid bilayers, but most of these do not show convincing equilibration of ions near the lipids. Two interesting papers by Bockmann et al. show that equilibration of sodium occurs on a tens of nanosecond time scale (Bockmann *et al.*, 2003), while binding of calcium requires equilibration times of over 100 ns (Bockmann and Grubmuller, 2004). Although not directly relevant for lipid–protein interactions, these sobering conclusions on simpler ions suggest extreme caution with the interpretation of simulations of *e.g.* peptide binding to a membrane on a shorter time scale.

Simulations of simple pentapeptides Ac-W-L-X-L-L in POPC bilayers illustrate this point (Aliste *et al.*, 2003; Aliste and Tieleman, 2005). These peptides are largely hydrophobic, and rapidly 'bind' to a POPC bilayer on a 10 ns simulation time scale, most of the time. However, the motions of these peptides are so slow that the distribution of different side chains does not accurately equilibrate on a 50 ns time scale. When two of the same peptides are present in one simulation, one bound to each side the membrane, the distribution of side chains on both sides is similar, especially at a coarse level like the center of mass of the peptide relative to the center of mass of the membrane. However, detailed distributions of individual side chains do not average accurately on a 50 ns time scale, which we think would be an essential requirement to calculate a hydrophobicity scale for the X residue using free energy simulations that can be compared to experiment (see below).

In simulations of integral membrane proteins the property of interest might not be directly associated with lipids, but in all cases lipids provide the environment for such proteins, and in several cases lipid and protein interact in a functionally important way (Lee, 2004). The G-protein coupled receptor rhodopsin is an excellent example of a protein that shows both interesting internal phenomena and important interactions with lipids. Based on an interesting study of rhodopsin at an unprecedented degree of sampling (Grossfield *et al.*, 2006), Grossfield et al., have shown that 26 independent simulations of rhodopsin run over 100 ns each were not sufficient to sample correctly the structure, especially the loops between transmembrane segments (Grossfield *et al.*, 2007), although averaging over all 26 simulations allowed good agreement with available experimental data.

B. Improving Sampling

Although complete sampling will remain impossible for most interesting systems, there are several shortcuts that may be helpful.

Straightforward molecular dynamics simulations may be improved by careful construction of starting structures that minimize the amount of equilibration time required. Starting structures in lipid–protein simulations are a key concern because, except in some published studies in which peptides bind to the membrane starting from solution (Mátyus *et al.*, 2007), they determine much of the outcome of a simulation. There are several algorithms for creating starting structures, which we recently reviewed (Kandt *et al.*, 2007). However, these algorithms minimize the time required for local equilibration of lipids around a peptide or protein; if a protein is inserted in a wrong orientation or at the wrong depth, they do not enable sufficient sampling to automatically correct the problem.

In practice, it is advisable to explore multiple starting structures, and multiple copies of the same starting structure with different initial velocities or solvent configurations. A key recent paper showed that even in extremely long (by current standards) simulations of rhodopsin, convergence in a single simulation is not observed, and a large number of runs is required to identify irrelevant fluctuations and compare to experiment (Grossfield *et al.*, 2007). A second interesting example are the simulations of pore-formation by magainin in a lipid bilayer by Leontiadou and Marrink (Leontiadou *et al.*, 2006).

A more sophisticated approach is to bias sampling towards interesting regions of phase space. There are many ways to achieve this. If a clear reaction path can be identified, umbrella sampling or a similar approach may be useful. This is commonly used in studies of ion permeation in ion channels (Allen *et al.*, 2006) or the distribution of small molecules in a lipid bilayer (MacCallum and Tieleman, 2006). Enhanced sampling algorithms like replica exchange molecular dynamics (Scheraga *et al.*, 2007) and conformational flooding (Bockmann and Grubmuller, 2002) may allow enhanced exploration of phase space, each with their own technical difficulties and limitations. Simulations of proteins may also be biased in particular directions. One recent example is a simulation of the vitamin B12 receptor BtuB, in which the authors applied force to a part of the protein that is thought to move under the influence of a second protein (Gumbart *et al.*, 2007). A second example are our simulations of the ABC transporter BtuCD, in which we derived general directions between domains from crystal structures in a related protein and used those to bias the motions of BtuCD (Sonne *et al.*, 2007b).

In some cases, it may also be possible to use coarse-grained simulations to explore large parts of conformational space. This is an active area of development, but examples include the coarse-grained MARTINI force field of Marrink *et al.* (Marrink *et al.*, 2007; Periole *et al.*, 2007) in which approximately 4 non-hydrogen atoms are replaced by a single interaction site. This force field allows simulations on a microsecond to millisecond time scale of lipids and proteins, at the expense of a loss of computational detail. A second example of coarse-graining is the approach by Voth *et al.* to combine different levels of resolution in simulations (Ayton and Voth, 2007). In principle, it is possible to move between atomistic and coarse-grained detailed. To study the interactions between lipids and antimicrobial peptides, for instance, one could start with coarse-grained simulations to equilibrate mixtures of lipids and peptides, and then translate the structures back into atomistic structures for further simulations to explore the details of interactions in the system.

V. VALIDATION OF LIPID–PROTEIN SIMULATIONS

Computer simulations are able to monitor the detailed motion of each molecule of a system, which is not possible by experiment. However, since they are based on models, they require validation before the results of simulations are useful. In principle, MD simulations are based on a transferrable description of interactions between different types of atoms, but as described above it is critical to make sure that interactions between protein and lipid, as well as lipid and water, are accurately described. This is surprisingly difficult, primarily because experiments that reach atomistic detail on membranes are relatively rare and because the connection between experiments and simulations is not trivial. We highlight several types of experimental data that are useful for testing lipid–protein simulations.

A. Free Energies of Transfer as Basis for Parameterization

Free energies of solvation in water and transfer between water and alkane are a useful experimental information to assess the accuracy of parameters describing the thermodynamics of interactions between water and the hydrocarbon interior of a bilayer. Free energies of transfer of small-molecule side-chain analogues have been accurately measured (Wolfenden, 2007) and can be calculated with a moderate amount of computational effort (MacCallum and Tieleman, 2003; Shirts and Pande, 2005; Villa and Mark, 2002). An example is shown in Table III, which compares water to cyclohexane transfer free energies for different force field modifications (Xu *et al.*, 2007). Tieleman's calculation shows the OPLS-AA + Berger combination is slightly more hydrophobic than the OPLS-AA set, at least for residues with a large aliphatic group. The difference between the 3 sets remains small relative to the magnitude of the errors.

TABLE III

Cyclohexane to Water Transfer Free Energies of the OPLS-AA Force Field (Tieleman *et al.*, 2006). All Energies Are in $kJ \, mol^{-1}$. Calculated Energies Are Shown as Averages with Standard Errors. The Error Columns Correspond to the Difference Between the Average Calculated Value and the Experimental Value. Experimental Value Have Been Determined by Radzicka and Wolfenden (1988). OPLS-AA Solvation Energies Are Taken from (MacCallum and Tieleman, 2003)

Residue analogue	Experiment	OPLS-AA		OPLS-AA + Berger lipids		OPLS-AA (modified) + Berger lipids	
		ΔG	Error	ΔG	Error	ΔG	Error
Asn	−27.7	−21.4 ± 1.0	6.3	−21.0 ± 0.9	6.7	−25.6 ± 0.6	2.1
Arg	−24.2	−23.6 ± 2.1	0.6	−19.8 ± 1.8	4.4		
Gln	−22.9	−13.0 ± 1.9	9.9	−19.4 ± 1.6	3.5	−20.2 ± 0.8	2.7
His	−18.7	−7.5 ± 1.3	11.2	−8.2 ± 1.2	10.5	−31.1 ± 1.2	−12.4
Asp	−18.6	−17.3 ± 1.0	1.3	−17.3 ± 0.9	1.3	−13.1 ± 0.6	5.5
Ser	−14.2	−14.4 ± 0.9	−0.2	−14.1 ± 0.8	0.1		
Glu	−13.0	−13.7 ± 1.1	−0.7	−13.2 ± 0.8	−0.2	−8.9 ± 0.7	4.1
Thr	−11.1	−12.1 ± 1.1	−1.0	−9.8 ± 0.9	1.3		
Lys	−1.6	5.2 ± 1.5	6.8	10.0 ± 1.3	11.6		
Tyr	−0.8	4.8 ± 1.7	5.7	5.8 ± 1.4	6.6	−0.6 ± 1.6	0.2
Cys	5.2	8.1 ± 1.0	3.0	7.6 ± 0.9	2.4		
Ala	7.7	8.4 ± 0.7	0.7	10.4 ± 0.6	2.7		
Trp	9.5	15.5 ± 1.8	6.0	14.5 ± 1.7	5	4.3 ± 1.2	−5.2
Met	9.7	9.1 ± 2.7	−0.6	10.3 ± 2.1	0.6		
Phe	14.1	18.8 ± 1.8	4.7	19.8 ± 1.6	5.7		
Val	16.7	17.8 ± 1.4	1.1	20.6 ± 0.9	3.9		
Ile	20.2	20.7 ± 1.4	.5	26.4 ± 1.1	6.2		
Leu	20.5	22.8 ± 1.4	2.4	26.2 ± 1.2	5.7		
Average error			3.2		4.3		2.4[a]

[a] Average error includes error from OPLS-AA residues that were not modified.

A drawback of transfer free energies is that bilayers are quite different from cyclohexane. Although it is possible to calculate potentials of mean force of the side-chain analogues as a function of depth in the bilayer (MacCallum *et al.*, 2007), there is no experimental information to validate these PMFs.

Wimley and White have measured 'hydrophobicity' scales, using the host-guest peptides AcWLXLL (with X any of the 20 natural amino acids) in water/octanol and in POPC vesicles. Reproducing these scales under the same conditions as in the experiments would be an excellent test of force fields. Aliste and Tieleman simulated the peptides in a lipid bilayer and some in octanol (Aliste *et al.*, 2003; Aliste and Tieleman, 2005), but on a 50 ns time scale equilibration in a bilayer is insufficient to expect much accuracy in free energy perturbation simulations aimed at reproducing the POPC scale. We will address this again in the near future.

B. Structural and Dynamic Information

Powerful biophysical methods to study lipid–protein interactions are X-ray and neutron diffraction and NMR, EPR, IR, and fluorescence spectroscopy (*e.g.* in (Bechinger *et al.*, 2004; Benz *et al.*, 2005; Klauda *et al.*, 2006; Vigh *et al.*, 2005)). Such experiments, particularly diffraction and NMR, play key roles in parameterizing and validating lipid force fields, but the interpretation of experiments on systems containing peptides is typically more complicated, and often provides information about a protein or peptide that is not directly useful in validating simulations. In many cases, experimental information is relatively global, such as in EPR data on binding of a protein to a membrane. While this is useful information, it is not normally critical enough to distinguish between force fields. Other important information is available from relaxation measurements using NMR (Pastor *et al.*, 2002) and NMR measurements on model peptides in model bilayers (Killian and Nyholm, 2006). As simulations begin to approach time scales where relaxation properties that can be measured by NMR can be calculated accurately, and the interpretation of NMR experiments becomes clearer, a closer connection between NMR and simulation may be a promising avenue for validating simulations.

VI. CONCLUSION AND PERSPECTIVE

Despite major progress in force fields, sampling and computer power, simulating membrane proteins and lipid–protein interactions remains challenging. Although there are several viable parameter sets to study such systems, they all have technical drawbacks that require addressing. Lipid parameters themselves have proven difficult to develop, as lipid bilayer properties critically depend on a perfect balance between lipid/lipid and water/lipid interactions. We discussed several technical issues related to combining different parameter sets, but feel this will continue to be an important method until substantial and successful further development of new parameters. In addition to concerns about parameters, the long relaxation times involved in any lipid system cause serious sampling problems. Although simulations of the order of 100–250 ns may seem 'long' compared to current literature standards, they are in many cases barely adequate for the problems that are currently being studied. The use of sampling-enhancing methods, even as simple as running multiple copies of a simulation, is highly recommended for many problems. Lastly, we emphasized the lack of experimental data that may be used for a critical validation of simulations of mixed lipid/protein systems. Thermodynamic data on small molecules is a powerful tool, but not a direct link to real peptides and proteins, and additional data will be necessary.

Acknowledgements

This work is supported by the Natural Sciences and Engineering Research Council (NSERC). DPT is an Alberta Heritage Foundation for Medical Research Senior Scholar and Canadian Institute of Health Research New Investigator. We thank Dr. L. Monticelli for his comments.

References

Aliste, M.P., MacCallum, J.L., Tieleman, D.P. (2003). *Biochemistry* **42**, 8976–8987.
Aliste, M.P., Tieleman, D.P. (2005). *BMC Biochem.* **6**, 30.
Allen, T.W., Andersen, O.S., Roux, B. (2006). *Biophys. Chem.* **124**, 251–267.
Anézo, C., de Vries, A.H., Höltje, H.D., Tieleman, D.P., Marrink, S.J. (2003). *J. Phys. Chem. B* **107**, 9424–9433.
Ash, W.L., Zlomislic, M.R., Oloo, E.O., Tieleman, D.P. (2004). *Biochim. Biophys. Acta (BBA) Biomembranes* **1666**, 158–189.
Ayton, G.S., Voth, G.A. (2007). *J. Struct. Biol.* **157**, 570–578.
Bandyopadhyay, S., Tarek, M., Klein, M.L. (1999). *J. Phys. Chem. B* **103**, 10075–10080.
Bechinger, B., Aisenbrey, C., Bertani, P. (2004). *Biochim. Biophys. Acta (BBA) Biomembranes* **1666**, 190.
Benz, R.W., Castro-Roman, F., Tobias, D.J., White, S.H. (2005). *Biophys. J.* **88**, 805–817.
Berendsen, H.J.C., Postma, J.P.M., van Gunsteren, W.F., Hermans, J. (1981). Interaction models for water in relation to protein hydration. In: Pullman, B. (Ed.), "Intermolecular Forces". Reidel, Dordrecht, pp. 331–342.
Berger, O., Edholm, O., Jahnig, F. (1997). *Biophys. J.* **72**, 2002–2013.
Betanzos, M., Chiang, C.S., Guy, H.R., Sukharev, S. (2002). *Nat. Struct. Mol. Biol.* **9**, 704–710.
Bockmann, R.A., Grubmuller, H. (2002). *Nat. Struct. Mol. Biol.* **9**, 198–202.
Bockmann, R.A., Grubmuller, H. (2004). *Angew. Chem. Int. Ed. Engl.* **43**, 1021–1024.
Bockmann, R.A., Hac, A., Heimburg, T., Grubmuller, H. (2003). *Biophys. J.* **85**, 1647–1655.
Case, D.A., Cheatham, T.E., Darden, T., Gohlke, H., Luo, R., Merz, K.M., Onufriev, A., Simmerling, C., Wang, B., Woods, R.J. (2005). *J. Comput. Chem.* **26**, 1668–1688.
Chandrasekhar, I., Kastenholz, M., Lins, R.D., Oostenbrink, C., Schuler, L.D., Tieleman, D.P., van Gunsteren, W.F. (2003). *Eur. Biophys. J.* **32**, 67–77.
Chiu, S.W., Clark, M.M., Jakobsson, E., Subramaniam, S., Scott, H.L. (1999). *J. Phys. Chem. B* **103**, 6323–6327.
Forrest, L.R., Sansom, M.S. (2000). *Curr. Opin. Struct. Biol.* **10**, 174–181.
Grossfield, A., Feller, S.E., Pitman, M.C. (2006). *Proc. Natl. Acad. Sci.* **103**, 4888–4893.
Grossfield, A., Feller, S.E., Pitman, M.C. (2007). *Proteins Struct. Funct. Bioinform.* **67**, 31–40.

Gumbart, J., Wang, Y., Aksimentiev, A., Tajkhorshid, E., Schulten, K. (2005). *Curr. Opin. Struct. Biol.* **15**, 423–431.

Gumbart, J., Wiener, M.C., Tajkhorshid, E. (2007). *Biophys, J.* **93**, 496–504.

Hess, B., van der Vegt, N.F.A. (2006). *J. Phys. Chem. B* **110**, 17616–17626.

Jakobsson, E. (1997). *Trends Biochem. Sci.* **22**, 339–344.

Jorgensen, W.L., Chandrasekhar, J., Madura, J.D., Impey, R.W., Klein, M.L. (1983). *J. Chem. Phys.* **79**, 926–935.

Jorgensen, W.L., Maxwell, D.S., Tirado-Rives, J. (1996). *J. Am. Chem. Soc.* **118**, 11225–11236.

Kandt, C., Ash, W.L., Tieleman, D.P. (2007). *Methods* **41**, 475–488.

Killian, J.A., Nyholm, T.K. (2006). *Curr. Opin. Struct. Biol.* **16**, 473–479.

Klauda, J.B., Kucerka, N., Brooks, B.R., Pastor, R.W., Nagle, J.F. (2006). *Biophys. J.* **90**, 2796–2807.

Lee, A.G. (2004). *Biochim. Biophys. Acta* **1666**, 62–87.

Lennard-Jones, J.E. (1924). *Proc. R. Soc. London A* **106**, 463–477.

Leontiadou, H., Mark, A.E., Marrink, S.J. (2006). *J. Am. Chem. Soc.* **128**, 12156–12161.

MacCallum, J.L., Bennett, W.F., Tieleman, D.P. (2007). *J. Gen. Physiol.* **129**, 371–377.

MacCallum, J.L., Tieleman, D.P. (2003). *J. Comput. Chem.* **24**, 1930–1935.

MacCallum, J.L., Tieleman, D.P. (2006). *J. Am. Chem. Soc.* **128**, 125–130.

MacKerell Jr., A.D., Bashford, D., Bellott, D., Dunbrack Jr., R.L., Evanseck, J.D., Field, M.J., Fischer, S., Gao, J., Guo, H., Ha, S., Joseph-McCarthy, D., Kuchnir, L., Kuczera, K., Lau, F.T.K., Mattos, C., Michnick, S., Ngo, T., Nguyen, D.T., Prodhom, B., Reiher III, W.E., Roux, B., Schlenkrich, M., Smith, J.C., Stote, R., Straub, J., Watanabe, M., Wiorkiewicz-Kuczera, J., Yin, D., Karplus, M. (1998). *J. Phys. Chem. B* **102**, 3586–3616.

MacKerell Jr., A.D., Feig, M., Brooks III, C.L. (2004). *J. Am. Chem. Soc.* **126**, 698–699.

Marrink, S.J., Risselada, H.J., Yefimov, S., Tieleman, D.P., de Vries, A.H. (2007). *J. Phys. Chem. B* **111**, 7812–7824.

Mátyus, E., Kandt, C., Tieleman, D.P. (2007). *Curr. Med. Chem.* **14**, 2789–2798.

Pastor, R.W., Venable, R.M., Feller, S.E. (2002). *Acc. Chem. Res.* **35**, 438–446.

Periole, X., Huber, T., Marrink, S.J., Sakmar, T.P. (2007). *J. Am. Chem. Soc.* **129**, 10126–10132.

Pertsin, A., Platonov, D., Grunze, M. (2007). *Langmuir* **23**, 1388–1393.

Ponder, J.W., Case, D.A. (2003). *Protein Simul.* **66**, 27–85.

Radzicka, A., Wolfenden, R. (1988). *Biochemistry* **27**, 1664–1670.

Ryckaert, J.P., Bellemans, A. (1978). *Faraday Discuss. Chem. Soc.* **66**, 95–106.

Scheraga, H.A., Khalili, M., Liwo, A. (2007). *Ann. Rev. Phys. Chem.* **58**, 57–83.

Scott, W.R.P., Hünenberger, P.H., Tironi, I.G., Mark, A.E., Billeter, S.R., Fennen, J., Torda, A.E., Huber, T., Krüger, P., van Gunsteren, W.F. (1999). *J. Phys. Chem. A* **103**, 3596–3607.

Shirts, M.R., Pande, V.S. (2005). *J. Chem. Phys.* **122**, 144107.

Shirts, M.R., Pitera, J.W., Swope, W.C., Pande, V.S. (2003). *J. Chem. Phys.* **119**, 5740, doi:10.1063/11587119.

Sonne, J., Jensen, M.O., Hansen, F.Y., Hemmingsen, L., Peters, G.H. (2007a). *Biophys. J.* **92**, 4157–4167.

Sonne, J., Kandt, C., Peters, G.H., Hansen, F.Y., Jensen, M.O., Tieleman, D.P. (2007b). *Biophys. J.* **92**, 2727–2734.

Tieleman, D.P., Berendsen, H.J., Sansom, M.S. (1999a). *Biophys. J.* **76**, 1757–1769.

Tieleman, D.P., Hess, B., Sansom, M.S.P. (2002). *Biophys. J.* **83**, 2393–2407.

Tieleman, D.P., MacCallum, J.L., Ash, W.L., Kandt, C., Xu, Z., Monticelli, L. (2006). *J. Phys. Condens. Matter* **18**, S1221–S1234.

Tieleman, D.P., Marrink, S.J., Berendsen, H.J. (1997). *Biochim. Biophys. Acta* **1331**, 235–270.

Tieleman, D.P., Sansom, M.S., Berendsen, H.J. (1999b). *Biophys. J.* **76**, 40–49.

van Gunsteren, W.F., Billeter, S.R., Eising, A.A., Hünenberger, P.H., Krüger, P., Mark, A.E., Scott, W.R.P., Tironi, I.G. (1996). "Biomolecular Simulation: The GROMOS96 Manual and User Guide". Vdf Hochschulverlag AG an der ETH Zürich, Zürich.

Vigh, L., Escriba, P.V., Sonnleitner, A., Sonnleitner, M., Piotto, S., Maresca, B., Horvath, I., Harwood, J.L. (2005). *Prog. Lipid Res.* **44**, 303–344.

Villa, A., Mark, A.E. (2002). *J. Comput. Chem.* **23**, 548–553.

Wohlert, J., Edholm, O. (2006). *J. Chem. Phys.* **125**, 204703.

Wolfenden, R. (2007). *J. Gen. Physiol.* **129**, 357–362.

Xu, Z.T., Luo, H.H., Tieleman, D.P. (2007). *J. Comput. Chem.* **28**, 689–697.

CHAPTER 5

Implicit Modeling of Membranes

Alan Grossfield *,†

*IBM T. J. Watson Research Center, 1101 Kitchawan Road, Yorktown Heights, NY 10598, USA
†Present address: University of Rochester Medical Center, Department of Biochemistry and Biophysics, 601 Elmwood Ave, Box 712, Rochester, NY 14642, USA

I. Introduction
II. Classes of Models
   A. Solute-Focused
   B. Membrane-Focused
III. Interesting Problems in Implicit Membrane Modeling
   A. Antimicrobial Peptides
   B. Protein–Protein Association
IV. Conclusion
   References

I. INTRODUCTION

Biological lipid membranes are central to many biological processes. They form selectively permeable barriers, allowing cells to control their contents and create concentration gradients. However, in contrast to the totally passive view espoused by standard undergraduate texts, biological membranes also actively modify cell behavior by altering the function of membrane proteins, modulating the stability of protein–protein associations, and altering the binding and distribution of small molecules including salts and osmolytes (Jensen and Mouritsen, 2004; Brown, 1994; White and Wimley, 1999; Epand, 1998, 2003; Mouritsen and Bloom, 1993; Nyholm et al., 2007).

Membrane composition varies significantly in different tissues within a given organism, emphasizing that the distribution of specific lipid species is not a matter of simple abundance. For example, rod outer segment disk membranes, found in the mammalian visual system, contain roughly 50% polyunsaturated ω-3 fatty acids; this is an enormous enrichment considering their natural abundance is more like 5% (Boesze-Battaglia and Albert, 1989; Boesze-Battaglia et al., 1989). Since

Current Topics in Membranes, Volume 60
1063-5823/08 $35.00
DOI: 10.1016/S1063-5823(08)00005-7

humans are unable to synthesize ω-3s, this implies the body must be specifically trafficking them to the disk membranes. Polyunsaturated lipids have been shown to significantly enhance the function of rhodopsin (which constitutes more than 90% of the protein in the disk membranes) (Mitchell *et al.*, 2001; Niu *et al.*, 2001; Grossfield *et al.*, 2006), clearly demonstrating that the cell is manipulating membrane content in order to optimize function. The story gets even more interesting when the concentration of cholesterol, which inhibits rhodopsin function, is considered: the concentration is high (25–30 mol%) in immature disks found near the bottom of the rod outer segment stack, and is greatly reduced (5 mol%) in mature disks (Mitchell *et al.*, 2001; Niu *et al.*, 2001, 2002; Pitman *et al.*, 2005).

The last 20 years have seen a significant increase in attempts to model membranes, and in particular to model their interactions with proteins and other permeants. Conceptually, the simplest approach is to perform all-atom molecular dynamics simulations; to our knowledge, the first example of a protein modeled in an explicit lipid membrane was gramicidin in a DMPC bilayer, conducted by Woolf and Roux (1994). Recent improvements in computer speed have significantly changed the landscape of this field; where simulations were once run for 100s of picoseconds, present technology allows us to run explicit membrane systems for hundreds of nanoseconds (Grossfield *et al.*, 2006) or even microseconds (Martínez-Mayorga *et al.*, 2006; Grossfield *et al.*, 2007). However, in order to reach these timescales, the calculations require extremely powerful supercomputers (Allen *et al.*, 2001), and even then, running long enough to generate statistical convergence is difficult (Grossfield *et al.*, 2007; Faráldo-Gomez *et al.*, 2004).

As a result, many interesting calculations can be expected to remain out of reach of all-atom molecular dynamics for the foreseeable future. These include simulations of membrane protein folding and insertion, dimerization (or oligomerization) of integral membrane proteins, and membrane poration by antimicrobial peptides. In each case, the time- and length-scales involved would require prohibitively long simulations. In principle, one could trade temporal information for improved convergence by using enhanced sampling tools such as replica exchange dynamics (Okamoto, 2004). However, while some such calculations have been reported for membranes (Nymeyer *et al.*, 2005), this technique is difficult to apply to bilayers because of the higher temperatures tend to disrupt bilayer structure.

One approach which has garnered significant interest in recent years is the use of coarse-grained molecular models, where the number of atoms per molecular is strategically reduced and the interaction potential simplified, dramatically diminishing the computational cost of the calculations and concomitantly increasing the feasible simulation size and time. However, these methods will be discussed extensively elsewhere in this volume, and so we will not explore them here.

An alternative approach is to combine a continuum representation of the membrane with a atomic representations of the rest of the system. This scenario allows us to focus our computational effort on the portion of we are most interested in, for example the membrane protein whose folding we wish to explore. Representing the membrane implicitly has a number of significant advantages, the most important of which is computational efficiency: as a rule, implicit models are dramatically less expensive per energy evaluation than the equivalent all-atom systems. Moreover, these models generally produce an approximation to the solvation free energy of the system, as opposed to simply the potential energy as computed in an all-atom system. As a result, no additional sampling of environmental degrees of freedom is required for a fixed solute structure. This savings is significant when one considers the nanosecond to microsecond relaxation and reorganization times of explicit lipid membranes. Finally, the absence of explicit "solvent" molecules simplifies conformational searching and enhances the power of sampling techniques like replica exchange.

The primary tradeoff in using implicit membrane models is their presumed lack of high-resolution accuracy; almost by definition, replacing all-atom models with analytic formulas sacrifices a certain level of detail. Thus, the key question becomes: what physical characteristics of the membrane must be reproduced in our membrane models to yield physically correct behavior of membrane permeants? The answer, of course, depends on the details of the system being considered, and the scientific questions asked. A model could simultaneously be well suited for some circumstances and wholly inadequate in others.

II. CLASSES OF MODELS

Lipid membranes self-assemble in an attempt to isolate hydrophobic acyl chains from the surrounding aqueous environment. Under appropriate conditions, this leads to the formation of stable bilayers, with a hydrophobic core, an interfacial region containing a mixture of polar headgroups and water, and the surrounding aqueous medium. Despite this apparent simplicity, biological membranes are capable of remarkable diversity of structure and dynamics. Although many models focus on a single property—the thickness of the hydrophobic core—real membranes have a broad range of physical characteristics which vary with lipid composition. These include structural quantities such as surface area per lipid, chain order, intrinsic curvature, and pressure profile, as well as dynamic properties such as the dielectric profile, diffusion coefficients, and chain and headgroup reorientation relaxation times.

Moreover, these properties are not independent of each other. For example, changing the hydration levels of model membranes modulates the surface area per lipid, and thus the chain order parameters. However, lipid surface area can also be controlled by varying the headgroup type, the length and degree of unsaturation of the chains, and the presence of other membrane permeants, such as cholesterol. The surface area per lipid in turn affects the magnitude of headgroup–headgroup interactions and, particularly in the case of charged headgroups, the distribution of ions near the membrane surface. All of these properties can in principle affect the binding, conformation, stability, and oligomerization of peptides and proteins.

Thus, one critical question is, which of these membrane properties must be included to generate a successful implicit membrane model? Unfortunately, there is no simple answer to this question, as the answer depends largely on the model's intended use.

For example, if one plans to model a membrane protein using an all-atom representation, then effectively including the electrostatic effects of the membrane is probably paramount. If instead one wishes to use a simpler rigid-cylinder model for an embedded alpha helix then other issues become more important. As a direct result of this diversity, many different approaches to implicit modeling have emerged in the literature. For purposes of discussion, we have divided them into two broad classes: *Solute-focused* and *Membrane structure-focused*. Obviously, these labels represent something of a simplification, but we think this classification is on the whole helpful, in that it provides a context for understanding recent work.

A. Solute-Focused

This section will describe models best characterized as "solute-focused". By this, we mean models where the solute—typically a protein, peptide, or small molecule—is considered with atomic or near-atomic resolution, and the membrane is largely a backdrop intended to provide an appropriate venue. As a rule, these models neglect details of membrane structure other than the thickness of the hydrophobic low-dielectric core, and contain few provisions to account for the solute's disruption of membrane structure. Instead, these methods tend to focus on the membrane–solute interactions and the ways in which the membrane modifies solute–solute interactions, especially via electrostatics.

Although membrane electrostatics will be discussed in a separate chapter of this book, one cannot adequately introduce this class of implicit membrane models without first reviewing electrostatic and dielectric theory as applied to membranes. Indeed, one of the most pervasive concepts in membrane modeling is the notion of a low dielectric slab embedded between semi-inifinite regions of high dielectric. For this reason, we begin by considering the simplest circumstance, a spherical charge in an infinite uniform dielectric. The charging free energy for a charge q in a sphere of radius a embedded in a region of dielectric ϵ can be computed by integrating the electric field over all volume outside the sphere

$$\Delta G = \frac{1}{8\pi} \int \epsilon E^2 \, dV = \frac{1}{8\pi} \int_a^\infty \epsilon \left(\frac{q}{\epsilon r^2}\right)^2 4\pi r^2 \, dr = -\frac{q^2}{2\epsilon a} \tag{1}$$

generating the familiar Born equation (Born, 1920). This derivation can be generalized for the case of a permanent (Bell, 1931) or polarizable (Bonner, 1951) dipole (the so-called Onsager equation), or for an arbitrary charge distribution (Kirkwood, 1939).

However, the situation becomes somewhat more complex when the environment itself becomes heterogeneous, as is the case in a membrane. The electrostatic field due to a point charge approaching the barrier between two semi-infinite dielectric slabs is easily computed using the method of images (Jackson, 1962), and from this one can compute the charging energy. However, this solution contains an unphysical divergence as the charge approaches the dielectric interface due to the point charge approximation. This divergence, which appears repeatedly in the development of membrane models, was resolved by Ulstrup and coworkers by converting the volume integrals into surface integrals and directly accounting for the intersection between the ion surface and dielectric interface (Kharkats and Ulstrup, 1991). The result for an ion a distance $h > a$ from the interface is

$$\Delta G = -\frac{q^2}{8\epsilon_1 a}\left\{4 + \left(\frac{\epsilon_1 - \epsilon_2}{\epsilon_1 + \epsilon_2}\right)\left(\frac{2}{h/a}\right)\right.$$
$$\left. + \left(\frac{\epsilon_1 - \epsilon_2}{\epsilon_1 + \epsilon_2}\right)^2\left[\frac{2}{1 - (h/a)^2} + \frac{1}{2h/a}\ln\frac{(2h/a) + 1}{(2h/a) - 1}\right]\right\} \qquad (2)$$

while the result for $0 \leqslant h \leqslant a$, where the ion overlaps the dielectric interface, is

$$\Delta G = -\frac{q^2}{8\epsilon_1 a}\left\{\left(2 + \frac{2h}{a}\right) + \left(\frac{\epsilon_1 - \epsilon_2}{\epsilon_1 + \epsilon_2}\right)\left(4 - \frac{2h}{a}\right)\right.$$
$$\left. + \left(\frac{\epsilon_1 - \epsilon_2}{\epsilon_1 + \epsilon_2}\right)^2\left[\frac{(1 + h/a)(1 - h/a)}{1 + 2h/a} + \frac{1}{2h/a}\ln(1 + 2h/a)\right]\right\}$$
$$+ \frac{q^2}{4\epsilon_2 a}\left(\frac{2\epsilon_2}{\epsilon_1 + \epsilon_2}\right)\left(1 - \frac{h}{a}\right). \qquad (3)$$

The same authors also derived analytic solutions for the charging free energy for a spherical charge in the presence of a slab of low dielectric surrounded by semi-infinite regions of higher dielectric, albeit without the finite ion size corrections (Iversen *et al.*, 1998). However, this solution involves several infinite sums, and is thus cumbersome to implement computationally, although Flewelling and Hubbell devised an efficient approximate solution (Flewelling and Hubbell, 1986). Krishtalik took this approach one step further, deriving an analytical solution for the case where there are 5 distinct dielectric regions (2 high-dielectric water regions, 2 moderate-dielectric interfacial regions, 1 low dielectric core) (Krishtalik, 1996). Previously, Parsegian published analytic solutions to several simple problems related to ion permeation through membranes (Parsegian, 1969).

Since analytic approaches are only readily applicable to simple geometries such as spheres, numerical methods are necessary in order to treat more biologically-relevant systems. The most obvious approach is to numerically solve the Poisson equation (or, if salt effects are to be included, the Poisson–Boltzmann equation)

(Schnitzer and Lambrakis, 1991; Sharp and Honig, 1990; Murray et al., 1997; Lin et al., 2002)

$$\nabla \cdot \epsilon(\vec{r}) \nabla \Phi(\vec{r}) = -4\pi \rho(\vec{r}), \tag{4}$$

where Φ is the electrostatic potential, $\epsilon(\vec{r})$ is the position-dependent dielectric constant, and ρ is the charge density (typically represented by a finite number of point charges q_i). Once this equation has been solved, the electrostatic free energy can be computed as

$$\Delta G_{\text{elec}} = \frac{1}{2} \int \rho(\vec{r}) \Phi(\vec{r}) \, dV = \frac{1}{2} \sum_i^{\text{charges}} q_i \Phi_i, \tag{5}$$

where Φ_i is the electrostatic potential at the location of the ith charge. There is an extensive literature on the application of this formalism to biomolecular problems, so instead of reviewing it here we will simply suggest readers consult the recent review by Baker and the references cited there (Baker, 2005).

Several groups have directly applied the Poisson–Boltzmann approach to membrane-protein association thermodynamics. For example, Ben-Tal et al. used it to examine the thermodynamics of α-helix insertion, representing the membrane as a simple low dielectric slab (Ben-Tal et al., 1996). Murray and coworkers explicitly included the lipid headgroups in their calculations (Murray et al., 1997); this was particularly important in later work examining the association of basic peptides with anionic lipid bilayers (Murray et al., 1999).

Computing the electrostatic solvation free energy via the Poisson equation has a number of distinct advantages: it is directly seated in electrostatic theory, and within the limits of the dielectric assumption and numerical accuracy it is correct. However, there are a number of drawbacks. Historically, the finite-difference approaches typically used are relatively expensive, and computing forces sufficiently accurate for use in molecular dynamics was difficult. Recently, progress has been made in some of these areas (Lu et al., 2005a, 2005b; Feig et al., 2004), but for many applications rigorous Poisson electrostatics are still prohibitively expensive. Moreover, unless headgroups and some waters are explicitly included, this approach does not reproduce the correct sign of the electrostatic potential at the center of the lipid bilayer, which is thought to be crucial in the thermodynamics of many membrane permeants (Lin et al., 2002).

As a result, significant effort has been invested in developing faster, if more approximate, methods for computing electrostatic energies in dielectric media. Many of the most commonly used methods are variants of the generalized Born approach originally developed by Still and coworkers (Still et al., 1990). Although these developments have been the subject of several recent reviews (Bashford and Case, 2000; Feig et al., 2004), the underlying techniques and assumptions become relevant when the formalism is expanded to cover membranes, so we will discuss it here as well.

Most generalized Born methods are built around the empirical solvation free energy expression suggested by Still et al. (Still *et al.*, 1990)

$$\Delta G = -\left(1 - \frac{1}{\epsilon}\right) \sum_{i,j}^{N_{\text{atoms}}} \frac{q_i q_j}{\sqrt{r_{ij}^2 + \alpha_i \alpha_j \exp(-r_{ij}^2/4\alpha_i \alpha_j)}}, \tag{6}$$

where q_i is the partial charge on the ith atom, the sum is over all atom pairs (including self-interaction) and α_i is the ith generalized Born radius. The radii are constructed by computing the electrostatic free energy to solvate each charge individually in the protein, and then plugging that free energy into the Born equation (Eq. (1)) to extract an effective radius. The free energy is computed assuming the protein is a region of low dielectric, usually 1, and only the atom under consideration is charged. As a rule, the Coulomb field approximation is invoked to simplify the calculation; that is, the electric field due to the charge is presumed to be undistorted by the surrounding dielectric boundary. The result is volume integral over all space excluding the atom itself. This can be converted to a difference between two volume integrals, one over all space outside the atom, and the other only over the volume outside the atom but inside the protein

$$\frac{1}{\alpha_i} = \frac{1}{R_i} - \frac{1}{4\pi} \int_{\text{solute},r>R_i} \frac{1}{r^4} \, dV. \tag{7}$$

Although Eq. (6) appears to contain only pairwise interactions, many-body effects are included implicitly via the volume integral in Eq. (7).

The key to the effectiveness of the generalized Born method is the calculation of the effective Born radii; recent work has shown that if "perfect" radii are used—the electrostatic free energy of each atom is computed numerically using a standard Poisson solver—then Eq. (6) does an excellent job reproducing the molecular solvation energies computed using the Poisson equation (Lee *et al.*, 2002), which in turn does a good job reproducing the electrostatic portion of the solute–solvent interaction from explicit solvent simulations (Wagoner and Baker, 2004). The original Still formulation used a numerical integration over the protein volume, which was expensive and was ill-suited to computing forces suitable for molecular dynamics calculations. Many groups developed better approaches to performing this integral, including pairwise approximations (Bashford and Case, 2000; Feig *et al.*, 2004), various numerical schemes (Srinivasan *et al.*, 1999; Lee *et al.*, 2002, 2003; Grycuk, 2003; Tjong and Zhou, 2007), and a reformulation as a surface integral (Ghosh *et al.*, 1998; Gallicchio *et al.*, 2002). Some groups also added corrections intended to improve on the Coulomb field assumption (Lee *et al.*, 2003; Grycuk, 2003; Tjong and Zhou, 2007)

The situation becomes significantly more complex when one considers a membrane environment, because in that instance energies and forces depend not only on the relative position of the atoms but on their absolute location in the membrane. To our knowledge, the first membrane model explicitly based on the generalized Born formulation was due to Spassov et al. (Spassov *et al.*, 2002). Their approach, which they call GB/IM, includes the heterogeneous dielectric environment by considering the membrane interior to have the same dielectric as the protein, with the result that the volume integral over the embedded portion is replaced by an integral over the whole of the membrane interior, which is approximated by an analytic function fit to Poisson–Boltzmann results. The remaining protein volume is integrated using the efficient pairwise method of Dominy and Brooks (Dominy and Brooks, 1999). They applied their methodology rigid structures of bacteriorhodopsin and rhodopsin, and performed a short dynamics simulation of the influenza fusion peptide bound to the membrane.

Im *et al.* (2003a) took a related approach. They extended previous work from the Brooks group (Im *et al.*, 2003b), where numeric behavior of the volume integration was improve by use of a smoothing function, and, like Spassov *et al.* (2002), considered the membrane to be part of the protein interior. Their approach contained analytic corrections to the Coulomb field assumption originally designed for soluble proteins. They validated their results by examining the behavior of several membrane-binding peptides, including melittin, M2-TMP, and the glycophorin A dimer, comparing against Poisson–Boltzmann calculations and experimental structural information. The same model was later used to explore the folding and insertion of several designed helical transmembrane peptides (Im and Brooks, 2005).

Although both of these methods appear to perform well in practice, the assumption that the membrane and protein have the same dielectric is troubling. Because the protein charges are explicitly represented, one should represent the continuum dielectric inside the protein as 1, as is done for simulations where both solute and solvent are explicitly represented; the force field parameters are chosen with this application in mind. By contrast, the membrane interior has a dielectric of 2–4.

Feig and coworkers introduced a formalism to explicitly handle multiple dielectric environments (Feig *et al.*, 2004), and later applied it membrane modeling (Tanizaki and Feig, 2005, 2006). This model, called the heterogeneous dielectric generalized Born or HDGB, contains several notable technical advancements. First, the membrane representation is improved: the chemical heterogeneity of the membrane–water interface (Jacobs and White, 1989) is explicitly included in the calculation by modeling the membrane as a series of dielectric slabs, rather than just two regions. They computed the free energy profile for a test charge in this model using the Poisson–Boltzmann equation and used the results to spline-fit an effective dielectric constant profile to be used in the simulations. While the physical meaning of a bulk quantity like the dielectric varying smoothly on atomic lengthscales is unclear, the result is a formalism which accurately recapitulates a more realistic model for membrane electrostatics. Moreover, this method is built on top of a rigorous volume integration scheme (Lee *et al.*, 2003), and has been very carefully parameterized and characterized.

However, this careful characterization revealed some unfortunate complications when applying this model to larger molecules, such as a bacteriorhodopsin monomer or trimer. Tanizaki and Feig found that the results were very sensitive to the long range electrostatics cutoff, oscillating over a range of hundreds of kcal/mol (Tanizaki and Feig, 2006); this can result in almost comic failures, where setting the electrostatic cutoff at a seemingly reasonable 16 Å causes the bacteriorhodopsin monomer to be most stable in a horizontal orientation, with the helices lying in the plane of the membrane and the loops embedded in the membrane core. These effects go away with a sufficiently long cutoff, in the range of 36–38 Å, but the result is a dramatic increase in the computational cost. Although this problem has not been reported with the other methods discussed here, it seems likely that the underlying mechanisms will be present in all of them.

To this point, we have focused entirely on the electrostatic components of these models. This is of course incomplete; all of these models, whether intended for bulk solvent or specific to membrane modeling, contain at least one additional term representing non-electrostatic effects. Most follow the traditional approach from bulk solvent modeling and assume that these interactions can be related to the solvent accessible surface area. This approximation makes intuitive sense, and has some theoretical basis in the scaled-particle theory of hard sphere solvation (Pierotti, 1976). However, several groups have argued that nonpolar solvation, which includes terms from cavitation, hydrophobic effects, and favorable van der Waal's interactions, requires a somewhat more subtle treatment (Gallicchio *et al.*, 2002; Levy *et al.*, 2003). Wagoner and Baker showed that a significant fraction of the error in continuum methods, when compared to explicit solvent calculations, was due to the treatment of the non-electrostatic components, and that including terms to explicitly account for volume effects and attractive solute–solvent interactions greatly improved the situation (Wagoner and Baker, 2006).

The situation is—at least in principle—far more complex in the context of a lipid bilayer. Lipid acyl chains are far larger than water molecules, and unlike bulk solvent, the chains have a net orientation. As a result, cavitation effects should arguably have some additional shape and location dependence. Furthermore, the largely anhydrous environment means that hydrophobic interactions should likely be neglected in the membrane core, but other nonpolar terms, such as favorable solute–solvent dispersion interactions and solvent entropy remain. Finally, molecules permeating the lipid bilayer feel a lateral pressure profile; this pressure varies significantly not only with location in the membrane but also with membrane composition (Carrillo-Tripp and Feller, 2005; Ollila *et al.*, 2007; Cantor, 1999a), and may have functional implications (Cantor, 1997a, 1997b, 1999b; Pitman *et al.*, 2005).

Still, most of the methods discussed above model nonpolar interactions strictly on the basis of solvent accessible surface area, usually parameterized to reproduce partitioning free energies from water to liquid alkane and combined with a position-dependent scaling factor which turns the interactions off in the membrane interior. However, even within this genre, there are some interesting variations. Tanizaki and Feig (2005) attempted to capture position dependent effects by fitting the effective surface tension in the membrane to the potential of mean force of O_2 permeating an explicit lipid bilayer (Marrink and Berendsen, 1996). As a result, this parameterization accounts for variations in cavitation and dispersion with membrane depth.

Ben-Tal *et al.* (1996) took a different approach in their calculations exploring rigid helix insertion into lipid bilayers; while they did use an area term, they also included a term to account for lipid disruption, based on chain statistics calculations from Fattal and Ben-Shaul (1995). However, the inclusion of this term relies on the rigid-body nature of the calculation, since it is derived by treating the helix as a featureless cylinder.

Lazaridis took an entirely different approach in developing his implicit membrane model. IMM1 (Lazaridis, 2003), which is a generalization of the EFF1 solvation model (Lazaridis and Karplus, 1999) to a membrane environment. EFF1 consists of a distance-dependent dielectric combined with a pairwise, distance-dependent solvation term. IMM1 adds an explicit position-dependent atomic potential, parameterized to reproduce liquid-hydrocarbon transfer free energies of model compounds, combined with an enhancement of the electrostatic interactions in the bilayer interior. Although there is little theoretical justification underlying the functional forms of the model, it is easy to compute, and dynamics trajectories on systems such as the glycophorin A dimer, a helix isolated from bacteriorhodopsin, and several membrane-binding peptides all produced qualitatively reasonable results. In a later paper, IMM1 was further generalized to represent the internal water in a simulation of a transmembrane β-barrel protein (Lazaridis, 2005).

Several groups have also developed purely empirical implicit membrane models. For example, IMPALA model of Ducarme *et al.* (1998) applies simple atom-restraints, parameterized to reproduce partitioning experiments, independent of molecular context. This method is essentially free computationally, but has a number of unphysical implications, most notably that atoms of a given type feel exactly the same membrane forces whether on the surface of the molecule or buried in the interior. As a result, such a method is completely incapable of reproducing basic phenomena such as the stabilization of helical structure by the membrane environment. The same complaint can be made about the model from Sanders and Schwonek, although that model succeeds admirably in its stated goal of reproducing binding thermodynamics of small rigid molecules (Sanders and Schwonek, 1993).

Efremov et al. (1999a, 1999b) developed a model based entirely on solvent accessible surface area, generalizing the atomic solvation parameters of model (Eisenberg and McLachlan, 1986). Initially, their models were parameterized to reproduce low dielectric bulk solvent like hexane (Efremov *et al.*, 1999a) and octanol (Efremov *et al.*, 1999b), although later studies introduced spatial heterogeneity (Efremov *et al.*, 2002; Vereshaga *et al.*, 2007). These models produce qualitatively correct behavior, for example stabilizing α-helices, but as with the original atomic solvation parameters are not quantitatively accurate.

Of course, conceptually simple models are not to be disdained solely on that account. Rather, the goals and assumptions of the calculation must always be considered. For example, Pappu et al. showed they were able to find the correctly packed dimer structure of glycophorin A by representing the membrane as an infinite dielectric (no electrostatics at all) and a spring to prevent helix flipping (Pappu *et al.*, 1999).

B. Membrane-Focused

In contrast to the methods discussed in Section II.A, the methods presented in this chapter are largely focused on understanding the effects of membrane structure on the behavior of bound molecules. As a rule, the solute representations are not as detailed, but more care is taken to retain information about the membrane. In general, these models are intended in large part to describe the variation of membrane-solute interactions as a function of membrane composition, phenomena that are largely neglected by the models described previously.

The present class of models can be further divided into two subclasses: continuum models and chain models. The former class represents the membrane using some form of continuum mechanics based on some bulk property, *e.g.* hydrophobic thickness, while the latter attempts to build toward macroscopic predictions via a microscopic consideration of chain statistics.

The best known of the continuum models is the "mattress model" of Mouritsen and Bloom (Mouritsen and Bloom, 1984; Jensen and Mouritsen, 2004). In this approach, the lipids (and any additional membrane components, such as transmembrane proteins) are represented primarily as coupled springs with variable hydrophobic thickness. By assigning equilibrium thicknesses to the lipids and protein in specific phases, the elastic energy can be computed as

$$H_{\text{Elastic}}^{\alpha} = n_L^{\alpha} A_L^{\alpha} \left(d_L^{\alpha} - d_L^{0,\alpha} \right)^2 + n_P^{\alpha} A_P^{\alpha} \left(d_P^{\alpha} - d_P^{0,\alpha} \right)^2, \tag{8}$$

where n_i^{α} is the number of molecules of i in phase α, A is the effective force constant for thickness deformations, d_i is the hydrophobic thickness of molecules of type i, and $d_i^{0,\alpha}$ is the equilibrium thickness of molecules of type i in phase α. Specific protein–lipid interactions can also be included, such as hydrophobic mismatch

$$H_{\text{hydro}}^{\alpha} = \frac{n_L^{\alpha} n_P^{\alpha}}{n_L^{\alpha} + n_P^{\alpha}} B_{LP}^{\alpha} \left| d_L^{\alpha} - d_P^{\alpha} \right| \tag{9}$$

and favorable adhesion

$$H_{\text{adhes}}^{\alpha} = \frac{n_L^{\alpha} n_P^{\alpha}}{n_L^{\alpha} + n_P^{\alpha}} C_{LP}^{\alpha} \min(d_L^{\alpha}, d_P^{\alpha}), \tag{10}$$

where B and C are the positive and negative interaction coefficients for the respective terms. These energy terms constitute the excess enthalpy for the system, which can then be combined with the free energy for an ideal mixture to compute the free energies for different states. Thus, one can use the mattress model examine the effects of membrane permeants on lipid structure (and *vice versa*). For example, the original paper focuses on the effects of different "proteins" on the lipid phase diagram (Mouritsen and Bloom, 1984). In part because of its lack of atomic-level details, this model has been very successful in interpreting and suggesting experiments, particularly those involving designed single transmembrane helices such as the WALP and KALP families (Nyholm *et al.*, 2007).

However, the mattress model is neither the only nor the first continuum model for lipid–protein interactions. To pick one representative example, we consider the work of Owicki and McConnell, who used Landau–de Gennes theory to consider lipid–protein interactions in terms of order parameters related to the lipid gel–liquid phase transition (Owicki *et al.*, 1978; Owicki and McConnell, 1979). Their model describes mechanisms by which different lipid species could alter protein–lipid and even protein–protein interactions. However, the utility of the work is somewhat limited by its mandate that the protein be evenly distributed in the membrane, and by its focus on the notion of an annulus of boundary lipids surrounding the protein.

Brown and coworkers have proposed an alternative continuum formulation. Inspired by the unusual lipid composition of retinal rod outer segment disk membranes, with high concentrations of non-lamellar-forming lipids, they focused on spontaneous curvature of the membrane rather than simple hydrophobic matching (Gibson and Brown, 1993; Brown, 1994, 1997). When applied to the Meta-I/Meta-II equilibrium of photoactivated rhodopsin (Endress *et al.*, 2002), this model suggests that the lipid-composition dependent portion of the free energy change can be written as

$$\Delta G^0 = \kappa\left[\left(H_{\text{MII}}^L - H_0^L\right)^2 - \left(H_{\text{MI}}^L - H_0^L\right)^2\right] + \gamma_{\text{LP}}\left(A_{\text{MII}}^P - A_{\text{MI}}^P\right), \tag{11}$$

where H_i^L is the mean curvature of the membrane with the protein in state i, γ_{LP} is the lipid–protein surface tension, and A^P is the exposed area of the protein, believed to increase upon formation of Meta-II. More recently, this model was used to argue for the role of spontaneous curvature in modulating rhodopsin aggregation as well as function (Botelho *et al.*, 2006). This work proposes that increased protein concentration and decreasing bilayer thickness alter rhodopsin's properties via the same mechanism, a competition between curvature strain and hydrophobic matching.

By contrast, another subclass of models focused on the statistical physics of the lipid chains. In fairness, these two subclasses are not as distinct as they appear: chain states are invoked in the derivation and parameterization of the continuum models, and the results derived from the chain models, especially in the mean-field approximation, point back toward quantities used in the continuum methods.

A useful starting place to review chain-based models is the work of Marčelja (1974, 1976). His model considers the membrane as a set of hexagonal lattice sites, each of which contains a single lipid molecule. Each lipid is characterized by a molecular order parameter

$$\eta_j = \left\langle \frac{1}{n} \sum_m \left(\frac{3}{2} \cos^2 v_m - \frac{1}{2} \right) \right\rangle_j, \tag{12}$$

where n is the number of carbon segments in the chain v_m is the orientation of the mth segment. Lipid–lipid interaction is represented in the molecular field approximation, summing over nearest neighbors

$$\Phi_i = \frac{1}{6} \sum_{j=1}^6 V_0 \phi_j, \tag{13}$$

where

$$\phi_j = \left\langle \frac{n_{tr}}{n^2} \sum_m \left(\frac{3}{2} \cos^2 v_m - \frac{1}{2} \right) \right\rangle_j \tag{14}$$

and n_{tr}/n is the fraction of trans states, V_0 is the coupling constant, and Φ_i describes the strength of the molecular field acting to orient the molecule at site i. If there is a protein molecule at a neighboring site, a term in Eq. (13) is replaced by the lipid–protein interaction V_{lp} (Marčelja, 1976). Thus, the total energy for the ith position in the lattice is

$$E_i(\Phi_i, P) = E_{int} - \Phi_i \left(n_{tr}/n^2 \right) \sum_m \left(\frac{3}{2} \cos^2 v_m - \frac{1}{2} \right) + PA \tag{15}$$

is dependent on the chain's internal energy E_{int}, the cross-sectional area A and lateral pressure P. The partition function for the ith chain is

$$Z_i = \sum_{confs} \exp\left[-E_i(\Phi_i, P)/k_B T\right]. \tag{16}$$

Thus, the average orientations for all chains can be calculated by solving the following set of coupled non-linear differential equations

$$\phi_i = \frac{1}{Z_i} \sum_{confs} \frac{n_{tr}}{n} \sum_m \left(\frac{3}{2} \cos^2 v_m - \frac{1}{2} \right) \exp\left[-E_i(\Phi_i, P)/k_B T\right]. \tag{17}$$

The total internal energy thus becomes

$$U = \sum_i \left[E_{int} - n\Phi_i \phi_i / 2 \right] + V_{lp} \sum_{lp \text{ pairs}} n\phi_i / 12, \tag{18}$$

where the latter term is necessary to correctly account for protein–lipid interactions in the hexagonal lattice.

Writing the total system partition function as the product of the individual partition function in Eq. (16), and combining with Eq. (18), it is straightforward to write the system's entropy and Gibbs free energy.

This formalism can be used to investigate the effects of a protein on lipid structure as a function of temperature, protein size, and concentration. Most interestingly, one can compute the effective protein–protein potential of mean force; in this manner, a Marčelja-type model can be used to explore the effects of lipid composition on protein oligomerization and aggregation. Several other groups have explored similar models, differing primarily in the details of lipid chain representation and the manner in which the resulting equations are solved (Meraldi and Schlitter, 1981; Pink and Chapman, 1979).

More recently, Ben-Shaul and coworkers have developed a comprehensive membrane model which explicitly account for chain statistics (Fattal and Ben-Shaul, 1993, 1994, 1995). They begin by expressing the system's free energy as a sum of three terms

$$F = 2N(f_t + f_s + f_h), \tag{19}$$

where N is the number of lipids per leaflet (the present equation assumes a symmetric planar bilayer), f_t is the free energy per lipid chain, f_s the surface free energy due water–chain interaction, and f_h is the free energy due to headgroup–headgroup and water–headgroup interactions. Describing individual chain conformations using the rotational isomeric approximation combined with an overall tilt vector, the probability distribution P over all conformations α can be constructed, and the chain free energy computed as

$$f_t = \sum_\alpha P(\alpha)\epsilon(\alpha) + k_B T \sum_\alpha P(\alpha) \ln P(\alpha), \tag{20}$$

where $\epsilon(\alpha)$ is the internal energy of conformation α. $P(\alpha)$ is constrained to obey

$$\sum_\alpha P(\alpha) = 1, \tag{21}$$

$$\sum_\alpha P(\alpha)\left[\phi(z_i; \alpha) + \phi(-z_i; \alpha) \right] = a\rho, \quad \text{for all } z_i. \tag{22}$$

The first equation represents a simple normalization of probability. In the latter, $\phi(z_i; \alpha)$ is the atomic number density of conformation α in the membrane slice z_i, a is the area per chain, and ρ is density of the bilayer interior. Equation (22) explicitly assumes constant density in the hydrophobic interior of a symmetric planar bilayer, but can be easily generalized to account for heterogeneous density drawn from experiment, or to deal with more complex geometries such as curved bilayers or micelles.

Minimizing Eq. (20) with respect to Eqs. (21) and (22) gives

$$P(\alpha) = \frac{\exp[\beta\epsilon(\alpha) - \beta \sum_{z_i} \pi(z_i)\phi(z_i;\alpha)\Delta z]}{\sum_{\alpha} \exp[\beta\epsilon(\alpha) - \beta \sum_{z_i} \pi(z_i)\phi(z_i;\alpha)\Delta z]}, \tag{23}$$

where $\beta \equiv 1/k_B T$, the denominator is the chain partition function Z, and $\pi(z_i)$ are a series of Lagrange multipliers physically corresponding to the lateral pressure profile along the membrane normal. Substituting back into Eq. (20) generates

$$f_t = k_B T \ln Z - a\rho \sum_{z_i} \pi(z_i)\Delta z. \tag{24}$$

Thus, calculating any chain property amounts to computing the appropriate $\pi(z_i)$ values for the system. This model has a number of desirable properties, most notably the direct dependence of the chain thermodynamics on the headgroup type, via the surface area per chain a. This area can in turn be considered as a variable, which is where the latter two terms in Eq. (19) come into play.

In addition to considering alternative membrane geometries, this model can also be generalized to include the effects of other molecules included in the bilayer. Ben-Shaul and coworkers considered simple protein models such as impermeable walls and cylinders (Fattal and Ben-Shaul, 1995), while other groups have used analogous approaches to consider protein–protein interaction (May and Ben-Shaul, 2000; Bohinc et al., 2003). The lateral pressure profile computed in this model can be directly connected to the intrinsic curvature described in the models from Brown and coworkers, in that heterogeneity in the pressure leads directly to a preference for intrinsic curvature. As we will see below, this is a repeating theme in chain-based models of membranes.

Frink and Freschknecht (2005a, 2005b) have used a simple coarse-grained chain representation combined with density functional theory to compute lipid bilayer properties. Although the details of their formulation are too complex to describe here, it is interesting to note that once again the lateral pressure profile plays a central role. They have applied their model to explore the effects of alcohols on lipid structure (Frischknecht and Frink, 2006), and to explore pore-formation due to the binding of rigid helices (Frink and Frischknecht, 2006). By contrast to the models described in Section II.A, their model allowed them to compare different modes of pore formation.

Over the last 15 years, Cantor has presented a series of papers describing another chain-based model focused on the lateral pressure profile (Cantor, 1993, 1996, 1997a, 1997b, 1999a, 1999b, 2002). His approach uses a lattice model to account for chain conformations, with a constant density assumption similar to that of Ben-Shaul. He has used his model to explore the effects of chain length and unsaturation on the equilibrium width, area fluctuations, and lateral pressure profile of planar lipid bilayers (Cantor, 1999a). He has also proposed that lateral pressure is a common mechanism by which bilayer composition can be used to regulate protein function (Cantor, 1999b, 2002). Specifically, if a protein has two states with different shapes in the membrane, then the free energy difference between the two states is

$$\Delta G_{12} = \Delta G_{12}^0 - \int \pi(z)\big(A_2(z) - A_1(z)\big)\,dz, \qquad (25)$$

where ΔG_{12}^0 is the intrinsic free energy difference between the states, excluding bilayer effects, $A_i(z)$ is the area profile for the protein in state i, and $\pi(z)$ is the lateral pressure profile. The last term in Eq. (25) provides a direct mechanistic coupling between the lateral pressure profile (and thus lipid composition) and the protein's conformation equilibrium even in the total absence of specific lipid–protein interactions. Cantor has also proposed that this model provides a simple framework for understanding the mechanism of most general anesthetics (Cantor, 1997b, 2001).

III. INTERESTING PROBLEMS IN IMPLICIT MEMBRANE MODELING

There has been a great deal of work to develop models for implicitly modeling membrane–protein interactions. This work spans a broad range of different approaches, each with different strengths and weaknesses. In particular, the methods described in Section II.A have the advantage of using atomic descriptions of the solute of interest; this means that subtle differences, such as the effects of mutations or chemical modifications, can be directly examined. However, the membrane is usually represented in a very simple manner, and non-electrostatic effects in particular are not included in detail. As a result, these models do not as a rule capture the effects of lipid composition, except crudely via the hydrophobic thickness of the membrane. Moreover, if solute binding is correlated with disruption of membrane structure, these models will not capture it, since these models explicitly assume membrane structure is invariant. By contrast, the models in II.B represent membrane-bound solutes in far less detail, but do far more to include the effects of lipid chain structure. However, they typically lack atomic detail and tend to represent lipid–protein interactions phenomenologically. As a result, they cannot easily be used to resolve questions which depend on the details of solute structure. This means that there are many interesting problems which will require new approaches combining the strengths of the existing solute-centric models with better representations of membrane structure.

A. Antimicrobial Peptides

Antimicrobial peptides (AMPs) are an ancient immune mechanism, ubiquitous in multicellular organisms and even found in some bacteria (Zhang *et al.*, 2000; Risso, 2000; Zasloff, 2002a, 2002c, 2002b). In humans, they are found mostly on exposed organs, such as the eyes, skin, and mouth. Unlike the rest of the immune system, these peptides generally act in a non-inflammatory way. This is critical, as organs such as the eyes are constantly exposed to pathogens, and permanent inflammation would seriously degrade their performance.

AMPs exhibit a broad diversity of structures, ranging from single helices to β-strands to small globules (Zasloff, 2002c). However, the vast majority of them share two characteristics: amphipathic structure and positive charge. The former quality encourages binding to the membrane–water interface, while the latter enhances selectivity toward bacterial membranes, which tend to be enriched in anionic lipids compared to the zwitterionic lipids most common in mammalian cells. Interestingly, transformed cancer cells also have a higher concentration of anionic lipids, and some AMPs have been shown to have antitumor activity (Jacob and Zasloff, 1994; Mader and Hoskins, 2006). Lipopeptides have also found use in the development of vaccines (BenMohamed *et al.*, 2002). The biophysics of AMPs binding to lipid membranes have been extensively reviewed (Epand and Vogel, 1999; Shai, 1999; Huang, 2000, 2006; Doherty *et al.*, 2006; Chan *et al.*, 2006).

Unlike most other classes of drugs, AMPs do not inhibit enzymatic pathways, or indeed specifically bind any proteins in their targets. Rather, they operate by binding to and disrupting the membrane (Boman *et al.*, 1994). As a direct result, pathogens such as bacteria, fungi and viruses are far less likely to evolve immunity to them, because doing so would require changing the lipid composition of their membranes and likely disrupting many of their own native proteins.

Despite this, AMPs have not for the most part found much use as antibiotic drugs, because they are relatively hard to synthesize and tend to break down rapidly in the body. However, in recent years, scientists have borrowed from the basic properties of AMPs to design new potential drugs, for example using peptide mimetics (see for example Ishitsuka *et al.*, 2006). Shai and coworkers have had remarkable successes by combining two strategies: including D-amino acids to foil peptidases (Avrahami and Shai, 2003), and conjugating shorter peptides to fatty acids (Avrahami *et al.*, 2001; Avrahami and Shai, 2002, 2004; Makovitzki *et al.*, 2006; Makovitzki and Shai, 2005; Malina and Shai, 2005). The essential insight into the value of lipidization is that hydrophobicity is relatively non-specific. That is, most of the sequence in AMPs is devoted to making them hydrophobic enough to bind to membranes, as opposed to lysing them after binding. Thus, one can dispense with most of the sequence if the peptide is attached to an acyl chain; Shai and coworkers found that sequences as short as 4 amino acids had strong antimicrobial activity against fungi and bacteria, without significantly damaging human cells (Makovitzki *et al.*, 2006).

Unsurprisingly, there has been a great deal of interest in modeling the binding of AMPs to model lipid membranes (La Rocca *et al.*, 1999). Indeed, membrane-disrupting peptides such as gramicidin, alamethicin, and melittin have become standard test cases in the development of implicit membrane models. In fact, much of the interest in modeling isolated helices and helical aggregates in membranes derives from the classic "barrel-stave" model, where amphipathic helical peptides initially bind to the membrane interface, then cooperatively associate and fully insert to form pores; this process has been explored via theoretical means (Frink and Frischknecht, 2005a; Bohinc *et al.*, 2003) and by explicit molecular dynamics simulations (Tieleman *et al.*, 1999b, 1999a).

However, there is significant evidence that many if not most AMPs do not operate via the "barrel-stave" mechanism. Rather, the "carpet model" appears more prevalent (Chan *et al.*, 2006); in this view, peptides bind interfacially and stabilize highly curved lipid structures, leading to the formation of toroidal pores ("wormholes") and even micellization of the membrane.

This case is a difficult one to treat computationally. Poration due to lipid binding occurs on too long a time scale for all-atom molecular dynamics; the calculations which have been done have typically begun by preforming a particular pore-forming oligomeric structure (Tieleman *et al.*, 1999b; La Rocca *et al.*, 1999). In principle, solute-focused implicit membrane models can reach the necessary time- and length-scales for spontaneous oligomerization, and as described above have had some successes in describing barrel-stave-type pore formation. However, these models cannot represent the kinds of membrane disruption expected according to the carpet model, and so cannot be used to elucidate which if either model applies to a given solute. The only example we are aware of where an implicit membrane model was used to examine the mechanism of poration is the work of Frink and Frischknecht, described above (Frink and Frischknecht, 2005a, 2005b), which applied density functional theory. In this model, however, protein helices are represented as impermeable hard cylinders, neglecting other phenomena such as electrostatics and amphipathicity. As such, these calculations, while highly instructive, cannot be used to reveal the binding mode of. for example, a particular amino acid sequence.

An implicit membrane model which could successfully attack this problem would most likely need the following properties:

- Atomic- or near-atomic-level solute description.
- Accurate electrostatics, including both dielectric effects and effects due to headgroup charge and dipoles.
- Membrane which responds to solute structure.

Several models from Section II.A have the first property, and could readily be extended to have the second. However, none of the existing models can readily meet the third requirement, which should make this an interesting future research problem.

B. Protein-Protein Association

Although biophysical experiments frequently focus on the behavior of isolated proteins, biologically many if not most membrane proteins function as part of complexes (Alberts *et al.*, 1994). Of these complexes, homodimers are most easily studied by biophysical techniques, since only a single protein needs to be overexpressed. In particular, there is great interest in the dimerization of G protein-coupled receptors (GPCRs) (Park *et al.*, 2004), the largest superfamily of proteins in the human genome. These proteins are responsible for a broad array of physiological processes involving signaling (Bockaert and Pin, 1999), and as a result are commonly targetted in drug development (Ma and Zemmel, 2002). Among GPCRs, only one protein, rhodopsin, has been structurally resolved at atomic resolution (Edwards *et al.*, 2004; Li *et al.*, 2004; Okada *et al.*, 2002, 2004; Palczewski *et al.*, 2000; Schertler *et al.*, 1993). In recent years, significant controversy has erupted over the oligomeric state of rhodopsin in its native membrane environment. Several groups have demonstrated that rhodopsin dimerizes in non-native membranes (Kota *et al.*, 2006; Mansoor *et al.*, 2006), and Palczewski and coworkers showed striking images from atomic force microscopy showing ordered rows of rhodopsin dimers (Fotiadis *et al.*, 2003a). However, others have argued that these dimers are artifacts of sample preparation conditions (Chabre *et al.*, 2003; Fotiadis *et al.*, 2003b), and have argued that the functional form of the protein is most likely monomeric (Chabre and le Maire, 2005).

In principle, one way resolve this controversy would be an unambiguous determination of the dimeric structure of rhodopsin. However, this seems unlikely, because any crystal structure could be countered by the argument that the dimer was stabilized only by crystal packing. Instead, various groups have attempted to map a generic GPCR dimer interface using mutagenesis studies (Javitch, 2004; Guo *et al.*, 2005; Fanelli and De Benedetti, 2005; Filizola *et al.*, 2002; Filizola and Weinstein, 2005b).

Ideally, this sort of strategy would be complemented by molecular-level simulation, to flesh out the details and validate the interactions. Indeed, there have been several efforts along these lines (see for example Filizola *et al.*, 2006; Filizola and Weinstein, 2005a). However, such efforts are complicated by the very long time scales necessary to sample large-scale rearrangement of protein–protein interfaces; it is to be expected that even a totally incorrect protein–protein docking would be stable on the 10–100 ns timescale readily accessible by all-atom molecular dynamics. Using existing implicit membrane models is more appealing, since conventional molecular dynamics simulations could be abandoned in favor more efficient searching and sampling techniques. However, the best of the existing solute-focused are expensive for larger systems; for example, Feig's work on bacteriorhodopsin trimers showed that the need for very long electrostatics cutoffs greatly increased the computational cost (Tanizaki and Feig, 2006). Moreover, these models are not capable of capturing the effects of specific lipid species. This could be critical to assessing the stability of dimers, since we know that rhodopsin is both highly sensitive to and capable of perturbing its lipid environment (Brown, 1994; Botelho *et al.*, 2006; Polozova and Litman, 2000). As such, we once again reach a point where new models will be needed in order to resolve the problem.

IV. CONCLUSION

Lipid membranes are critically important biologically, both as passive barriers and as active participants in membrane protein function. Molecular modeling has already made significant contributions to our understanding of their roles, and can be expected to be even more valuable as structures of more membrane proteins become available. However, for many applications, explicit all-atom calculations are prohibitively expensive, and will remain so for the foreseeable future. In this context, the development of new models for representing protein–lipid interactions implicitly becomes extremely important. A great deal of impressive work has been done, but still more remains.

References
Alberts, B., Bray, D., Lewis, J., Raff, M., Roberts, K., Watson, J.D. (1994). "Molecular Biology of the Cell", 3rd ed. Garland Publishing, Inc., New York.
Allen, F., Almasi, G., Andreoni, W., Beece, D., Berne, B.J., Bright, A., Brunheroto, J., Cascaval, C., Castanos, J., Coteus, P., Crumley, P., Curioni, A., Denneau, M., Donath, W., Eleftheriou, M., Fitch, B., Fleisher, B., Georgiou, C.J., Germain, R., Giampapa, M., Gresh, D., Gupta, M., Haring, R., Ho, H., Hochschild, P., Hummel, S., Jonas, T., Lieber, D., Martyna, G., Maturu, K., Moreira, J., Newns, D., Newton, M., Philhower, R., Picunko, T., Pitera, J., Pitman, M., Rand, R., Royyuru, A., Salapura, V., Sanomiya, A., Shah, R., Sham, Y., Singh, S., Snir, M., Suits, F., Swetz, R., Swope, W.C., Vishnumurthy, N., Ward, T.J.C., Warren, H., Zhou, R. (2001). Blue Gene: A vision for protein science using a petaflop supercomputer. *IBM Syst. J.* **40**, 310.
Avrahami, D., Shai, Y. (2002). Conjugation of a magainin analogue with lipophilic acids controls hydrophobicity, solution assembly, and cell selectivity. *Biochemistry* **41** (7), 2254–2263.
Avrahami, D., Shai, Y. (2003). Bestowing antifungal and antibacterial activities by lipophilic acid conjugation to D,L-amino acid-containing antimicrobial peptides: A plausible mode of action. *Biochemistry* **42** (50), 14946–14956. http://dx.doi.org/10.1021/bi035142v.
Avrahami, D., Shai, Y. (2004). A new group of antifungal and antibacterial lipopeptides derived from non-membrane active peptides conjugated to palmitic acid. *J. Biol. Chem.* **279**, 12277–12285.
Avrahami, D., Oren, Z., Shai, Y. (2001). Effect of multiple aliphatic amino acids substitutions on the structure, function, and mode of action of diastereomeric membrane active peptides. *Biochemistry* **40** (42), 12591–12603.

Baker, N.A. (2005). Improving implicit solvent simulations: A Poisson-centric view. *Curr. Opin. Struct. Biol.* **15**, 137–143.

Bashford, D., Case, D.A. (2000). Generalized Born models of macromolecular solvation effects. *Annu. Rev. Phys. Chem.* **51**, 129–152.

Bell, R.P. (1931). The electrostatic energy of dipole molecules in different media. *Trans. Faraday Soc.* **27**, 797–802.

Ben-Tal, N., Ben-Shaul, A., Nicholls, A., Honig, B. (1996). Free energy determinants of α-helix insertion into lipid bilayers. *Biophys. J.* **70**, 1803–1812.

BenMohamed, L., Wechsler, S.L., Nesburn, A.B. (2002). Lipopeptide vaccines—yesterday, today, and tomorrow. *Lancet Infect. Dis.* **2**, 425–431.

Bockaert, J., Pin, J.P. (1999). Molecular tinkering of G protein-coupled receptors: An evolutionary success. *EMBO J.* **18**, 1723–1729.

Boesze-Battaglia, K., Albert, A.D. (1989). Fatty acid composition of bovine rod outer segment plasma membrane. *Exp. Eye Res.* **49** (4), 699–701.

Boesze-Battaglia, K., Hennessey, T., Albert, A.D. (1989). Cholesterol heterogeneity in bovine rod outer segment disk membranes. *J. Biol. Chem.* **264** (14), 8151–8155.

Bohinc, K., Kralj-Iglic, V., May, S. (2003). Interaction between two cylindrical inclusions in a symmetric lipid bilayer. *J. Chem. Phys.* **119**, 7435–7444.

Boman, H.G., Marsh, J., Goode, J.A. (Eds.) (1994). Antimicrobial peptides. In "Ciba Foundation Symposium", vol. **186**. Wiley, Chichester, pp. 1–272.

Bonner, W.B. (1951). The electrostatic energy of molecules in solution. *Trans. Faraday Soc.* **47**, 1143–1152.

Born, M. (1920). Volumen und hydratationswarme der ionen. *Z. Phys.* **1**, 45.

Botelho, A.V., Huber, T., Sakmar, T.P., Brown, M.F. (2006). Curvature and hydrophobic forces drive oligomerization and modulate activity of rhodopsin in membranes. *Biophys. J.* **91**, 4464–4477.

Brown, M.F. (1994). Modulation of rhodopsin function by properties of the membrane bilayer. *Chem. Phys. Lipids* **73** (1–2), 159–180.

Brown, M.F. (1997). Influence of nonlamellar-forming lipids on rhodopsin. In: Epand, R.M. (Ed.), Lipid Polymorphism and Membrane Properties. In: "Current Topics in Membranes", vol. **44**. Academic Press, San Diego, pp. 285–356.

Cantor, R.S. (1993). Statistical thermodynamics of curvature elasticity in surfactant monolayer films: A molecular approach. *J. Chem. Phys.* **99**, 7124–9149.

Cantor, R.S. (1996). Theory of lipid monolayers comprised of mixtures of flexible and stiff amphiphiles in athermal solvents: Fluid phase coexistence. *J. Chem. Phys.* **104**, 8082–8095.

Cantor, R.S. (1997a). Lateral pressure in cell membranes: A mechanism for modulation of protein function. *J. Phys. Chem. B* **101**, 1724–1725.

Cantor, R.S. (1997b). The lateral pressure profile in membranes: A physical mechanism of general anesthesia. *Biochemistry* **36**, 2339–2344.

Cantor, R.S. (1999a). Lipid composition and the lateral pressure profile in bilayers. *Biophys. J.* **76**, 2625–2639.

Cantor, R.S. (1999b). The influence of membrane lateral pressures on simple geometric models of protein conformational equilibria. *Chem. Phys. Lipids* **101**, 45–56.

Cantor, R.S. (2001). Breaking the Meyer–Overton rule: Predicted effects of varying stiffness and interfacial activity on the intrinsic potency of anesthetics. *Biophys. J.* **80**, 2284–2297.

Cantor, R.S. (2002). Size distribution of barrel-stave aggregates of membrane peptides: Influence of the bilayer lateral pressure profile. *Biophys. J.* **82**, 2520–2525.

Carrillo-Tripp, M., Feller, S.E. (2005). Evidence for a mechanism by which omega-3 polyunsaturated lipids may affect membrane protein function. *Biochemistry* **44**, 10164–10169.

Chabre, M., le Maire, M. (2005). Monomeric G-protein-coupled receptor as a functional unit. *Biochemistry* **44** (27), 9395–9403. http://dx.doi.org/10.1021/bi0507200.

Chabre, M., Cone, R., Saibil, H. (2003). Is rhodopsin dimeric in native retinal rods? *Nature* **426**, 30–31.

Chan, D.I., Prenner, E.J., Vogel, H.J. (2006). Tryptophan- and arginine-rich antimicrobial peptides: Structures and mechanisms of action. *Biochim. Biophys. Acta* **1758**, 1184–1202.

Doherty, T., Waring, A., Hong, M. (2006). Peptide–lipid interactions of the beta-hairpin antimicrobial peptide tachyplesin and its linear derivatives from solid-state NMR. *Biochim. Biophys. Acta* **1758**, 1285–1291.

Dominy, B.N., Brooks, C.L. (1999). Development of a generalized born model parameterization for proteins and nucleic acids. *J. Phys. Chem. B* **103**, 3765–3773.

Ducarme, P., Rahman, M., Brasseur, R. (1998). IMPALA: A simple restraint field to simulate the biological membrane in molecular structure studies. *Proteins: Struct. Funct. Gen.* **30**, 357–371.

Edwards, P.C., Li, J., Burghammer, M., McDowell, J.H., Villa, C., Hargrave, P.A., Schertler, G.F.X. (2004). Crystals of native and modified bovine rhodopsins and their heavy atom derivatives. *J. Mol. Biol.* **343** (5), 1439–1450. http://dx.doi.org/10.1016/j.jmb.2004.08.089.

Efremov, R.G., Nolde, D.E., Vergoten, G., Arseniev, A.S. (1999a). A solvent model for simulations of peptides in bilayers. I. Membrane-promoting alpha-helix formation. *Biophys. J.* **76**, 2448–2459.

Efremov, R.G., Nolde, D.E., Vergoten, G., Arseniev, A.S. (1999b). A solvent model for simulations of peptides in bilayers. II. Membrane spanning alpha-helices. *Biophys. J.* **76**, 2460–2471.

Efremov, R.G., Volynsky, P.E., Nolde, D.E., Dubovskii, P.V., Arseniev, A.S. (2002). Interaction of cardiotoxins with membranes: A molecular modeling study. *Biophys. J.* **83** (1), 144–153.

Eisenberg, D., McLachlan, A.D. (1986). Solvation energy in protein folding and binding. *Nature* **319**, 199–203.

Endress, E., Heller, H., Casalta, H., Brown, M.F., Bayerl, T.M. (2002). Anisotropic motion and molecular dynamics of cholesterol, lanosterol, and ergosterol in lecithin bilayers studied by quasi-elastic neutron scattering. *Biochemistry* **41** (43), 13078–13086. 0006-2960 J. Art.

Epand, R. (1998). Lipid polymorphism and protein–lipid interactions. *Biochim. Biophys. Acta* **1376**, 353–368.

Epand, R.M. (2003). Fusion peptides and the mechanism of viral fusion. *Biochim. Biophys. Acta* **1614**, 116–121.

Epand, R.M., Vogel, H.J. (1999). Diversity of antimicrobial peptides and their mechanism of action. *Biochim. Biophys. Acta* **1462**, 11–28.

Fanelli, F., De Benedetti, P.G. (2005). Computational modeling approaches to structure-function analysis of G protein-coupled receptors. *Chem. Rev.* **105**, 3297–3351.

Faráldo-Gomez, J.D., Forrest, L.R., Baaden, M., Bond, P.J., Domene, C., Patargias, G., Cutherbertson, J., Sansom, M.S.P. (2004). Conformational sampling and dynamics of membrane proteins from 10-nanosecond computer simulations. *Proteins* **57**, 783–791.

Fattal, D.R., Ben-Shaul, A. (1993). A molecular model for lipid–protein interactions in membranes: The role of hydrophobic mismatch. *Biophys. J.* **65**, 1795–1809.

Fattal, D.R., Ben-Shaul, A. (1994). Mean-field calculations of chain packing and conformational statistics in lipid bilayers: Comparison with experiments and molecular dynamics studies. *Biophys. J.* **67**, 983–995.

Fattal, D.R., Ben-Shaul, A. (1995). Lipid chain packing and lipid–protein interaction in membranes. *Physica A* **220**, 192–216.

Feig, M., Im, W., Brooks III, C.L. (2004). Implicit solvation based on generalized Born theory in different dielectric environments. *J. Chem. Phys.* **120**, 903–911.

Feig, M., Onufriev, A., Lee, M.S., Im, W., Case, D.A., Brooks III, C.L. (2004). Poisson methods in the calculation of electrostatic solvation free energies for protein structures. *J. Comput. Chem.* **25**, 265–284.

Filizola, M., Weinstein, H. (2005a). The structure and dynamics of GPCR oligomers: A new focus in models of cell-signaling mechanisms and drug design. *Curr. Opin. Drug Discov. Devel.* **8** (5), 577–584.

Filizola, M., Weinstein, H. (2005b). The study of G-protein coupled receptor oligomerization with computational modeling and bioinformatics. *FEBS J.* **272** (12), 2926–2938. http://dx.doi.org/10.1111/j.1742-4658.2005.04730.x.

Filizola, M., Olmea, O., Weinstein, H. (2002). Prediction of heterodimerization interfaces of G-protein coupled receptors with a new subtractive correlated mutation method. *Protein Eng.* **15** (11), 881–885.

Filizola, M., Wang, S.X., Weinstein, H. (2006). Dynamic models of G-protein coupled receptor dimers: Indications of asymmetry in the rhodopsin dimer from molecular dynamics simulations in a popc bilayer. *J. Comput. Aided Mol. Des.* **20** (7–8), 405–416. http://dx.doi.org/10.1007/s10822-006-9053-3.

Flewelling, R.F., Hubbell, W.L. (1986). The membrane dipole potential in a total membrane potential model. *Biophys. J.* **49**, 541–552.

Fotiadis, D., Liang, Y., Filipek, S., Saperstein, D.A., Engel, A., Palczewski, K. (2003a). Rhodopsin dimers in native disc membranes. *Nature* **421**, 127–128.

Fotiadis, D., Liang, Y., Filipek, S., Saperstein, D.A., Engel, A., Palczewski, K. (2003b). Is rhodopsin dimeric in native retinal rods? (reply). *Nature* **426**, 31.

Frink, L.J.D., Frischknecht, A.L. (2005a). Density functional theory approach for coarse-grained lipid bilayers. *Phys. Rev. E* **72**, 041923.

Frink, L.J.D., Frischknecht, A.L. (2005b). Comparison of density functional theory and simulation of fluid bilayers. *Phys. Rev. E* **72**, 041924.

Frink, L.J.D., Frischknecht, A.L. (2006). Computational investigations of pore forming peptide assemblies in lipid bilayers. *Phys. Rev. Lett.* **97**, 208701–208704.

Frischknecht, A.L., Frink, L.J.D. (2006). Alcohols reduce lateral membrane pressures: Predictions from molecular theory. *Biophys. J.* **91**, 4081–4090.

Gallicchio, E., Zhang, L.Y., Levy, R.M. (2002). The SGB/NP hydration free energy model based on the Surface Generalized Born solvent reaction field and novel nonpolar hydration free energy estimators. *J. Comput. Chem.* **23**, 517–529.

Ghosh, A., Rapp, C.S., Friesner, R.A. (1998). Generalized Born model based on a surface integral formulation. *J. Phys. Chem. B* **102**, 10983–10990.

Gibson, N.J., Brown, M.F. (1993). Lipid headgroup and acyl chain composition modulate the MI–MII equilibrium of rhodopsin in recombinant membranes. *Biochemistry* **32**, 2438–2454.

Grossfield, A., Feller, S.E., Pitman, M.C. (2006). A role for direct interactions in the modulation of rhodopsin by omega-3 polyunsaturated lipids. *Proc. Natl. Acad. Sci. USA* **103**, 4888–4893.

Grossfield, A., Feller, S.E., Pitman, M.C. (2007). Convergence of molecular dynamics simulations of membrane proteins. *Proteins: Struct. Funct. Bioinf.* **67**, 31–40.

Grossfield, A., Pitman, M.C., Feller, S.E., Soubias, O., Gawrisch, K., The role of water in the activation of the G protein-coupled receptor rhodopsin. *JMB*, in press.

Grycuk, T. (2003). Deficiency of the Coulomb-field approximation in the generalized Born model: An improved formula for Born radii evalutation. *J. Chem. Phys.* **119**, 4817–4826.

Guo, W., Shi, L., Filizola, M., Weinstein, H., Javitch, J.A. (2005). Crosstalk in G protein-coupled receptors: Changes at the transmembrane homodimer interface determine activation. *Proc. Natl. Acad. Sci. USA* **102** (48), 17495–17500. http://dx.doi.org/10.1073/pnas.0508950102.

Huang, H.W. (2000). Action of antimicrobial peptides: Two-state model. *Biochemistry* **39**, 8347–8352.

Huang, H.W. (2006). Molecular mechanism of antimicrobial peptides: The origin of cooperativity. *Biochim. Biophys. Acta* **1758**, 1292–1302.

Im, W., Brooks III, C.L. (2005). Interfacial folding and membrane insertion of designed peptides studied by molecular dynamics simulations. *Proc. Natl. Acad. Sci. USA* **102**, 6771–6776.

Im, W., Feig, M., Brooks III, C.L. (2003a). An implicit membrane generalized Born theory for the study of structure, stability, and interactions of membrane proteins. *Biophys. J.* **85**, 2900–2918.

Im, W., Lee, M.S., Brooks III, C.L. (2003b). Generalized Born model with a simple smoothing function. *J. Comput. Chem.* **24**, 1691–1702.

Ishitsuka, Y., Arnt, L., Majewski, J., Frey, S., Ratajczek, M., Kjaer, K., Tew, G.N., Lee, K.Y.C. (2006). Amphiphilic poly(phenyleneethynylene)s can mimic antimicrobial peptide membrane disordering effect by membrane insertion. *J. Am. Chem. Soc.* **128** (40), 13123–13129. http://dx.doi.org/10.1021/ja061186q.

Iversen, G., Kharkats, Y.I., Ulstrup, J. (1998). Simple dielectric image charge models for electrostatic interactions in metalloproteins. *Mol. Phys.* **94**, 297–306.

Jackson, J.D. (1962). "Classical Electrodynamics". Wiley, New York.

Jacob, L., Zasloff, M. (1994). Potential therapeutic applications of magainins and other antimicrobial agents of animal origin. *Ciba Found. Symp.* **186**, 197–223.

Jacobs, R.E., White, S.H. (1989). The nature of the hydrophobic binding of small peptides at the bilayer interface: Implications for the insertion of transbilayer helices. *Biochemistry* **28**, 3421–3437.

Javitch, J.A. (2004). The ants go marching two by two: Oligomeric structure of G-protein coupled receptors. *Mol. Pharmacol.* **66**, 1077–1082.

Jensen, M.Ø., Mouritsen, O.G. (2004). Lipids do influence protein function—the hydrophobic matching hypothesis revisited. *Biochim. Biophys. Acta* **1666**, 205–226.

Kharkats, Y.I., Ulstrup, J. (1991). The electrostatic Gibbs energy of finite-size ions near a planar boundary between two dielectric media. *J. Electroanal. Chem.* **308**, 17–26.

Kirkwood, J.G. (1939). The dielectric polarization of polar liquids. *J. Chem. Phys.* **7**, 911–919.

Kota, P., Reeves, P.J., Rajbhandary, U.L., Khorana, H.G. (2006). Opsin is present as dimers in COS1 cells: Identification of amino acids at the dimeric interface. *Proc. Natl. Acad. Sci. USA* **103** (9), 3054–3059. http://dx.doi.org/10.1073/pnas.0510982103.

Krishtalik, L.I. (1996). Intramembrane electron transfer: Processes in the photosynthetic reaction center. *Biochim. Biophys. Acta Bioenergetics* **1273**, 139–149.

La Rocca, P., Biggin, P.C., Tieleman, D.P., Sansom, M.S. (1999). Simulation studies of the interaction of antimicrobial peptides and lipid bilayers. *Biochim. Biophys. Acta* **15**, 185–200.

Lazaridis, T. (2003). Effective energy function for proteins in lipid membranes. *Proteins: Struct. Funct. Gen.* **52**, 176–192.

Lazaridis, T. (2005). Structural determinants of transmembrane β-barrels. *J. Chem. Theory Comput.* **23**, 1090–1099.

Lazaridis, T., Karplus, M. (1999). Effective energy function for proteins in solution. *Proteins: Struct. Funct. Gen.* **35**, 133–152.

Lee, M.S., Salsbury, F.R., Brooks III, C.L. (2002). Novel generalized Born methods. *J. Chem. Phys.* **116**, 10606–10614.

Lee, M.S., Feig, M., Salsbury Jr., F.R., Brooks III, C.L. (2003). New analytic approximation to the standard molecular volume definition and its application to generalized born calculations. *J. Comput. Chem.* **24**, 1348–1356.

Levy, R.M., Zhang, L.Y., Gallicchio, E., Felts, A.K. (2003). On the nonpolar hydration free energy of proteins: Surface area and continuu, solvent models for the solute–solvent interaction energy. *J. Am. Chem. Soc.* **125**, 9523–9530.

Li, J., Edwards, P.C., Burghammer, M., Villa, C., Schertler, G.F.X. (2004). Structure of bovine rhodopsin in a trigonal crystal form. *J. Mol. Biol.* **343** (5), 1409–1438.

Lin, J.-H., Baker, N.A., McCammon, J.A. (2002). Bridging implicit and explicit solvent approaches for membrane electrostatics. *Biophys. J.* **83**, 1374–1379.

Lu, B., Cheng, X., Hou, T., McCammon, J.A. (2005a). Calculation of the maxwell stress tensor and the Poisson–Boltzmann force on a solvated molecular surface using hypersingular boundary integrals. *J. Chem. Phys.* **123** (8), 084904. http://dx.doi.org/10.1063/1.2008252.

Lu, B., Zhang, D., McCammon, J.A. (2005b). Computation of electrostatic forces between solvated molecules determined by the Poisson–Boltzmann equation using a boundary element method. *J. Chem. Phys.* **122** (21), 214102. http://dx.doi.org/10.1063/1.1924448.

Ma, P., Zemmel, R. (2002). Value of novelty? *Nat. Rev. Drug Discov.* **1**, 571–572.

Mader, J.S., Hoskins, D.W. (2006). Cationic antimicrobial peptides as novel therapeutic agents for cancer treatment. *Expert. Opin. Investig. Drugs* **15**, 933–946.

Makovitzki, A., Shai, Y. (2005). pH-dependent antifungal lipopeptides and their plausible mode of action. *Biochemistry* **44** (28), 9775–9784. http://dx.doi.org/10.1021/bi0502386.

Makovitzki, A., Avrahami, D., Shai, Y. (2006). Ultrashort antibacterial and antifungal lipopeptides. *Proc. Natl. Acad. Sci. USA* **103** (43), 15997–16002. http://dx.doi.org/10.1073/pnas.0606129103.

Malina, A., Shai, Y. (2005). Conjugation of fatty acids with different lengths modulates the antibacterial and antifungal activity of a cationic biologically inactive peptide. *Biochem. J.* **390** (Pt 3), 695–702. http://dx.doi.org/10.1042/BJ20050520.

Mansoor, S.E., Palczewski, K., Farrens, D.L. (2006). Rhodopsin self-associates in asolectin liposomes. *Proc. Natl. Acad. Sci. USA* **103** (9), 3060–3065. http://dx.doi.org/10.1073/pnas.0511010103.

Marčelja, S. (1974). Chain ordering in liquid crystals. II. Structure of bilayer membranes. *Biochim. Biophys. Acta* **367**, 165–176.

Marčelja, S. (1976). Lipid–mediate protein interactions in membranes. *Biochim. Biophys. Acta* **455**, 1–7.

Marrink, S.J., Berendsen, H.J.C. (1996). Permeation process of small molecules across lipid membranes studied by molecular dynamics simulations. *J. Phys. Chem.* **100**, 16729–16738.

Martínez-Mayorga, K., Pitman, M.C., Grossfield, A., Feller, S.E., Brown, M.F. (2006). Retinal counterion switch mechanism in vision evaluated by molecular simulation. *J. Am. Chem. Soc.* **128**, 16502–16503.

May, S., Ben-Shaul, A. (2000). A molecular model for lipid-mediated interaction between proteins and membranes. *Phys. Chem. Chem. Phys.* **2**, 4494–4502.

Meraldi, J.P., Schlitter, J. (1981). A statistical mechanical treatment of fatty acyl-chain order in phospholipid bilayers and correlation with experimental data A: Theory. *Biochim. Biophys. Acta* **645**, 183–192.

Mitchell, D.C., Niu, S.-L., Litman, B.J. (2001). Optimization of receptor—G protein coupling by bilayer lipid composition I. *J. Biol. Chem.* **276** (46), 42801–42806.

Mouritsen, O.G., Bloom, M. (1984). Mattress model of lipid–protein interactions in membranes. *Biophys. J.* **46**, 141–153.

Mouritsen, O.G., Bloom, M. (1993). Models of lipid–protein interactions in membranes. *Ann. Rev. Biophys. Biomol. Struct.* **22**, 145–171.

Murray, D., Ben-Tal, N., Honig, B., McLaughlin, S. (1997). Electrostatic interaction of myristolated proteins with membranes: Simple physics, complicated biology. *Structure* **15**, 985–989.

Murray, D., Arbuzova, A., Hangyás-Mihályné, G., Gambhir, A., Ben-Tal, N., Honig, B., McLaughlin, S. (1999). Electrostatic properties of membranes containing acidic lipids and adsorbed basic peptides: Theory and experiment. *Biophys. J.* **77**, 3176–3188.

Niu, S.-L., Mitchell, D.C., Litman, B.J. (2001). Optimization of receptor—G protein coupling by bilayer composition II. *J. Biol. Chem.* **276** (46), 42807–42811.

Niu, S.-L., Mitchell, D.C., Litman, B.J. (2002). Manipulation of cholesterol levels in rod disk membranes by methyl–b-cyclodextrin. *J. Biol. Chem.* **277**, 20139–20145.

Nyholm, T.K.M., Özdirekcan, S., Killian, J.A. (2007). How protein transmembrane segments sense the lipid environment. *Biochemistry* **46**, 1457–1465.

Nymeyer, H., Woolf, T.B., Garcia, A.E. (2005). Folding is not required for bilayer insertion: Replica exchange simulations of an alpha-helical peptide with an explicit lipid bilayer. *Proteins* **59** (4), 783–790.

Okada, T., Fujiyoshi, Y., Silow, M., Navarro, J., Landau, E.M., Shichida, Y. (2002). Functional role of internal water molecules in rhodopsin revealed by X-ray crystallography. *Proc. Natl. Acad. Sci. USA* **99** (9), 5982–5987.

Okada, T., Sugihara, M., Bondar, A.N., Elstner, M., Entel, P., Buss, V. (2004). The retinal conformation and its environment in rhodopsin in light of a new 2.2 angstrom crystal structure. *J. Mol. Biol.* **342**, 571–583.

Okamoto, Y. (2004). Generalize-ensemble algorithms: Enhanced sampling techniques for Monte Carlo and molecular dynamics simulations. *J. Mol. Graphics Modell.* **22**, 425–439.

Ollila, S., Hyvönen, M.T., Vattulainen, I. (2007). Polyunsaturation in lipid membranes: Dynamic properties and lateral pressure profiles. *J. Phys. Chem. B* **111** (12), 3139–3150.

Owicki, J.C., McConnell, H.M. (1979). Theory of protein–lipid and protein–protein interactions in bilayer membranes. *Proc. Natl. Acad. Sci. USA* **76** (10), 4750–4754.

Owicki, J.C., Springgate, M.W., McConnell, H.M. (1978). Theoretical study of protein–lipid interactions in bilayer membranes. *Proc. Natl. Acad. Sci. USA* **75** (4), 1616–1619.

Palczewski, K., Kumasaka, T., Hori, T., Behnke, C.A., Motoshima, H., Fox, B.J., Le Trong, I., Teller, D.C., Okada, T., Stenkamp, R.E., Yamamoto, M., Miyano, M. (2000). Crystal structure of rhodopsin: A G protein-coupled receptor. *Science* **289**, 739–745.

Pappu, R.V., Marshall, G.M., Ponder, J.W. (1999). A potential smoothing algorithm accurately predicts transmembrane helix packing. *Nat. Struct. Biol.* **6**, 50–55.

Park, P.S.-H., Filipek, S., Wells, J.W., Palczewski, K. (2004). Oligomerization of G protein-coupled receptors: Past, present and future. *Biochemistry* **43**, 15643–15656.

Parsegian, A. (1969). Energy of an ion crossing a low dielectric membrane: Solutions to four relevant electrostatic problems. *Nature* **221**, 844–846.

Pierotti, R.A. (1976). A scaled particle theory of aqueous and nonaqueous solutions. *Chem. Rev.* **76**, 717–726.

Pink, D., Chapman, D. (1979). Protein–lipid interactions in bilayer membranes: A lattice model. *Proc. Natl. Acad. Sci. USA* **76**, 1542–1546.

Pitman, M.C., Grossfield, A., Suits, F., Feller, S.E. (2005). Role of cholesterol and polyunsaturated chains in lipid–protein interactions: Molecular dynamics simulations of rhodopsin in a realistic membrane environment. *J. Am. Chem. Soc.* **127**, 4576–4577.

Polozova, A., Litman, B.J. (2000). Cholesterol dependent recruitment of di22:6-PC by a G protein-coupled receptor into lateral domains. *Biophys. J.* **79**, 2632–2643.

Risso, A. (2000). Leukocyte antimicrobial peptides: Multifunctional effector molecules of innate immunity. *J. Leukoc. Biol.* **68**, 785–792.

Sanders II, C.R., Schwonek, J.P. (1993). An approximate model and empirical energy function for solute interactions with a water–phosphatidylcholine interface. *Biophys. J.* **65**, 1207–1218.

Schertler, G.F., Villa, C., Henderson, R. (1993). Projection structure of rhodopsin. *Nature* **362**, 770–772.

Schnitzer, J.E., Lambrakis, K.C. (1991). Electrostatic potential and Born energy of charged molecules interacting with phospholipid membranes: Calculation via 3-D numerical solution for the full Poisson equation. *J. Theor. Biol.* **152**, 203–222.

Shai, Y. (1999). Mechanism of the binding, insertion and destabilization of phospholipid bilayer membranes by alpha-helical antimicrobial and cell non-selective membrane-lytic peptides. *Biochim. Biophys. Acta* **1462**, 55–70.

Sharp, K.A., Honig, B. (1990). Electrostatic interactions in macromolecules: Theory and applications. *Annu. Rev. Biophys. Biophys. Chem.* **19**, 301–332.

Spassov, V.Z., Yan, L., Szalma, S. (2002). Introducing an implicit membrane in generalized-Born/solvent accessibility continuum solvent models. *J. Phys. Chem. B* **106**, 8726–8738.

Srinivasan, J., Trevathan, M.W., Beroza, P., Case, D.A. (1999). Application of a pairwise generalized Born model to proteins and nucleic acids: Inclusion of salt effects. *Theor. Chem. Acc.* **101**, 426–434.

Still, W.C., Tempczyk, A., Hawley, R.C., Hendrickson, T. (1990). Semianalytical treatment of solvation for molecular mechanics and dynamics. *J. Am. Chem. Soc.* **112**, 6127–6129.

Tanizaki, S., Feig, M. (2005). A generalized Born formalism for heterogeneous dielectric environments: Application to the implicit modeling of biological molecules. *J. Chem. Phys* **122**, 12706–12713.

Tanizaki, S., Feig, M. (2006). Molecular dynamics simulations of large integral membrane proteins with an implicit membrane model. *J. Phys. Chem. B* **110**, 548–556.

Tieleman, D.P., Sansom, M.S.P., Berendsen, H.J.C. (1999a). Alamethicin helices in a bilayer and in solution: Molecular dynamics simulations. *Biophys. J.* **76**, 40–49.

Tieleman, D.P., Sansom, M.S.P., Berendsen, H.J.C. (1999b). An alamethicin channel in a lipid bilayer: Molecular dynamics simulations. *Biophys. J.* **76**, 1757–1769.

Tjong, H., Zhou, H.-X. (2007). GBr[6]: A parameterization-free, accurate, analytical generalized Born method. *J. Phys. Chem. B* **111**, 3055–3061.

Vereshaga, Y.A., Volynsky, P.E., Nolde, D.E., Arseniev, A.S., Efremov, R.G. (2007). Helix interactions in membranes: Lessons from unrestrained Monte Carlo simulations. *J. Chem. Theory Comput.* **1**, 1252–1264.

Wagoner, J., Baker, N.A. (2004). Solvation forces on biomolecular structures: A comparison of explicit solvent and Poisson–Boltzmann models. *J. Comput. Chem.* **25**, 1623–1629.

Wagoner, J.A., Baker, N.A. (2006). Assessing implicit models for nonpolar mean solvation forces: The importance of dispersion and volume terms. *Proc. Natl. Acad. Sci. USA* **103**, 8331–8336.

White, S.H., Wimley, W.C. (1999). Membrane protein folding and stability: Physical principles. *Annu. Rev. Biophys. Biomol. Struct.* **28**, 319–365.

Woolf, T.B., Roux, B. (1994). Molecular dynamics simulation of the gramicidin channel in a phospholipid bilayer. *Proc. Natl. Acad. Sci. USA* **91**, 11631–11635.

Zasloff, M. (2002a). Innate immunity, antimicrobial peptides, and protection of the oral cavity. *Lancet* **360**, 1116–1117.

Zasloff, M. (2002b). Antimicrobial peptides in health and disease. *N. Engl. J. Med.* **347**, 1199–1200.

Zasloff, M. (2002c). Antimicrobial peptides of multicellular organisms. *Nature* **415**, 389–395.

Zhang, G., Ross, C.R., Blecha, F. (2000). Porcine antimicrobial peptides: New prospects for ancient molecules of host defense. *Vet. Res.* **31**, 277–296.

CHAPTER 6

Blue Matter: Scaling of N-Body Simulations to One Atom per Node

Blake G. Fitch, Aleksandr Rayshubskiy, Maria Eleftheriou,
T.J. Christopher Ward, Mark Giampapa, Michael C. Pitman, Jed Pitera,
William C. Swope and Robert S. Germain

IBM Research Division, IBM Thomas J. Watson Research Center, 1101 Kitchawan, Rd Route 134, Yorktown Heights, NY 10598, USA

Abstract

N-body simulations present some of the most interesting challenges in the area of massively parallel computing, especially when the object is to improve the total time to solution for a fixed size problem. The Blue Matter molecular simulation framework has been developed specifically to address these challenges in order to explore programming models for massively parallel machine architectures in a concrete context and to support the scientific goals of the IBM Blue Gene project. This paper reviews the key issues involved in achieving ultra-strong scaling of methodologically correct biomolecular simulations, in particular, the treatment of the long range electrostatic forces present in simulations of proteins in water and membranes. Blue Matter computes these forces using the

1063-5823/08 $35.00
DOI: 10.1016/S1063-5823(08)00006-9

Particle–Particle–Particle–Mesh Ewald (P3ME) method which breaks the problem up into two pieces, one of which requires the use of three-dimensional Fast Fourier Transforms with global data dependencies and the other which involves computing interactions between pairs of particles within a cut-off distance. We will summarize our exploration of the parallel decompositions used to compute these finite-ranged interactions carried out as part of the Blue Matter development effort, describe some of the implementation details involved in these decompositions, and present the evolution in (strong scaling) performance achieved over the course of this exploration along with evidence for the quality of simulation achieved.

I. INTRODUCTION

The IBM Blue Gene Membrane Protein Science Effort provided a two-fold opportunity: Focus an unprecedented computational resource to gain insights into membrane proteins such as G-protein Coupled Receptors (GPCRs) through methodologically rigorous all-atom molecular dynamics simulation, and secondly, advance the state of the art in massively parallel molecular dynamics with the goal of reducing total time to solution (strong scaling). The effort to gain insights into GPRCs continues to yield fascinating results, from early insights into pure membranes (Pitman *et al.*, 2005b; Suits *et al.*, 2005) and membranes with cholesterol (Pitman *et al.*, 2004), to lipid–protein interactions (Grossfield *et al.*, 2006a, 2006b; Pitman *et al.*, 2005a), to insights into protein activation (Martinez-Mayorga *et al.*, 2006; Grossfield *et al.*, 2007). Advances in strong scaling of Blue Matter enabled microsecond scale membrane protein simulations on a practical timescale, and represent a decisive step forward in the history of massively parallel molecular dynamics applications. This chapter describes the Blue Matter strong scaling effort, which supports massively parallel, methodologically rigorous, all-atom molecular dynamics of complex biological systems such as membrane proteins in a native-like environment.

A. The Problem

Numerical simulation of molecular systems can yield unique insights into the details of the structure and dynamics of biomolecules (Karplus and McCammon, 2002). Such simulations are used to sample the configurations assumed by the biomolecule at specified temperature and also to study the dynamical evolution of the system under some specified set of conditions. Among the many challenges facing the biomolecular simulation community, increasing the time-scales probed by simulation in order to better make contact with physical experiment is the one that stresses computer systems to the utmost. Even sampling techniques that do not themselves yield up kinetic information can benefit from increased computation rate for a single trajectory.

B. N-Body Problem

Classical biomolecular simulation includes both Monte Carlo and molecular dynamics (Frenkel and Smit, 1996). The focus of our work has been on molecular dynamics although the Replica Exchange or Parallel Tempering Method (Sugita and Okamoto, 1999) which combines Molecular Dynamics with Monte Carlo-style moves has been implemented in Blue Matter as well (Eleftheriou et al., 2006b). Classical molecular dynamics is an N-body problem in which the evolution of the system is computed by numerical integration of the classical equations of motion. At each time step, forces on particles are computed and then the equations of motion are integrated to update the velocities and positions of the particles.

The forces on the particles can be classified as follows:

Bonded Forces These forces act between covalently bonded atoms and include bond stretches, angle bends, and torsions.

Non-bonded Forces These forces act between all pairs of particles and include the hard core repulsion and van der Waals interactions that are typically modeled by a Lennard-Jones 6-12 potential of the form $V_{ij}^{\text{L-J}}(r_{ij}) = 4 \cdot \epsilon_{ij} \cdot [(\sigma_{ij}/r)^{12} - (\sigma_{ij}/r)^6]$ as well as the electrostatic forces which have a potential energy of the form $V_e(r_{ij}) = q_i \cdot q_j / r_{ij}$.

The forces induced by the Lennard-Jones potential drop off rapidly with distance (varying as $1/r_{ij}^{13}$ and $1/r_{ij}^7$) and can be modeled as finite-ranged forces. The electrostatic forces fall off much more slowly with distance (varying as $1/r_{ij}^2$ and cannot be approximated simply by neglecting interactions between pairs of particles beyond some cut-off distance (even when a smooth switching function is used) (Bader and Chandler, 1992).

Ideally, a simulation would model a protein in an infinite volume of water, but this is not practical. Instead, the usual approach is to use periodic boundary conditions so that the simulation actually models an infinite array of identical cells containing the biomolecules under study along with the water. This is generally preferred over simulating a single finite box in a vacuum (or some dielectric) because it eliminates the interface at the surface of the simulation cell. The choice of simulation cell size is important. Too large, and unnecessary computations are being done. Too small, and the interactions between biomolecules in different cells of the periodic array can introduce artifacts. The most commonly used techniques for handling the long range interactions with periodic boundary conditions are based on the Ewald summation method and particle mesh techniques that divide the electrostatic force evaluation into a real-space portion that can be approximated by a potential with a finite range cut-off and a reciprocal space portion that involves a convolution of the charge distribution with an interaction kernel (Hockney and Eastwood, 1988; Darden *et al.*, 1993; Essmann *et al.*, 1995; Deserno and Holm, 1998a). This convolution is evaluated using a Fast Fourier Transform (FFT) method in all of the particle-mesh techniques including the Particle–Particle–Particle–Mesh (P3ME) technique (Hockney and Eastwood, 1988; Deserno and Holm, 1998a) used by Blue Matter. In this case, the $O(n^2)$ dependence on particle number n is reduced to $O(n \log n)$. The evaluation of the three dimensional FFT (3D-FFT) and its inverse on every time step imposes a global data dependency on the program. That is, the result depends on the position and value of *every* charge in the system.

Various algorithmic techniques for increasing the effective rate at which the kinetics of the systems evolves have been explored including kinetic acceleration techniques (Voter, 1998; Shirts and Pande, 2001) and multiple time-stepping algorithms (Tuckerman *et al.*, 1992). Issues like the appropriate identification of "transitions" within a simulation have raised concerns about the applicability of the kinetic acceleration techniques to biomolecular simulation. In our experience, even with a correct splitting of the electrostatic forces (Zhou *et al.*, 2001), the use of multiple time-stepping leads to significantly larger drifts in the total energy for NVE (constant particle number, volume, and energy) simulations over that obtained using the velocity Verlet integrator (Swope *et al.*, 1982). Our view is that direct kinetic simulation is still the most reliable technique for accessing long time-scale dynamical information in spite of the challenge of scaling of a fixed size N-body problem with some tens of thousands of particles onto a parallel computer with many thousands of nodes (strong scaling). The scale of this challenge can be envisioned by considering that a simulation rate of one microsecond per two weeks of wall clock time requires each molecular dynamics (MD) time-step to complete within 1.2 milliseconds, or fewer than one million processor clock cycles on Blue Gene/L when using a time-step size of 1 femtosecond (10^{-15} s). Carrying out such long time-scale simulation also potentially exposes "correctness" issues with a particular application's implementation of MD.

II. BACKGROUND ON MASSIVELY PARALLEL BIOMOLECULAR SIMULATION

The highest degree of strong scaling in the published literature prior to the availability of the Blue Gene/L hardware platform (Gara, 2005), was achieved by the NAMD (Kale *et al.*, 1999) package which demonstrated continued speed-up through about 60 atoms/processor (and a time per time-step of about 15 ms without multiple time-stepping) on the Pittsburgh Supercomputing Center Lemieux system using 1536 processors (Phillips *et al.*, 2002). At that time, the NAMD code used a combination of volume and force decompositions (Plimpton, 1995; Plimpton and Hendrickson, 1996) for the evaluation of the real-space forces that allowed them to create a large number of units of work that could be distributed for load balancing ($14\times$ the number of volume elements in the system where the dimension of each volume element was larger than a cutoff radius). Subsequently, the NAMD developers incrementally increased the number of units of work available by effectively "splitting" volume elements in half along a selected axis or axes. Also, the parallel decomposition used for the three dimensional Fast Fourier Transform (3D-FFT) in the NAMD code is a "slab" decomposition which limits the distribution of work for that module to N for a $N \times N \times N$ FFT. Much of the focus in the NAMD work has been on enabling simulations of large biomolecular systems rather than on taking smaller systems to longer time-scales.

The intellectual point of departure for much of the work on highly scalable parallel decompositions of the N-body problem (the real space portion at least) is the work of Plimpton and Hendrickson (Hendrickson and Plimpton, 1995; Plimpton, 1995), who demonstrated a force decomposition method for which the number of communicating partners of a single node scales like $O(\sqrt{p})$, where p is the number of nodes. Algorithms that attempted to achieve comparable theoretical scaling behavior within a volume decomposition have been published by Snir (Snir, 2004) and Shaw (Shaw, 2005) although to our knowledge, no implementation of either of these algorithms has been published. Subsequent to the publication of our initial description and performance characterization of the Blue Matter "V4" technique in September 2005 (Fitch *et al.*, 2005; Germain *et al.*, 2005), which will be described below, the D.E. Shaw team published a description of essentially the same technique (Bowers *et al.*, 2006b). An implementation of this algorithm on a commodity cluster was described in a subsequent publication (Bowers *et al.*, 2006a), but performance was reported using aggressive approximations including single precision arithmetic and multiple time-stepping with a reported energy drift of 7×10^{-4} K/ns ($100\times$ larger than the worst energy drift measured on Blue Matter under production or benchmarking conditions).

A. Background on Blue Matter

Blue Matter has provided a concrete context in which to explore algorithmic techniques and programming models required to exploit massively parallel machine architectures like Blue Gene as well as providing the capability required to execute one of the primary goals of IBMs Blue Gene project (Allen, 2001): To use the unprecedented computational resource developed during the course of the project to attack grand challenge life sciences problems such as advancing our understanding of biologically important processes, in particular, the mechanisms behind protein folding. Blue Matter has been used in production by computational scientists on the Blue Gene project since 2003, initially on the SP2 platform and later on Blue Gene/L (Swope *et al.*, 2004; Pitman *et al.*, 2004, 2005a, 2005b; Suits *et al.*, 2005; Grossfield *et al.*, 2006a; Eleftheriou *et al.*, 2006a; Martinez-Mayorga *et al.*, 2006).

Through use of the real space parallelization techniques described below and the highly scalable 3D-FFT developed as part of this project (Eleftheriou *et al.*, 2005), the Blue Matter molecular dynamics code has demonstrated continued speed-up through approximately one atom per node on 16,384 Blue Gene/L nodes and a time per time-step of under 2 ms for a 43,222 atom solvated membrane protein system. All of this was achieved with methodologically rigorous methods. Our success in meeting the challenges of strong scaling has enabled detailed atomistic simulations of biologically interesting systems at time-scales and with ensemble sizes that were previously unattainable including 26 trajectories of 100 nanoseconds each of a G-Protein Coupled Receptor (GPCR), Rhodopsin, in a realistic membrane environment (Grossfield *et al.*, 2006a) and multiple microsecond scale simulations of that system (Martinez-Mayorga *et al.*, 2006) and others. Furthermore, since the path to increased hardware performance now seems to lie more along the path of increasing concurrency (multiple CPU cores per chip and increased parallelism) rather than increasing clock speed (Greer, 2005), future work with even very large molecular systems with hundreds of thousands of atoms may require scalability to small ratios of atoms per node. While there have been some theoretical studies of scaling in this limit (Taylor *et al.*, 1997), the Blue Matter classical biomolecular simulation application running on Blue Gene/L represents the first demonstration of strong scaling of such a code to this degree (Fitch *et al.*, 2005; Germain *et al.*, 2005; Fitch *et al.*, 2006b, 2006a). With access to such time-scales comes increased concern about whether the simulations are valid and Blue Matter has also demonstrated the ability to generate trajectories with excellent energy conservation over microsecond time-scales.

III. INHERENT CONCURRENCY OF MOLECULAR DYNAMICS

We describe molecular dynamics as comprised of four major modules:

- real-space non-bonded (finite range pair interactions),
- K-space (FFT-based),
- bonded (graph-based),
- integration (per particle).

Before attempting to scale an algorithm onto many thousands of nodes, it is useful to estimate how much concurrency (potential for parallelism) is inherent in various components of that algorithm. First, consider the anatomy of a molecular dynamics time step starting with the availability of the coordinates and velocities of all of the particles in the system (r_i, v_i):

- Compute forces on each particle due to bonded (intramolecular) interactions
  - bond stretches,
  - angle bends,
  - torsions.
- Compute forces on each particle due to non-bond interactions (assuming periodic boundary conditions)
  - Hard core repulsive and van der Waals forces (usually represented by a Lennard-Jones 6-12 potential function that is smoothly switched off beyond some cutoff distance r_c).
  - Electrostatic forces ($1/r^2$ forces that are most commonly evaluated using the Ewald summation technique or its mesh-based variants (Deserno and Holm, 1998a)).
- Accumulate the total force on each particle and use that force along with the current position and velocity of particle to propagate the motion of the particle forward in time by some small increment.

It is possible to view a molecular dynamics time step (or any parallel computation) as the successive materialization of distributed data structures on which local computation takes place. Given a choice of granularity below which no parallelism will be attempted and taking into account the data dependencies in the algorithm, one can estimate the number of (data) independent computations required at each phase. An example of such an analysis is pictured in Fig. 1 for a molecular dynamics simulation using the P3ME method to treat the long-range electrostatic forces. One can afford a lack of concurrency in components that impose very little computation or communication burden, but eventually even these will become bottlenecks if they are not parallelized as well. This is an instance of Amdahl's Law, which states that the amount of speedup possible for a fixed size problem is limited by the fraction of the problem that is non-parallelizable (Amdahl, 1967). More precisely, the *PotentialSpeedup* $= 1/(f + (1 - f)/N)$ where f is the fraction of time consumed by serial operations (or ones that are replicated on every node) and N is the number of nodes.

In the limit of very strong scaling, the P3ME convolution step is expected to be the limiting factor assuming that a good distribution of work can be achieved for the bonded and real-space non-bonded force computations. We conjecture that this would be the case even for an architecture with full bisectional bandwidth because of latencies due to the hardware and software overheads in the successive communication phases required by P3ME shown in Fig. 1. The development of a highly scalable three-dimensional Fast Fourier Transform (3D-FFT) was essential for Blue Matter and has been reported on in detail previously (Eleftheriou *et al.*, 2006a).

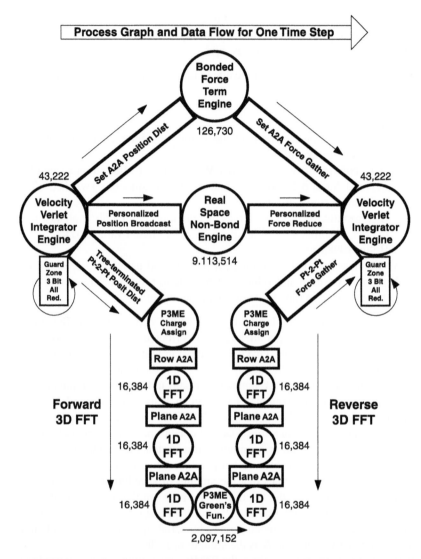

FIGURE 1 Diagram showing the data dependencies and opportunities for concurrency in various stages of the non-bonded force calculations in a molecular dynamics time step using the P3ME method. For each step, the actual number of independent work items are displayed for one of the molecular systems, Rhodopsin, benchmarked in this paper. Three separate threads of computation are shown: finite-ranged pair interactions, bonded interactions, and long-ranged electrostatic interactions computed via the P3ME method. The center circle at the bottom of the figure represents the convolution step which is evaluated by first Fourier-transforming the meshed charge distribution, then multiplying by a kernel (Green's function), and then inverse Fourier-transforming the result to obtain the meshed electrostatic potential. A detailed explanation of the P3ME method shown here can be found in (Deserno and Holm, 1998a).

IV. PARALLEL DECOMPOSITIONS

A. Introduction

Our explorations of parallel decompositions have included replicated data methods that leverage the hardware facilities of Blue Gene/L to globalize and reduce data structures (Fitch *et al.*, 2003; Germain *et al.*, 2005) and spatial/interaction decompositions (Fitch *et al.*, 2005; Germain *et al.*, 2005; Fitch *et al.*, 2006a). This paper focuses on the two spatial decompositions used in Blue Matter for which we have strong scaling data. We refer to these two decompositions, which will be described in more detail below, as V4 and V5. Our investigations began with the realization that for finite-range pair potentials, given a three-dimensional domain decomposition of the simulation cell onto a set of nodes, it is possible to limit the broadcast of positions by any originating node to those nodes containing a portion of simulation space within half a cut-off radius of the boundary of the originating node. The first spatial decomposition deployed in Blue Matter, V4, uses geometric criteria to determine where the real-space non-bonded interaction between two particles will be computed, namely on whichever node contains the point half-way between the two particles. This provides a large number of distinct units of computational work that can be distributed by partitioning space using an Optimal Recursive Bisection (ORB) scheme (Fitch *et al.*, 2005; Germain *et al.*, 2005; Fitch *et al.*, 2006b). The implementation of this method entails the broadcast of a particle's position to a sphere with radius R_{eff}. Nominally R_{eff} is be half the molecular dynamics cut-off distance for real-space non-bond interactions for both V4 and V5. However, enabling the preservation of particle assignments to nodes over several time-steps requires the introduction of a guard zone that increases the R_{eff} beyond half the molecular dynamics cut-off. The size of the guard zone is a tuning parameter. The use of this guard zone in V4 also ensures that the particle positions required by the K-space and bonded force modules are available.

The most recent parallel decomposition, referred to as V5, uses geometry primarily as a heuristic to prime the set-based optimization process that follows (Fitch *et al.*, 2006a). Where V4 managed data distribution (of particle positions) and reduction (of forces) via a data "push" and caching module, V5 specializes communications between the integrator module and the three force computation modules described above. Where the V4 push/cache method required a distinct communications phase, V5 allows overlap of one module's communication and/or computation with another module's communication and/or computation. On Blue Gene/L, which has two processors per node, modules are scheduled to cores to maximize overlap. Currently the scheduling is static and we place the longest-running module on its own core.

Achieving a reduction in the number of communicating partners per node (for the real-space non-bond computations) was the major driver behind the evolution from V4 to V5. In Blue Matter V4, each node ("home" node) broadcasts the positions of the particles homed on that node to each node containing a volume element that intersects the volume enclosed by an imaginary shell extending half a cut-off distance (with some additional guard distance) away from the volume owned by the "home" node as shown in Fig. 2a. This local volumetric broadcast and the corresponding reduction of forces back to the "home" node involve a number of nodes that scales like $O(p)$, the same scaling obtained with a classic volumetric decomposition (though broadcasting to only $1/8$ of the volume in simulation space). The advantage of a technique like that used by Blue Matter V4 on a mesh interconnect topology is that the number of hops for each message is minimized.

The next incremental step in improving the V4 algorithm was taken by noting that all interactions can still be computed if the local broadcast only goes to nodes containing a portion of the imaginary shell that forms the boundary of volumetric broadcast used in the V4 technique as shown in Fig. 2b. This improves the scaling of the number of communicating partners per node to $O(p^{2/3})$. In this method, the interaction between two particles can be computed on any of the nodes that contain the intersection of the broadcast shells for the nodes containing the two particles. In contrast to the V4 technique, there is no simple geometric construction (analogous to the choice of the mid-point in V4) to select the node that should compute the interaction.

The insight that the real space non-bond algorithm can be cast as an optimization problem leads to the method actually implemented in Blue Matter V5 (Fitch *et al.*, 2006a). The full optimization problem is to minimize the average execution time per MD time-step and this is beyond our ability to solve at present. Attempting to minimize the number of communicating partners for each node subject to the constraint of load balance a much more tractable problem, particularly given a good heuristic (such as the broadcast to a shell described above) for starting this optimization process.

We begin a description of the optimization process used in V5 with a set of definitions and initializations:

1. Begin by defining the set of nodes to which each node will *potentially* broadcast particle coordinates. In the case of Blue Matter V5, this is the set of nodes which contain some portion of the surface in simulation space defined by the locus of points that are exactly a broadcast distance r_{eff} away from the surface of the simulation space volume assigned to the originating or "central" node i. We call this the "Candidate Send To Node Set" or in the case of V5, the "Surface Node Set" which we will denote by C_i. Of course, the optimization process could use the set of nodes defined by the broadcast volume in V4 or some other set of nodes as a starting point.

2. For each pair of nodes **i** and **j**, define the "Interaction Assignment Option Set" to be the intersection of the corresponding Candidate Send To Node Sets. In principle, the task of computing the interactions between particles

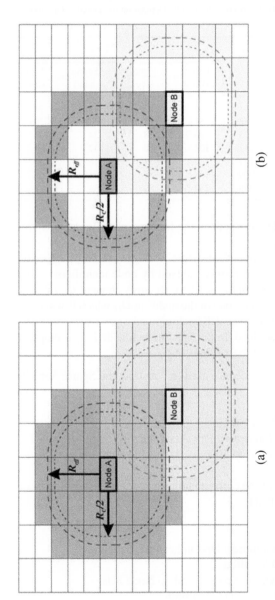

(a)

(b)

FIGURE 2   (a) Broadcast to nodes containing any portion of simulation space within R_{eff} of the boundary of the central node (V4). (b) Broadcast to nodes containing any point in simulation space which is exactly R_{eff} away from the boundary of the central node (V5 full skin before optimization). These figures illustrate the basic geometric ideas behind the spatial decompositions used by Blue Matter. They represent a two-dimensional slice of the simulation space with the domain decomposition onto nodes shown by the array of rectangular cells superimposed on the figures. Node A broadcasts positions (of particles within the volume of simulation space that it manages) to nodes shaded "salmon" or brown. Node B broadcasts positions to nodes shaded cyan or brown. The coarse dashed lines are drawn at a distance R_{eff} from the central nodes A and B. R_{eff} is somewhat larger than half the cut-off distance, R_c used for the real space non-bond forces to provide for a "guard zone". The cells shaded brown receive positions from both node A and node B and can therefore compute forces between particles. While the figure on the left shows all of the possible nodes that could compute the interactions between nodes A and B in brown, the Blue Matter V4 implementation actually assigns the interaction between two particles to the node containing the point in simulation space halfway between the two particles. The figure on the right shows the starting point for the optimization process used in Blue Matter V5. The nodes shown in brown in this figure are part of the "Interaction Option Set" for the node pair (A, B) whose role in the V5 optimization procedure is described in the text.

TABLE I

Communicating Partner Counts as a Function of Partition Size for V4 Volumetric Broadcast, "Full Skin" Broadcast, and V5 Set-Based Optimization. The V5 Results Give the Minimum and Maximum Values as Well as the Average Since the V5 Algorithm Results Depend on the Detailed Distribution of Particles

Node count	Partition			V4	V5			
	$\mathbf{P}_x$	$\mathbf{P}_y$	$\mathbf{P}_z$		Full skin	Sparse (avg.)	Sparse (min.)	Sparse (max)
512	8	8	8	45	42	22	19	24
1024	8	8	16	63	58	27	24	30
2048	8	16	16	105	90	38	34	40
4096	16	16	16	147	122	50	47	54
4096	8	32	16	147	122	51	46	52
8192	16	32	16	209	162	65	61	67
16384	16	32	32	349	242	89	84	91

in **i** and **j** can be assigned to any node in the Interaction Assignment Option Set which we will call O_{ij}.

3. The "Interacting Pair Assignment" structure is defined to be an upper triangular two-dimensional integer array indexed by node identifiers **i** and **j**. When the optimization procedure is finished, the array element I_{ij} will contain the node identifier of exactly one of the nodes in O_{ij} which is where the interactions between the node pair will be computed.

4. We also define the "Sparse Send To Set" S_i for each node **i**. S_i is empty at the start, but at the end S_i will be a subset of C_i.

Next, we outline the iterative procedure used to construct the Sparse Send To Sets S_i:

{Construct Sparse Send To Node Set}
Initialize all elements of the Interacting Pair Assignment structure I_{ij} to -1
Let sequence $P = \{(i, j) \in \mathcal{P} \times \mathcal{P} | (i < j)\}$ where $\mathcal{P}$ is the set of node identifiers
Sort P according to the size of the corresponding Interaction Assignment Option Set $\|O_{ij}\|$ {smallest first}
for $k = 0$ to $\|P\| - 1$ **do**
   $(i, j) = P_k$
   $D = O_{ij}$ {Interaction Assignment Option Set}
   if $\exists a \in \mathcal{P} | a \in (D \cap S_i \cap S_j)$ **then**
      {No need add any nodes to Sparse Send To Node Sets}
   else if $\exists a \in \mathcal{P} | a \in (D \cap S_i) \vee (D \cap S_j)$ **then**
      Choose the element a that appears in the smallest number of Sparse Send To Node Sets S_n and append it to S_i and to S_j {One of these appends will be a no-op because $(a \in S_i) \vee (a \in S_j)$ already}

else
> From $b \in D$ choose $b|b$ appears in the smallest number of Sparse Send
> To Sets S_m and append it to S_i and to S_j

end if
end for

The results of this optimization process in terms of the communicating partner count are shown in Table I. In fact, scaling plots of the communicating partner count as a function of node count (Fitch *et al.*, 2006a) indicate that the Blue Matter V5 algorithm equals and sometimes betters the $O(\sqrt{p})$ scaling of the number of communicating partners achieved by the Plimpton–Hendrickson technique.

V. RESULTS

A. Performance Data

Figure 3 plots the computational throughput in time steps per second versus the number of atoms per node. Plotting the data as a function of atoms per node

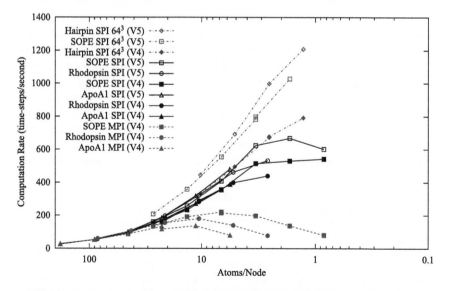

FIGURE 3 This plot shows computational rates on a number of molecular systems as a function of the number of atoms per node. This facilitates comparisons between systems of different sizes and also explicitly shows the degree of strong scaling achieved. Results for Blue Matter using the V5 method running on the low level communications System Programming Interface (SPI) provided by the Blue Gene/L Advanced Diagnostic Environment (Giampapa *et al.*, 2005) are plotted along with results from the V4 decomposition implemented on both MPI and SPI. Details on the systems studied are provided in Table II.

TABLE II

Details About the Systems Benchmarked with Blue Matter. All Runs Were Made with the Velocity Verlet Integrator (Swope *et al.*, 1982), Used the Particle–Particle–Particle–Mesh (P3ME) Technique to Handle Long Range Electrostatic Interactions, and Were Constant Particle Number, Volume, and Energy (NVE) Simulations. Rigid Water Models Were Used and All Heavy-Atom to Hydrogen Bonds in Non-Water Molecules Were Constrained Using Rattle (Andersen, 1983). All Runs Performed the P3ME Calculation on Every Time-Step. Except for the SOPE (64^3) Data, These Choices Are Those Used in Production Scientific Work (Hairpin, Rhodopsin) or Attempt to Match (or Exceed) Benchmarking Conditions Reported Elsewhere (ApoA1)

System	Total atoms	Cutoff/ switch (Å)	P3ME mesh	Time step (fs)
Hairpin	5239	9.0/1.0	64^3	1
SOPE	13,758	9.0/1.0	64^3, 128^3	1
Rhodopsin	43,222	9.0/1.0	128^3	2
ApoA1	92,224	10.0/2.0	128^3	1

provides some degree of normalization for system size when the real-space non-bonded interactions are the dominant contribution to the iteration time. One can observe that the scalability plots in the left-hand portion of Fig. 3 (corresponding to larger values of atoms/node and lower node counts) seem to lie on a "universal" curve. This would be even more "universal" if we plotted the number of real-space non-bond interactions per node rather than atoms per node on the horizontal axis. Figure 3 includes the Blue Matter V5 (SPI only) data as well the V4 data on both SPI and MPI (Fitch *et al.*, 2006b).

Table III provides both the time per time-step and the amount of time required for the neighborhood broadcast and reduce required by the V4 and V5 real-space non-bond algorithms. The effects of the reduction in numbers of communicating partners from V4 to V5 shown in Table I for Rhodopsin is manifest in the broadcast and reduction data in Table III. These data also show the dramatic difference

TABLE III

Performance Data for V4 and V5 Algorithms. Results for Two 4096 Node Partition Geometries Are Shown: 4096^r is $8 \times 32 \times 16$ and 4096^c is $16 \times 16 \times 16$

(a) β-Hairpin

Node count	Total (ms)		Broadcast (ms)		Reduce (ms)		Atoms per node
	V4 SPI	V5 SPI	V4 SPI	V5 SPI	V4 SPI	V5 SPI	
512	3.01	2.25	0.25	0.11	0.24	0.11	10.2
1024	2.02	1.45	0.26	0.11	0.25	0.10	5.1
2048	1.48	1.00	0.25	0.12	0.23	0.10	2.6
4096^r	1.52	1.12	0.26	0.14	0.23	0.17	1.3
4096^c	1.26	0.83	0.24	0.12	0.22	0.08	1.3

(b) SOPE

Node count	Total (ms)				Broadcast (ms)				Reduce (ms)				Atoms per node
	V4		V5		V4		V5		V4		V5		
	MPI	SPI	SPI	SPI 64^3	MPI	SPI	SPI	SPI 64^3	MPI	SPI	SPI	SPI 64^3	
512	7.47	6.81	6.22	4.83	0.56	0.35	0.16	0.13	0.44	0.35	0.19	0.16	26.9
1024	5.25	4.30	3.89	2.79	0.63	0.31	0.14	0.10	0.53	0.30	0.16	0.12	13.4
2048	4.66	2.81	2.45	1.80	0.88	0.25	0.12	0.09	0.86	0.23	0.15	0.10	6.7
4096^r	5.61	2.57	2.11	1.28	1.38	0.25	0.14	0.09	1.33	0.24	0.13	0.10	3.4
4096^c	5.08	1.95	1.60	1.25	1.40	0.22	0.12	0.08	1.31	0.21	0.15	0.08	3.4
8192	7.31	1.89	1.50	0.97	2.49	0.23	0.14	0.08	2.21	0.21	0.11	0.08	1.7

(c) Rhodopsin

Node count	Total (ms)			Broadcast (ms)			Reduce (ms)			Atoms per node
	V4		V5	V4		V5	V4		V5	
	MPI	SPI	SPI	MPI	SPI	SPI	MPI	SPI	SPI	
512	16.77	16.82	17.52	0.77	0.47	0.39	0.54	0.51	0.44	84.4
1024	9.42	9.50	9.48	0.77	0.39	0.24	0.59	0.39	0.26	42.2
2048	6.46	5.58	5.07	0.95	0.35	0.19	0.77	0.33	0.19	21.1
4096^r	5.83	3.55	3.21	1.44	0.28	0.19	1.23	0.26	0.15	10.6
4096^c	5.56	3.47	3.11	1.42	0.29	0.18	1.16	0.27	0.15	10.6
8192	7.17	2.51	2.16	2.47	0.24	0.20	2.05	0.23	0.13	5.3
16384	12.88	2.28	1.88	5.30	0.25	0.28	4.52	0.24	0.20	2.6

(d) ApoA1

Node count	Total (ms)			Broadcast (ms)			Reduce (ms)			Atoms per node
	V4		V5	V4		V5	V4		V5	
	MPI	SPI	SPI	MPI	SPI	SPI	MPI	SPI	SPI	
512	35.56	36.37	38.42	1.05	1.08	0.66	0.68	0.97	0.90	180.1
1024	19.29	19.18	18.95	1.02	0.74	0.40	0.71	0.76	0.51	90.1
2048	11.48	10.68	9.97	1.14	0.60	0.26	0.88	0.55	0.30	45.0
4096^r	8.57	6.33	5.39	1.66	0.51	0.22	1.54	0.47	0.23	22.5
4096^c	7.55	5.97	5.44	1.37	0.50	0.19	1.23	0.48	0.21	22.5
8192	7.34	3.68	3.14	2.22	0.43	0.15	2.15	0.38	0.16	11.3
16384	12.58	2.57	2.09	4.83	0.40	0.13	4.80	0.34	0.13	5.6

in scalability between MPI and SPI for V4 that has been previously reported (Fitch *et al.*, 2007).

The application-based tracing facility used by Blue Matter allows us to collect timing data from all of the nodes in the system. Trace points in this facility are placed in start/finish pairs within the source code and are turned on by compile time macro definitions. For a single node, a timing diagram could be constructed

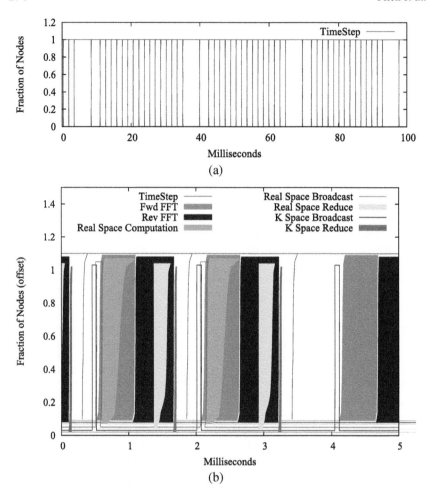

FIGURE 4 (a) Time-step start/finish shown. (b) Magnified view showing a few time-steps with additional trace point start/stop pairs shown. "Timing" diagram style plot in which the distribution function for the start and finish trace points, corresponding to the start and finish of a time step on each node, is computed. For each time step, the distribution function, $F_{start}(x)$, of the "start" trace points are plotted first, followed by the distribution, $F_{finish}(x)$, for the "finish" trace points. Some of the traces are shown with fills and the individual traces have been slightly offset in the vertical direction for clarity.

by setting an indicator variable equal to 1 when the start trace point is encountered and resetting it to 0 after the stop trace point reached. In order to create something analogous to a timing diagram that contains aggregated data from all the nodes, we compute the distribution function of time stamps (over nodes) for every time step. The distribution function is a function of time giving the fraction of nodes which have already "encountered" the specified trace point at the time passed to

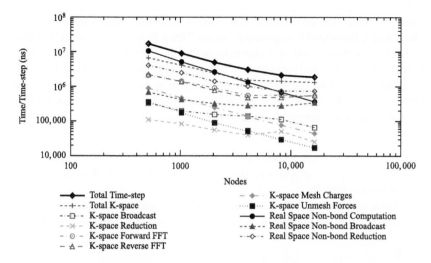

Nodes

- ◆— Total Time-step
- --+-- Total K-space
- --□-- K-space Broadcast
- --✕-- K-space Reduction
- -◉- K-space Forward FFT
- -△- K-space Reverse FFT
- -◆- K-space Mesh Charges
- ▪ K-space Unmesh Forces
- ●— Real Space Non-bond Computation
- --▲-- Real Space Non-bond Broadcast
- --◇-- Real Space Non-bond Reduction

FIGURE 5 This figure shows the contributions of selected components to the total time-step as a function of node count for the Rhodopsin system run using Blue Matter V5. Each major force engine has its own communication driver, hence the separate broadcast/reduce operations for Real-space and K-space (as well as for the Bonded forces which are not shown). One should note that there are periods during a time-step when both cores are making use of the torus simultaneously as can be seen in Fig. 4b. This decreases the performance of the 3D-FFT operations, but the overlapping of communications operations gives an improvement in overall performance.

the distribution function as an argument. This information is used to construct the plots containing aggregated data from all 16,384 nodes shown in Fig. 4 for Blue Matter V5 running the Rhodopsin system on Blue Gene/L.

In Fig. 4a, only the distribution functions of the start/finish trace point pairs that bracket the entire time step are shown. The periodic "long" time step is caused by the need to migrate particles from one node to another as their positions shift. We avoid migration on every time step by keeping track of particles that are somewhat farther away than half the cut-off distance (the additional amount is the "guard zone" mentioned earlier) and also monitoring the drift of each particle since the last migration. When one or more particles drifts far enough, an alarm is raised using the fast barrier facilities available on BG/L and a new migration phase follows.

Figure 4b zooms in on a few time steps and plots distribution functions for a number of important components of a time step. It is obvious from this plot that the forward and reverse 3D-FFT operations dominate the total time-step. Also, the scheduling of communication and computation tasks on both CPUs on each node can also be seen, e.g. the overlap of the reverse 3D-FFT and the real space reduce.

A detailed breakdown of the time required for the various operations required to complete a time-step is provided for the Rhodopsin system in Fig. 5. From this figure, it is clear that the dominant contribution to the time-step switches from the "Real Space Non-bond Computation" to the "Total K-space" above 2048 nodes. At the highest node count, it is also evident that the time required for the various localized broadcasts and reductions is increasing and even if the "Total K-space" contribution could be reduced, the reductions, in particular, would become the limiting factors for scalability. One should also keep in mind that the total time per time-step is not simply the sum of the various contributions shown in Fig. 5 since:

- Both cores are used on each node and so, for example, the "Real Space Non-bond Computation" can overlap with portions of the "Total K-space" work as seen in Fig. 4b.
- Some of the quantities plotted are already aggregations of other quantities in the plot, e.g. the various "K-space" contributions and "Total K-space".

B. Energy Drift in Long NVE Simulations

In general, a computational scientist will want to use the largest time-step size possible consistent with "correctness" in order to maximize throughput. Other performance-critical simulation parameters affecting simulation accuracy and stability include the FFT mesh spacing for Particle–Particle–Particle–Mesh (P3ME) methods (Deserno and Holm, 1998b) and the force-splitting scheme and time-step ratios chosen for symplectic multiple time-stepping methods (Sexton and Weingarten, 1992; Tuckerman *et al.*, 1992; Zhou *et al.*, 2001). Determining the optimal parameters for simulations enabled by multi-teraflop and larger machines that involve billions or tens of billions of time-steps provides a considerable challenge. Figure 6 shows plots of the change in the total energy in a simulation of a 66,728 atom solvated protein system, the Lambda Repressor, for a series of time-step sizes. Previously, we have reported a measured energy drift of 6×10^{-4} K/ns (over a 1.6 µs simulation) for the 43,222 atom Rhodopsin system using a 2 fs time-step (Fitch *et al.*, 2006a; Fitch *et al.*, 2007) which is consistent with the drift of 5.1×10^{-4} K/ns seen in the 2 fs time-step data for the Lambda Repressor.

VI. SUMMARY AND CONCLUSIONS

We have described the progression of parallel decompositions for the real-space non-bond forces explored as part of the Blue Matter effort. The progression started with non-geometric replicated data decompositions, continued with a spatial/interaction hybrid decomposition with geometric assignment of workload (V4), and culminated in the V5 decomposition that uses geometry only as a heuristic that defines the starting point of the set-based optimization procedure

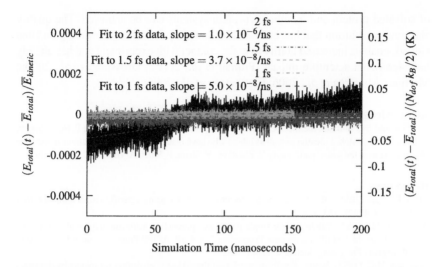

FIGURE 6 Energy drift for different values of Verlet integration time-step. The left axis shows the deviation from the average total energy normalized by the average kinetic energy while the right axis shows the deviation from the average total energy in equivalent temperature units (simply obtained by multiplying the left axis by the average "instantaneous temperature" of 394 K). The fitted slopes in terms of equivalent temperature per unit time are 2×10^{-5} K/ns 1.5×10^{-5} K/ns and 5.1×10^{-4} K/ns for time-step sizes of 1 fs, 1.5 fs, and 2 fs, respectively.

for interaction assignment. As implemented in Blue Matter, these algorithms have made methodologically rigorous simulations of biologically interesting systems on the microsecond scale routine on the Blue Gene Watson facility (see Table IV). Furthermore, we have demonstrated that excellent energy conservation over microsecond scale NVE (constant particle number, volume, and energy) simulations

TABLE IV

The Performance Results for Blue Matter V5 Expressed as the Number of Days Required to Simulate One Microsecond for a Specified Time-Step Size

	β-Hairpin	SOPE 64^3	SOPE 128^3	Rhodopsin	ApoA
Time-step (fs)	1.0	1.0	1.0	2.0	1.0
Partition size	Days required to reach 1 µs				
512	26.0	55.9	72.0	101.4	444.7
1024	16.8	32.4	45.0	54.9	219.3
2048	11.6	20.8	28.4	29.3	115.4
4096^r	13.0	14.9	24.4	18.6	63.0
4096^c	9.6	14.6	18.5	18.0	62.4
8192		11.2	17.4	12.5	36.3
16,384			19.2	10.9	24.2

of solvated protein and membrane/protein systems can be achieved. The quality and time-to-solution that has been demonstrated by Blue Matter running on Blue Gene/L enables increased ability to make contact with experiment and has already had significant scientific impact (Pitman *et al.*, 2005a; Grossfield *et al.*, 2006a; Martinez-Mayorga *et al.*, 2006; Eleftheriou *et al.*, 2006a).

Acknowledgements

We would like to thank the members of the Blue Gene hardware team including P. Heidelberger, A. Gara, M. Blumrich, J. Sexton as well as others who have made use of and contributions to the Blue Matter code over the years, particularly Y. Zhestkov, Y. Sham, F. Suits, A. Grossfield, and R. Zhou.

References

Allen, F., *et al.* (2001). Blue Gene: A vision for protein science using a petaflop supercomputer. *IBM Systems J.* **40** (2), 310–327.

Amdahl, G.M. (1967). Validity of the single-processor approach to achieving large scale computing capabilities. In: "AFIPS Conference Proceedings", vol. **30**. AFIPS Press, American Federation of Information Processing Societies, pp. 483–485.

Andersen, H.C. (1983). Rattle: A velocity version of the, SHAKE algorithm for molecular dynamics calculations. *J. Comput. Phys.* **52** (1), 24–34.

Bader, J.S., Chandler, D. (1992). Computer simulation study of the mean forces between ferrous and ferric ions in water. *J. Phys. Chem.* **96** (15), 6423–6427.

Bowers, K.J., Chow, E., Xu, H., Dror, R.O., Eastwood, M.P., Gregersen, B.A., Klepeis, J.L., Kolossvary, I., Moraes, M.A., Sacerdoti, F.D., Salmon, J.K., Shan, Y., Shaw, D.E. (2006a). Molecular dynamics—scalable algorithms for molecular dynamics simulations on commodity clusters. In: "SC '06: Proceedings of the 2006 ACM/IEEE Conference on Supercomputing". ACM Press, New York, NY, USA, p. 84.

Bowers, K.J., Dror, R.O., Shaw, D.E. (2006b). The midpoint method for parallelization of particle simulations. *J. Chem. Phys.* **124** (18), 184109.

Darden, T., York, D., Pedersen, L. (1993). Particle mesh Ewald: An $n \log(n)$ method for Ewald sums in large systems. *J. Chem. Phys.* **98** (12), 10089–10092.

Deserno, M., Holm, C. (1998a). How to mesh up, Ewald sums. I. A theoretical and numerical comparison of various particle mesh routines. *J. Chem. Phys.* **109** (18), 7678–7693.

Deserno, M., Holm, C. (1998b). How to mesh up, Ewald sums. II. An accurate error estimate for the particle–particle–particle–mesh algorithm. *J. Chem. Phys.* **109** (18), 7694–7701.

Eleftheriou, M., Fitch, B., Rayshubskiy, A., Ward, T.J.C., Germain, R.S. (2005). Performance measurements of the 3d, FFT on the Blue Gene/L supercomputer. In: Cunha, J.C., Medeiros, P.D. (Eds.), "Euro-Par 2005 Parallel Processing: 11th International Euro-Par Conference, Lisbon, Portugal, August 30–September 2, 2005". In: "Lecture Notes in Computer Science", vol. **3648**. Springer-Verlag, New York, pp. 795–803.

Eleftheriou, M., Germain, R.S., Royyuru, A.K., Zhou, R. (2006a). Thermal denaturing of mutant lysozyme with both the oplsaa and the charmm force fields. *J. Am. Chem. Soc.* **128** (41), 13388–13395.

Eleftheriou, M., Rayshubskiy, A., Pitera, J.W., Fitch, B.G., Zhou, R., Germain, R.S. (April 2006b). Parallel implementation of the replica exchange molecular dynamics algorithm on Blue Gene/L. In: "Fifth IEEE International Workshop on High Performance Computational Biology".

Essmann, U., Perera, L., Berkowitz, M.L., Darden, T., Lee, H., Pedersen, L.G. (1995). A smooth particle mesh, Ewald method. *J. Chem. Phys.* **103** (19), 8577–8593.

Fitch, B.G., Germain, R.S., Mendell, M., Pitera, J., Pitman, M., Rayshubskiy, A., Sham, Y., Suits, F., Swope, W., Ward, T.J.C., Zhestkov, Y., Zhou, R. (2003). Blue Matter, an application framework for molecular simulation on Blue Gene. *J. Parallel Distrib. Comput.* **63**, 759–773.

Fitch, B.G., Rayshubskiy, A., Eleftheriou, M., Christopher Ward, T.J., Giampapa, M., Zhestkov, Y., Pitman, M.C., Suits, F., Grossfield, A., Pitera, J., Swope, W., Zhou, R., Germain, R.S., Feller, S. (August 2005). Blue Matter: Strong scaling of molecular dynamics on Blue Gene/L. Research Report RC23688, IBM Research Division.

Fitch, B.G., Rayshubskiy, A., Eleftheriou, M., Christopher Ward, T.J., Giampapa, M., Pitman, M.C., Germain, R.S. (2006a). Molecular dynamics—Blue Matter: Approaching the limits of concurrency for classical molecular dynamics. In: "SC'06: Proceedings of the 2006 ACM/IEEE Conference on Supercomputing". ACM Press, New York, NY, USA, p. 87.

Fitch, B.G., Rayshubskiy, A., Eleftheriou, M., Christopher Ward, T.J., Giampapa, M., Zhestkov, Y., Pitman, M.C., Suits, F., Grossfield, A., Pitera, J., Swope, W., Zhou, R., Feller, S., Germain, R.S. (2006b). Blue Matter: Strong scaling of molecular dynamics on Blue Gene/L. In: Alexandrov, V., van Albada, D., Sloot, P., Dongarra, J. (Eds.), "International Conference on Computational Science (ICCS, 2006)". In: "Lecture Notes in Computer Science", vol. **3992**. Springer-Verlag, New York, pp. 846–854.

Fitch, B.G., Rayshubskiy, A., Eleftheriou, M., Christopher Ward, T.J., Giampapa, M., Pitman, M.C., Germain, R.S. (2007). Progress in scaling biomolecular simulations to petaflop scale platforms. In: Lehner, W., Meyer, N., Streit, A., Stewart, C. (Eds.), "Euro-Par 2006 Workshops: Parallel Processing". In: "Lecture Notes in Computer Science", vol. **4375**. Springer-Verlag, New York, pp. 279–288.

Frenkel, D., Smit, B. (1996). "Understanding Molecular Simulation". Academic Press, San Diego, CA.

Gara, A., et al. (2005). Overview of the, Blue Gene/L system architecture. *IBM J. Res. Develop.* **49** (2/3), 195–212.

Germain, R.S., Fitch, B., Rayshubskiy, A., Eleftheriou, M., Pitman, M.C., Suits, F., Giampapa, M., Christopher Ward, T.J. (2005). Blue Matter on Blue Gene/L: Massively parallel computation for biomolecular simulation. In: "CODES+ISSS '05: Proceedings of the 3rd IEEE/ACM/IFIP International Conference on Hardware/Software Codesign and System Synthesis". ACM Press, New York, NY, USA, pp. 207–212.

Germain, R.S., Zhestkov, Y., Eleftheriou, M., Rayshubskiy, A., Suits, F., Ward, T.J.C., Fitch, B.G. (2005). Early performance data on the, Blue Matter molecular simulation framework. *IBM J. Res. Develop.* **49** (2/3), 447–456.

Giampapa, M.E., Bellofatto, R., Blumrich, M.A., Chen, D., Dombrowa, M.B., Gara, A., Haring, R.A., Heidelberger, P., Hoenicke, D., Kopcsay, G.V., Nathanson, B.J., Steinmacher-Burow, B.D., Ohmacht, M., Salapura, V., Vranas, P. (2005). Blue Gene/L advanced diagnostics environment. *IBM J. Res. Develop.* **49** (2/3), 319–332.

Greer, D. (2005). Industry trends: Chip makers turn to multicore processors. *IEEE Computer* **38** (5), 11–15.

Grossfield, A., Feller, S.E., Pitman, M.C. (2006a). A role for direct interactions in the modulation of rhodopsin by ω-3 polyunsaturated lipids. *Proc. Nat. Acad. Sci.* **103** (13), 4888–4893.

Grossfield, A., Feller, S.E., Pitman, M.C. (2006b). Contribution of omega-3 fatty acids to the thermodynamics of membrane protein solvation. *J. Phys. Chem. B* **110** (18), 8907–8909.

Grossfield, A., Feller, S.E., Pitman, M.C. (2007). Convergence of molecular dynamics simulations of membrane proteins. *Proteins: Structure, Function, and Bioinformatics* **67** (1), 31–40.

Hendrickson, B., Plimpton, S. (1995). Parallel many-body simulations without all-to-all communication. *J. Parallel Distrib. Comput.* **27** (1), 15–25.

Hockney, R.W., Eastwood, J.W. (1988). "Computer Simulation Using Particles". Institute of Physics Publishing.

Kale, L., Skeel, R., Bhandarkar, M., Brunner, R., Gursoy, A., Krawetz, N., Phillips, J., Shinozaki, A., Varadarajan, K., Schulten, K. (1999). Namd2: Greater scalability for parallel molecular dynamics. *J. Comput. Phys.* **151** (1), 283–312.

Karplus, M., McCammon, J.A. (2002). Molecular dynamics simulations of biomolecules. *Nature Struct. Biol.* **9** (9), 646–652.

Martinez-Mayorga, K., Pitman, M.C., Grossfield, A., Feller, S.E., Brown, M.F. (2006). Retinal counterion switch mechanism in vision evaluated by molecular simulations. *J. Am. Chem. Soc.* **128** (51), 16502–16503.

Phillips, G., Zheng, J.C., Kumar, S., Kale, L.V. (2002). NAMD: Biomolecular simulation on thousands of processors. In: "Supercomputing '02: Proceedings of the 2002 ACM/IEEE Conference on Supercomputing". IEEE Computer Society Press, Los Alamitos, CA, USA, pp. 1–18.

Pitman, M.C., Suits, F. Jr., MacKerell, A.D., Feller, S.E. (2004). Molecular-level organization of saturated and polyunsaturated fatty acids in a phosphatidylcholine bilayer containing cholesterol. *Biochemistry* **43** (49), 15318–15328.

Pitman, M.C., Grossfield, A., Suits, F., Feller, S.E. (2005a). Role of cholesterol and polyunsaturated chains in lipid–protein interactions: Molecular dynamics simulation of rhodopsin in a realistic membrane environment. *J. Am. Chem. Soc.* **127** (13), 4576–4577.

Pitman, M.C., Suits, F., Gawrisch, K., Feller, S.E. (2005b). Molecular dynamics investigation of dynamical properties of phosphatidylethanolamine lipid bilayers. *J. Chem. Phys.* **122** (24), 244715.

Plimpton, S., Hendrickson, B. (1996). A new parallel method for molecular dynamics simulation of macromolecular systems. *J. Comput. Chem.* **17** (3), 326–337.

Plimpton, S. (1995). Fast parallel algorithms for short-range molecular dynamics. *J. Comput. Phys.* **117** (1), 1–19.

Sexton, J.C., Weingarten, D.H. (1992). Hamiltonian evolution for the hybrid, Monte Carlo algorithm. *Nucl. Phys. B* **380**, 665–677.

Shaw, D.E. (2005). A fast, scalable method for the parallel evaluation of distance-limited pairwise particle interactions. *J. Comput. Chem.* **26** (13), 1318–1328.

Shirts, M.R., Pande, V.S. (2001). Mathematical analysis of coupled parallel simulations. *Phys. Rev. Lett.* **86** (22), 4983–4987.

Snir, M. (2004). A note on N-body computations with cutoffs. *Theory of Computing Systems* **37**, 295–318. http://dx.doi.org/10.1007/s00224-003-1071-0.

Sugita, Y., Okamoto, Y. (1999). Replica-exchange molecular dynamics method for protein folding. *Chem. Phys. Lett.* **314**, 141–151.

Suits, F., Pitman, M.C., Feller, S.E. (2005). Molecular dynamics investigation of the structural properties of phosphatidylethanolamine lipid bilayers. *J. Chem. Phys.* **122** (24).

Swope, W.C., Andersen, H.C., Berens, P.H., Wilson, K.R. (1982). A computer simulation method for the calculation of equilibrium constants for the formation of physical clusters of molecules: Application to small water clusters. *J. Chem. Phys.* **76**, 637–649.

Swope, W.C., Pitera, J.W., Suits, F., Pitman, M., Eleftheriou, M., Fitch, B.G., Germain, R.S., Rayshubskiy, A., Ward, T.J.C., Zhestkov, Y., Zhou, R. (2004). Describing protein folding kinetics by molecular dynamics simulations. 2. Example applications to alanine dipeptide and a β-hairpin peptide. *J. Phys. Chem. B* **108** (21), 6582–6594.

Taylor, V.E., Stevens, R.L., Arnold, K.E. (1997). Parallel molecular dynamics: Implications for massively parallel machines. *J. Parallel Distrib. Comput.* **45** (2), 166–175.

Tuckerman, M., Berne, B.J., Martyna, G.J. (1992). Reversible multiple time scale molecular dynamics. *J. Chem. Phys.* **97** (3), 1990–2001.

Voter, A.F. (1998). Parallel replica method for dynamics of infrequent events. *Phys. Rev. B* **57**, 13985.

Zhou, R., Harder, E., Xu, H., Berne, B.J. (2001). Efficient multiple time step method for use with, Ewald and particle mesh Ewald for large biomolecular systems. *J. Chem. Phys.* **115** (5), 2348–2358.

CHAPTER 7

Multiscale Simulation of Membranes and Membrane Proteins: Connecting Molecular Interactions to Mesoscopic Behavior

Gary S. Ayton, Sergei Izvekov, W.G. Noid and Gregory A. Voth

Center for Biological Modeling and Simulation, and Department of Chemistry, University of Utah, 315 S. 1400 E. Rm 2020, Salt Lake City, UT 84112-0850, USA

Abstract

Current progress in the field of the multiscale simulation of membranes is presented. A general survey of various simulation methods, and how they can relate to the overall multiscale spatial and temporal problem associated with membranes is given. An overview of our current simulation methodology in this area is summarized, as well as a set of selected research applications.

Current Topics in Membranes, Volume 60   1063-5823/08 $35.00
DOI: 10.1016/S1063-5823(08)00007-0

I. MULTISCALE ASPECTS OF MEMBRANE SIMULATION: A BRIEF REVIEW

This chapter will focus on recent developments in the multiscale simulation of membranes. Specifically, an overview of our efforts in this area will be given (Izvekov and Voth, 2005a, 2006a; Shi *et al.*, 2006; McWhirter *et al.*, 2004; Ayton and Voth, 2002, 2004a, 2004b, 2007; Ayton *et al.*, 2001a, 2001b, 2002a, 2002b, 2004, 2005a, 2005b, 2007; Chang *et al.*, 2005; Blood *et al.*, 2005; Shi and Voth, 2005; Noid *et al.*, 2007). From a biological perspective, membranes serve a number of key roles in living systems, including, for example, compartmentalization (cells) and mechanical support (membrane bound proteins). In real cells, the membrane is far from homogeneous and includes a variety of lipids, non-lipid molecules (e.g., cholesterol) and membrane bound proteins. The lipid bilayer membrane can also remodel into new structures with the aid of proteins (e.g., BAR domains (Peter *et al.*, 2004; Blood and Voth, 2006)).

The general structure of a membrane is remarkable in that its thickness is microscopic (on the order of three to four nanometers) while its area can extend up to near macroscopic dimensions (μm^2). Furthermore, its material properties are that of an elastic material in terms of uniform dilations and bends, while it behaves like a two dimensional fluid in terms of in-plane shears (Evans and Needham, 1987; Sackmann, 1994; Lipowsky and Sackmann, 1995). The fluid nature of the lipid bilayer can be seen in the finite diffusion constant of the lipids; they are not in a solid phase. Furthermore, under an in-plane shear, the bilayer exhibits a viscous response characterized by a shear viscosity (Evans and Needham, 1987).

The combination of the structure of the bilayer (i.e., microscopic thickness, macroscopic area), its material properties (elastic bulk and bending moduli, viscous shear), and its highly inhomogeneous composition (domains, membrane bound proteins) results in a distinct multiscale character. For example, small changes in lipid structure, or the inclusion of non-lipid molecules like cholesterol in the bilayer, can alter macroscopically observed material properties (Ayton *et al.*, 2002b). In turn, long wavelength thermal membrane undulations can be perturbed via changes in the bending modulus of the bilayer.

Examining membranes with computer simulation has been an ongoing study for almost twenty years now. A large effort has been at the atomistic-level with molecular dynamics (MD) simulation (e.g., see reviews in Refs. (Anezo *et al.*, 2003; Tieleman *et al.*, 1997; Pastor and Feller, 1996; Feller, 2000; Pastor, 1994; Scott, 2002)). Coarse-grained (CG) simulation, where lipid molecules are modeled with simplified quasi-molecular structures, has recently gained considerable attention (Izvekov and Voth, 2005a, 2006a; Marrink *et al.*, 2004; Faller and Marrink, 2004; Shelley *et al.*, 2001a, 2001b; Nielsen *et al.*, 2005a; Lopez *et al.*, 2005; Nielsen *et al.*, 2005b, 2004; Faller, 2004; Sun and Faller, 2005; Meyer *et al.*, 2000; Stevens, 2004, 2005; Bond and Sansom, 2006; Shih *et al.*, 2006; Brannigan and Brown, 2004, 2005, Brannigan *et al.*, 2006; Cooke and Deserno, 2006, 2005; Cooke *et al.*, 2005; Farago, 2003; Farago and Pincus, 2004;

denOtter and Briels, 2003; Goetz *et al.*, 1999; Goetz and Lipowsky, 1998). Other membrane models operate at even longer length and timescales from the mesoscopic (McWhirter *et al.*, 2004; Ayton and Voth, 2004a, 2004b; Ayton *et al.*, 2005b, 2007; Lin and Brown, 2004; Brown, 2003; Lin and Brown, 2005; Lin *et al.*, 2006) up to macroscopic scales (Ayton *et al.*, 2002b, 2005a). Clearly, from the number of different simulation methodologies currently employed to examine bilayers at different length and time-scales, ranging from the atomistic (nm, ps), to the mesoscopic (μm, μs), different emergent phenomena can be examined.

However, in real bilayers, some degree of "communication" or coupling between the different scales takes place, and this critical aspect is very difficult to readily address within the current simulation methodologies that have been developed for bilayers. Atomistic-level simulations are still too small in both size and duration to examine long wavelength phenomena, while mesoscopic simulations lack the atomistic-level detail to allow the system to fully sample different configurations. As such, the search for a fully multiscale simulation methodology for bilayers is still an ongoing effort.

One reoccurring theme in a multiscale simulation approach is taking atomistic-level information, as determined from MD simulation, and using that information, in some averaged sense, at higher length and timescales. This can be seen in Fig. 1, which graphically illustrates the approach: Panel (a) is a snapshot of a membrane bound ion channel in a fully hydrated lipid bilayer at the atomistic-level. The system size is such that long wavelength thermal undulation modes cannot form, and the timescale is on the order of multiple nanoseconds. Panel (b) gives the CG representation of (a) where the lipids (and other non-lipid molecules) are transformed into simpler representations. As will be discussed in this chapter, a new methodology for implementing this type of CG transformation is called "multiscale coarse-graining" (MS-CG) (Izvekov and Voth, 2005a, 2005b, 2006a, 2006b; Shi *et al.*, 2006; Wang *et al.*, 2006; Zhou *et al.*, 2007) and essentially gives a systematic means whereby CG models can be derived from atomistic-level simulations. Panel (c) is a snapshot of a mesoscopic bilayer/solvent methodology (McWhirter *et al.*, 2004; Ayton and Voth, 2004a, 2002; Ayton *et al.*, 2005b, 2007) denoted the Elastic Membrane Version 1 and 2 (EM, EM2). The "particle-like" representation of the image is intentional; a "quasi-particle" discretization of the governing field theory mesoscopic model is employed, resulting in a mesoscopic membrane that can have any shape, and even remodel into new topologies, all within a multiscale mesoscopic solvent (Ayton *et al.*, 2004, 2005b) that gives the correct hydrodynamic behavior. Panel (d) shows the highest lengthscale mesoscopic model (in this case showing domain structures on a Giant Unilamellar Vesicle (Ayton *et al.*, 2005a)), where the domains can be modeled, e,g., via a Landau model for composition (McWhirter *et al.*, 2004; Jiang *et al.*, 2000) using Smooth Particle Applied Mechanics (SPAM) (Kum *et al.*, 1995; Hoover and Hoover, 2003; Hoover and Posch, 1996; Lucy, 1977; Monaghan, 1992).

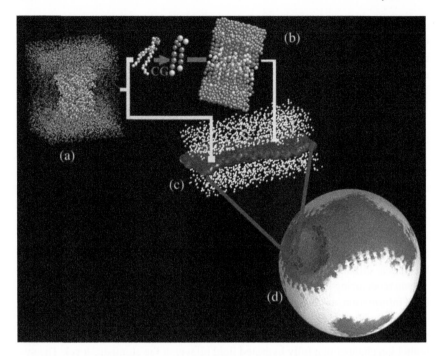

FIGURE 1 An illustration of the overall multiscale simulation methodology for membranes. Panel (a) begins at the atomistic-level. Panel (b) represents the coarse-grained MS-CG lipid representation, while panel (c) is the first mesoscopic representation just above the accessible length and time scales of the atomistic and CG domains. Panel (d) gives the second mesoscopic regime typically characterized by lengthscales of tens to hundreds of μm and timescales of μs.

A. Molecular Dynamics

The discussion will be restricted to issues concerning the overall spatial and temporal restrictions of atomistic-level MD simulation. Clearly, with massively parallel MD simulations such as with the computer program NAMD, the system sizes that are currently being accessed are ever increasing (e.g., see Ref. (Blood and Voth, 2006)). However the goal here is not to review the largest MD bilayer simulations, but to focus on a few simulations that address key multiscale issues. In that regard, some fairly recent atomistic-level MD simulations are noteworthy of mention.

Presently, large-scale atomistic molecular dynamics (MD) simulations can reach near mesoscopic spatial with length-scales on the order of 20 nm (Hofsab *et al.*, 2003; Lindahl and Edholm, 2000; deVries *et al.*, 2004; Marrink and Mark, 2001, 2003a, 2003b; Tieleman *et al.*, 2003; Marrink and Tieleman, 2002; Pandit *et al.*, 2004a, 2004b) using, e.g., the GROMACS force field (Spoel *et al.*, 1996). The simulation time that is accessed depends on the number of available processors and CPU time. For example, Edholm (Hofsab *et al.*, 2003) examined systems with 1024 lipids and the accompanying hydration layer up to 10 ns. The noteworthy work of deVries *et al.* (2004) examined a system of 1017 lipids for a total of 90 ns and were able to observe the formation of small lipid vesicles. Pandit et al. has examined nanoscopic domain formation in sphingomyelin/cholestrol/dioleylphosphatidylcholine (DOPC) mixtures (Pandit *et al.*, 2004a, 2004b). In one case, the system was composed of 1424 DOPC lipids, 122 Cholesterol, 266 sphingomyelin, and 62,561 water molecules examined for upwards of 20 ns (Pandit *et al.*, 2004a). Another study employed 100 DOPC, 100 sphingomyelin, and 5000 water molecules examined for 200 ns (Pandit *et al.*, 2004b).

The diffusion constant for bilayers, as defined from the slope of the mean square displacement (Pastor and Feller, 1996), is a key quantity that experimentally exhibits multiscale behavior. A wealth of experimental studies have aimed at measuring the lateral diffusion coefficient in lipid bilayers, and these can be loosely divided into short-range diffusion methods such as quasielastic neutron scattering (Lipowsky and Sackmann, 1995; Shin *et al.*, 1991; Pfeiffer *et al.*, 1988), long-range diffusion methods such as fluorescence (Korlach *et al.*, 1999), and magnetic resonance (Filippov *et al.*, 2003; Oradd *et al.*, 2002). Depending on whether short or long-range methods are used, different measurements of the lateral diffusion coefficient have been obtained (Vaz and Almeida, 1991).

The origin of this multiscale behavior is believed to originate from short-time "free volume" displacements (Vaz and Almeida, 1991) versus long-time lipid motion. Experimentally, the short-time diffusion coefficient can be two orders of magnitude larger than the corresponding long-time diffusion coefficient. Simulations by Essmann of dipalmitoylphosphatidylcholine (DPPC) bilayers (Essmann and Berkowitz, 1999), employing 10 ns trajectories, calculated the lateral diffusion coefficient from the slope of the mean square displacement and no discrepancy between the short and long-time diffusion coefficients was observed. The value of the lateral diffusion coefficient was found to be in reasonable agreement with neutron scattering experiments (Pfeiffer *et al.*, 1988). Simulations over much longer times (100 ns) (Lindahl and Edholm, 2001), for the same system, indicate that the long-time diffusion coefficient is indeed smaller, and more in agreement with fluorescence techniques. However, large system size simulations (Hofsab *et al.*, 2003) (1024 lipids compared to 64 in the previous work) examined over 10 ns exhibit a lateral diffusion similar to that found in Ref. (Essmann and Berkowitz, 1999). The main conclusion is that there is a distinct coupling between the time examined (length of simulation) and the accessible non-periodic translational space (system size). That is, even though the system size was significantly larger in Ref. (Hofsab *et al.*, 2003), the lipid displacements occurred on the same time-scale as in Ref. (Essmann and Berkowitz, 1999).

As the length and time-scales that can be reached with MD are intimately tied with the current available computer power, it is not surprising that the accessible system sizes, along with corresponding simulation times, have steadily increased. However, and this is an important point, at no time in the near future will simulations with atomistic detail spanning length-scales of 10 or 20 μm (about the size of a cell), and time-scales of microseconds (a typical time-span for many biological processes), likely be performed, nor are they warranted given the natural reduction in resolution required across the physical scales.

B. Coarse-Grained MD

Coarse-graining can be thought of as the process of reducing the number of degrees of freedom that must be taken into account in a complex system that is initially characterized by a large number of atomistic-scale degrees of freedom. Coarse-grained MD simulations of membranes (Izvekov and Voth, 2005a, 2006a; Marrink *et al.*, 2004; Faller and Marrink, 2004; Shelley *et al.*, 2001a, 2001b; Nielsen *et al.*, 2005a, 2005b, 2004; Lopez *et al.*, 2005; Faller, 2004; Sun and Faller, 2005; Meyer *et al.*, 2000; Stevens, 2004, 2005; Bond and Sansom, 2006; Shih *et al.*, 2006; Brannigan and Brown, 2004, 2005; Brannigan *et al.*, 2006; Cooke and Deserno, 2006, 2005; Cooke *et al.*, 2005; Farago, 2003; Farago and Pincus, 2004; denOtter and Briels, 2003; Goetz *et al.*, 1999; Goetz and Lipowsky, 1998) have been increasingly utilized as a means of examining long length-scale/time-scale membrane phenomena.

In the CG approach, an entire lipid molecule is "mapped" onto a reduced-resolution object which qualitatively resemble a lipid, but is much simpler, and usually consists of a number of spheres linked with "bonds". If the parameters are chosen carefully (Marrink *et al.*, 2004; Shelley *et al.*, 2001a, 2001b) they can form bilayers (Goetz and Lipowsky, 1998) or even reproduce a number of membrane material properties such as the bending modulus and area compressibility (Marrink *et al.*, 2004; Stevens, 2004). Some CG bilayer models have been parameterized such that they are "solvent-free" (Brannigan and Brown, 2004, 2005; Brannigan *et al.*, 2006; Cooke and Deserno, 2006, 2005; Cooke *et al.*, 2005; Farago, 2003); given that a large computational cost is the surrounding solvent, these CG bilayer models can potentially access phenomena at even longer length and time-scales.

Although the CG approach has been quite successful in reproducing static quantities (i.e., the membrane thickness and bending modulus), there are some issues that should be addressed when CG dynamics are considered. In particular, CG models typically exhibit an increased diffusion coefficient. For example, in Ref. (Marrink and Mark, 2003b) it was noted that both the CG water and lipids diffuse approximately 4 times faster than their atomistic-level counterparts. To give a meaningful time-scale, the solution was to employ an effective time (i.e., the real time multiplied by a factor of 4). In doing so, the lateral diffusion constant for the lipids agreed with the experimental estimate

(Marrink *et al.*, 2004). In fact, the phenomena of increased diffusion is a characteristic result of coarse-graining, and has it origins in the fact that the high frequency interactions that were "averaged out". This emerging scheme of arbitrarily re-scaling time in CG simulations is in many ways problematic. In the process of coarse-graining a system, high frequency modes are averaged over, and the result is that the dynamics at the new coarse-grained level should be governed by a Langevin equation (Izvekov and Voth, 2006b; Chandler, 1987; Kubo, 1966) where, above and beyond the conservative interactions between coarse-grained particles, there is also a random and frictional force component.

C. Mesoscopic Membrane Models

Mesoscopic models for membranes can employ a particle-based (Yamamoto *et al.*, 2002; Shillcock and Lipowsky, 2002a, 2002b; Groot and Rabone, 2001; Kumar and Rao, 1998; Kumar *et al.*, 2001; Laradji and Kumar, 2004; Khelashvili *et al.*, 2005; Pandit *et al.*, 2004c) or continuum-level (Lin and Brown, 2004, 2005; Brown, 2003; Lin *et al.*, 2006) formulation and are designed to work at length- and time-scales well beyond the molecular, or CG, level. Some of the particle-based schemes typically employ Dissipative Particle Dynamics (DPD) (Groot and Warren, 1997; Espanol and Warren, 1995) where the lipids are modeled via soft repulsive potentials and the particle dynamics are augmented with the characteristic DPD random-drag forces. However, the interpretation of a DPD "lipid" model remains is still unclear.

One continuum approach in particular is quite novel and models the membrane in Fourier space (Lin and Brown, 2004, 2005; Brown, 2003; Lin *et al.*, 2006) (employing Fourier Space Brownian Dynamics). This model takes advantage of the functional form of the Helfrich bending energy (Helfrich, 1973) along with the fact that the hydrodynamic interaction between the membrane and viscous solvent is well defined (Seifert, 1997; Granek, 1997; Brochard and Lennon, 1975). The result is a unified dynamical membrane model whose key parameterization is essentially only the bending modulus. This approach works quite well in mostly homogeneous systems; however, the method has also been employed to examine, for example, pinned membranes (Lin and Brown, 2004, 2005).

II. NEW MULTISCALE METHODS FOR MEMBRANE SIMULATIONS

This section will discuss the theoretical and methodological foundations of a new multiscale simulation framework for membranes. The overall methodology can be divided into two components: Multiscale coarse-graining (MS-CG) and mesoscopic modeling. Referring to Fig. 1, MS-CG applies to panel (b), while mesoscopic modeling applies to both panel (c) (the mesoscopic scales just above the atomistic and CG level) to panel (d) (the mesoscopic scales approaching the macroscopic regime).

A. Multiscale Coarse-Graining (MS-CG)

Arguably, the key issue in CG modeling is how to accurately determine the *effective interactions* between the CG sites, and as one possible solution, the multiscale coarse-graining (MS-CG) method has been developed (Izvekov and Voth, 2005a, 2005b, 2006a, 2006b; Shi *et al.*, 2006; Wang *et al.*, 2006; Zhou *et al.*, 2007). The MS-CG method utilizes atomistic-level MD simulation data within a rigorous variational principle to systematically generate "exact" CG force-fields that can, in turn, be employed in subsequent CG-MD simulations at much larger length and time-scales. The MS-CG approach allows key molecular-scale information to be rigorously propagated upward in scale to the CG model.

In the next two subsections, the theoretical framework and related concepts behind the MS-CG method will be discussed.

1. Background

The statistical mechanical foundation of MS-CG can be formulated within the framework of field theoretic coarse-graining (McWhirter *et al.*, 2004; Ayton *et al.*, 2005a; Chaikin and Lubensky, 1995) and liquid state theory (Noid *et al.*, 2007; Hansen and McDonald, 1986). In this section, some key concepts will be considered. Suppose that the coordinates of the n atoms of the system are given by $\mathbf{r}^n = \{\mathbf{r}_1, \mathbf{r}_2, \mathbf{r}_3, \ldots, \mathbf{r}_n\}$, while the N coarse-grained coordinates are given by $\mathbf{R}^N = \{\mathbf{R}_1, \mathbf{R}_2, \mathbf{R}_3, \ldots, \mathbf{R}_N\}$, with $N < n$. In the subsequent analysis, upper case letters will refer to the CG representation while lower case letters apply to the atomistic system. A mapping operator relates the high resolution atomically detailed system to the low resolution CG representation, and is given by $\mathbf{M}_{\mathbf{R}}^N = \{\mathbf{M}_{\mathbf{R}_1}, \mathbf{M}_{\mathbf{R}_2}, \mathbf{M}_{\mathbf{R}_3}, \ldots, \mathbf{M}_{\mathbf{R}_N}\}$. For example, if a center of mass transformation is desired, then

$$\mathbf{M}_{\mathbf{R}_I}\left(\mathbf{r}^n\right) = \frac{\sum_{i=1}^n P_{Ii} m_i \mathbf{r}_i}{\sum_{i=1}^n P_{Ii} m_i}, \tag{1}$$

where $P_{Ii} = 1$ if atom i is part of the CG site I and is zero otherwise. The mass of atom i is m_i, and the mass of the CG site is $M_I = \sum_{i=1}^n P_{Ii} m_i$, which in this case is just the sum of the masses of the atoms that belong to CG site I. It can be shown that the coarse-grained potential, $U(\mathbf{R}^N)$, is related to the potential energy function of the atoms $u(\mathbf{r}^n)$ in the following way:

$$U\left(\mathbf{R}^N\right) = -k_{\mathrm{B}} T \ln\left[\frac{Z_N}{z_n} \int d\mathbf{r}^n \prod_{I=1}^N \delta\left(\mathbf{R}_I - \mathbf{M}_{\mathbf{R}_I}\left(\mathbf{r}^n\right)\right) e^{-\beta u(\mathbf{r}^n)} \right]. \tag{2}$$

In the previous expression, k_B is Boltzmann's constant, T is the thermodynamic temperature, $\beta = 1/k_B T$, Z_N is the CG configurational integral, and z_n is the atomistic configurational integral. The two configurational integrals are given by, respectively,

$$Z_N = \int d\mathbf{R}^N e^{-\beta U(\mathbf{R}^N)},$$

$$z_n = \int d\mathbf{r}^n e^{-\beta u(\mathbf{r}^n)}. \tag{3}$$

A corresponding canonical transformation can also be employed for Eq. (2) (Noid *et al.*, 2007).

The coarse-grained potential $U(\mathbf{R}^N)$ is formally a many-dimensional potential of mean force (PMF) and it cannot be obtained directly from computer simulations of the atomic system. However, it can be shown that its derivatives, which are, in effect, forces that act on the coarse-grained coordinates as given by

$$\mathbf{F}_I(\mathbf{R}^N) \equiv -\nabla_{\mathbf{R}_I} U(\mathbf{R}^N), \tag{4}$$

can, in principle, be obtained from equilibrium MD simulations of the atomic system since they can be expressed as canonical ensemble averages of well defined functions of the atomic coordinates. In the simplest case, $\mathbf{F}_I$ is equal to the total force from the set of atoms that contribute to the CG site at $\mathbf{R}_I$, averaged over all those states in a long equilibrium MD run for which the coarse-grained coordinates are equal to $\mathbf{R}^N$, i.e., $\mathbf{F}_I(\mathbf{R}^N) = \langle \mathbf{f}_I(\mathbf{r}^n) \rangle_{\mathbf{R}^N}$, where $\mathbf{f}_I = \sum_{i=1}^n P_{Ii} \mathbf{f}_i$, $\mathbf{f}_i$ is the total force on atom i.

In practice, in order to obtain the coarse-grained forces from an atomistic MD trajectory, it is necessary to assume a simple, but flexible, functional form for $U(\mathbf{R}^N)$ and then variationally determine the parameters in the functional form so that the analytic derivatives match the appropriate numerical average of the forces observed in the MD data. It is most convenient to express $\mathbf{U}(\mathbf{R}^N)$ as a sum of pairwise additive central potentials whose distance dependence is represented by a spline function, for example. Algorithms have been developed in our group for this "force matching" (FM) process and have already been carried out for a number of systems (Izvekov and Voth, 2005a, 2005b, 2006a, 2006b; Shi *et al.*, 2006; Wang *et al.*, 2006). This scheme is quite different from the Inverse Monte Carlo method (Lyubartsev and Laaksonen, 1997, 1995, 1999); the difference lies in the fact that with MS-CG the CG forces are derived directly from the atomistic force data and include three-body correlations (Noid *et al.*, 2007). Non-bonded and bonded forces are also usually treated separately in the MS-CG approach (Izvekov and Voth, 2005a, 2006a) but this is not a necessity.

The force matching methodology, in principle, is completely general; in cases where some indication of the underlying functional form is known beforehand (e.g., strong elastic interactions are present), then this information can be incorporated to simplify the MS-CG scheme (Chu and Voth, 2006). The result in the latter case is similar in spirit to an elastic network type of approach, but with the important distinction that the underlying atomistic-level thermally averaged MD trajectory information is used to construct and parameterize the MS-CG model.

2. Force Matching Specifics

The success of the MS-CG method relies on decomposing the total CG force, $\mathbf{F}_I$ over the surrounding CG sites. The process of force matching (FM) involves modeling the total CG force with an effective model that is linearly dependent on its parameters (Izvekov and Voth, 2005a). For example, the CG force can be approximated by a pairwise interaction over the surrounding CG sites as

$$\mathbf{F}_I^{\mathrm{MS}} = \sum_{J \neq I}^{N} \mathbf{F}_{IJ}^{\mathrm{MS}}, \tag{5}$$

where $\mathbf{F}_I^{\mathrm{MS}}$ is a fitted force and $\mathbf{F}_{IJ}^{\mathrm{MS}}$ is an effective pair force between CG site I and CG site J. The fitted force is assumed to have a pre-specified form that is linearly dependent on N_k parameters, which are expressed as the vector $\mathbf{l} = \{l_1, l_2, l_3, \ldots, l_{N_k}\}$.

The means by which this decomposition is accomplished is found from force matching (FM) which relies on the variational minimization of a residual function, χ_{MS}^2, to give the "best fit" for $\mathbf{F}_{IJ}^{\mathrm{MS}}$ subject to the constraints on the pair interaction model. This scheme can be applied to the case that the pair interaction is assumed to be radial (i.e., along the inter CG-site vector, $\mathbf{R}_{IJ} = \mathbf{R}_I - \mathbf{R}_J$), and then the residual is defined as

$$\chi_{\mathrm{MS}}^2(\mathbf{l}) = \frac{1}{3n_t N} \sum_{t=1}^{n_t} \sum_{i=1}^{N} |\mathbf{f}_I(\mathbf{r}_t^n) - \mathbf{F}_I^{\mathrm{MS}}(\mathbf{M}_{\mathbf{R}}^N(\mathbf{r}_t^n; \mathbf{l}))|^2, \tag{6}$$

where n_t is the total number of configurations sampled, N is the number of CG sites, and $\mathbf{r}_t^n$ corresponds to the tth atomically detailed configuration. The best solution, in practice, resides at the minimum of the residual which can be found from the location where the gradient is zero, i.e., $\nabla_{\mathbf{l}} \chi_{\mathrm{MS}}^2 = \mathbf{0}$; the fact that the residual is quadratic guarantees that only one minimum can exist. The end result is a tabulated radial force that can be used in a subsequent MS-CG simulation. The situation becomes more complicated when complex systems with many components and bonded interactions are considered (e.g., a lipid bilayer or protein) (Izvekov and Voth, 2005a, 2006a); however, the overall scheme still applies.

There exists a further theoretical foundation of the MG-CG method that is not immediately apparent from Eqs. (2), (4), and (6), and current research is aimed at exploring these aspects. In particular, it can be shown that the set of normal linear equations found from evoking the $\nabla_{\mathbf{l}} \chi_{\mathrm{MS}}^2 = \mathbf{0}$ condition can be mapped over to the Yvon–Born–Green (YBG) equation (Noid *et al.*, 2007; Hansen and McDonald, 1986) in certain cases. As such, the MS-CG method is much more than a convenient algorithm by which to coarse-grain complex systems. It is, in fact, a powerful scheme to indirectly include information from the three particle distribution function; a quantity that is exceedingly difficult to calculate. These three body correlations can be critical in order to correctly model a CG force field.

B. Mesoscopic Models of Membranes

Beyond the length- and time-scales of coarse-grained MD simulation lies the mesoscopic domain. At this scale the concept of individual atomic sites and their molecular frames become blurred and in many ways the latter level of description is no longer necessary. The mesoscopic scale is a scale dominated by "domain motions", elastic deformations, and thermal "fluctuations". From the theoretical and computational point of view, this scale may in many ways be the most challenging because it effectively involves billons of atoms for tens to hundreds of microseconds of dynamics (or beyond). CG MD (e.g., the MS-CG method) can provide a bridge between the emergent mesoscopic scale and the atomistic scale, but it alone is not likely to model these phenomena entirely. At the mesoscale, new approaches are required, utilizing the notions of "fields" and "collective variables." Within statistical mechanics such approaches formally fall into the class of field theories (Chaikin and Lubensky, 1995). In our group, new quasi-particle methods have been employed to solve the field theoretic descriptions of membranes (McWhirter et al., 2004; Ayton and Voth, 2002, 2004a, 2004b; Ayton et al., 2004, 2005a, 2005b, 2007). In turn, new methodological tools have also been developed to bridge MS-CG MD data into these models (Shi and Voth, 2005), along with "multiscale coupling" methods, wherein atomistically detailed systems can be effectively embedded in a mesoscopic membrane model (Chang et al., 2005; Ayton and Voth, 2007) (e.g., atomistically detailed transmembrane proteins coupled to a mesoscopic membrane). In the next subsections these new mesoscopic simulation methodologies will be reviewed.

1. Homogeneous Membranes

The EM2 model (Ayton et al., 2005b, 2007) takes a step forward from the original EM model (Ayton and Voth, 2004a, 2002) by establishing a connection with the well known Helfrich Hamiltonian for the mean bending energy of a membrane (Helfrich, 1973) for specific deformations. As such, the differences between the EM and EM2 model lie primarily in the derivation of the final expressions. The inclusion of an explicit mesoscopic solvent (Ayton et al., 2004) is also a key refinement. With the inclusion of a mesoscopic solvent (Ayton et al., 2004), the dampened membrane undulation dynamics as predicted by hydrodynamic theory (Ayton et al., 2005b; Seifert, 1997; Granek, 1997; Brochard and Lennon, 1975) can be recovered.

The EM2 model originates from the expression for the continuum elastic energy of a membrane with thickness h. This energy expression can be split into a mean bending, a deviatoric, and a bulk expansion/contraction energy contribution as

$$F = F_{k_c} + F_D + F_\lambda$$

$$= \int dA_m \frac{k_c}{2} \left[\frac{1}{R_1} + \frac{1}{R_2} \right]^2 + \int dA_m \frac{B_a}{2} \left[\left| \frac{1}{R_1} - \frac{1}{R_2} \right| \right]^2$$

$$+ \int dA_0 \frac{\lambda h}{2} [2\varepsilon]^2, \tag{7}$$

where the first term, F_{k_c}, accounts for the mean bending energy, the second term, F_D, accounts for deviatoric (Gaussian) bending energies (Ayton *et al.*, 2007; Fischer, 1992a), and the final term, F_λ, accounts for local in-plane membrane plane strains, 2ε. The bending modulus is given by k_c and $R_{1,2}$ are the two local principal radii of curvature, while B_a is the deviatoric modulus (Ayton *et al.*, 2007; Fischer, 1992a), and the bulk modulus is given by λ (Ayton *et al.*, 2002a). The energy expression in Eq. (7) gives the free energy difference between the membrane in a deformed and a reference state. More specifically, for a perfectly flat membrane (i.e., when $R_{1,2}^{-1} = 0$) with no local in-plane strains ($2\varepsilon = 0$), F is zero. The differential, dA_m, represents the area of a material element on the membrane evaluated in some local membrane frame/coordinate system. The differential, dA_0, in Eq. (7) represents the area of a material element on the undeformed reference state.

Equation (7) describes the mean bending and plane strain energies associated with small deformations. However, in practice, for a real membrane (i.e., a vesicle or liposome) it is difficult to determine the two radii of curvature and the local plane strains. The problem involves defining a local coordinate system on the membrane that can be transformed back into the lab reference frame. However, this problem can be avoided with a "quasi-particle" discretization of the continuum-level energy density (Ayton *et al.*, 2005a, 2005b, 2007).

This idea of transforming continuum-level energy expressions into a discretized quasi-particle analog will be a reoccurring theme in the mesoscopic modeling described here. With EM2, a specific type of deformation is employed to initially parameterize the model. Briefly, in order to parameterize the mean bending contribution, a reference state composed of an initially flat configuration of EM2 quasi-particles evenly distributed over a surface in the xy plane is considered. The two principle curvatures can then be defined to be along the x and y axes. Then, a deformation in the form of a spherical "cup" is imposed on the system while the local area is conserved. It can be shown that a deviatoric curvature contribution also arises (Ayton *et al.*, 2007). Next, in order to parameterize the bulk response, a uniform expansion is imposed on the flat plane of EM2 quasi-particles.

Before continuing further with the discussion, it is important to review some key concepts that will be employed in subsequent quasi-particle decompositions, as this scheme will be revisited in later sections. The first step in arriving at the final EM2 equations from Eq. (7) involves discretizing the original field integral. This is done by defining the local density of the previous flat membrane in terms of "macroscopic" delta functions $\delta_M(\mathbf{r} - \mathbf{r}_i)$ as (Evans and Morriss, 1990)

$$\rho(\mathbf{r}) = \sum_{i=1}^{N} \delta_M(\mathbf{r} - \mathbf{r}_i), \tag{8}$$

which has the dimensions of $1/\text{length}^3$. In this expression, N is the number of quasi-particles that will be used to model the membrane. The fundamental length-scale of discretization of the membrane, L_{EM2}, arises via the definition of the macroscopic delta function: $\delta_M(\mathbf{r} - \mathbf{r}_i) \approx 1/hL_{EM2}^2$ for $\mathbf{r}_i$ within some volume hL_{EM2}^2 centered about $\mathbf{r}$, and $\delta_M(\mathbf{r}-\mathbf{r}_i) = 0$ otherwise. A normalization condition consistent with this definition is given by $\int d\mathbf{r}\delta_M(\mathbf{r} - \mathbf{r}_i) = 1$. Alternatively, a smooth weight function as will introduced shortly (Ayton et al., 2005a) could also be employed; the idea and resulting expressions are essentially unaltered. The average density is defined as $\rho_0 = N/V$, where V is the relevant volume required to characterize the system.

The properties of the quasi-particle membrane can then be related to those in the continuum membrane such that the continuum-level properties are averaged across length-scales specified by the fundamental lengthscale, L_{EM2}, of the macroscopic delta function. In this way, a general single quasi-particle property a_i is defined though its continuum-level counterpart $a(\mathbf{r})$ through

$$a(\mathbf{r})\rho(\mathbf{r}) \equiv \sum_{i=1}^{N} a_i \delta_M(\mathbf{r} - \mathbf{r}_i), \tag{9}$$

where $a(\mathbf{r})$ is the continuum (field) variable. The relationship between a_i and $a(\mathbf{r})$ can be found from noting that

$$\int d\mathbf{r}\, a(\mathbf{r})\rho(\mathbf{r}) = \int d\mathbf{r}\, a(\mathbf{r}) \left[\sum_{i=1}^{N} \delta_M(\mathbf{r} - \mathbf{r}_i) \right] = \sum_{i=1}^{N} \bar{a}_{\mathbf{r}_i}, \tag{10}$$

where $\bar{a}_{\mathbf{r}_i}$ is the average value of $a(\mathbf{r})$ evaluated in the non-zero region of $\delta_M(\mathbf{r} - \mathbf{r}_i)$. Equation (10) can also be re-expressed with the definition of a_i in Eq. (9), such that

$$\int d\mathbf{r}\, a(\mathbf{r})\rho(\mathbf{r}) = \int d\mathbf{r} \left[\sum_{i=1}^{N} a_i \delta_M(\mathbf{r} - \mathbf{r}_i) \right] = \sum_{i=1}^{N} a_i. \tag{11}$$

Comparing each term in the summations in Eqs. (10) and (11), it is observed that $a_i = \bar{a}_{\mathbf{r}_i}$, that is, the associated particle property gives an average of $a(\mathbf{r})$ over the pre-selected discretization lengthscale.

The key step of discretizing Eq. (7) (and other similar continuum-level expressions) involves evoking a uniform density approximation, expressed as $\rho(\mathbf{r})/\rho_0 \simeq 1$. If this expression is inserted into Eq. (7), then the final EM2 energy can be expressed as (Ayton *et al.*, 2005b)

$$H_{\text{eff}}(\{\mathbf{r}_{ij}, \hat{\boldsymbol{\Omega}}_i, \hat{\boldsymbol{\Omega}}_j\}) = H_{k_c,\text{eff}}(\{\mathbf{r}_{ij}, \hat{\boldsymbol{\Omega}}_i, \hat{\boldsymbol{\Omega}}_j\}) + H_{\lambda,\text{eff}}(\{\mathbf{r}_{ij}\}), \quad (12)$$

where

$$H_{k_c,\text{eff}}(\{\mathbf{r}_{ij}, \hat{\boldsymbol{\Omega}}_i, \hat{\boldsymbol{\Omega}}_j\}) = \frac{1}{2}\sum_{i=1}^{N}\sum_{j\neq i}^{N_{c,i}} \frac{8k_c}{\rho_A N_{c,i}} \frac{[(\hat{\boldsymbol{\Omega}}_i \cdot \hat{\mathbf{r}}_{ij})^2 + (\hat{\boldsymbol{\Omega}}_j \cdot \hat{\mathbf{r}}_{ij})^2]}{r_{ij}^2}, \quad (13)$$

and

$$H_{\lambda,\text{eff}}(\{\mathbf{r}_{ij}\}) = \frac{1}{2}\sum_{i=1}^{N}\sum_{j\neq i}^{N_{c,i}} \frac{2\pi (r_{ij}^0)^2 h\lambda}{N_{c,i}^2}\left[2\left(\frac{r_{ij} - r_{ij}^0}{r_{ij}^0}\right)\right]^2. \quad (14)$$

In both of these expressions, the "eff" subscripts stand for "effective" and reflects the fact that the resultant EM2 expressions are discretized approximations to the original continuum level expressions (i.e., Eq. (7)), and are parameterized via a specific set of initial deformations.

From Eq. (13), is it seen that the additional quasi-particle property of orientation is required, where $\hat{\boldsymbol{\Omega}}_i$ is a unit vector designating the orientation of quasi-particle "i". The orientation vectors, $\hat{\boldsymbol{\Omega}}_i$ and $\hat{\boldsymbol{\Omega}}_j$, roughly correspond to the membrane normal vectors that would occur at positions "i" and "j".

The expression in Eq. (12) maps over to the mean bending energy as given in Eq. (7) to second order in $\delta\theta_{ij}$, where $\delta\theta_{ij}$ is the angle formed between $\hat{\boldsymbol{\Omega}}_i$ and $\hat{\boldsymbol{\Omega}}_j$. The density ρ_A corresponds to the reduced density ρ^* via $\rho^* = NL_{\text{EM2}}^2/A = \rho_A L_{\text{EM2}}^2$, which is a constant that is evaluated in the initial flat configuration, while $N_{c,i}$ corresponds to the number of neighboring "j" particles found within a distance r_c of the "i" particle. In Eq. (14), r_{ij}^0 corresponds to the initial distance found between the two quasi-particles in the initially flat, unstressed membrane. Moreover, it can be shown that the EM2 decomposition gives a deviatoric modulus of $B_a \cong k_c/2$ (Ayton *et al.*, 2007).

In the form of Eq. (12), the EM2 quasi-particles are "bonded" together in a highly cross-linked mesh. If, for example, membrane remodeling (Ayton *et al.*, 2007; Blood and Voth, 2006) was to be examined, the EM2 model can be modified such that the quasi-particle "bonds" are "unhooked"; the elastic bonds as in Eq. (14) can be replaced with a soft repulsive-attractive inverse power potential. However, in "unhooking" the EM2 bulk elastic bond network, the fine control over the exact value of the bulk modulus is lost. This non-bonded form of EM2 is simply denoted as EM2-S, where S stands for "self-assembling".

A spontaneous curvature can also be included in the EM2-S model (Ayton *et al.*, 2007). The mean curvature is defined by $H = (1/R_1 + 1/R_2)/2$. In this case, the free energy difference, F, appearing in Eq. (7) is expressed as

$$F = \int dA_m \frac{k_c}{2}(2H - nC_0)^2 + F_D, \tag{15}$$

where the plane strain term, F_λ is not explicitly included. The plane strain contribution now arises from the soft repulsive interactions of the EM2-S model, and is mostly likely not strictly elastic. The desired spontaneous curvature is given by C_0 (Fischer, 1992b), and $n = 2$ is for an isotropic spontaneous curvature.

Following the quasi-particle transformation that was done with EM2, the EM2-S model employs a new discretized mesoscopic Hamiltonian, F^{eff}. Again, the spherical cup deformation is employed to initially parameterize the model. In this case, F^{eff} is expressed as (Ayton *et al.*, 2005b)

$$F^{\text{eff}} = \frac{k_c}{2\rho_A} \sum_{i=1}^{N}(2H_i - nC_0)^2, \tag{16}$$

where the mean curvature at the ith EM2 quasi-particle is given by H_i. Equation (16) is then expressed as a pairwise sum of interacting EM2 quasi-particles as

$$F^{\text{eff}} = (1/2) \sum_{i=1}^{N} \sum_{j \neq i, r_{ij} \leqslant r_c}^{N} \Delta u_{ij}, \tag{17}$$

where

$$\Delta u_{ij} = 4\varepsilon_\phi \phi_{ij}(L_{\text{EM2}}/r_{ij})^2, \tag{18}$$

$r_{ij} = |\mathbf{r}_i - \mathbf{r}_j|$, r_c is a pre-set cut-off radius, and L_{EM2} was previously defined. The summation in F^{eff} reflects the fact that since the EM2-S quasi-particles are not bonded together, different pairs of interacting quasi-particles will occur during the course of the simulation. The term Δu_{ij} contains the spontaneous bending contribution via the orientationally dependent ϕ_{ij} term (Ayton *et al.*, 2005b). Other variations for the functional form of ϕ_{ij} could be devised; however, mirroring EM2 (Ayton *et al.*, 2005b), ϕ_{ij} is chosen to be

$$\phi_{ij} = \left(\hat{\mathbf{\Omega}}_i \cdot \hat{\mathbf{r}}_{ij} - \frac{\gamma r_{ij}}{2}\right)^2 + \left(\hat{\mathbf{\Omega}}_j \cdot \hat{\mathbf{r}}_{ij} + \frac{\gamma r_{ij}}{2}\right)^2, \tag{19}$$

where it can be shown that $\hat{\mathbf{\Omega}}_i \cdot \hat{\mathbf{r}}_{ij} \cong \delta\theta_{ij}/2$; $\hat{\mathbf{\Omega}}_j \cdot \hat{\mathbf{r}}_{ij} \cong -\delta\theta_{ij}/2$. It can also be shown that $\gamma = nC_0/2$. It should be noted that different types of spontaneous curvatures can be incorporated into EM2-S (Ayton *et al.*, 2007), including isotropic $n = 2$ and anisotropic $n = 1$ forms. These will be discussed further in Subsection III.B.3.

2. Domains and Membrane Bound Proteins

Membrane fluid domains have been experimentally observed in Giant Unilamellar Vesicles (GUVs) composed of ternary lipid/cholesterol mixtures (see, e.g., Refs. (Baumgart *et al.*, 2003; Veatch and Keller, 2003)). At the mesoscopic level, this type of phenomena can be modeled using a Landau model for composition (McWhirter *et al.*, 2004; Ayton *et al.*, 2005a; Jiang *et al.*, 2000) using either Landau–Ginzburg (LG) or Cahn–Hilliard (CH) composition dynamics (Metiu *et al.*, 1976; Cahn and Hilliard, 1958). The idea is that the domains can be characterized by a single composition variable, ϕ, where ϕ is bound between -1 (representing one phase) and $+1$ (representing the other phase). These domains then can move/form over the surface of the vesicle. The phases can further couple to curvature and other deformations/structures in the bilayer (e.g., a membrane bound protein). With this composition-based formulation, the "rate" that a domain, or phase, moves about the membrane surface does not directly relate to how the actual molecules that constitute the phase move; rather, the dynamics should be considered more of an exploration of free energy minima.

At first glance, this general composition-based approach to mesoscopic simulation appears quite appealing; however, it can very quickly become computationally challenging, or even intractable, due to the complex transformations, Jacobians, and boundary conditions that are involved in solving the overall problem. As a result, this approach has generally been restricted to fairly simple geometries and in the absence of thermal fluctuations (Jiang *et al.*, 2000).

The quasi-particle decomposition as discussed in the previous subsection can instead also be applied here. In this case, Smooth Particle Applied Mechanics (SPAM) (Kum *et al.*, 1995; Hoover and Hoover, 2003, 1996; Lucy, 1977; Monaghan, 1992) can directly be utilized. SPAM was originally designed as a Lagrangian scheme to solve hydrodynamic problems; however, it can also be used to solve more involved problems such as composition dynamics. In a fashion similar to the way in which the Helfrich bending energy was transformed into a discrete EM2 quasi-particle system, the Landau composition model can be transformed into a discrete set of quasi-particles. In the same way that EM2 quasi-particles can possess the additional quasi-particle property of orientation, the SPAM quasi-particles exhibit the additional particle property of membrane composition such that they can exchange composition between themselves. As such, one SPAM quasi-particle can "give" composition to another according to the governing dynamics (i.e., either LG or CH) and the corresponding composition free energy model.

The discussion below will be restricted to applying SPAM to domain formation and dynamics. More general applications of SPAM can be found elsewhere (Kum *et al.*, 1995; Hoover and Hoover, 2003, 1996; Lucy, 1977; Monaghan, 1992). A detailed description of the method to solve composition dynamics is given in Ref. (Ayton *et al.*, 2005a); here, the focus will be on some key highlights. A general Landau model for composition on an undulating membrane surface can be expressed as

$$F = F_{k_c} + F_\lambda + F_\phi + F_{\phi,H}, \tag{20}$$

where F_{k_c} and F_λ are given in Eq. (7), F_ϕ gives the composition dependent contribution, and $F_{\phi,H}$ couples the composition to the local curvature on the membrane. Only closed surfaces will be considered; F_D is not included. A standard Landau model for composition is used for F_ϕ and is expressed as

$$F_\phi = \int dA \left[\frac{\zeta^2}{2} |\nabla \phi|^2 + V(\phi) \right], \tag{21}$$

where the first term on the right hand side is non-local and drives the system to a uniform composition state (i.e., $\phi = 0$ over the membrane surface), while the second term is a double well potential that drives the system to phase separate. In this case, $V(\phi) = a\phi^n/n - b\phi^m/m$, where $n > m$ and are both positive. More complex forms can also be employed (McWhirter et al., 2004). The composition coupling to curvature is given by

$$F_{\phi,H} = \int dA \, \Lambda \phi H^2, \tag{22}$$

where H is the mean curvature as defined in the previous section. This form of curvature coupling arises from assuming that the bending modulus is linearly dependent on composition, i.e., the material can get stiffer or softer depending on the local composition. At this point, a feedback between the composition, and the membrane material properties takes place.

Composition dynamics can take the form of LG or CH dynamics. A Lagrangian formulation of LG dynamics is given by

$$\frac{d\phi}{dt} = -\Gamma \left(\frac{\delta F[\phi, H]}{\delta \phi} - \mu^* \right) + \mathbf{u} \cdot \nabla \phi - \alpha, \tag{23}$$

where $F[\phi, H] = F_\phi + F_{\phi,H}$ and μ^* is a target chemical potential. The local flow field is given by $\mathbf{u}$. This equation of motion drives the composition field chemical potential, $\mu = \delta F[\phi, H]/\delta\phi$, to the target. Once this is reached, the dynamics stops. The mean composition is not conserved under these dynamics, and as such, a "composition-stat" α is included. CH dynamics are given by

$$\frac{d\phi}{dt} = M\nabla^2 \frac{\delta F[\phi, H]}{\delta \phi} + \mathbf{u} \cdot \nabla \phi, \tag{24}$$

which conserves composition, but allows the chemical potential to fluctuate.

SPAM enters into these dynamics in much the same way as in the EM2 model. However, in the former case, the mass density is defined via smooth weight functions as

$$\rho(\mathbf{r}) = \sum_{j=1}^{N} m W \left(|\mathbf{r} - \mathbf{r}_j| \right), \tag{25}$$

where m is the mass (which can be set to unity), W is the Lucy function (Lucy, 1977), and is expressed as

$$W(r) = \begin{cases} D\left[1 - \dfrac{r}{\sigma}\right]^3 \left[1 + \dfrac{3r}{\sigma}\right], & r \leqslant \sigma, \\ 0, & \text{otherwise.} \end{cases} \tag{26}$$

Here, σ is a fundamental lengthscale, much like L_{EM2}. The composition at any point can thus be represented as a sum over SPAM quasi-particle compositions as

$$\phi(\mathbf{r})\rho(\mathbf{r}) = \sum_j^N m\phi_j W(|\mathbf{r} - \mathbf{r}_j|). \tag{27}$$

With this definition, it is possible to express both Eqs. (23) and (24) in a SPAM quasi-particle representation. The challenge is in evaluating $\nabla^2\phi$ and $\nabla^4\phi$ as required by Eqs. (23) and (24). Normally, evaluating these quantities could be exceedingly difficult in complex geometries (i.e., a vesicle with large thermal undulations). However, SPAM recasts the problem in an entirely local reference frame that can be decomposed into pairwise interactions between SPAM particles.

SPAM quasi-particles carry with them certain particle properties. For example, particle position, $\mathbf{r}_i$, is a fairly obvious property. SPAM quasi-particles also carry with them their intrinsic acceleration, $\mathbf{a}_i$ (Kum *et al.*, 1995; Hoover and Hoover, 2003; Hoover and Posch, 1996), which can be found from the resulting EM2 or SPAM forces (Ayton *et al.*, 2005a). Integration of the SPAM quasi-particle acceleration in turn defines the SPAM quasi-particle velocity, $\mathbf{u}_i$. Including composition dynamics into the scheme adds an additional particle property, that being composition, ϕ_i, which is updated by integrating the equation of motion for the SPAM quasi-particle's composition either by LG or CH dynamics.

For $a = b = \Lambda = 0$, the resulting SPAM version of LG dynamics is given by

$$\frac{d\phi_i}{dt} = \Gamma \zeta^2 \sum_j^N \frac{m}{\rho_{ij}} [\boldsymbol{\chi}_j + \boldsymbol{\chi}_i] \cdot \nabla_i W_{ij} + \mathbf{u}_i \cdot \nabla\phi_i - \alpha_i. \tag{28}$$

A number of terms in this expression need to be defined. First, the symmetrized mass density is given by $\rho_{ij} = (\rho_i + \rho_j)/2$, where $\rho_i \equiv \rho(\mathbf{r}_i) = \sum_j m W_{ij}$, and the shorthand notation $W_{ij} = W(|\mathbf{r}_i - \mathbf{r}_j|)$ is introduced here. The gradient of the composition at SPAM quasi-particle i is given by $\boldsymbol{\chi}_i = (\nabla\phi)_i = \sum_j (m/\rho_{ij})[\phi_j - \phi_i]\nabla_i W_{ij}$. Finally, the SPAM-stat, α_i has the equation of motion $\dot{\alpha}_i = (1/Q)[\sum_j \phi_j]$ where Q is an arbitrary constant. Therefore, under LG dynamics the SPAM quasi-particles also carry with them the gradient of composition, as well as the SPAM-stat as particle properties.

The CH dynamics take the following form:

$$\frac{d\phi_i}{dt} = M \sum_j^N \frac{m}{\rho_{ij}} [\boldsymbol{\psi}_j + \boldsymbol{\psi}_i] \cdot \nabla_i W_{ij} + \mathbf{u}_i \cdot \nabla \phi_i, \tag{29}$$

where

$$\boldsymbol{\psi}_i = \sum_j^N \frac{m}{\rho_{ij}} [\mu_j - \mu_i] \nabla_i W_{ij}, \tag{30}$$

and

$$\mu_i = -\zeta^2 \sum_j^N \frac{m}{\rho_{ij}} [\boldsymbol{\chi}_j + \boldsymbol{\chi}_i] \cdot \nabla_i W_{ij}. \tag{31}$$

Under CH dynamics, the SPAM quasi-particles carry with them their chemical potential, as well as the gradient of their chemical potential as particle properties. The SPAM composition equations of motion can be implemented in the same manner as any pair-wise "force field" in an MD simulation. In principle, the flexibility of SPAM means that any number of field-theoretic models can be re-cast in a SPAM representation.

C. Multiscale Coupling (MSC)

The final mesoscopic simulation methodology to be described here for membranes is denoted multiscale coupling (MSC) (Chang et al., 2005). This is a true "concurrent" multiscale framework wherein an ensemble atomistic-level MD simulations are coupled to a single EM2 mesoscopic membrane model (McWhirter et al., 2004; Ayton and Voth, 2004a, 2007; Ayton et al., 2004, 2005a, 2005b; Chang et al., 2005). The result, in a qualitative sense, is an atomistically detailed "window" in the mesoscopic-level model and thus specific regions of interest (e.g., a transmembrane protein) can be examined with full atomistic-level detail while the rest of the system is modeled at a mesoscopic level. The coupling between the two scales takes the form of altered boundary conditions at the MD level. This scenario is depicted in Fig. 2 where an ensemble of atomistic-level systems (upper panel) is correlated to a single mesoscopic region (shown as the small square on the EM2 membrane). The idea of coupling atomistic-level systems to continuum-level models has been attempted before (Delgado-Buscalioni and Coveney, 2003; Kopelevich et al., 2005; Hyman, 2005); however, in these cases long-range electrostatics were not present and a "buffer region" was required in order to maintain momentum conservation. Furthermore, in this previous work the MD system was explicitly "embedded" in the continuum model; here, the ensemble of MD systems each evolve under their intact periodic boundaries.

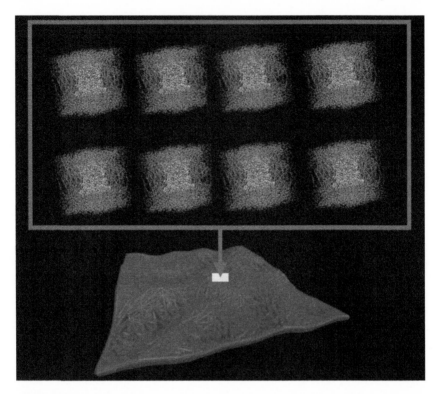

FIGURE 2 An illustration of multiscale coupling (MSC). An ensemble of atomistic level systems (shown in the top box) are coupling to a single region on the mesoscopic membrane (shown as the small square on the mesoscopic membrane). The coupling between the two occurs via modulated boundary conditions on the atomistic-level MD ensemble.

MSC is implemented by selecting a small area on the mesoscopic EM2 membrane with length-scales on par with the ensemble of atomistic-level MD systems (shown as the small region in Fig. 2). The material properties are found from preliminary non-equilibrium MD (NEMD) simulations (Ayton and Voth, 2004a; Ayton *et al.*, 2002a, 2001a, 2001b, 2002b). The two simulations are coupled via the plane stress (or surface tension), such that the ensemble of atomistic-level simulations are coupled to the longer wavelength EM2 mesoscopic-level plane stresses.

MSC can be implemented within an MPI-2 parallel scheme where the EM2 code "spawns" the ensemble of atomistic-level DL_POLY simulations (Chang *et al.*, 2005). The atomistic-level and mesoscopic simulations are matched in time; however, since the mesoscopic EM2 code is much faster, due to its larger time-step ($\delta t_{EM2} = 0.04$ ps versus $\delta t_{DL\_POLY} = 0.002$ ps) and the short-ranged EM2 interactions, it actually "waits" in the background until the atomistic-level simulation catches up.

In the initial formulation of MSC (Chang *et al.*, 2005), the local mesoscopic surface tension was employed to couple the mesoscopic and atomistic-level systems. Alternatively, the plane stress as evaluated through the underlying constitutive relation governing the EM2 model can be employed. The two simulations are coupled every 25 mesoscopic (or 500 atomistic MD) timesteps, although this coupling is in principle adjustable.

III. SELECTED APPLICATIONS

In the following subsections, a select set of applications of our overall multiscale methodology for bilayers will be presented. These particular studies clearly demonstrate key aspects of the methodology. Other related studies will be referenced within.

A. Multiscale Coarse-Grained Simulations

1. Pure Lipid Bilayers

The MS-CG method was first employed (Izvekov and Voth, 2005a) to obtain a CG model of a dimyristoylphosphatidylcholine (DMPC) lipid bilayer (Smondyrev and Berkowitz, 1999). A single atomistic MD simulation of the bilayer, was used to generate the required atomistic trajectories for the force matching (as described in Subsection II.A.2). More details are found in Ref. (Izvekov and Voth, 2005a).

Water molecules were mapped into a single CG interaction site associated with their geometrical center, while DMPC molecules were coarse-grained in a similar manner to other CG lipid models (Marrink *et al.*, 2004; Shelley *et al.*, 2001a, 2001b). All the non-bonded interactions, and their corresponding potentials, were tabulated on a fine mesh. The intra-lipid CG sites were linked by harmonic bonds with their force constants determined using a least-squares linear fit of the respective FM force profiles; further details involving the separation of bonded and non-bonded forces can be found in Ref. (Izvekov and Voth, 2005a). The final CG scheme is shown in Fig. 3 and some resulting selected MS-CG force profiles are shown in Fig. 4.

One of the tests of the MS-CG method is to employ the resulting CG force field in a new CG MD simulation of the same system with exactly the same dimensions. The resulting radial distribution functions of the CG sites can then be directly compared to the original atomistic-level MD counterparts. As seen in Fig. 5, the MS-CG method is able to quantitatively reproduce the membrane structure. It should be noted, however, that the structural properties of the bilayer are also

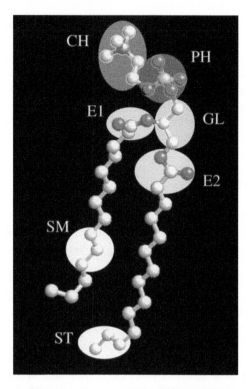

FIGURE 3   The MS-CG decomposition of a DMPC molecule. Single coarse-grained (CG) interaction sites were associated with centers-of-mass of the choline (CH), the phosphate (PH), the glycerol
backbone (GL: CH_2–CH–CH_2), the ester groups (E1 and E2: O_2CCH_2), the triplets of carbon atoms
of alkane chains [SM: $(CH_2)_3$], and the tail of alkane chains [ST: $(CH_2)_2$–CH_3].

sensitive to the CG water–water and CG water–lipid interactions. If, for example,
an insufficient interaction cut-off is employed in the FM algorithm, the resulting
force fields can eventually lead to bilayer destabilization.

In Ref. (Izvekov and Voth, 2005a), no in depth analysis of the CG dynamics
was performed. As was discussed previously in Subsection I.B, CG dynamics
can only give meaningful dynamical information when the appropriate drag and
random forces are included in the dynamics in a manner consistent with the
First and Second Fluctuation-Dissipation Theorems (Izvekov and Voth, 2006b;
Chandler, 1987; Kubo, 1966). In Ref. (Izvekov and Voth, 2005a), the simulation
was essentially classical dynamics employing the MS-CG potential such that a
more efficient exploration of the many dimensional CG PMF was obtained. However the resulting dynamics cannot reflect the actual dynamical behavior of the
CG groups as would be seen in an exact atomistic MD simulation. This issue can
be addressed within the MS-CG method (Izvekov and Voth, 2006b).

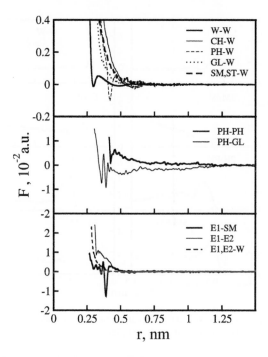

FIGURE 4   Effective pairwise forces between selected coarse-grained (CG) interaction sites as a function of intersite separation as calculated by MS-CG method. The CG sites are shown in Fig. 3. The "W" stands for CG water.

2. Mixed Lipid Bilayers Containing Cholesterol

The MS-CG method has also been employed to readily develop CG models for a 50 mol% mixture of DMPC and cholesterol (Izvekov and Voth, 2006a). This application was especially challenging and a clear demonstration of the power of the MS-CG method since it has been very difficult to develop CG models for mixed bilayers using simpler CG schemes. The DMPC molecules were modeled as before (Smondyrev and Berkowitz, 1999) and the cholesterol molecules were modeled with a modified AMBER force field (Ayton *et al.*, 2002b). Again, a one-site MS-CG model based on the TIP3P potential was used for water, with the interaction site placed at the molecular geometrical center. The atomistic system consisted of 6336 total atoms (32 lipids, 32 cholesterol molecules, and 1312 water molecules) and was integrated with a time step of 2 fs.

The centers of mass of the underlying atomic groups were employed as the CG sites for lipid and cholesterol molecules. Four and seven site CG representations of the cholesterol molecule were employed, where the latter seven site cholesterol model differed in the CG bonding scheme. The two cholesterol CG schemes are shown in Fig. 6 where the different labels for the CG sites are shown. In the four-site representation, the tetracyclic ring structure is split between CA and CB

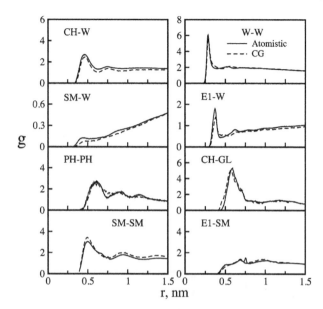

FIGURE 5   Selected CG site-site RDFs from atomistic MD simulation (solid lines) compared to those from the MS-CG MD simulation (dashed lines). The peaks from the bonded intramolecular sites are not shown.

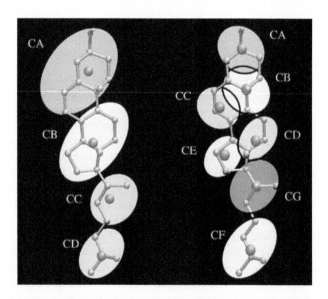

FIGURE 6   Coarse-grained representation of cholesterol. The left panel is the 4-site cholesterol MS-CG model while the right panel is the 7-site MS-CG model. The various labels represent the CG sites.

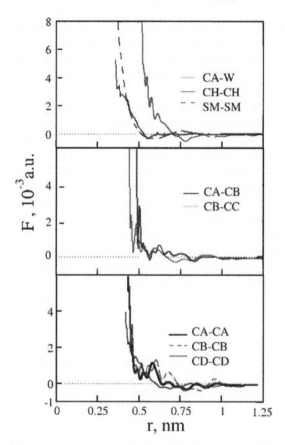

FIGURE 7 Effective pairwise nonbonded forces between selected MS-CG interaction sites smoothed by a polynomial fit as a function of inter-site separation. The approximate error bars in these curves are about 5–10% of the force value. Labels refer to the CG schemes in Fig. 6.

sites. The CA site includes the polar hydroxyl group, while the CC-CD site pair represents the aliphatic side chain attached to the ring structure. In the four-site cholesterol representation, the arrangement of the CG sites is close to linear. In the seven-site representation, three more sites were added to better reflect the planar geometry of the ring structure. Details involving the intramolecular bonding schemes within the FM algorithm can be found elsewhere (Izvekov and Voth, 2006a).

The resulting MS-CG force fields were then employed in a CG simulation under constant NPT conditions, and an accurate reproduction of the structural properties of the bilayer was observed. However, MS-CG force fields should be carefully treated under constant pressure (NPT) conditions. Under constant volume (NVT) conditions, they can reasonably reproduce bulk phase structural properties; however, the internal pressure may be off, and as a result, under constant NPT conditions, the density is typically underestimated. One option to correct this includes a virial constraint within the FM algorithm (Izvekov and Voth, 2006a). Other more rigorous, constant NPT MS-CG methods are currently under development.

The FM algorithm employed a 5 ns atomistic-level simulation under constant volume, isothermal conditions, where the starting system density was found from the equilibrium volume obtained from a constant NPT MD simulation. The FM trajectory data, along with the instantaneous virial was then sampled every 1 ps; the instantaneous atomistic velocities along with virial itself are necessary to implement the virial constraint. The trajectories, atomistic velocities, and force data were applied to the CG sites in the FM algorithm with an explicit separation of bonds and the virial constraint. The FM force field was represented by a spline over a mesh with a grid spacing of approximately 0.01 nm for nonbonded and 0.0025 nm for bonded interactions. For nonbonded forces, the mesh was extended to 0.13 nm. Some of the resulting non-bonded FM force fields are shown in Fig. 7; the different CG sites are as labeled in the previous figure. Clearly, the nature of the resulting force fields is not trivial (e.g., a simple power law decay). As shown in Fig. 8, the resulting radial distribution functions are also in good agreement with the atomistic-level MD simulation results.

3. CG Bilayers with All-Atom Resolution Membrane Bound Proteins

The MS-CG method can also be used to construct mixed resolution all-atom and coarse-grained (AA-CG) models (Shi *et al.*, 2006). This idea was used to build a mixed all-atom and coarse-grained (AA-CG) model of the gramicidin A (gA) ion channel embedded in a fully solvated dimyristoylphosphatidylcholine (DMPC) membrane. The gA peptide was described in full atomistic detail, while the lipid and water molecules were described using MS-CG representations; the entire set of atom-CG and CG-CG interactions were determined using MS-CG. The resulting AA-CG model was employed in an MD simulation, and the results from simulations of the AA-CG model compare very favorably to those from all-atom MD simulations of the entire system (Shi *et al.*, 2006). Since the general framework of the MS-CG method can readily be extended to the AA-CG scenario; this approach has the potential to be extended to many other systems.

A 10 ns constant NVT simulation of the peptide-membrane system was performed using the AA-CG force field where the initial AA-CG configuration was found from an equilibrated all-atom configuration. Selected force profiles of the atom-CG interactions are plotted in Fig. 9. Both the CG membrane and the AA gA helical dimer and remained stable during the 10 ns AA-CG MD simulation. The gA-gA and gA-lipid radial distribution functions (RDFs) are compared to the all-atom MD result in Fig. 10 where it can be seen that the agreement between the gA-gA RDFs from AA-CG and all atom simulation is quite good. This trend

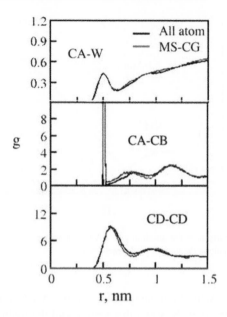

FIGURE 8 Selected cholesterol–water, cholesterol–DMPC, and cholesterol–cholesterol site–site RDFs from the all-atom MD simulation (dark lines) compared to those from the MS-CG MD simulation (light lines) using the four-site representation of the cholesterol molecule.

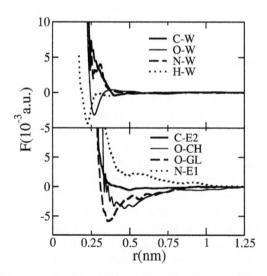

FIGURE 9 Effective pairwise nonbonded forces for selected AA-CG interactions. Shown in the figure are smoothed fits of the effective forces, approximate error bars are about 5–10% of the force value. C, O, N, H correspond to the peptide backbone atoms, W is coarse-grained water molecule, and CH, GL, E1, and E2 are coarse-grained choline, glycerol backbone, and ester groups, respectively.

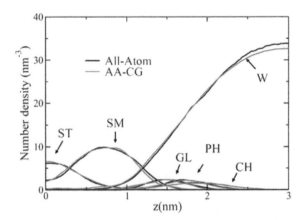

FIGURE 10   Comparison of the all-atom and AA-CG density profile for several CG sites [cf. caption to Figs. 3 and 9] along the normal to the membrane.

continues with the gA-lipid tail, suggesting that the AA-CG model successfully reproduces the interactions between the peptide and the hydrophobic lipid tails. A snapshot of the final configuration of the system is shown in Fig. 11. Finally, the calculated RMSD of the peptide backbone atoms in the AA-CG simulation as compared to the all-atom MD simulation is quite good considering the degree to which the lipid and solvent molecules have been coarse-grained.

B. Mesoscopic Simulations

1. Correlations of Close Pairs of Pure Lipid Bilayers

The EM2 model was employed to examine the entropic repulsion between two mesoscopic bilayers. In confined geometries, both the structure and dynamics of a membrane are altered compared to when it is only surrounded by a viscous solvent. In general, confined geometries includes smectic membranes (deJeu *et al.*, 2003; Helfrich, 1978), bilayers against surfaces (Helfrich, 1978; Gov *et al.*, 2004; Groves *et al.*, 1997; Groves and Boxer, 2002), and bilayer pairs at close distances (Sackmann, 1994; Lipowsky and Sackmann, 1995; Brochard and Lennon, 1975; Helfrich, 1978). The case of two bilayers separated by a small distance d is particularly interesting as an entropic repulsion that scales like $1/d^2$ is predicted theoretically (Lipowsky and Sackmann, 1995; Helfrich, 1978) and has been observed experimentally by reflection interference contrast microscopy (Radler and Sackmann, 1993). The narrow channel of solvent between the two bilayers couples both their structure and dynamics via the membrane's thermal undulations (Brochard and Lennon, 1975;

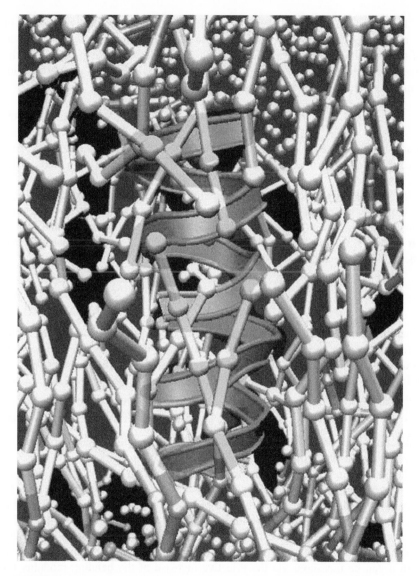

FIGURE 11 Configuration of the atomistic gA peptide in MS-CG DMPC lipid and water after 10 ns AA-CG MD simulation. The all-atom gA molecule can be seen in ribbon format in the center of the image.

Helfrich, 1978). At very far distances, the two membranes behave as if they were isolated; however, when $qd < 1$ (where q is a wavevector) the two membranes couple.

The correlation in the membrane undulations at close proximity is predicted to have a specific scaling, dependent on the average distance d of the separation of the two bilayers. For example, Helfrich (1978) derived an expression for the free energy difference, ΔF, when the undulations of one of the membranes are dampened by the close presence of the other and is given by

$$\Delta F \approx \frac{(k_B T)^2}{k_c} \frac{1}{d^2}, \tag{32}$$

that is, the free energy is a repulsive mean potential that scales like $1/d^2$. This prediction has proven exceedingly hard to test through simulation models at the all atom level.

The free energy difference for two membranes separated by a distance d can be explicitly evaluated with the EM2 model by examining the effective bending modulus found when the two membranes are at close proximity, k_c^{eff}, and then at far distances, k_c. It can be shown (Ayton *et al.*, 2005b) that the total entropic free energy difference for the two membranes can be calculated from

$$\Delta F = \sum_q^{N_q} k_B T \ln \frac{k_c^{\text{eff}}(\mathbf{q})}{k_c} = N_q k_B T \ln \frac{k_c^{\text{eff}}}{k_c}, \tag{33}$$

where the sum is over all N_q sampled wavevectors $\mathbf{q}$, and $k_c^{\text{eff}} = k_c^{\text{eff}}(\mathbf{q})$ is the effective bending modulus of the membranes that differs from k_c due to their coupling. The second equality in Eq. (33) assumes that $k_c^{\text{eff}}(\mathbf{q})$ is independent of $\mathbf{q}$ for the modes that are sampled. The effective bending modulus is determined via equipartition, i.e.,

$$A k_c^{\text{eff}} q^4 \langle |u_{\mathbf{q}}^d|^2 \rangle = k_B T, \tag{34}$$

where the superscript "d" refers to the fact that the undulations are dampened in one membrane due to the presence of the other.

It should be noted that in order to map the simulation results over to the Helfrich theory, the distance, d, is found from $d = d' - L_{\text{EM2}}$, where d' is the average separation or distance between the mid-planes of the two membranes. Indeed it was found from the EM2 simulations that over a range of separations ($d = 1.5$ to 3 nm) that ΔF scales like $1/d^2$ (Ayton *et al.*, 2005b). The origin of this entropic repulsion originates from the undulation modes being dampened, and, in turn it can be measured via the effective increase of the bending modulus for each of the two membranes. A snapshot of the EM2 mesoscopic system is shown in Fig. 12, where it can be seen that the membrane undulations must result in substantial interstitial solvent flows. The result seems to be a distinct lack of correlation between the structures of the two membranes and a "smashing out" of thermal membrane undulation magnitudes. In fact, it is this lack of undulation that results in an increase in the apparent bending modulus due to the loss of energy, and hence the $1/d^2$ scaling of the free energy as a function of intermembrane distance.

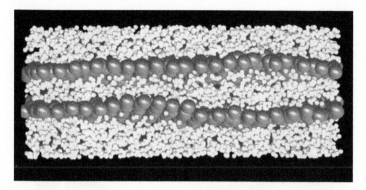

FIGURE 12   A simulation snapshot of the EM2 system for $d = 2.9$ nm. The mesoscopic solvent particles are drawn smaller than their actual size for clarity. Note the frustration of the undulations of the mesoscopic membranes.

2. Domain Formation on a Giant Unilamellar Vesicle (GUV)

By virtue of our multiscale simulation methods, domain formation can be modeled on the surface of a GUV using a Landau field theory model for phase coexistence (Eq. (21)) which is further coupled to elastic deformation mechanics (e.g., membrane curvature). As introduced previously, Smooth Particle Applied Mechanics (SPAM) (Ayton et al., 2005a; Kum et al., 1995; Hoover and Hoover, 2003; Hoover and Posch, 1996; Lucy, 1977; Monaghan, 1992), a form of smoothed particle continuum mechanics, is used to solve either the time-dependent Landau–Ginzburg or Cahn–Hilliard (Metiu et al., 1976; Cahn and Hilliard, 1958) form of the composition dynamics. At the same time, the underlying elastic membrane is modeled using SPAM, resulting in a unified computational scheme capable of treating the response of the composition fields to arbitrary vesicle deformations (Ayton et al., 2005a).

The SPAM framework was described previously in Subsection II.B.2; here, the focus will be on the results. In Ref. (McWhirter et al., 2004) domain formation was examined at relatively short mesoscopic length-scales, where thermal undulations played an important role. However, in order to examine GUVs, the required length scales are now on the order of μm, far beyond the typical wavelengths associated with thermally induced membrane undulations (Evans and Rawicz, 1997). As such, membrane deformations at these scales usually take the form of some external perturbation (i.e., a mechanical rather than thermal perturbation). For example, an external deformation could mirror micromanipulation experiments (Rawicz et al., 2000; Olbrich et al., 2000) where, in essence, the GUV is externally "poked", resulting in an indentation (cf. Fig. 13). In terms of the composition coupling to curvature (i.e., via Eq. (22)), this scenario could give insight into the details of the coupling from an experimental viewpoint. Of course, if the composition-curvature coupling is small, or even non-existent, then employing

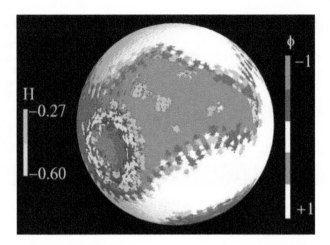

FIGURE 13 Dynamic composition coupling to induced curvature in the mesoscopic SPAM-composition simulation. The color-coding for the composition is given in the bar on the left of the image, where green corresponds to $\phi = -1$ and white corresponds to $\phi = +1$. The orange regions are the local curvature field superimposed on the images.

this type of external deformation experimentally would most likely not perturb the underlying domain structure.

With the present SPAM LG/CH-membrane model as described in Subsection II.B.2, it was possible to externally impose a small deformation on an otherwise spherical GUV and then allow the composition fields to adjust to the new membrane surface. Furthermore, if a composition dependent bending modulus was incorporated into the model, then the feedback became more complex (i.e., via Eq. (22)). Since the SPAM representation of the composition dynamics as given in Eqs. (23) and (24) is naturally resolved in a local reference frame, the imposition of an external deformation is automatically incorporated into the underlying composition dynamics and requires no additional formulation.

A 20 μm SPAM GUV was annealed such that the domains could weakly couple to small surface irregularities (details involving the initial preparation of the GUV can be found in Ref. (Ayton *et al.*, 2005a)). It was then symmetrically "dinted" along the x-axis such that the shape of the "dint" mimicked the shape of an external experimental probe 5 to 10 μm in diameter. The composition dynamics (LG and CH) were allowed to couple to this new deformation, and a reorganization of the domain structure was observed.

A snapshot of the domain structure on the externally dented GUV is shown in Fig. 13. Small surface irregularities (indicated by orange regions corresponding to small mean curvatures) act as nucleating sites for domain structures with $\phi \approx -1$. The effect of deforming the GUV is clearly evident in Fig. 13, as the "ring" of large negative curvature, at the perimeter of the dent, is now embedded within a

$\phi \approx -1$ domain that was not present in the original GUV. These current results may also be pertinent to the formation and evolution of lipid domain "rafts" in real cellular membranes especially during important processes such as vesicle budding.

On a related note, a previous study of domain formation using CG simulation (Shi and Voth, 2005) was able to bridge in a multiscale fashion some of the key quantities that appear in Eq. (21). A CG model based on the work of Marrink (Marrink *et al.*, 2004) was used, but in the future the MS-CG approach as was described in Subsection II.A, will be employed to provide a more accurate CG description. In this way, the MS-CG simulation will act as a bridge between the atomistic and mesoscopic levels, as described in Section I.

3. Mesoscopic Membrane Remodeling

Liposome remodeling processes (e.g., vesiculation and tubulation) due to N-BAR domain interactions with a liposome can be explored with the EM2-S model (Ayton *et al.*, 2007). Some key results from atomistic-level MD simulations of membrane binding to the concave face of N-BAR domains (Blood and Voth, 2006) can also be employed with an EM2-S liposome simulation to examine how the spontaneous curvature fields generated by N-BAR domains result in membrane remodeling.

However, before proceeding, a brief summary of some experimental observations and background material is in order, while a more detailed account can be found in Ref. (Ayton *et al.*, 2007) and in references therein. It is widely accepted that proteins play an key role in affecting the geometry of membranes by altering the local curvature (Peter *et al.*, 2004; Zimmerberg and Kozlov, 2006; Ford *et al.*, 2002; Lee *et al.*, 2002; McMahon and Gallop, 2005; Razzaq *et al.*, 2001; Masuda *et al.*, 2006). Of particular interest is the crescent-shaped N-BAR (Bin/amphiphysin/Rvs) domain dimer protein module that has been experimentally observed to induce local curvature in liposomes *in vitro* over relatively short (atomistic) timescales. It should be noted that the membrane remodeling process occurs on macroscopic timescales (seconds or even minutes). The structure of the N-BAR domain dimer (as determined from the crystal structure) contains both the BAR domain, as well as two N-terminal amphipathic helices along with a high density of positively charged residues on its concave surface (Peter *et al.*, 2004). It is believed that both the curved shape and the charge distribution are the key molecular-scale components responsible for the bending of a bilayer.

Experimentally (Peter *et al.*, 2004), the results can be summarized as the following: At low to medium N-BAR concentrations (20 µM), liposomes were remodelled into *tubules*, while at high N-BAR concentrations (40 µM), liposome *vesiculation* was observed, where the liposome was remodelled into a large number of smaller shaped liposomes with a variety of shapes and sizes. The present working hypothesis is that whether a liposome is tubulated or vesiculated is not only dependent on the molecular-level details of the N-BAR domain, but on how the N-BAR domains arrange themselves both spatially and orientationally, on average, on the liposome surface. As such, a mesoscopic EM2-S simulation can directly examine the "smoothed out" averaged interaction picture. In particular, the EM2-S model can directly incorporate both spontaneous curvature as well as remodeling within its mesoscopic quasi-particle formulation. As such, this approach was employed to examine how different kinds of spontaneous curvature fields (that represent different averaged spatial distributions of N-BAR domains on the liposome surface) can affect liposome remodeling (Ayton *et al.*, 2007).

Two different kinds of N-BAR domain spontaneous curvature fields were considered (Ayton *et al.*, 2007): The first corresponded to an isotropic averaged spatial distribution of N-BAR domains on the liposome surface. The second modeled an *anisotropic* N-BAR domain density, where some degree of N-BAR domain orientational order on the liposome surface occurs. In the second case, the N-BAR domains arrange themselves such that they "line up" to some degree on the surface of the liposome. The resulting spontaneous curvature field will have a preferred directionality that is correlated with the alignment of the N-BAR domains. It results not only from the averaged collective interaction between N-BAR domain proteins themselves, but also from an indirect collective interaction where the local curvature on the liposome couples, in a non-local fashion, to the curvature in nearby regions. Details of the implementation can be found elsewhere (Ayton *et al.*, 2007).

The governing free energy functional was given in Eq. (15); the EM2-S quasi-particle decomposition approximates this with Eq. (16). Isotropic spontaneous curvature fields have $n = 2$, while the anisotropic curvature fields have $n = 1$. In this latter case the directionality of the anisotropic curvature needs to be considered. Details of how the anisotropic fields are modeled within the EM2-S model are given in Ref. (Ayton *et al.*, 2007), and involve specifying the local, in-plane directionality of the spontaneous curvature. As such, each EM2-S quasi-particle has two degrees of orientational freedom: the membrane normal vector, $\hat{\boldsymbol{\Omega}}_i$, and the in-plane spontaneous curvature field direction vector, $\hat{\mathbf{n}}_i^T = n_{ix}^T \hat{\mathbf{i}} + n_{iy}^T \hat{\mathbf{j}}$, where $\hat{\mathbf{i}}, \hat{\mathbf{j}}$ are the local in-plane axes.

Regardless of whether an isotropic ($n = 2$) or anisotropic ($n = 1$) spontaneous curvature field is employed, the strength of the spontaneous curvature, C_0, as appearing in Eqs. (15) and (16), is given as the product of the averaged N-BAR domain density, ρ, an area element δA, and the intrinsic curvature of a single N-BAR domain as found from atomistic MD simulation, i.e., $C_0 = \rho \delta A H_{\text{N-BAR}}$, where $H_{\text{N-BAR}} \sim 0.15 \text{ nm}^{-1}$ (Blood and Voth, 2006).

In the mesoscopic simulation, an EM2-S liposome composed of 4000 EM2-S quasi-particles with an initial radius of 130 nm was equilibrated with no spontaneous curvature field at 308 K. The resulting final structure is shown in panel (a) of Fig. 14; thermal undulations are still quite pronounced at this lengthscale, as shown by the deformations in the liposome surface. The two different N-BAR domain spontaneous curvature fields [isotropic ($n = 2$), anisotropic ($n = 1$)]

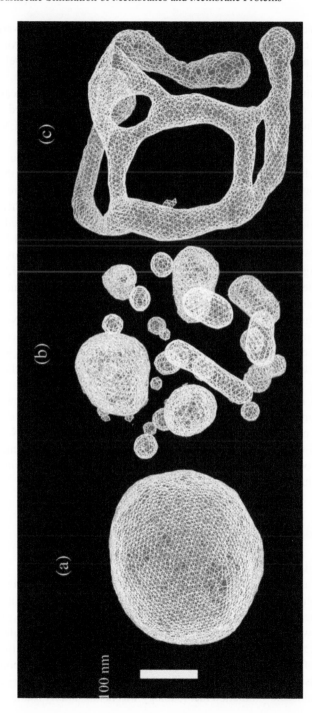

FIGURE 14 (a) An EM2-S liposome with an initial radius of 130 nm and no N-BAR spontaneous curvature, i.e., $C_0 = 0$ nm^{-1}. (b) The remodeled liposome as in (a) subjected to an isotropic ($n = 2$) spontaneous curvature field with a value of $C_0 = 0.14$ nm^{-1}. (c) The remodeled liposome as in (a) subjected to an anisotropic ($n = 1$) spontaneous curvature field with a value of $C_0 = 0.08$ nm^{-1}. The scale bar corresponds to 100 nm.

were then examined at various curvature field strengths corresponding to various possible N-BAR concentrations on the membrane surface, where the maximum spontaneous curvature was that calculated from the original atomistic-level MD simulation (Blood and Voth, 2006), with $H_{\text{N-BAR}} = 0.15 \text{ nm}^{-1}$.

The main results are summarized in panels (b) and (c) of Fig. 14. Recalling the experimental observation (Peter *et al.*, 2004) that tubulation occurred only with low to medium N-BAR concentrations, it was observed from the EM2-S simulation that a similar result was *only* obtained with an anisotropic ($n = 1$) spontaneous curvature field (shown in panel (c) of Fig. 14). That is, tubulation can only occur when there is some degree of orientational correlation in the N-BAR domain spontaneous curvature field. In fact, the anisotropic curvature field actually "wraps" around the tubules, in a sense holding the structure together.

Conversely, again from experiment (Peter *et al.*, 2004) is was observed that *vesiculation* only occurred at very high N-BAR domain densities. The EM2-S model only exhibited vesiculation with an *isotropic* N-BAR spontaneous curvature field at high densities. Matching the experimental observation with the EM2-S results, the conclusion is that at high densities the spatial arrangement of the N-BAR domains is isotropic in nature. This observation, at first glance, almost seems counter-intuitive in that one might assume that at high densities the N-BAR domains would "pack" like a bunch of bananas. However, that does not seem to be the case; if the N-BAR domains lined up, on average, the liposome would not undergo vesiculation, but more likely it would exhibit tubulation (panel (c) in Fig. 14).

This work has resulted in new insight into membrane remodeling; however, there is still a future MS-CG component that is required in order to complete the picture. What is now required is a MS-CG simulation of multiple CG N-BAR domains on a CG membrane, so that the average spatial structure and orientation correlations that occur in membranes with various densities of N-BAR domains on them can be elucidated.

C. Multiscale Coupling

1. Pure Lipid Bilayer

MSC was first applied to a pure DMPC membrane (Chang *et al.*, 2005). From a structural viewpoint, under MSC, both the 2D pair correlation function of the centers of mass of the DMPC lipids, as well as lipid chain order parameter S_{CD} were almost identical when compared to an isolated NPT atomistic-level simulation. These results suggest that the structural lipid correlations are not overly affected by the inclusion of mesoscopic plane stress modes. However, it should be pointed out that the atomistic-level bilayer remained stable under the MSC boundary conditions, so the multiscale simulation methodology was demonstrably stable.

Although the lipid structural properties were not overly sensitive to the MSC boundary conditions, some dynamical properties, such as the mean square displacement (MSD), were altered. For example, although the lateral diffusion of single lipids did not show a significant change whether or not the MD system was isolated or under MSC, the perpendicular diffusion component strongly depends on the MSC coupling. Lipid diffusion under MSC consists of two components: the local lipid diffusion in the atomistic MD system and then membrane fluctuations arising from the mesoscopic EM model. In fact, after 1 ns, the z-component of the lipid mean square displacement in the MSC simulation was almost two orders of magnitude larger than that of an isolated NPT system. This striking difference is due to the long wavelength undulation modes in the mesoscopic model effectively "dragging" the entire atomistic bilayer simulation up and down with the thermal modes. The small lengthscales of the isolated MD simulation, along with the small periodic lattice, inhibits these motions.

2. Bilayers with Trans-Membrane Proteins

MSC has been also employed to examine changes in atomistic-level protein structure due to long wavelength membrane undulations and plane stress fields (Ayton and Voth, 2007). In this case, an ensemble of atomistic-level simulations of a model of a transmembrane influenza A virus M2 proton channel (Wu and Voth, 2005; Smondyrev and Voth, 2002) in a DMPC bilayer was coupled to a corresponding mesoscopic EM2 model of a DMPC bilayer (Ayton et al., 2005b) in an explicit mesoscopic solvent (BLOBs). The aim was to examine any structural variations in the key proton gating His[37] residues of the M2 channel, as it is these residues that are believed to be responsible for the free energy barrier to proton transport.

The overall goal was to explore whether or not the structure of a protein is coupled to long wavelength membrane phenomena, which takes a step beyond the issue of how the local lipid environment affects protein structure/function (Grossfield et al., 2006). Briefly, the transmembrane domain of the M2 protein is about 19 residues, and it is composed of a tetrameric bundle of α-helical transmembrane peptides (Duff and Ashley, 1992; Kovacs and Cross, 1997). A column of water molecules within the channel is believed to be interrupted by a ring of four His[37] residues (Pinto et al., 1997), which play a key role in proton gating and transport (Wu and Voth, 2005; Smondyrev and Voth, 2002). Two states have been proposed: In the closed state, the imidazole side chains of the four His[37] residues are pointed towards the lumen, thus occluding the pore (Wu and Voth, 2005). Alternatively, a possible open state can occur with three ε protonated and one or more doubly protonated His[37] residues (Smondyrev and Voth, 2002), where the proposed "shuttle mechanism" (Pinto et al., 1997) suggests that the proton relay system intrinsically contains the doubly protonated His[37] residue. This scenario contrasts with the electrostatic repulsion dominated "shutter mechanism" (Sansom et al., 1997).

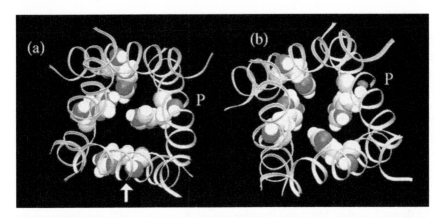

FIGURE 15 A snapshot of the open M2 protein channel His[37] residues under (a) MSC and (b) standard constant NPT MD simulation conditions. The "P" corresponds to the doubly protonated His[37] residue. Under MSC, the doubly protonated His[37] residue moves into the center of the channel, displacing the His[37] residue to its left (i.e., the His[37] residue indicated by the arrow clockwise one over from the doubly protonated one denoted by "P"). This structure results in an overall larger effective volume, and hence a lower density.

The influenza A virus M2 channel has been examined previously with atomistic-level MD simulation (Wu and Voth, 2005; Smondyrev and Voth, 2002; Pinto *et al.*, 1997; Sansom *et al.*, 1997; Zhong *et al.*, 1998; Forrest *et al.*, 2000; Schweighofer and Pohorille, 2000). A review on the overall topic of MD simulation of membrane proteins can be found in Ref. (Ash *et al.*, 2004). However, in all cases the length-scales were such that only a relatively small bilayer region surrounded the protein. MSC can effectively include longer wavelength phenomena, and these long wavelength thermally induced modes can, in principle, change the trans-membrane protein structure.

Details concerning the MSC implementation for this study can be found elsewhere (Ayton and Voth, 2007). The important result was that MSC can indeed alter the structure of the His[37] residues in both the open and closed states. However, how the structural alteration is manifest must be considered carefully. First, the effect of the MSC coupled mesoscopic stress fields on the mean square displacement (MSD) of the His[37] residues in the xy plane was negligible, as the MSD is largely dominated by relatively fast frequency atomistic-level fluctuations. However, the average effective volume of the His[37] residues, V_{eff}, did increase by about 4% under MSC. The increase in *volume* of the open channel under MSC reflects the frustrated open state of the His[37] residues; the MSC plane stress field acts to decrease the density of the His[37] complex in all directions. However, the more organized structure of the closed channel responds differently in that area changes are coupled to the plane stress, and thus its effective volume remains essentially constant.

Figure 15 shows the final simulation snapshots of the His[37] residues in the open state under both MSC (a) and constant NPT (b) conditions. It is observed that with MSC the conformation of the doubly protonated His[37] residue (designated with a "P") is directed more towards the channel center such that the residue to its left (the bottom residue in Fig. 15) is displaced out of the channel. This structure is consistent with the slightly larger V_{eff} and is most likely due to a more vertical orientation of the His[37] residues. Conversely, under constant NPT atomistic MD conditions (b), the protrusion of the doubly protonated residue is not as great, and residue to the left is not displaced. Thus the structure in (b) corresponds to a slightly higher effective density.

IV. SUMMARY

This review has discussed our current efforts in the multiscale simulation of membranes. An ongoing theme of this work is to attempt to systematically connect the different spatial and temporal scales that exist in membranes, ranging from the atomistic-level to the nearly macroscopic. A large part of the initial effort has been to develop computational methodologies and simulation capabilities that can eventually be employed in an overall multiscale simulation methodology. Our current research directions aim to continue to develop these new methodologies and to further bridge them together such that a more unified multiscale picture for membranes and membrane proteins can be developed.

Acknowledgements

This research was supported by the National Institutes of Health (5 R01 GM063796). We thank Dr. Jhih-Wei Chu for valuable discussions and assistance.

References

Anezo, C., Vries, A.H.d., Holtje, H.D., Tieleman, D.P., Marrink, S.J. (2003). Methodological issues in lipid bilayer simulation. *J Phys. Chem. B* **107**, 9424–9433.

Ash, W.L., Zlomislic, M.R., Oloo, E.O., Tieleman, D.P. (2004). Computer simulations of membrane proteins. *Biochem. Biophys. Acta* **1666**, 158–189.

Ayton, G., Voth, G.A. (2002). Bridging microscopic and mesoscopic simulations of lipid bilayers. *Biophys. J.* **83**, 3357–3370.

Ayton, G.S., Voth, G.A. (2004a). The simulation of biomolecular systems at multiple length and timescales. *Int. J. Mult. Comp. Eng.* **2**, 291–311.

Ayton, G.S., Voth, G.A. (2004b). Mesoscopic lateral diffusion in lipid bilayers. *Biophys. J.* **87**, 3299–3311.

Ayton, G.S., Voth, G.A. (2007). Multiscale simulation of membrane bound proteins. *J. Struct. Biol.* **157**, 570–578.

Ayton, G., Bardenhagen, S., McMurtry, P., Sulsky, D., Voth, G.A. (2001a). Interfacing molecular dynamics with continuum dynamics in computer simulation: Towards an application to biological membranes. *IBM J. Res. Dev.* **45**, 417–426.

Ayton, G., Bardenhagen, S., McMurtry, P., Sulsky, D., Voth, G.A. (2001b). Interfacing continuum and molecular dynamics: An application to lipid bilayers. *J. Chem. Phys.* **114**, 6913–6924.

Ayton, G., Smondyrev, A.M., Bardenhagen, S., McMurtry, P., Voth, G.A. (2002a). Calculating the bulk modulus for a lipid bilayer using non-equilibrium molecular dynamics. *Biophys. J.* **82**, 1226–1238.

Ayton, G., Smondyrev, A.M., Bardenhagen, S., McMurtry, P., Voth, G.A. (2002b). Interfacing molecular dynamics and macro-scale simulations for lipid bilayer vesicles. *Biophys. J.* **83**, 1026–1038.

Ayton, G.S., Tepper, H., Mirijanian, D., Voth, G.A. (2004). A new perspective on the coarse-grained dynamics of fluids. *J. Chem. Phys.* **120**, 4074–4088.

Ayton, G.S., McWhirter, J.L., McMurtry, P., Voth, G.A. (2005a). Coupling field theory with continuum mechanics: A simulation of domain formation in giant unilamellar vesicles. *Biophys. J.* **88**, 3855–3869.

Ayton, G.S., McWhirter, J.L., Voth, G.A. (2005b). A second generation mesoscopic lipid bilayer model: Connections to field-theory descriptions of membranes and nonlocal hydrodynamics. *J. Chem. Phys.* **124**, 064906.

Ayton, G.S., Blood, P.D., Voth, G.A. (2007). Membrane remodeling from N-BAR domain interactions: Insights from multiscale simulation. *Biophys. J.* **92**, 3595–3602.

Baumgart, T., Hess, S.T., Webb, W.W. (2003). Imaging coexisting fluid domains in biomembrane models coupling curvature and line tension. *Nature* **425**, 821–824.

Blood, P.D., Voth, G.A. (2006). Direct observation of BAR domain-induced membrane curvature via molecular dynamics simulation. *Proc. Natl. Acad. Sci.* **103**, 15068–15072.

Blood, P.D., Ayton, G.S., Voth, G.A. (2005). Probing the molecular-scale lipid bilayer response to shear flow using nonequilibrium molecular dynamics. *J. Phys. Chem. B* **109**, 18673–18679.

Bond, P.J., Sansom, M.S.P. (2006). Insertion and assembly of membrane proteins via simulation. *J. Am. Chem. Soc.* **128**, 2697–2704.

Brannigan, G., Brown, F.L.H. (2004). Solvent-free simulations of membrane bilayers. *J. Chem. Phys.* **120**, 1059–1071.

Brannigan, G., Brown, F.L.H. (2005). Composition dependence of bilayer elasticity. *J. Chem. Phys.* **122**, 074905.

Brannigan, G., Lin, L.C.L., Brown, F.L.H. (2006). Implicit solvent simulation models for biomembranes. *Eur. Biophys. J.* **35**, 104–124.

Brochard, F., Lennon, J.F. (1975). Frequency spectrum of the flicker phenomenon in erythrocytes. *J. Phys. (France)* **36**, 1035–1047.

Brown, F.L.H. (2003). Regulation of protein mobility via thermal membrane undulations. *Biophys. J.* **84**, 842–853.

Cahn, J.W., Hilliard, J.E. (1958). Free energy of a nonuniform system I. Interfacial free energy. *J. Chem. Phys.* **28**, 258–267.

Chaikin, P.M., Lubensky, T.C. (1995). "Principles of Condensed Matter Physics". University Press, Cambridge.

Chandler, D. (1987). "Introduction to Modern Statistical Mechanics". Oxford University Press, New York.

Chang, R., Ayton, G.S., Voth, G.A. (2005). Multiscale coupling of mesoscopic and atomistic-level lipid bilayer simulations. *J. Chem. Phys.* **122**, 244716.

Chu, J., Voth, G.A. (2006). Coarse-grained modeling of the actin filament derived from atomistic-scale simulations. *Biophys. J.* **90**, 1572–1582.

Cooke, I.R., Deserno, M. (2005). Solvent-free model for self-assembling fluid bilayer membranes: Stabilization of the fluid phase based on broad attractive tail potentials. *J. Chem. Phys.* **123**, 224710.

Cooke, I.R., Deserno, M. (2006). Coupling between lipid shape and membrane curvature. *Biophys. J.* **91**, 487–495.

Cooke, I.R., Kremer, K., Deserno, M. (2005). Tunable generic model of fluid bilayer membranes. *Phys. Rev. E* **72**, 011506.

deJeu, W.H., Ostrovskii, B.I., Shalaginov, A.N. (2003). Structure and fluctuations of smectic membranes. *Rev. Mod. Phys.* **75**, 181–235.

Delgado-Buscalioni, R., Coveney, P.V. (2003). Continuum-particle hybrid coupling for mass, momentum, and energy transfers in unsteady fluid flow. *Phys. Rev. E* **67**, 046704.

denOtter, W.K., Briels, W.J. (2003). The bending rigidity of an amphiphilic bilayer from equilibrium and nonequilibrium molecular dynamics. *J. Chem. Phys.* **118**, 4712–4720.

deVries, A.H., Mark, A.E., Marrink, S.J. (2004). Molecular dynamics simulation of the spontaneous formation of a small DPPC vesicle in water in atomistic detail. *J. Am. Chem. Soc.* **126**, 4488–4489.

Duff, K.C., Ashley, R.H. (1992). The transmembrane domain of influenza A M2 protein forms amantadine-sensitive proton channels in planar lipid bilayer. *Virology* **190**, 485–489.

Espanol, P., Warren, P.B. (1995). Statistical-mechanics of dissipative particle dynamics. *Europhys. Lett.* **30**, 191–196.

Essmann, U., Berkowitz, M.L. (1999). Dynamical properties of phospholipid bilayers from computer simulation. *Biophys. J.* **76**, 2081–2089.

Evans, D.J., Morriss, G.P. (1990). "Statistical Mechanics of Nonequilibrium Liquids". Academic Press, London.

Evans, E., Needham, D. (1987). Physical properties of surfactant bilayer membranes: Thermal transitions, elasticity, rigidity, cohesion and colloidal interactions. *J. Phys. Chem.* **91**, 4219–4228.

Evans, E., Rawicz, W. (1997). Elasticity of "fuzzy" membranes. *Phys. Rev. Lett.* **79**, 2379–2382.

Faller, R. (2004). Automatic coarse graining of polymers. *Polymer* **45**, 3869–3876.

Faller, R., Marrink, S.J. (2004). Simulation of domain formation on DLPC–DSPC mixed bilayers. *Langmuir* **20**, 7686–7693.

Farago, O. (2003). "Water-free" computer model for fluid bilayer membranes. *J. Chem. Phys.* **119**, 596–605.

Farago, O., Pincus, P. (2004). Statistical mechanics of bilayer membrane with a fixed projected area. *J. Chem. Phys.* **120**, 2934–2950.

Feller, S.E. (2000). Molecular dynamics simulations of lipid bilayers. *Curr. Opin. Colloid Interface Sci.* **5**, 217–223.

Filippov, A., Oradd, G., Lindblom, G. (2003). The effect of cholesterol on the lateral diffusion of phospholipids in oriented bilayers. *Biophys. J.* **84**, 3079–3086.

Fischer, T.M. (1992a). Bending stiffness of lipid bilayers. III. Gaussian curvature. *J. Phys. II (France)* **2**, 337–343.

Fischer, T.M. (1992b). Bending stiffness of lipid bilayers. II. Spontaneous curvature of the monolayers. *J. Phys. France II* **2**, 327–336.

Ford, M.G., Mills, I.G., Peter, B.J., Vallis, Y., Praefcke, G.J., Evans, P.R., McMahon, H.T. (2002). Curvature of clathrin-coated pits driven by epsin. *Nature* **419**, 361–366.

Forrest, L.R., Kukol, A., Arkin, I.T., Tieleman, D.P., Sansom, M.S.P. (2000). Exploring models of the influenza A M2 channel: MD simulations in a phospholipid bilayer. *Biophys. J.* **78**, 55–69.

Goetz, R., Lipowsky, R. (1998). Computer simulations of bilayer membranes: Self-assembly and interfacial tension. *J. Chem. Phys.* **108**, 7397–7409.

Goetz, R., Gompper, G., Lipowsky, R. (1999). Mobility and elasticity of self-assembled membranes. *Phys. Rev. Lett.* **82**, 221–224.

Gov, N., Zilman, A.G., Safran, S. (2004). Hydrodynamics of confined membranes. *Phys. Rev. E* **70**, 011104.

Granek, R. (1997). From semi-flexible polymers to membranes: Anomalous diffusion and reptation. *J. Phys. (France)* **7**, 1761–1788.

Groot, R.D., Rabone, K.L. (2001). Mesoscopic simulation of cell membrane damage, morphology change and rupture by nonionic surfactants. *Biophys. J.* **81**, 725–736.

Groot, R.D., Warren, P.B. (1997). Dissipative particle dynamics: Bridging the gap between atomistic and mesoscopic simulation. *J. Chem. Phys.* **107**, 4423–4435.

Grossfield, A., Feller, S.E., Pitman, M.C. (2006). A role for direct interactions in the modulation of rhodopsin by w-3 polyunsaturated lipids. *Proc. Natl. Acad. Sci.* **103**, 4888–4893.

Groves, J.T., Boxer, S.G. (2002). Micropattern formation in supported lipid membranes. *Acc. Chem. Res.* **35**, 149–157.

Groves, J.T., Ulman, N., Boxer, S.G. (1997). Micropatterning fluid lipid bilayers on solid supports. *Science* **275**, 651–653.

Hansen, J.P., McDonald, I.R. (1986). "Theory of Simple Liquids", 2 ed. Academic Press, San Diego.

Helfrich, W. (1973). Elastic properties of lipid bilayers: Theory and possible experiments. *Z. Naturforsch (c)* **28**, 693–703.

Helfrich, W. (1978). Steric interaction of fluid membranes in multilayer systems. *Z. Naturforsch (a)* **33a**, 305–315.

Hofsab, C., Lindahl, E., Edholm, O. (2003). Molecular dynamics simulations of phospholipid bilayers with cholesterol. *Biophys. J.* **84**, 2192–2206.

Hoover, W.G., Hoover, C.G. (2003). Links between microscopic and macroscopic fluid mechanics. *Mol. Phys.* **101**, 1559–1573.

Hoover, W.G., Posch, H.A. (1996). Numerical heat conductivity in smooth particle applied mechanics. *Phys. Rev. E* **54**, 5142–5145.

Hyman, J.M. (2005). Patch dynamics for multiscale problems. *Comput. Sci. Eng.* **7**, 47–53.

Izvekov, S., Voth, G.A. (2005a). A multiscale coarse-graining method for biomolecular systems. *J. Phys. Chem. B* **109**, 2469–2473.

Izvekov, S., Voth, G.A. (2005b). Multiscale coarse-graining of liquid state systems. *J. Chem. Phys.* **123**, 134105.

Izvekov, S., Voth, G.A. (2006a). Multiscale coarse-graining of mixed phospholipid/cholesterol bilayers. *J. Chem. Theor. Comp.* **2**, 637–648.

Izvekov, S., Voth, G.A. (2006b). Modeling real dynamics in the coarse-grained representations of condensed phase systems. *J. Chem. Phys.* **125**, 151101.

Jiang, Y., Lookman, T., Saxena, A. (2000). Phase separation and shape deformation of two-phase membranes. *Phys. Rev. E* **61**, R57–R60.

Khelashvili, G.A., Pandit, S.A., Scott, H.L. (2005). Self-consistent mean-field model based on molecular-dynamics: Application to lipid–cholesterol bilayers. *J. Chem. Phys.* **123**, 034910.

Kopelevich, D.I., Panagiotopoulos, A.Z., Kevrekidis, I.G. (2005). Coarse-grained computations for a micellar system. *J. Chem. Phys.* **122**, 044907.

Korlach, J., Schwille, P., Webb, W.W., Feigenson, G.W. (1999). Characterization of lipid bilayer phases by confocal microscopy and fluorescence correlation spectroscopy. *Proc. Natl. Acad. Sci. USA* **96**, 8461–8466.

Kovacs, F.A., Cross, T.A. (1997). Transmembrane four-helix bundle of influenza A M2 protein channel: Structural implications from helix tilt and orientation. *Biophys. J.* **73**, 2511–2517.

Kubo, R. (1966). The Fluctuation-Dissipation Theorem. *Rep. Prog. Phys.* **29**, 255–284.

Kum, O., Hoover, W.G., Posch, H.A. (1995). Viscous conducting flows with smooth-particle applied mechanics. *Phys. Rev. E* **52**, 4899–4907.

Kumar, P.B.S., Rao, M. (1998). Shape instabilities in the domains of a two-component fluid membrane. *Phys. Rev. Lett.* **80**, 2489–2492.

Kumar, P.B.S., Gompper, G., Lipowsky, R. (2001). Budding dynamics of multicomponent membranes. *Phys. Rev. Lett.* **86**, 3911–3914.

Laradji, M., Kumar, P.B.S. (2004). Dynamics of domain growth in self-assembled fluid vesicles. *Phys. Rev. Lett.* **93**, 198105.

Lee, E., Marcucci, M., Daniell, L., Pypaert, M., Weisz, O.A., Ochoa, G.C., Farsad, K., Wenk, M.R., De Camilli, P. (2002). Amphiphysin 2 (Bin1) and T-tubule biogenesis in muscle. *Science* **297**, 1193–1196.

Lin, L.C.-L., Brown, F.L.H. (2004). Dynamics of pinned membranes with application to protein diffusion on the surface of red blood cells. *Biophys. J.* **86**, 764–780.

Lin, L.C.L., Brown, F.L.H. (2005). Dynamic simulations of membranes with cytoskeletal interactions. *Phys. Rev. E* **72**, 011910.

Lin, L.C.L., Gov, N., Brown, F.L.H. (2006). Nonequilibrium membrane fluctuations driven by active proteins. *J. Chem. Phys.* **124**, 074903.

Lindahl, E., Edholm, O. (2000). Mesoscopic undulations and thickness fluctuations in lipid bilayers from molecular dynamics simulations. *Biophys. J.* **79**, 426–433.

Lindahl, E., Edholm, O. (2001). Molecular dynamics simulations of NMR relaxation rates and slow dynamics in lipid bilayers. *J. Chem. Phys.* **115**, 4938–4950.

Lipowsky, R., Sackmann, E. (1995). "Structure and Dynamics of Membranes", vol. **1A**. North-Holland, Amsterdam.

Lopez, C.F., Nielsen, S.O., Ensing, B., Moore, P.B., Klein, M.L. (2005). Structure and dynamics of model pore insertion into a membrane. *Biophys. J.* **88**, 3083–3094.

Lucy, L.B. (1977). A numerical approach to the testing of the fission hypothesis. *Astrophys. J.* **82**, 1013–1024.

Lyubartsev, A.P., Laaksonen, A. (1995). Calculation of effective interaction potentials from radial distribution functions: A reverse Monte Carlo approach. *Phys. Rev. E* **52**, 3730–3737.

Lyubartsev, A.P., Laaksonen, A. (1997). Osmotic and activity coefficients from effective potential for hydrated ions. *Phys. Rev. E* **55**, 5689–5696.

Lyubartsev, A.P., Laaksonen, A. (1999). Effective potentials for ion–DNA interactions. *J. Chem. Phys.* **111**, 11207–11215.

Marrink, S.J., Mark, A.E. (2001). Effect of undulations on surface tension in simulated bilayers. *J. Phys. Chem. B* **105**, 6122–6127.

Marrink, S.J., Mark, A.E. (2003a). The mechanism of vesicle fusion as revealed by molecular dynamics simulations. *J. Am. Chem. Soc.* **125**, 11144–11145.

Marrink, S.J., Mark, A.E. (2003b). Molecular dynamics simulation of the formation, structure, and dynamics of small phospholipid vesicles. *J. Am. Chem. Soc.* **125**, 15233–15242.

Marrink, S.J., Tieleman, D.P. (2002). Molecular dynamics simulation of spontaneous membrane fusion during a cubic-hexagonal phase transition. *Biophys. J.* **83**, 2386–2392.

Marrink, S.J., deVries, A.H., Mark, A.E. (2004). Coarse Grained Model for semiquantitative lipid simulations. *J. Phys. Chem. B* **108**, 750–760.

Masuda, M., Takeda, S., Sone, M., Ohki, T., Mori, H., Kamioka, Y., Mochizuki, N. (2006). Endophilin BAR domain drives membrane curvature by two newly identified structure-based mechanisms. *EMBO J.* **25**, 2889–2897.

McMahon, H.T., Gallop, J.L. (2005). Membrane curvature and mechanisms of dynamic cell membrane remodelling. *Nature* **438**, 590–596.

McWhirter, J.L., Ayton, G.S., Voth, G.A. (2004). Coupling field theory with mesoscopic dynamical simulations of multi-component lipid bilayers. *Biophys. J.* **87**, 3242–3263.

Metiu, H., Kitahara, K., Ross, J. (1976). A derivation and comparison of two equations (Landau–Ginzburg and Cahn) for the kinetics of phase transitions. *J. Chem. Phys.* **65**, 393–396.

Meyer, H., Biermann, O., Faller, R., Reith, D., Muller-Plathe, F. (2000). Coarse graining of nonbonded inter-particle potentials using automatic simplex optimization to fit structural properties. *J. Chem. Phys.* **113**, 6264–6275.

Monaghan, J.J. (1992). Smoothed particle hydrodynamics. *Annu. Rev. Astron. Astrophys.* **30**, 543–574.

Nielsen, S.O., Lopez, C.F., Srinivas, G., Klein, M.L. (2004). Coarse-grain models and the computer simulation of soft materials. *J. Phys. Condens. Matter* **16**, R481–R512.

Nielsen, S.O., Srinivas, G., Klein, M.L. (2005a). Incorporating a hydrophobic solid into a coarse grain liquid framework: Graphite in an aqueous amphiphilic environment. *J. Chem. Phys.* **123**, 124907.

Nielsen, S.O., Srinivas, G., Lopez, C.F., Klein, M.L. (2005b). Modeling surfactant adsorption on hydrophobic surfaces. *Phys. Rev. Lett.* **94**, 228301.

Noid, W.G., Chu, J.-W., Ayton, G.S., Voth, G.A. (2007). Multiscale coarse-graining and structural correlations: Connections to liquid state theory. *J. Phys. Chem. B* **111**, 4116–4127.

Olbrich, K., Rawicz, W., Needham, D., Evans, E. (2000). Water permeability and mechanical strength of polyunsaturated lipid bilayers. *Biophys. J.* **79**, 321–327.

Oradd, G., Lindblom, G., Westerman, P.W. (2002). Lateral diffusion of cholesterol and dimyristoylphosphatidylcholine in a lipid bilayer measured by pulsed field gradient NMR spectroscopy. *Biophys. J.* **83**, 2702–2704.

Pandit, S.A., Vasudevan, S., Chiu, S.W., Mashi, R.J., Jakobsson, E., Scott, H.L. (2004a). Sphingo-myelin-cholesterol domains in phospholipid membranes: Atomistic simulation. *Biophys. J.* **87**, 1092–1100.

Pandit, S.A., Jakobsson, E., Scott, H.L. (2004b). Simulation of the early stages of nano-domain formation in mixed bilayers of sphingomyelin, cholesterol and dioleylphosphatidylcholine. *Biophys. J.* **87**, 3312–3322.

Pandit, S., Khelashvili, G., Chu, S.W., Jackobsson, E., Mashl, R.J., Scott, H.L. (2004c). Modeling mixed lipid bilayers: Atomic and Ginzburg–Landau simulations. *Biophys. Soc. Discuss. Probing Membrane Microdomains.* http://www.biophysics.org/discussions/2007/scott.pdf.

Pastor, R.W. (1994). Molecular-dynamics and Monte-Carlo simulations of lipid bilayers. *Curr. Opin. Struct. Biol.* **4**, 486–492.

Pastor, R.W., Feller, S.E. (1996). Time scales of lipid dynamics and molecular dynamics. *Biol. Membr.* **1**, 4–29.

Peter, B.J., Kent, H.M., Mills, I.G., Vallis, Y., Butler, P.J.G., Evans, P.R., McMahon, H.T. (2004). BAR domains as sensors of membrane curvature: The amphiphysin BAR structure. *Science* **303**, 495–499.

Pfeiffer, W., Schlossbauer, G., Knoll, W., Farago, B., Steyer, A., Sackmann, E. (1988). Ultracold neutron scattering study of local lipid mobility in bilayer membranes. *J. Phys. (France)* **49**, 1077–1082.

Pinto, L.H., Dieckmann, G.R., Gandhi, C.S., Papworth, C.G., Braman, J., Shaughnessy, M.A., Lear, J.D., Lamb, R.A., DeGrado, W.F. (1997). A functionally defined model for the M2 proton channel of influenza A virus suggests a mechanism for its ion selectivity. *Proc. Natl. Acad. Sci.* **94**, 11301–11306.

Radler, J., Sackmann, E. (1993). Imaging optical thicknesses and separation distances of phospholipid vesicles at solid surfaces. *J. Phys. II (France)* **3**, 727–748.

Rawicz, W., Olbrich, K.C., McIntosh, T., Needham, D., Evans, E. (2000). Effect of chain length and unsaturation on elasticity of lipid bilayers. *Biophys. J.* **79**, 328–339.

Razzaq, A., Robinson, I.M., McMahon, H.T., Skepper, J.N., Su, Y., Zelhof, A.C., Jackson, A.P., Gay, N.J., O'Kane, C.J. (2001). Amphiphysin is necessary for organization of the excitation-contraction coupling machinery of muscles, but not for synaptic vesicle endocytosis in Drosophila. *Genes Dev.* **15**, 2967–2979.

Sackmann, E. (1994). Membrane bending energy concept of vesicle- and cell-shapes and shape-transitions. *FEBS Lett.* **346**, 3–16.

Sansom, M.S., Kerr, I.D., Smith, G.R., Son, H.S. (1997). The influenza A virus M2 channel: A mole-cular modeling and simulation study. *Virology* **233**, 163–173.

Schweighofer, K.J., Pohorille, A. (2000). Computer simulation of ion channel gating: The M2 channel of influenza A virus in a lipid bilayer. *Biophys. J.* **78**, 150–163.

Scott, H.L. (2002). Modeling the lipid component of membranes. *Curr. Opin. Struct. Biol.* **12**, 495–502.

Seifert, U. (1997). Configurations of fluid membranes and vesicles. *Adv. Phys.* **46**, 13–137.

Shelley, J.C., Shelley, M.Y., Reeder, R.C., Bandyopadhyay, S., Klein, M.L. (2001a). A coarse grain model for phospholipid simulation. *J. Phys. Chem. B* **105**, 4464–4470.

Shelley, J.C., Shelley, M.Y., Reeder, R.C., Bandyopadhyay, S., Klein, M.L. (2001b). Simulation of phospholipids using a coarse grain model. *J. Phys. Chem. B* **105**, 9785–9792.

Shi, Q., Izvekov, S., Voth, G.A. (2006). Mixed atomistic and coarse-grained molecular dynamics: Simulation of a membrane bound ion channel. *J. Phys. Chem. B* **110**, 15045–15048.

Shi, Q., Voth, G.A. (2005). Multiscale modeling of phase separation in mixed lipid bilayers. *Biophys. J.* **89**, 2385–2394.

Shih, A.Y., Arkhipov, A., Freddolino, P.L., Schulten, K. (2006). Coarse grained protein–lipid model with applications to lipoprotein particles. *J. Phys. Chem. B* **110**, 3674–3684.

Shillcock, J.C., Lipowsky, R. (2002a). Dissipative particle dynamics simulations of planar amphiphilic bilayers. *NIC Symp. Proc.* **9**, 407–417.

Shillcock, J.C., Lipowsky, R. (2002b). Equilibrium structure and lateral stress distribution of amphiphilic bilayers from dissipative particle dynamics. *J. Chem. Phys.* **117**, 5048–5061.

Shin, Y.K., Ewert, U., Budil, D.E., Freed, J.H. (1991). Microscopic versus macroscopic diffusion in model membranes by electron spin resonance spectral-spatial imaging. *Biophys. J.* **59**, 950–957.

Smondyrev, A.M., Berkowitz, M.L. (1999). United atom force field for phospholipid membranes: Constant pressure molecular dynamics simulation of dipalmitoylphosphatidylcholine/water system. *J. Comput. Chem.* **20**, 531–545.

Smondyrev, A.M., Voth, G.A. (2002). Molecular dynamics simulation of proton transport through the Influenza A Virus M2 channel. *Biophys. J.* **83**, 1987–1996.

Spoel, D.V.D., Buuren, A.R.V., Apol, E., Meulenhoff, P.J., Tieleman, D.P., Sijbers, A.L.T., Drunen, R.V., Berendsen, H.J.C. (1996). http://rugmdo.chem.rug.nl/gmx.

Stevens, M.J. (2004). Coarse-grained simulations of lipid bilayers. *J. Chem. Phys.* **121**, 11942–11948.

Stevens, M.J. (2005). Complementary matching in domain formation within lipid bilayers. *J. Am. Chem. Soc.* **127**, 15330–15331.

Sun, Q., Faller, R. (2005). Systematic coarse-graining of atomistic models for simulation of polymeric systems. *Comput. Chem. Eng.* **29**, 2380–2385.

Tieleman, D.P., Marrink, S.J., Berendsen, H.J.C. (1997). A computer perspective of membranes: Molecular dynamics studies of lipid bilayer systems. *Biochim. Biophys. Acta* **1331**, 235–270.

Tieleman, D.P., Leontiadou, H., Mark, A.E., Marrink, S.J. (2003). Simulation of pore formation in lipid bilayers by mechanical stress and electric fields. *J. Am. Chem. Soc.* **125**, 6382–6383.

Vaz, W.L.C., Almeida, P.F. (1991). Microscopic versus macroscopic diffusion in one-component fluid phase lipid bilayer membranes. *Biophys. J.* **60**, 1553–1554.

Veatch, S.L., Keller, S.L. (2003). Separation of liquid phases in giant vesicles of ternary mixtures of phospholipids and cholesterol. *Biophys. J.* **85**, 3074–3083.

Wang, Y., Izvekov, S., Yan, T., Voth, G.A. (2006). Multiscale coarse-graining of ionic liquids. *J. Phys. Chem. B* **110**, 3564–3575.

Wu, Y., Voth, G.A. (2005). A computational study of the closed and open states of the influenza A M2 proton channel. *Biophys. J.* **89**, 2402–2411.

Yamamoto, S., Maruyama, Y., Hyodo, S. (2002). Dissipative particle dynamics study of spontaneous vesicle formation of amphiphilic molecules. *J. Chem. Phys.* **116**, 5842–5849.

Zhong, Q., Husslein, T., Moore, P.B., Newns, D.M., Pattnaik, P., Klein, M.L. (1998). The M2 channel of influenza A virus: A molecular dynamics study. *FEBS. Lett.* **434**, 265–271.

Zhou, J., Thorpe, I.F., Izvekov, S., Voth, G.A. (2007). Coarse-grained peptide modeling using a systematic multiscale approach. *Biophys. J.* **92**, 4289–4303.

Zimmerberg, J., Kozlov, M.M. (2006). How proteins produce cellular membrane curvature. *Nat. Rev. Mol. Cell. Biol.* **7**, 9–19.

CHAPTER 8

Interactions between Small Molecules and Lipid Bilayers

Justin L. MacCallum and D. Peter Tieleman

Dept. of Biological Sciences, University of Calgary, 2500 University Dr NW, Calgary AB, T2N 1N4, Canada

Abstract

The cell membrane forms a selectively permeable barrier that enables cells to maintain a distinct chemical environment from their surroundings, making cell membranes one of the key components required for life. In this chapter we review the interactions between small molecules and lipid bilayers from a computational perspective, focusing on the permeation of small molecules and how small molecules affect the properties of lipid bilayers. We organize the molecules based on broad chemical characteristics such as hydrophobicity, polarity, and charge. We also discuss three specific classes of interest: anesthetics, sugars, and carbon nanoparticles.

Current Topics in Membranes, Volume 60
1063-5823/08 $35.00
DOI: 10.1016/S1063-5823(08)00008-2

ABBREVIATIONS

DOPC: dioleoylphosphatidylcholine

POPC: palmitoyloleoylphosphatidylcholine

POPE: palmitoyloleoylphosphatidylethanolamine

DOPS: dioleoylphosphatidylserine

DMPC: dimyristoylphosphatidylcholine

PMF: potential of mean force

MD: molecular dynamics

PCP: pentachlorophenol

GMO: glycerol-1-monoleate

TFE: (1,1,2)-trifluorethane

HFE: hexafluorethane

FCS: fluorescence correlation spectroscopy

I. INTRODUCTION

A. Biological Membranes and Small Molecules

The membrane surrounding every cell serves the critical biological function of compartmentalization, which allows cells to maintain an environment within the volume enclosed by the membrane that is distinct from the environment outside. Higher organisms also contain many membrane bound organelles. In this case the membrane still serves the same function of compartmentalization, allowing the cell to maintain a distinct environment within the organelle. The cell membrane is able to fulfill its role of compartmentalization by forming a selectively permeable barrier. Without the assistance of membrane proteins, only small nonpolar molecules are able to permeate the bilayer at a substantial rate, while biologically important molecules such as ions, and most nutrients are unable to cross the membrane at significant rates. Thus, it is important to understand the basic physical chemistry underlying the permeation of small molecules across the lipid bilayer.

The cell membrane is also host to a wide variety of membrane proteins, with important functions ranging from transport to signaling. Because small molecules affect the properties of the lipid bilayer, they may also indirectly affect the properties of the membrane proteins contained within (Cantor, 1997, 1998; Sharp *et al.*, 2001; Brink-van der Laan *et al.*, 2004; van den Brink-van der Laan *et al.*, 2004a, 2004b). For example, it has been proposed that the mechanism of general anesthesia may involve modulation of bilayer properties, which in turn affect the conformational equilibrium of membrane proteins (Cantor, 1997, 1998). Small molecules may also modify the properties of the lipid bilayer, allowing cells to survive extremes in temperature or hydration, for example.

Despite this importance, our knowledge of the interactions between small molecules and lipid bilayers is still rather limited. Experiments on lipid bilayer systems are complicated by the disordered, "soft" nature of the system. Computer simulations provide a complementary view to experiments, revealing a level of detail that is often impossible or very difficult to achieve experimentally. At the same time, macroscopic observables can be calculated from the simulations—allowing for comparison between simulation and experiment.

In this chapter we will review the available literature on computer simulations of the interactions between small molecules and lipid bilayers. The chapter will be loosely organized based on the type of small molecule, although many molecules could fit into more than one class. We will focus on the permeation and distribution of these molecules, as well as how these molecules affect the properties of the lipid bilayer. There are two previous reviews on similar subject matter. First, a comprehensive review of lipid bilayer simulations from Tieleman *et al.* (1997) also includes a section on the interaction between small molecules in lipid bilayers. Second, a review from Xiang and Anderson focused on the partitioning of small molecules as it pertains to liposomal drug transport (Xiang and Anderson, 2006).

B. Limitations

Before beginning a detailed examination of the interactions between small molecules and lipid bilayers, we will first pause briefly to review the limitations of molecular dynamics simulations. The principal limitations of molecular dynamics simulations for understanding the interactions between small molecules and lipid bilayers are the limited length and time scales accessible to simulation, and the force fields used to describe the interactions within the system.

Molecular dynamics simulations are limited in the length- and time-scales that they can access due to available computer power. A simulation of several hundred lipids for a few hundred nanoseconds would currently be considered state of the art. Although longer and larger simulations exist in the literature, they are still far from being routine. The time-scale limitation presents a serious problem for the calculation of interactions between many small molecules and lipid bilayers. For example, water permeates so slowly through a lipid bilayer that there is little chance of observing this phenomenon in a 100 ns simulation of 64 lipids. This necessitates the use of biased sampling techniques, such as umbrella sampling, which force the system to sample the event of interest, while still providing a rigorous method for unbiasing the simulation. However, even with biased simulations sufficient sampling over slow lipid motions is still required. Limitations of length-scale generally pose less of a problem in small molecule-lipid simulations. However, in some cases, for example those involving lipid phase changes or large changes in bilayer curvature, large systems may be required to avoid finite size effects and artificial stabilization of the bilayer due to boundary effects. As computers become faster, it is possible to simulate larger systems for longer times. Simulations that required enormous effort ten years ago can now be considered almost trivial in terms of required computer power.

The current force fields used to describe the interactions between lipids and small molecules also represent a potential limitation. In general, most modern force fields such as GROMOS (Oostenbrink *et al.*, 2004), OPLS-AA (Jorgensen *et al.*, 1996a, 1996b: Kaminski *et al.*, 2001), Amber (Cornell *et al.*, 1995), and CHARMM (MacKerell *et al.*, 1998) are parameterized, at least partially, on transfer free energies between water and nonpolar solvents. Thus, we can expect that free energies of transfer between water and the hydrocarbon region of the lipid bilayer should be relatively accurate. However, transfer to other environments has not been parameterized explicitly and may not be as accurate. Oostenbrink *et al.* (2004) were unable to accurately parameterize the amino acid side chains to simultaneously give accurate hydration free energies and water-alkane transfer free energies. They suggest this is likely due to the lack of polarizability in most current force fields. In practice, a change in polarization in moving from water to hydrocarbon is implicitly included in most force fields due to the parameterization on hydration or transfer free energies. For lipid simulations, the lack of polarizability in current force fields likely represents less of a problem than limitations of length and time scale, or overall force field accuracy. However, including polarizability in future force fields will surely be required to significantly improve force field accuracy beyond its current state.

Parameterization of the lipid model itself has also proven difficult as there is relatively little experimental data on which to parameterize. Some properties, such as area per lipid, are not sensitive enough to parameterize on as they do not do a good job of discriminating between sets of parameters (Anezo *et al.*, 2003). Other properties, such as phase transition temperature, are very computationally expensive and so have not been used to parameterize force fields to date. Many of the above mentioned limitations can be overcome in principal, although in practice this continues to be a challenge.

C. Four Region Model

In order to simplify the discussions to follow, we will first present an overview of the structure and chemical properties of the lipid bilayer. Figure 1 shows a snapshot and partial density profile for a DOPC bilayer. The bilayer is very inhomogeneous in the normal direction, with large gradients in density and polarity on a nanometer length scale. Similar to Marrink and Berendsen (1996), we will divide the bilayer into four arbitrary regions based on chemical properties. Region I almost exclusively contains lipid tails. This region has the lowest density, at the very center of the bilayer, with the density rapidly increasing towards the head groups. Region II contains a diverse mixture of functional groups. It contains the peak of the lipid tail density, the bulk of the carbonyl density, as well as a portion of the head group density, in addition to a small amount of water. This region is effectively the interface between the lipid tails and the

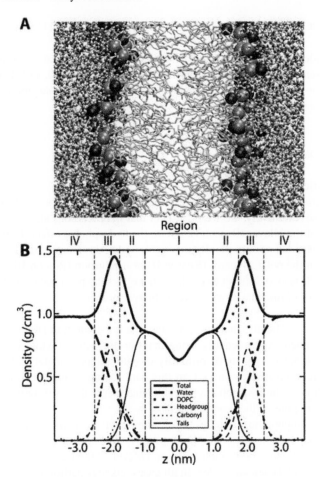

FIGURE 1 Snapshot and partial density profiles of a DOPC bilayer. (A) Snapshot of a DOPC bilayer system. Water is drawn as white and black tubes. The headgroups are represented as large spheres: the choline group is black, and the phosphate group is gray. The lipid tails are represented as grey tubes. (B) Partial density profile of a DOPC bilayer system. The normal of the membrane is parallel to the z-axis. The four regions of the bilayer are indicated (see text).

headgroups. Region III also contains a diverse mixture of functional groups, including a small portion of the carbonyl density, the bulk of the head group density, and a substantial amount of water. This region is probably best described as the interface between the head groups and water. Finally, Region IV consists mostly of bulk water, with very small amounts of head group density. This four region model will serve as a roadmap throughout the rest of this chapter.

II. PARTITIONING OF SMALL MOLECULES IN LIPID BILAYERS

We divide our review by class of molecule, focusing first on three classes based on generic chemical properties: hydrophobicity, polarity, and charge. Then we focus on several specific classes of interest: anesthetics, sugars, and carbon nanoparticles.

A. Non-Polar Molecules

1. Aliphatic Molecules

The interaction between hydrocarbons and lipid bilayers represents an interesting test case. Intuitively, we would expect that hydrocarbons should partition to the center of the bilayer (Region I), due to the hydrophobic effect. However, it is less clear what happens in the denser, highly ordered regions of the hydrocarbon chains, or in the polar glycerol (Region II) or headgroup (Region III) regions of the bilayer. It is also difficult to predict what effect the presence of hydrocarbons in the interior of the bilayer will have on bilayer properties.

a. Hexane. The distribution of hexane in a low-hydration DOPC bilayer has been determined experimentally using neutron diffraction (White *et al.*, 1981). As expected, hexane partitions primarily to within approximately 1.5 nm of the center of the bilayer (Region I). Can this partitioning be explained as a simple manifestation of the hydrophobic effect? What drives the partitioning of hexane in different regions of the lipid bilayer?

Several computational studies have attempted to address the distribution of short alkane molecules into lipid bilayer (Marrink and Berendsen, 1996; Pohorille and Wilson, 1996; Jedlovszky and Mezei, 2000; Bemporad *et al.*, 2004a; MacCallum and Tieleman, 2006). In all cases, alkanes partition preferentially to the center of the bilayer, in the disordered hydrocarbon tails of Region I. We have performed extensive umbrella sampling calculations to determine the distribution of hexane in a DOPC bilayer (MacCallum and Tieleman, 2006). By performing the simulations at different temperatures, we were able to determine the entropic and enthalpic components of the free energy (Fig. 2). As expected, partitioning from bulk water to the center of the bilayer is driven almost entirely by entropy, with a negligible enthalpic component—consistent with the hydrophobic effect. However, the situation rapidly changes when moving away from the center of the bilayer into the denser, ordered region of the lipid tails. Although the free energy increases only slightly, there are large changes in its enthalpic and entropic components. Comparing the partitioning to Region I and Region II from water, the signature has changed from being entirely entropy driven to enthalpy driven. The strong change in driving force can be understood in terms of molecular packing. Free volume is scarce in the more dense regions of the bilayer leading to a large

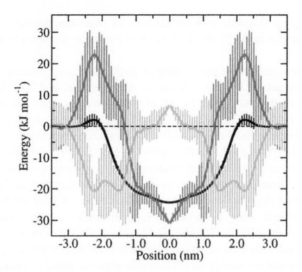

FIGURE 2 PMF for hexane in a DOPC bilayer. The free energy (black), and its enthalpic (ΔH, light gray) and entropic ($-T\Delta S$, dark gray) components are shown. Adapted from (MacCallum and Tieleman, 2006).

entropic penalty for partitioning a hexane to these regions. This is because the lipid chains must pack in such a way as to allow a cavity for the hexane molecule to fit; at high densities such configurations are rare and thus have low entropy. However, the high density in this region provides more opportunity for favorable van der Waal's interactions, leading to the favorable enthalpy in this region. In contrast, the very center of the lipid bilayer contains a great deal of free volume— regions where a hexane molecule could fit without overlapping with the lipid tails. Thus, there is no unfavorable entropy associated with partitioning to this region of the bilayer, and the entropy gains upon removing hexane from contact with water provide the driving force for partitioning.

b. Other Alkanes. Bemporand and co-workers calculated the distribution of ethane in a DPPC bilayer (Bemporad *et al.*, 2004a). Ethane partitions preferentially to Region I, with a maximum depth of −5.6 kJ/mol relative to water. There is a small barrier in Regions II and III, with a height of approximately 5 kJ/mol. This barrier is consistent with our studies on hexane partitioning, where we also found a barrier of approximately 5 kJ/mol. We were able to resolve the PMF into entropic and enthalpic components and found that the barrier is due to a combination of a very favorable enthalpic component and very unfavorable entropy, likely due to the difficulty in packing an alkane molecule in the very dense headgroup region. Pohorille *et al.* (1996) calculated the PMF for ethane in a GMO rather than DPPC bilayer. Ethane also partitions to Region I of a GMO bilayer, with a maximum depth of −5.3 kJ/mol. However, they did not observe a free energy barrier in Regions II and III, possibly because the interfacial density is not as high in GMO compared to DDPC due to the lack of a zwitterionic phosphatidylcholine group.

We have also studied the partitioning of methane, propane, butane, and isobutane as analogues of the side chains of Ala, Val, Ile, and Leu respectively (MacCallum et al., 2007a, 2007b). All of these molecules partition to Region I, with depths of -8.4, -13.8, -22.1, and -15.2 kJ/mol for methane, propane, butane, and isobutane. The value we obtain for methane is lower than that obtained for ethane in DPPC (-5.6 kJ/mol) or GMO (-5.3 kJ/mol), although direct comparison is difficult as the lipid compositions and temperatures of the simulations are different. There is a clear decrease in the PMF in Region I as the length of the hydrocarbon chain is increased, as would be expected from the hydrophobic effect. Interestingly, there is a relatively large difference in free energy between butane and isobutene. Butane, which is not branched, is able to better pack with the ordered lipid tails than isobutene, which is branched. When the partitioning between water and the unordered cyclohexane phase is calculated using the same force field, nearly identical results are obtained for both butane and isobutene. Thus, it appears that molecular shape is one determinant of partitioning into lipid bilayers.

c. Isoprene. Siwko et al. (2007) studied the effect of isoprene (2-methyl-1,3-butadiene) on POPC and DMPC membranes. Isoprene partitions to within approximately 1.5 nm of the center of the membrane, in good agreement with the results for alkanes. The presence of isoprene in the lipid bilayer results in an expansion in the area per lipid—up to 6% for DMPC containing 48 mole percent isoprene. The presence of isoprene also leads to an increase in order in the lipid tails, as measured by the deuterium order parameters of the tails. Interestingly, many species of trees, such as poplar and oak, as well as other plants, synthesize vast amounts of isoprene at high temperatures. The work of Siwko et al. supports the hypotheses that isoprene plays a thermal protective role, stabilizing the bilayer at high temperatures. Although only isoprene was studied, we speculate that most small alkane-like molecules will have a similar effect—effectively gluing the lipid tails together and providing enhanced order and increased stability.

2. Aromatic Molecules

Aromatic molecules have several interesting properties which may affect their partitioning. They are generally hydrophobic, although some may contain polar substitutions. Aromatic molecules are also generally flat, due to their rigid aromatic ring structures. The electrostatic properties of aromatic molecules are interesting; aromatic rings, such as benzene, have a net quadrupole moment. When such a ring is substituted, for example as in pentachlorophenol (PCP), the molecule may also have a net dipole moment. How does the combination of hydrophobicity, shape, and electrostatic properties determine the partitioning behavior of aromatic molecules?

a. Analogues of Aromatic Amino Acid Side Chains. In biology, it is well known that the aromatic residues Trp and Tyr are concentrated near the membrane interface, while Phe is found mostly in the center of the membrane (Yau *et al.*, 1998). Which properties of these molecules account for the observed differences?

Norman and Nymeyer (2006) studied the partitioning of benzene and indole in a POPC bilayer. Indole binds primarily to the interface region of the bilayer, near the carbonyl groups (Region II). There are also two weaker binding sites, one in the center of the bilayer (Region I), and the other near the choline moiety (Region III). Benzene also binds to the same three sites; however, the center of the membrane is the most stable location for benzene. By performing simulations at multiple temperatures, Norman and Nymeyer were able to decompose the PMF into enthalpic and entropic components. They found that partitioning of indole to the center of the membrane was an entropy driven process, consistent with the hydrophobic effect, while partitioning to the global minimum at the interface was strongly enthalpy driven. They attributed this favorable enthalpy to a combination of weak electrostatic forces including hydrogen bonding, cation–pi interactions, interactions between the indole dipole in the membrane electric fields, and other nonspecific electrostatic interactions. In contrast, benzene has fewer favorable electrostatic interactions because it does not have the capability to form hydrogen bonds or a net dipole to interact with the membrane electric field. Norman and Nymeyer also performed a set of simulations where the partial charges on indole were reduced to zero. In this case, indole partitions to the center of the bilayer as one would expect for a purely hydrophobic molecule. Increasing the partial charges on indole leads to an even stronger interfacial preference, while increasing the charges on benzene causes a shift in partitioning from the center of the membrane to the interface. These results support the notion that when aromatic molecules partition to the membrane interface, it is due to weak electrostatic forces. Norman and Nymeyer also studied the orientation of indole in the bilayer. They found a strong tendency for indole to align the plane of its aromatic rings with the lipid tails, while aligning its dipole moment with the membrane electrostatic field.

We also studied the partitioning of small molecule analogues of tryptophan (methylindole), tyrosine (*p*-cresol), and phenylalanine (toluene) (MacCallum *et al.*, 2007a, 2007b). The results for tryptophan are in good agreement with those of Norman and Nymeyer; tryptophan is found to partition preferentially to the interface, with a second minimum at the center of the bilayer (Fig. 3). The observed orientational preference is also consistent with the observations of Norman and Nymeyer. Tyrosine also displays an interfacial minimum; however, the partitioning of tyrosine to the center of the membrane is unfavorable compared to water,

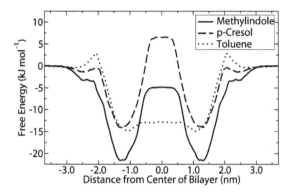

FIGURE 3   Potential of mean force for methylindole, *p*-cresol, and toluene in a DOPC bilayer at 300 K. Data from (MacCallum *et al.*, 2007a).

which can be attributed to the loss of hydrogen bonding to the polar hydroxyl group at the center of the membrane. When the hydroxyl group is located in the lipid chains, but still near in the interface, a water defect forms in the bilayer, which maintains hydration of the hydroxyl group. This phenomenon is similar to that observed for polar residues (see Section II.B.1.b below). In agreement with Norman and Nymeyer's results for benzene, we find that phenylalanine partitions throughout the lipid tails, without an interfacial preference.

b. Pyrene.   Hoff *et al.* (2005) studied the behavior of pyrene in a POPC bilayer through a combination of molecular dynamics simulations and solid-state NMR. Pyrene is of significant interest as a fluorescent probe for bilayer experiments, the interpretation of which may depend on the orientation and behavior of pyrene. They found that pyrene accumulates preferentially in the interfacial region of the lipid bilayer, similarly to indole. At this location of the bilayer, the plane of the pyrene ring orients parallel to the lipid tails. There is also a preference for the long axis of the pyrene molecule to run parallel to the lipid tails as well. Overall, the partitioning of pyrene to the interface region of the bilayer is somewhat surprising. Benzene, for example, partitions preferentially to the center of the bilayer. As pyrene is essentially four fused benzene rings, one might expect that it should partition to the center of the bilayer as well. Due to the large planar ring structure of pyrene, it might be entropically unfavorable to locate the pyrene in the center of the membrane due to the strong ordering effect it would have on the neighboring lipid tails. Previous very early molecular dynamics simulations of benzene observed a tendency towards partitioning to the interface region at higher temperatures (Bassolinoklimas *et al.*, 1995), which lends support to argument based on shape and chain entropy. It would be interesting to calculate the partitioning of pyrene with no partial charges in order to determine if electrostatic

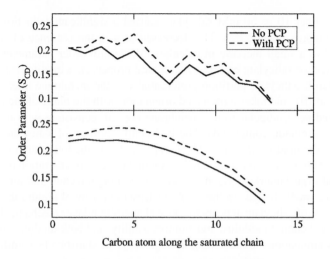

FIGURE 4 Average deuterium order parameters for POPC bilayer with and without the presence of 11 mol percent PCP. The upper and lower panels show the saturated and unsaturated chain, respectively. Adapted from (Mukhopadhyay *et al.*, 2004).

interactions play a role. Clearly, further study is needed to understand the effects of shape on partitioning in lipid bilayers.

c. Pentachlorophenol. Mukhopadhyay *et al.* (2004) studied the behavior of pentachlorophenol (PCP) in POPC and POPE bilayers. PCP accumulates in the interfacial region of both POPC and POPE bilayers, consistent with the presence of a polar hydroxyl group. As with benzene, indole, and pyrene, PCP also aligns the plane of its aromatic ring parallel to the lipid tails. In addition, there is a strong tendency for the polar hydroxyl group to be oriented towards the lipid head groups. The presence of PCP in the bilayer leads to a very small increase in the thickness of the bilayer and a corresponding increase in order. Large changes in the average chain tilt were observed upon the addition of PCP to both POPC and POPE bilayers, as shown in Fig. 4. The increasing order and large change in tilt angle is consistent with PCP driving the bilayer towards a more gel-like state. Interestingly, when bacteria are grown in an environment with high concentrations of PCP, they modify the composition of their cell membranes—perhaps to counteract the ordering effect of PCP (Lohmeier-Vogel *et al.*, 2004).

d. Salicylate. Salicylate is a non-steroidal anti-inflammatory drug that is closely related to aspirin. At high doses, salicylate can cause hearing loss and other hearing-related side effects that are related, in part, to changes in the membrane of outer hair cells in the cochlea. Song *et al.* (2005) have performed molecular dynamics simulations of salicylate interacting with a DPPC bilayer.

Including either 10 or 60 mM salicylate caused a significant reduction in area per lipid (approximately 2%). The decreased area in the presence of salicylate is coupled to a small increase in calculated deuterium order parameters. Overall, the effect of salicylate on bilayer structural properties is qualitatively similar to cholesterol. Salicylate partition preferentially to the membrane interface (Regions II and III), and takes a preferred orientation with the plane of the aromatic ring aligned perpendicular to the membrane normal, consistent with the results from other aromatic compounds. The hydrophilic portion of salicylate is coordinated by the lipid headgroups, which explains the reduction in area caused by salicylate. Salicylate also causes a reduction in the local concentration of chloride at the lipid–water interface. Such changes in lipid organization and ion distribution lead to small changes in the local electrostatic potential around the bilayer. The presence of salicylate appears to cause changes in bilayer mechanical properties, such as bending modulus, area compressibility, and bulk modulus; however, the 10 ns simulations are not long enough to yield statistically significant results.

3. Gases and Lennard-Jones Particles

Jedlovszky and Mezei (2000) studied the partitioning of O_2, CO, CO_2, and NO in a DMPC bilayer using the test cavity insertion technique. They observed that NO, CO, and O_2 have a free energy minimum in the hydrocarbon region of the bilayer (Region I)—consistent with their small, nonpolar nature. In contrast, for CO_2, the free energy and the hydrocarbon region is nearly identical to that in water. However, there is a large interfacial well near the location of the carbonyl groups (Region II). Jedlovszky and Mezei do not directly address the reasons for the difference between CO_2 and the other three molecules, but examination of the parameters used in their simulations reveals that the partial charges on CO_2 are much higher than those for the other solutes. Perhaps the higher partial charges, and corresponding opportunity for electrostatic interactions, drive partitioning towards interfaces—as may also be the case for aromatic molecules.

Marrink and Berendsen (1996) studied the partitioning behavior of O_2 and a series of Lennard-Jones molecules into DPPC bilayers. They found that O_2 partitions preferentially to the center of the bilayer, in agreement with the work of Jedlovszky and Mezei. They observed a strong correlation between the partitioning of oxygen and the available free volume within the bilayer. To further investigate the effects of solute size and available free volume on partitioning, they performed calculations on a series of Lennard-Jones molecules. Increasing the size of the Lennard-Jones particle leads to an increase in the preference for the center of the bilayer, a phenomenon which they attributed to the much higher available free volume at the center of the bilayer. To study the effects of asphericity on partitioning, they constructed a series of linear molecules composed of Lennard-Jones particles connected by rigid bonds. Increasing the asphericity of Lennard-Jones molecules tends to favor partitioning to the interface (Region II), consistent with the shape distribution of available free volume within the bilayer.

4. Summary: Non-Polar Molecules

Non-polar molecules partition to the lipid tails (Regions I and II). Aliphatic molecules partition to the center of the bilayer (Region I), driven by the hydrophobic effect. Aromatic molecules partition to either the center of the bilayers (Region I) or the dense portion of the lipid tails (Region II). The presence of polar groups or other electrostatic features, such as a weak dipole moment, appears to drive aromatic molecules towards the interface. Molecules lacking these electrostatic features, such as benzene, partition preferentially to the center of bilayer. However, pyrene partitions to the dense part of the lipid tails, despite a lack of polar groups or other electrostatic features. Also, there is a significant difference in partitioning between isobutene and butane. Clearly, further study is required to understand the role of molecular shape on the partitioning of non-polar molecules.

B. Polar and Charged Molecules

1. Polar Molecules

a. Ethanol. Feller *et al.* (2002) used a combination of NMR spectroscopy and simulations to deduce the distribution of ethanol molecules within the lipid bilayer. Using NMR spectroscopy, they measured cross-relaxation rates between ethanol and a POPC bilayer. However, such relaxation rates are not straightforward to interpret, as they depend on the distance between the two protons as well as the dynamics of proton dipole–dipole interactions. For example, although the strongest cross relaxation rates are observed between ethanol and the glycerol moiety, there are still significant cross relaxation rates observed between ethanol and the terminal methyl groups of the lipid tails. How do the observed cross relaxation rates correlate with the distribution of ethanol in the bilayer? By performing detailed molecular dynamics calculations, Feller *et al.* were able to calculate cross relaxation rates from their simulations, and achieved excellent agreement with experiment. In the simulations, they observe that the probability of finding and ethanol molecule at the center of the bilayer is extremely low. However, the cross relaxation rates between ethanol and the terminal methyl groups of the lipid tails, while small, are not negligible. In the simulation, they found that the cross relaxation between ethanol and the terminal methyl groups was due to disorder in the bilayer, where the terminal methyl groups were actually located near the carbonyl region, and thus able to interact with the ethanol. In addition, they were able to show that dynamic effects could account for up to 20% variation in observed cross relaxation rates. Using this combination of experimental and theoretical results, the authors were able to determine that the ethanol molecules partition preferentially to the interface region of the bilayer; the probability of finding an ethanol and the center of the bilayer was nearly 3 orders of magnitude lower than finding one and water. The authors attribute the interfacial partitioning of ethanol to a combination of the polar hydroxyl group and hydrophobic methyl group. The opposing partitioning tendencies—the polar hydroxyl group driving partitioning towards the polar phase, and the nonpolar methyl group driving partitioning to the lipid interior—leads the observed interfacial preference.

Patra *et al.* (2006) studied the interaction between two alcohols: ethanol and methanol, and two lipids: POPC and DPPC. The alcohols were added at concentrations of approximately 1 mol percent with respect to the number of water molecules, or about 40 mol percent with respect to the number of lipids. The addition of alcohol at these concentrations leads to a small increase in area per lipid, with a slightly larger increase for DPPC than for POPC. Both methanol and ethanol give similar increases in area per lipid. Preferential partitioning to the interface region is observed for both alcohols in combination with both bilayers. The bulk of the alcohol density is located below the lipid choline and phosphate groups, near the carbonyl region of the bilayer (Region II). The interfacial preference of ethanol appears to be much stronger than that of methanol. The density of alcohol at the center of the membrane is very low, consistent with the results of Feller *et al.* The concentration of alcohol used in this study leads to a slight increase in lipid chain order. However, experimental results show the opposite trend; incorporation of alcohol into the bilayer leads to a decrease in order parameter. The authors attempt to resolve this discrepancy by postulating that there are two separate contributions to the effects of alcohol bilayers: a local one that is the results of alcohols binding to individual lipid molecules, and a global one due to changes in surface tension due to the presence of alcohol. They postulate that this study was done in a regime where the second (global) effect is still increasing, whereas the experimental results are done at higher concentrations, and the first term now dominates. Support for this hypothesis comes from the simulation, where lipids which are bound to alcohol molecules have decreased order parameters compared to those that are not bound. Additional simulations performed at a wide variety of alcohol concentrations lead to increasing order parameters with increasing alcohol concentration, up to an alcohol-lipid ratio of approximately 1:1. At higher concentrations, the maximum order parameter was found to decrease with increasing alcohol concentration. Overall, these results appear to resolve a discrepancy between the simulated areas per lipid and the experimental ones. In the simulations, approximately 30 ethanol molecules are observed to cross the bilayer on a 40 ns time scale, while no crossing amounts are observed for methanol. This phenomenon can be explained by the much higher polarity of methanol, which contains one less carbon than ethanol. The addition of the single extra carbon of ethanol compared to methanol, leads to an increase in interfacial activity and the large increase in the observed crossing rate.

b. Small Molecule Analogues of Polar Amino Acid Side Chains.   We have calculated the distribution of small molecule analogues of the amino acid side chains obtained by truncating the side chain at the β-carbon and replacing the α-carbon by a proton (MacCallum *et al.*, 2007a, 2007b). All polar residues display a minimum near the carbonyl region of the bilayer (Region II) due to the amphipathic nature of the small molecules. The analogues of all polar side chains show an unfavorable free energy in the lipid tails compared to bulk water and a plateau in free energy within 0.3 to 0.6 nm from the center of the bilayer, as shown for Asn (acetamide) in Fig. 5. The free energy increases rapidly in the lipid tails, while at the same time a water defect forms that maintains partial hydration of the polar side chain (Fig. 6A). At some critical length the water defect spontaneously dissipates and the PMF becomes flat (Fig. 6B). There is an energetic balance between the cost of forming the defect, and the energy gained by hydration of the polar or charged residue. The energy gained from hydration is expected to remain relatively constant in the lipid tails, while the cost of forming the defect is expected to rise sharply as the defect becomes longer.

The PMFs for acetamide, acetic acid, methanol, and methylamine have been calculated by Bemporad and co-workers (Bemporad *et al.*, 2004a, 2004b). They observe a maximum at the center of the bilayer for all four molecules, with a plateau occurring for all molecules except methanol. A large water defect is observed when acetic acid is present at the center of the membrane, and all small molecules maintain non-zero hydration at the center of the membrane (Bemporad

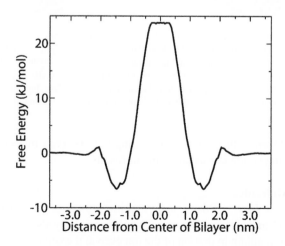

FIGURE 5   PMF for acetamide in a DOPC bilayer. Data from (MacCallum *et al.*, 2007a).

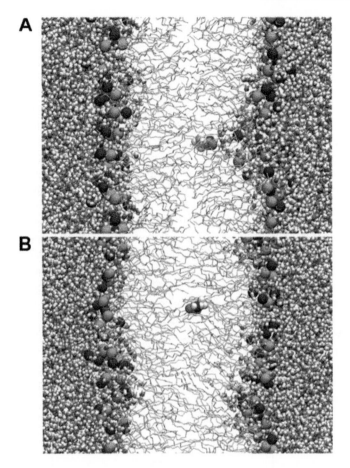

FIGURE 6 Snapshots of acetamide in a DOPC bilayer. Water is red and white tubes; choline groups are orange spheres; phosphate groups are blue speheres; the lipid tails are shown as grey tubes. The acetamide is shown in spacefilling (cyan—carbon, white—hydrogen, blue—nitrogen, red—oxygen). (A) Acetamide at 0.4 nm from the center of the bilayer. (B) Acetamide at 0.5 nm from the center of the bilayer. Adapted from (MacCallum *et al.*, 2007a). (For interpretation of the references to color in this figure legend, the reader is referred to the web version of this book.)

et al., 2005). Overall, these results, including the observed free energies at the center of the bilayer, are consistent with our results.

2. Charged Molecules

a. Monovalent Ions.   Wilson and Pohorille (1996) calculated the PMF for transport of Na^+ and Cl^- ions across a GMO bilayer. They also observe large water defects that maintain hydration of the ion even at the center of the membrane. Based on these calculations, the free energy of Na^+ at the center of the membrane is approximately 225 kJ/mol. While this value is very high, it is only about 60% of that obtained from continuum electrostatics—assuming a dielectric constant of 1.0 for the hydrophobic phase. However, alkanes actually have a dielectric constant of near two. In the continuum calculations, changing the dielectric constant from one to two lowers the free energy by approximately 50%. For the atomistic simulations, the force field does not explicitly include hydrogen from the lipid tails, thus the dielectric constant in the simulation is unity. To correct for this, the PMF is essentially scaled by a factor of two in the membrane interior. Although this method is crude, it does attempt to account for the discrepancy between the dielectric constant of real alkanes. Performing the scaling reduces the free energy for Na^+ at the center of the bilayer to approximately 112 kJ/mol. However, the rigor of this method is somewhat questionable as the presence of a large water defect at the center of the membrane likely makes the local dielectric constant much different from 1.0. These calculations point to the need for caution when assuming that the membrane is a fixed low dielectric slab. Using the free energy from the continuum calculations leads to a membrane permeability that is approximately 15 orders of magnitude too small, while the free energy calculated from the atomistic simulations, when corrected for the dielectric constant, yields a permeability that is within the range of experimental results.

b. Lipids. The PMF of a lipid in the membrane determines the cost of small local deformations in bilayer structure, which may be important for membrane-membrane interactions, such as membrane fusion, or interactions with membrane proteins, such as phospholipases. Tieleman and Marrink (2006) and calculated the PMF of a DPPC lipid in a DPPC bilayer. Pulling a lipid out of the bilayer into bulk water leads to a free energy change of approximately 80 kJ/mol, which compares favorably with the free energy derived from the critical micelle concentration of DPPC. Pulling the phosphate group of DPPC into the center of the bilayer leads to the formation of a large, two-sided water defect, with an energetic cost of approximately 80 kJ/mol. Tieleman and Marrink calculate the flux of water, lipids, and Na^+ through the bilayer by assuming that they pass through such pores, which occur spontaneously with a frequency depending on the free energy. Lipid flip-flop is calculated to occur on a timescale of approximately 30 h, which falls within the range of experimental values. The permeability of Na^+ is also in good agreement with experiment. In this case, the authors did not need to make any correction for the dielectric constant of the lipid tails, as was required by Wilson and Pohorille, to obtain good agreement with experiment. The source of this discrepancy is not clear; however, it is not straightforward to compare the simulations as the use different lipids (GMO versus DPPC) and different simulation protocols. Additionally, the calculations of Tieleman and Marrink were performed a decade later, and are based on a total simulation time of over 300 times that used by Wilson and Pohorille. The good agreement with experiment for the permeabilities of DPPC and Na^+ strongly suggests that the mechanism of basal membrane transport for these solutes is through transient water pores. The calculated permeability for water is approximately 7 orders of magnitude too slow compared to experiment, consistent with previous simulations that obtained the correct rate without the formation of large water pores (Marrink and Berendsen, 1994). Thus water permeation does not require the same large defects that lipid flip-flop and ion permeation do.

c. Charged Amino Acid Side Chains. We calculated (MacCallum *et al.*, 2007a, 2007b) the PMF's for the charged and neutral forms of the side chains of arginine (*n*-propylguanidinium), lysine (butylamine), aspartic acid (acetic acid), and glutamic acid (propionic acid) in a DOPC bilayer (Fig. 7). For the charged forms of all four molecules a large, lipid-lined water defect is observed that connects the charged molecule to bulk water (Fig. 8). This defect is stable on the timescale of the simulations ($\sim$100 ns), and forms spontaneously within a few nanoseconds. The calculated free energies at the center of the bilayer are approximately 80–90 kJ/mol for acetic and propionic acid, 50 kJ/mol for butylamine, and 60 kJ/mol for *n*-propylguanidinium. However, these molecules may undergo changes in ionization state as a function of the depth in the membrane. By calculating the PMF separately for the charged and neutral forms, one can determine the dominant species as a function of depth in the membrane. For the carboxylic acids, the neutral species becomes dominant within approximately 1.5 nm of the center of the bilayer. In transferring either acetic acid or propionic acid from water to the center of the membrane, the pK_a increases by nearly 10 units. Lys remains charged until within about 0.5 nm from the center of the membrane, while at the center of a DOPC membrane the free energies of the charged and neutral forms of *n*-propylguanidinium are nearly equal.

Dorairaj and Allen performed simulations of a long poly-leucine helix containing a single Arg at the central position (Dorairaj and Allen, 2007). When the Arg is placed in the center of a DPPC bilayer, a large water defect is formed consistent with our results. Several other recent studies of transmembrane helices containing charged residues have also observed the formation of a water defect and large deformations in bilayers structure (Monticelli *et al.*, 2004; Tieleman *et al.*, 2004; Freites *et al.*, 2005; Johansson and Lindahl, 2006). Dorairaj and Allen obtain a free energy of 71 kJ/mol, which is comparable to our calculated free energy for transferring a Leu side chain from the center of the membrane to water plus the free energy to transfer an Arg from water to the center of the membrane. Continuum electrostatics calculations yield free energies as high as 170 kJ/mol, depending on the assumptions used; clearly great care is required in performing and interpreting the results of continuum electrostatics calculations on lipid bilayers systems.

Carboxylic acids become protonated upon entering the lipid tails, at least in the absence of other stabilizing interactions such as salt bridge formation. On the other hand, it appears that basic residues are able to remain charged until

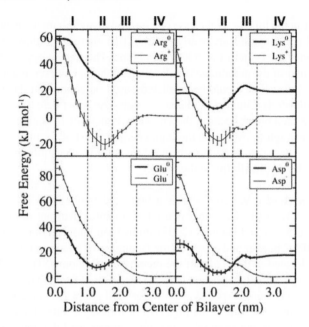

FIGURE 7 PMF's for small molecule analogues of the charged amino acid side chains. Both the charged (thin lines) and neutral (thick lines) species are shown. Adapted from (MacCallum *et al.*, 2007a).

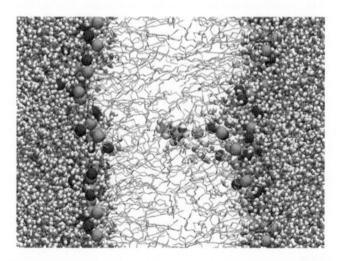

FIGURE 8 Snapshot of Arg at center of a DOPC bilayer. Water is red and white tubes; choline groups are orange spheres; phosphate groups are blue speheres; the lipid tails are shown as grey tubes. The acetamide is shown in spacefilling (cyan—carbon, white—hydrogen, blue—nitrogen, red—oxygen). Adapted from (MacCallum *et al.*, 2007a). (For interpretation of the references to color in this figure legend, the reader is referred to the web version of this book.)

much deeper in the bilayer. In the case of n-propylguanidinium, it appears that the charged and neutral forms are nearly equal in free energy, and thus nearly equal in population at the center of a DOPC membrane. The free energy for placing an ion at the center of the membrane is still relatively high; however, it appears to be significantly lower than would be predicted based on continuum theories which regard the bilayer as a static, low-dielectric slab. The water defect is stable at the center of the membrane because the energetic benefit of hydrating the charged molecule is enough to overcome the high energetic cost of forming such a large water defect. These results point to the need for caution when utilizing continuum theories that lack of flexibility in the bilayer.

d. Valproic Acid. Valproic acid is a branched fatty acid, $CH_3-CH_2-CH_2-CH(COOH)-CH_2-CH_2-CH_3$, that is used as an anticonvulsant in the treatment of epilepsy. Ulander and Haymet (2003) studied the permeation of valproic acid through a DPPC bilayer and observed the formation of a large water defect when valproate is located in the center of the membrane (Region I). The PMF shows a large barrier in Region I, consistent with the high cost of forming the water defect. Valproate is the dominant species from water until about 2.0 nm from the center of the bilayer; however, due to the high cost of forming a water defect it becomes cheaper to protonate the valproate—making valproic acid the dominant species in the bilayer interior (Region I). These results are consistent with those of the charged amino acid residues; in both cases there is a large free energy barrier at the center of the membrane and an associated water defect. The fact that valproate likely becomes protonated in the lipid tails is also consistent with the results for the acidic amino acid side chains.

3. Summary: Charged and Polar Molecules

The partitioning of polar and charged molecules the bilayer interior leads to the formation of large water defects. The formation of the water defect is associated with a steep increase in free energy, due to the high cost of forming the defect. For the charged molecules, this defect persists even when the ion is located in the center of the membrane. For polar molecules, the free energy plateaus and the water defect dissipates once the energetic cost of forming the defect becomes larger than desolvating the ion. The formation of water defects points to the deformability of the bilayer; caution must be used when interpreting continuum calculations that treat the bilayer as a rigid low-dielectric slab.

C. Anesthetics

Anesthesia is one of the backbones of modern medicine; without it most surgeries would be impossible to perform. Despite this importance, the mechanism of general anesthesia remains poorly understood. Over 100 years ago,

Meyer and Overton hypothesized that anesthetic potency is highly correlated with partitioning into non-polar media. Recently, this hypothesis has been revised to reflect the fact that there is a strong correlation between partitioning to the bilayer interface and anesthetic potency (Pohorille *et al.*, 1998). The most prominent current hypothesis proposes that anesthetic molecules bind to specific sites on integral membrane proteins involved in transmitting nerve signals, such as post-synaptic GABA receptors (Rudolph and Antkowiak, 2004; Franks, 2006). Alternatively, it has been proposed that rather than specific binding of anesthetics membrane proteins, there may be a membrane mediated mechanism whereby anesthetics affect the mechanical properties of the lipid bilayer, which in turn affects the conformational equilibrium of membrane proteins (Cantor, 1997, 1998), although this view remains controversial. Regardless of the exact mechanism of general anesthesia, the interfacial partitioning of anesthetic molecules is a reliable indicator of anesthetic potency.

1. Partitioning of Anesthetic Molecules

Pohorille *et al.* (1996) studied the partitioning of two closely related chemical compounds: TFE (1,1,2-trifluorethane) and HFE (hexafluorethane) between water and a GMO bilayer. TFE is a relatively potent general anesthetic and partitions to the membrane interface (Region II/III), while PFE is a nonanesthetic and partitions to the membrane interior (Region I). This is attributed to the interaction between the dipole moment of TFE and the strong dipole moment at the membrane interface. The strong dipole moment at the water–membrane interface arises from orientational anisotropy of both water molecules and the lipid headgroups. To provide further evidence for the correlation between the solute dipole moment and interfacial partitioning, Pohorille et al. also examined the orientation of the TFE molecule as a function of depth within the bilayer. They found that TFE was strongly oriented near the membrane interface, and that the point of strongest orientation corresponded to the free energy minimum.

Further evidence for the correlation between interfacial partitioning and solute dipole moment comes from the partitioning of a series of fluorinated methane compounds: CH_4, CFH_3, CH_2F_2, CHF_3, and CF_4 (Pohorille and Wilson, 1996). Those molecules possessing no net dipole—CH_4 and CF_4—partition to the center of the bilayer. On the other hand, CH_3F, CH_2F_2, and CHF_3, all of which possess a net dipole moment, partition to the membrane interface. By performing calculations with and without charge on the solute molecules, they were able to determine the nonpolar and electrostatic contributions to partitioning. Without charges, all of the solutes partition to the center of the bilayer, whereas with charges the molecules with a net dipole moment partition preferentially to the interface. The electrostatic contribution to the PMF presents a barrier of approximately 10 kJ/mol in the hydrophobic region of the bilayer (Region I). Both the electrostatic and nonpolar contributions to the PMF are monotonic; neither one displays an interfacial minimum. Thus, the interfacial partitioning of dipolar solutes is the result of a delicate balance between nonpolar and electrostatic forces. Neither profiles display a minimum at the interface—only their combination does.

Vemparala and co-workers calculated the PMF for halothane in a DPPC bilayer using steered molecular dynamics simulations (Vemparala and Klein, 2004; Vemparala *et al.*, 2006). They pulled a halothane molecule from its equilibrium position, near the carbonyl groups, into the water phase. For this process they obtained the free energy of approximately 20 kJ/mol, depending on the averaging method used. They did not pull the halothane molecule deeper into the hydrocarbon region, however. They also performed simulations of lipid bilayers containing embedded helices based on the α- and δ-subunits of the nicotinic acetylcholine receptor. Halothane was observed to bind to Tyr-277 of the alpha subunit, which is an experimentally determined binding site. This binding is observed to significantly alter the dynamics of the helix, which may play a role in anesthetic mechanism.

2. Effect of Anesthetic Molecules on Bilayer Properties

Koubi *et al.* (2000) studied the effect of halothane, at a 1:1 halothane to lipid ratio, on the structure of a DPPC bilayer. They found that halothane preferentially partitions to the interfacial region of the bilayer, with 95% of the halothane being located below the lipid carbonyl groups. The presence of this concentration of halothane in the bilayer leads to a slight expansion in the area per lipid, and the slight decrease in bilayer thickness. The increasing area per lipid results in an overall decrease in the acyl chain order parameters. However, the lipid segments near the preferred location of the halothane undergo a slight increase in order. These effects—increased area per lipid, and decreased order parameters—are opposite to those observed for hydrocarbons. The reorganization induced by halothane and the decrease in order results in changes to the lipid headgroup dipole orientations. At high anesthetic concentrations a larger fraction of lipid head groups is observed with their dipole pointing into the lipid bilayer, rather than out of it. This change in the distribution of headgroup dipoles also results in a change in the interfacial electric field. The authors speculate that this change in membrane electric field may cause changes in membrane protein function, as associated with anesthesia.

Koubi and co-workers also examined the effects of a non-anesthetic, hexafluorethane (HFE), on the structure of a DPPC bilayer (Koubi *et al.*, 2001, 2002). In contrast to halothane, HFE partitions preferentially to the center of the bilayer, as observed for GMO (Pohorille and Wilson, 1996). HFE causes a decrease in area per lipid, a slight increase in bilayer thickness, and increased order parameters— exactly the opposite effect of halothane. The presence of HFE causes only a slight change in the membrane potential, in contrast with the large change observed for halothane. The authors suggest that this difference may be the reason why halothane acts as an anesthetic while HFE does not.

3. Summary: Anesthetics

Regardless of the exact mechanism of general anesthesia, it seems very likely that the interfacial partitioning of anesthetics versus the membrane interior for nonanesthetics plays a role in determining anesthetic potency. Based on the simulations, a number of potential mechanisms have been put forward including changes in bilayer order, specific binding of anesthetics membrane proteins, and changes in membrane electric field. To date no molecular dynamics simulations have addressed the effect of anesthetics on the local lateral pressure profile, which may play a role in membrane protein function (Cantor, 1997, 1998; Brink-van der Laan *et al.*, 2004; van den Brink-van der Laan *et al.*, 2004a, 2004b; Bruno *et al.*, 2007).

D. Sugars

Many organisms are able to withstand extremes in temperature and hydration. Disaccharides, in particular trehalose, are produced in large quantities when these organisms undergo periods of extreme conditions. Experiments have shown that disaccharides can stabilize lipid membranes; however, the mechanism of stabilization is not clear. (Pereira *et al.*, 2004) have performed MD simulations of DPPC bilayers at temperatures of 325 and 475 K, and at three different concentrations of trehalose: 0, 0.6, and 1 M. At 325 K, the inclusion of trehalose leads to a very slight increase in area per lipid, with a corresponding decrease in order parameter. Trehalose partitions to the membrane interface, where it hydrogen bonds strongly with DPPC. Trehalose displaces water from the membrane surface, leading to a ~25% reduction in the number of hydrogen bonds between water and DPPC. In the absence of trehalose, the bilayer essentially falls apart when the temperature is increased to 475 K. Caution is required in interpreting the results at 475 K, however, as biomolecular force fields are not generally parameterized at the temperatures and it is questionable how accurate they are at a temperature this high. At 475 K, there is no longer a clear separation between the two leaflets of the bilayer and water is found throughout the bilayer. At a trehalose concentration of 0.6 M, the bilayer retains somewhat more structural integrity, although it is still significantly disrupted. 1 M trehalose appears to stabilize the bilayer to a great extent, with a structure that is clearly still a bilayer. Based on the calculated deuterium order parameters, the bilayer containing 1 M is clearly the most ordered at 475 K, although it is still less ordered than the bilayers at 325 K. The total number of hydrogen bonds in the system decreases by approximately 35% upon increasing the temperature to 475 K. Interestingly, although the total number of hydrogen bonds in the system is decreasing, the number of hydrogen bonds between trehalose and DPPC increases dramatically. Trehalose molecules are observed that form as many as eight hydrogen bonds with DPPC molecules. Many trehalose molecules form hydrogen bonds to multiple lipids—"bridging" them, with about 28% of the trehalose forming hydrogen bonds with two lipids, and 19% forming hydrogen bonds to three or more lipids.

Sum *et al.* (2003) reach similar conclusions based on their simulations of DPPC and either trehalose or sucrose. At high concentrations, trehalose replaces water in forming hydrogen bonds to DPPC. They also observe "bridging" trehalose molecules which form multiple hydrogen bonds with as many as three different lipids. These results appear to support the "water replacement" hypothesis, where disaccharides are able to stabilize membranes by replacing the hydrogen bonds formed by water that would be otherwise lost under extreme conditions.

Van den Bogaart *et al.* (2007) used a combination of fluorescence correlation spectroscopy (FCS) experiments and molecular dynamics simulations to study the effects of sugars on the lateral diffusion of lipids. A number of sugars were studied, including sucrose and trehalose in pure DOPC and DOPC/DOPS mixtures. FCS experiments show that sucrose causes the largest reduction in lipid mobility, even larger than trehalose, which has been shown to provide the best protection against the leakage and fusion of vesicles. All-atom simulations reveal that sucrose interacts with more phospholipid head groups simultaneously than trehalose does, although the 100 ns length is too short to accurately measure the difference in diffusion coefficients. Further simulations, using a coarse-grained model, were performed to determine the effect of sugars on the lateral diffusion of lipids. The diffusion coefficient was reduced by approximately a factor of 3 as the strength of the interactions between the sugar and the lipids was increased.

E. Carbon Nanoparticles

Carbon nanoparticles, such as nanotubes and fullerenes, have received significant attention in recent years, with many applications proposed. Studies on the toxicity of such nanoparticles are relatively scarce; in particular, very little is known about the interactions between nanoparticles and biological molecules, such as bilayers, at the microscopic level. Several groups have used computer simulations to study the interactions of carbon nanoparticles with lipid bilayers.

1. Combustion Generated Nanoparticles

The interactions between combustion generated nanoparticles and lipid bilayers have been studied by Chang and Violi (2006). In this study two different soot particles were used: one round particle ($C_{188}H_{53}$), which is roughly spherical although incomplete and unclosed; and a flat particle ($C_{50}H_{38}$). The interaction of these particles with DMPC bilayers was examined through both equilibrium and non-equilibrium (umbrella sampling) simulations. In equilibrium simulations the round particle appears to spend most of the time in the center of the lipid tails (Region I). In contrast, the flat particle spends most of the simulation in the dense region of the lipid tails (Region II). Marked differences in diffusion coefficient are also observed, with the flat particle diffusing nearly twice as fast as the round particle. The presence of either nanoparticle within the bilayer leads to the formation of a small water defect near the membrane surface, the effects of which are visible in the lipid-nanoparticle 2D pair-correlation function. Umbrella sampling simulations were performed to examine the permeation of the nanoparticles across the membrane. However, the simulation times are very short and significant hysteresis is observed—casting doubt on the reliability of the results. Much larger water defects are observed during the umbrella sampling simulations, although this is likely due to the non-equilibrium nature of the simulations. Overall, this paper presents the formation of water defects or pores as a novel mechanism of nanoparticle cytotoxicity. However, the simulations are quite short and the umbrella sampling simulations appear to be far from equilibrium. Further study is needed to validate the ability of soot particles to form water pores.

2. Fullerene and Fullerene Derivatives

Li *et al.* (2005) used atomistic simulations to study the interactions between fullerene molecules and a DMPC bilayer. Fullerene is resides preferentially in the dense region of the lipid tails (Region II), which the authors attribute to favorable dispersion interactions because of the high density in this region of the bilayer. This result is inconsistent with the behavior Chang and Violi observed for the round soot particle. The formation of fullerene dimers was studied by calculating fullerene–fullerene PMF's in both the lipid bilayer and in an isotropic alkane melt. In both cases, a regular pattern of peaks and valleys is observed with a spacing of around 0.4 nm due to the structure of the alkane tails or alkane melt. The interactions between fullerene molecules are noticeably less favorable in the lipid bilayer than in the alkane melt due to differences in density and the ordered non-isotropic nature of the bilayer compared to the alkane melt. When the two fullerene molecules are separated by large distances in the bilayer, they behave as essentially independent objects and reside primarily in the dense region of the bilayer (Region II). However, at small separations between the two fullerenes, particularly at 0.2–0.4 nm, the two fullerenes cause a much larger defect in lipid structure. This drives the fullerene dimer towards the center of the membrane (Region I) to minimize the disruption of the bilayer structure. Overall these results suggest that the solubility of fullerenes is enhanced in lipid bilayers compared to alkane melts, and thus the tendency to form fullerene aggregates or clusters is lower.

Qiao *et al.* (2007) examined the translocation of fullerene (C_{60}) and a polar derivative ($C_{60}[OH]_{60}$) through a DMPC bilayer. Pristine C_{60} is located preferentially in the dense region of the lipid tails (Region II), as was also observed by Li and co-workers. The free energy for transferring a single fullerene from water to a given location of the bilayer can be broken into two parts: the cost of forming a cavity to fit the fullerene into, and the energy gained through interactions between the fullerene and the surrounding lipids. The cost of cavity formation is higher in the more dense region of the lipid tails (Region II) than at the center of the bilayer (Region I), however, this is compensated for by the much more favorable fullerene-lipid interactions in Region II compared to Region I. This trend is opposite to that observed for small molecules, which are preferentially located at the center of the bilayer. When fullerene is located near the surface of the bilayer there is a large increase in the area per lipid for the lipids near the fullerene. This phenomenon is similar to the water pores observed in earlier studies. When the fullerene is buried deeper in the membrane, the changes in area per lipid are less significant, and it still remains unknown if the formation of water pores during fullerene permeation is an important mechanism of cytotoxicity. The polar derivative of fullerene ($C_{60}(OH]_{60}$) does not spontaneously move into the bilayer during a 48 ns simulation, as would be expected due to its highly hydrophilic surface. However, $C_{60}(OH)_{60}$ does adsorb to the surface of the membrane. PMF calculations reveal a free energy minimum of approximately -32 kJ/mol at the surface of the lipid bilayer (Region IV). The PMF increases steeply as the $C_{60}(OH)_{60}$ moves deeper into the lipid tails, similar to the PMF's observed for the polar analogues of the amino acid side chains. When $C_{60}(OH)_{60}$ is located at the minimum of the PMF (Region IV) the local area per lipid decreases by approximately 20%. Based on the local diffusion coefficients and the PMF's the mean translocation time was estimated at a few milliseconds for fullerene and approximately nine orders of magnitude longer for $C_{60}(OH)_{60}$, which is consistent with the high free energy barrier for locating $C_{60}(OH)_{60}$ in the center of the lipid bilayer.

3. Summary: Carbon Nanoparticles

Simulations of carbon nanoparticles have provided insight into several interesting phenomenon. It is possible that one mechanism of cytotoxicity may be the formation of water pores during the translocation of carbon nanoparticles. However, further study is required as this phenomenon was not observed in all simulations, and the simulations where it was observed were rather short. Interestingly, fullerene partitions to the dense region of the lipid tails, rather than to the center of the bilayer as would be expected of a hydrophobic molecule. Further study is needed to fully understand the driving forces behind partitioning and the effects of molecular shape, both for carbon nanoparticles and other non-polar molecules.

III. CONCLUSIONS

Computer simulations have been of tremendous value in understanding the interactions between small molecules and lipid bilayers. In particular, computer simulations have provided a level of spatial resolution that is impossible to achieve experimentally due to the "soft" nature of lipid bilayers. In most cases the experimental and calculated results are in good agreement. Reassuringly, the calculated results appear to be robust across different small molecule force fields and lipid models.

However, despite the success of computer simulations, our understanding of the physical chemistry underlying small molecule–bilayer interactions remains relatively poor. The exact location of the free energy minimum in the bilayer appears to be very sensitive to a number of molecular characteristics. Additionally, small changes in free energy mask large changes in enthalpy and entropy that are indicative of a change in molecular driving force in different regions of the bilayer. There is a very detailed balance between large opposing driving forces that is difficult to predict. More research is needed to fully understand the effects of molecular size, shape, and charge distribution on membrane partitioning. Additionally, few systematic studies have been undertaken to understand how the properties of the bilayer itself effect the partitioning of small molecules. Properties such as area per lipid, chain length, order parameter, headgroup, and chain saturation should be expected to have large effects on the partitioning of small molecules, but these effects are not well understood at this time.

The future of this area of research is bright. There are many interesting questions that remain unanswered. Small molecule–bilayer systems are generally very simple, making it easy test the effects of varying properties of either the small molecule or bilayer. Most importantly, the computational requirements for such studies are relatively modest, at least in comparison to other areas such as membrane protein or protein folding simulations. A large increase in computer power is not required to answer many of the interesting questions that remain.

Acknowledgements

This work is supported by the Natural Sciences and Engineering Research Council (NSERC). D.P.T. is an Alberta Heritage Foundation for Medical Research Senior Scholar and Canadian Institute of Health Research New Investigator. J.L.M. is supported by studentships from NSERC, Alberta Ingenuity, and Killam Trust.

References

Anezo, C., de Vries, A.H., Holtje, H.D., Tieleman, D.P., Marrink, S.J. (2003). Methodological issues in lipid bilayer simulations. *J. Phys. Chem. B* **107**, 9424–9433.

Bassolinoklimas, D., Alper, H.E., Stouch, T.R. (1995). Mechanism of solute diffusion through lipid bilayer-membranes by molecular-dynamics simulation. *J. Am. Chem. Soc.* **117**, 4118–4129.

Bemporad, D., Essex, J.W., Luttmann, C. (2004a). Permeation of small molecules through a lipid bilayer: A computer simulation study. *J. Phys. Chem. B* **108**, 4875–4884.

Bemporad, D., Luttmann, C., Essex, J.W. (2004b). Computer simulation of small molecule permeation across a lipid bilayer: Dependence on bilayer properties and solute volume, size, and cross-sectional area. *Biophys. J.* **87**, 1–13.

Bemporad, D., Luttmann, C., Essex, J.W. (2005). Behaviour of small solutes and large drugs in a lipid bilayer from computer simulations. *Biochim. Biophys. Acta Biomembranes* **1718**, 1–21.

Brink-van der Laan, E.V., Killian, J.A., de Kruijff, B. (2004). Nonbilayer lipids affect peripheral and integral membrane proteins via changes in the lateral pressure profile. *Biochim. Biophys. Acta Biomembranes* **1666**, 275–288.

Bruno, M.J., Koeppe, R.E., Andersen, O.S. (2007). Docosahexaenoic acid alters bilayer elastic properties. *Proc. Natl. Acad. Sci. USA* **104**, 9638–9643.

Cantor, R.S. (1997). The lateral pressure profile in membranes: A physical mechanism of general anesthesia. *Biochemistry* **36**, 2339–2344.

Cantor, R.S. (1998). The lateral pressure profile in membranes: A physical mechanism of general anesthesia. *Toxicol. Lett.* **101**, 451–458.

Chang, R., Violi, A. (2006). Insights into the effect of combustion-generated carbon nanoparticles on biological membranes: A computer simulation study. *J. Phys. Chem. B* **110**, 5073–5083.

Cornell, W.D., Cieplak, P., Bayly, C.I., Gould, I.R., Merz, K.M., Ferguson, D.M., Spellmeyer, D.C., Fox, T., Caldwell, J.W., Kollman, P.A. (1995). A 2nd generation force-field for the simulation of proteins, nucleic-acids, and organic-molecules. *J. Am. Chem. Soc.* **117**, 5179–5197.

Dorairaj, S., Allen, T.W. (2007). On the thermodynamic stability of a charged arginine sidechain in a transmembrane helix. *Proc. Natl. Acad. Sci. USA* **104**, 4943–4948.

Feller, S.E., Brown, C.A., Nizza, D.T., Gawrisch, K. (2002). Nuclear overhauser enhancement spectroscopy cross-relaxation rates and ethanol distribution across membranes. *Biophys. J.* **82**, 1396–1404.

Franks, N.P. (2006). Molecular targets underlying general anaesthesia. *Br. J. Pharmacol.* **147**, S72–S81.

Freites, J.A., Tobias, D.J., von Heijne, G., White, S.H. (2005). Interface connections of a transmembrane voltage sensor. *Proc. Natl. Acad. Sci. USA* **102**, 15059–15064.

Hoff, B., Strandberg, E., Ulrich, A.S., Tieleman, D.P., Posten, C. (2005). ^{2}H-NMR study and molecular dynamics simulation of the location, alignment, and mobility of pyrene in popc bilayers. *Biophys. J.* **88**, 1818–1827.

Jedlovszky, P., Mezei, M. (2000). Calculation of the free energy profile of H_2O, O_2, CO, CO_2, NO, and $CHCl_3$ in a lipid bilayer with a cavity insertion variant of the Widom method. *J. Am. Chem. Soc.* **122**, 5125–5131.

Johansson, A.C.V., Lindahl, E. (2006). Amino-acid solvation structure in transmembrane helices from molecular dynamics simulations. *Biophys. J.* **91**, 4450–4463.

Jorgensen, W.L., Maxwell, D.S., Tirado-Rives, J. (1996a). Development and testing of the OPLS all-atom force field on conformational energetics and properties of organic liquids. *J. Am. Chem. Soc.* **118**, 11225–11236.

Jorgensen, W.L., Maxwell, D.S., TiradoRives, J. (1996b). Development and testing of the OPLS all-atom force field on conformational energetics and properties of organic liquids. *J. Am. Chem. Soc.* **118**, 11225–11236.

Kaminski, G.A., Friesner, R.A., Tirado-Rives, J., Jorgensen, W.L. (2001). Evaluation and reparametrization of the OPLS-AA force field for proteins via comparison with accurate quantum chemical calculations on peptides. *J. Phys. Chem. B* **105**, 6474–6487.

Koubi, L., Tarek, M., Bandyopadhyay, S., Klein, M.L., Scharf, D. (2001). Membrane structural perturbations caused by anesthetics and nonimmobilizers: A molecular dynamics investigation. *Biophys. J.* **81**, 3339–3345.

Koubi, L., Tarek, M., Bandyopadhyay, S., Klein, M.L., Scharf, D. (2002). Effects of the nonimmobilizer hexafluorethane on the model membrane dimyristoylphosphatidylcholine. *Anesthesiology* **97**, 848–855.

Koubi, L., Tarek, M., Klein, M.L., Scharf, D. (2000). Distribution of halothane in a dipalmitoylphosphatidylcholine bilayer from molecular dynamics calculations. *Biophys. J.* **78**, 800–811.

Li, L.W., Bedrov, D., Smith, G.D. (2005). A molecular-dynamics simulation study of solvent-induced repulsion between C-60 fullerenes in water. *J. Chem. Phys.* **123**.

Lohmeier-Vogel, E.M., Ung, S., Turner, R.J. (2004). In Vivo ^{31}P nuclear magnetic resonance investigation of tellurite toxicity in Escherichia coli. *Appl. Environ. Microbiol.* **70**, 7342–7347.

MacCallum, J.L., Tieleman, D.P. (2006). Computer simulation of the distribution of hexane in a lipid bilayer: Spatially resolved free energy, entropy, and enthalpy profiles. *J. Am. Chem. Soc.* **128**, 125–130.

MacCallum, J.L., Bennett, W.F.D., Tieleman, D.P. (2007a). Distribution of amino acids in a lipid bilayer from computer simulations, *Biophys. J.*, in press.

MacCallum, J.L., Bennett, W.F.D., Tieleman, D.P. (2007b). Partitioning of amino acid side chains into lipid bilayers: Results from computer simulations and comparison to experiment. *J. Gen. Physiol.* **129**, 371–377.

MacKerell, A.D., Bashford, D., Bellott, M., Dunbrack, R.L., Evanseck, J.D., Field, M.J., Fischer, S., Gao, J., Guo, H., Ha, S., Joseph-McCarthy, D., Kuchnir, L., Kuczera, K., Lau, F.T.K., Mattos, C., Michnick, S., Ngo, T., Nguyen, D.T., Prodhom, B., Reiher, W.E., Roux, B., Schlenkrich, M., Smith, J.C., Stote, R., Straub, J., Watanabe, M., Wiorkiewicz-Kuczera, J., Yin, D., Karplus, M. (1998). All-atom empirical potential for molecular modeling and dynamics studies of proteins. *J. Phys. Chem. B* **102**, 3586–3616.

Marrink, S.J., Berendsen, H.J.C. (1994). Simulation of water transport through a lipid-membrane. *J. Phys. Chem.* **98**, 4155–4168.

Marrink, S.J., Berendsen, H.J.C. (1996). Permeation process of small molecules across lipid membranes studied by molecular dynamics simulations. *J. Phys. Chem.* **100**, 16729–16738.

Monticelli, L., Robertson, K.M., MacCallum, J.L., Tieleman, D.P. (2004). Computer simulation of the KvAP voltage-gated potassium channel: Steered molecular dynamics of the voltage sensor. *FEBS Lett.* **564**, 325–332.

Mukhopadhyay, P., Vogel, H.J., Tieleman, D.P. (2004). Distribution of pentachlorophenol in phospholipid bilayers: A molecular dynamics study. *Biophys. J.* **86**, 337–345.

Norman, K.E., Nymeyer, H. (2006). Indole localization in lipid membranes revealed by molecular simulation. *Biophys. J.* **91**, 2046–2054.

Oostenbrink, C., Villa, A., Mark, A.E., Van Gunsteren, W.F. (2004). A biomolecular force field based on the free enthalpy of hydration and solvation: The GROMOS force-field parameter sets 53A5 and 53A6. *J. Comput. Chem.* **25**, 1656–1676.

Patra, M., Salonen, E., Terama, E., Vattulainen, I., Faller, R., Lee, B.W., Holopainen, J., Karttunen, M. (2006). Under the influence of alcohol: The effect of ethanol and methanol on lipid bilayers. *Biophys. J.* **90**, 1121–1135.

Pereira, C.S., Lins, R.D., Chandrasekhar, I., Freitas, L.C.G., Hunenberger, P.H. (2004). Interaction of the disaccharide trehalose with a phospholipid bilayer: A molecular dynamics study. *Biophys. J.* **86**, 2273–2285.

Pohorille, A., Cieplak, P., Wilson, M.A. (1996). Interactions of anesthetics with the membrane–water interface. *Chem. Phys.* **204**, 337–345.

Pohorille, A., Wilson, M.A. (1996). Excess chemical potential of small solutes across water–membrane and water–hexane interfaces. *J. Chem. Phys.* **104**, 3760–3773.

Pohorille, A., Wilson, M.A., New, M.H., Chipot, C. (1998). Concentrations of anesthetics across the water–membrane interface; The Meyer–Overton hypothesis revisited. *Toxicol. Lett.* **101**, 421–430.

Qiao, R., Roberts, A.P., Mount, A.S., Klaine, S.J., Ke, P.C. (2007). Translocation of C-60 and its derivatives across a lipid bilayer. *Nano Lett.* **7**, 614–619.

Rudolph, U., Antkowiak, B. (2004). Molecular and neuronal substrates for general anaesthetics. *Nat. Rev. Neurosci.* **5**, 709–720.

Sharp, K.A., Madan, B., Manas, E., Vanderkooi, J.M. (2001). Water structure changes induced by hydrophobic and polar solutes revealed by simulations and infrared spectroscopy. *J. Chem. Phys.* **114**, 1791–1796.

Siwko, M.E., Marrink, S.J., de Vries, A.H., Kozubek, A., Uiterkamp, A., Mark, A.E. (2007). Does isoprene protect plant membranes from thermal shock? A molecular dynamics study. *Biochim. Biophys. Acta Biomembranes* **1768**, 198–206.

Song, Y.H., Guallar, V., Baker, N.A. (2005). Molecular dynamics simulations of salicylate effects on the micro- and mesoscopic properties of a dipalmitoylphosphatidylcholine bilayer. *Biochemistry* **44**, 13425–13438.

Sum, A.K., Faller, R., de Pablo, J.J. (2003). Molecular simulation study of phospholipid bilayers and insights of the interactions with disaccharides. *Biophys. J.* **85**, 2830–2844.

Tieleman, D.P., Marrink, S.J. (2006). Lipids out of equilibrium: Energetics of lipid desorption and pore mediated flip-flop. *J. Am. Chem. Soc.* **128**, 12462–12467.

Tieleman, D.P., Marrink, S.J., Berendsen, H.J.C. (1997). A computer perspective of membranes: Molecular dynamics studies of lipid bilayer systems. *Biochim. Biophys. Acta Rev. Biomembranes* **1331**, 235–270.

Tieleman, D.P., Robertson, K.M., MacCallum, J.L., Monticelli, L. (2004). Computer simulations of voltage-gated potassium channel KvAP. *Int. J. Quantum Chem.* **100**, 1071–1078.

Ulander, J., Haymet, A.D.J. (2003). Permeation across hydrated DPPC lipid bilayers: Simulation of the titrable amphiphilic drug valproic acid. *Biophys. J.* **85**, 3475–3484.

van den Bogaart, G., Hermans, N., Krasnikov, V., de Vries, A.H., Poolman, B. (2007). On the decrease in lateral mobility of phospholipids by sugars. *Biophys. J.* **92**, 1598–1605.

van den Brink-van der Laan, E., Chupin, V., Killian, J.A., de Kruijff, B. (2004a). Small alcohols destabilize the KcsA tetramer via their effect on the membrane lateral pressure. *Biochemistry* **43**, 5937–5942.

van den Brink-van der Laan, E., Chupin, V., Killian, J.A., de Kruijff, B. (2004b). Stability of KcsA tetramer depends on membrane lateral pressure. *Biochemistry* **43**, 4240–4250.

Vemparala, S., Klein, M.L. (2004). Interaction of inhalational anesthetic halothane with membrane protein: A molecular dynamics study. *Abstr. Pap. Am. Chem. Soc.* **228**, U181–U182.

Vemparala, S., Saiz, L., Eckenhoff, R.G., Klein, M.L. (2006). Partitioning of anesthetics into a lipid bilayer and their interaction with membrane-bound peptide bundles. *Biophys. J.* **91**, 2815–2825.

White, S.H., King, G.I., Cain, J.E. (1981). Location of hexane in lipid bilayers determined by neutron diffraction. *Nature* **290**, 161–163.

Wilson, M.A., Pohorille, A. (1996). Mechanism of unassisted ion transport across membrane bilayers. *J. Am. Chem. Soc.* **118**, 6580–6587.

Xiang, T.X., Anderson, B.D. (2006). Liposomal drug transport: A molecular perspective from molecular dynamics simulations in lipid bilayers. *Adv. Drug Delivery Rev.* **58**, 1357–1378.

Yau, W.-M., Wimley, W.C., Gawrisch, K., White, S.H. (1998). The preference of tryptophan for membrane interfaces. *Biochem.* **37**, 14713–14718.

CHAPTER 9

On the Nature of Lipid Rafts: Insights from Molecularly Detailed Simulations of Model Biological Membranes Containing Mixtures of Cholesterol and Phospholipids

Max Berkowitz

Department of Chemistry, University of North Carolina at Chapel Hill, Chapel Hill, NC 27599, USA

Abstract

This chapter describes molecular detailed simulations performed to study the nature of lipid raft domains that appear in model membranes. The simulations were performed on hydrated bilayers containing binary mixtures of cholesterol with phospholipids and also on ternary mixtures containing cholesterol, a phospholipid with a high main transition temperature T_m, and a phospholipid with a low transition temperature T_m. Although a large amount of data was obtained from these simulations, they still did not provide a clear molecular picture explaining existence of raft domains.

I. INTRODUCTION

Biological membranes are complicated entities containing self assembled mixtures of different lipids and proteins. Around 500 different lipid species are

Current Topics in Membranes, Volume 60
1063-5823/08 $35.00
DOI: 10.1016/S1063-5823(08)00009-4

identified in biomembranes (Mayor and Rao, 2004) and among all these lipids, one is playing quite a special role, cholesterol. The importance of cholesterol is partially related to the notion of "membrane rafts", a concept suggested in 1997 (Simons and Ikonen, 1997). The study of membrane rafts has undergone an explosion since their proposed existence. Lipid rafts participate in a multitude of cellular processes such as signal transduction (Draber *et al.*, 2001; Simons and Toomre, 2000; Baird *et al.*, 1999), protein and lipid sorting (Simons and Ikonen, 2000), cellular entry by toxins or viruses and viral budding (Brown and London, 1998a, 1998b) to name a few. The issues related to the nature and organization of rafts are far from being clarified and agreed upon (Hancock, 2006), although recently, a definition of what is a raft was established. According to the Keystone Symposium on Lipid Rafts and Cell Function in 2006, "Membrane rafts are small (10–200 nm) heterogeneous, highly dynamic, sterol- and sphingolipid-enriched domains that compartmentalize cellular processes" (Pike, 2006). The role of sterols in the rafts is therefore clearly defined, explaining the reason why it is studied so intensely (Silvius, 2003; Simons and Ehehalt, 2002; Ohvo-Rekila *et al.*, 2002).

Since biological membranes are complicated mixtures of phospholipids, sterols and proteins, it is very difficult to analyze them and understand the role every component plays in the functioning of the assembly. As a result, many investigations are performed on model membranes containing either pure components or well-controlled mixtures of two/three components (Simons and Vaz, 2004). A number of different physical techniques have been used to study lateral organization of lipid rafts in natural and model membranes (Bagatolli, 2003; Bagatolli and Gratton, 2001; Gaus *et al.*, 2003; Burgos *et al.*, 2003; Burns, 2003; Giocondi *et al.*, 2004; Ianoul *et al.*, 2003; Kahya *et al.*, 2004; Nicolini *et al.*, 2004; Subczynski and Kusumi, 2003; Yuan *et al.*, 2002). Although details about the structural properties and dimensions of lipid rafts in biological cells are not known yet, experimental work has demonstrated that physical properties of model membranes are related to the physical properties of natural membranes containing lipid rafts. In both natural and artificial membranes the raft domains are rich in cholesterol and saturated lipids. Therefore, it is very important to understand the nature of interactions between cholesterol and other lipids that are responsible for raft creation. While rafts in model membranes can be directly observed, not much is known about their nature, structure and dynamics and much still needs to be learned about the properties of lipid mixtures containing cholesterol.

A good starting point in the research of lipid mixtures is to consider bilayers containing binary mixtures of cholesterol and phospholipids. Phase diagrams for these mixtures depend on the value of the T_m, the temperature of the gel to liquid crystal phase transition.

For mixtures containing lipids with high T_m, such as saturated phosphatidylcholine (PC) or sphingomyline (SM), the assignment of phases in such a phase diagram is still not completely clarified, as different experimental techniques pro-

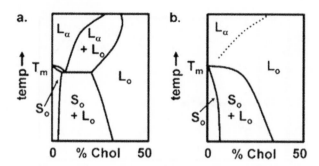

FIGURE 1 Two possible phase diagrams for the binary mixture of cholesterol and high T_m phospholipid. The figure is reproduced from the reference (Veatch and Keller, 2005b).

duce somewhat contradicting results (Veatch and Keller, 2005b). A recent review of Veatch and Keller (2005b) gives an excellent introduction into the subject of phase diagrams of two and three component mixtures in general and mixtures with cholesterol specifically. Fig. 1, reproduced from the review, presents two phase diagrams suggested for binary mixtures containing cholesterol and high T_m phospholipid. These diagrams demonstrate that three phases can exist in such mixtures. These are: L_α or L_d—liquid disordered phase in which both translational and conformational degrees of freedom for phospholipids molecules are disordered; L_o—liquid ordered phase that contains lipids that are translationally disordered and conformationally ordered; and finally, S_o phase—solid ordered phase where both translational and conformational degrees of freedom of lipids are ordered, as in a gel phase of a one component phospholipids bilayer. The phase diagram from panel 1a shows that at temperatures above T_m, regions of liquid–liquid coexistence are present in the two component lipid bilayers with cholesterol as one of the components, while diagram from Fig. 1b discards the coexisting regions. The nomenclature for the bilayer phases was established by Ipsen *et al.* (1987). They also demonstrated that it is possible to obtain the phase diagram shown on panel 1a from a specific model that can be considered to be an example of a "coarse-grained" model, i.e. a model containing only the gross features of the system. Since the model of Ipsen et al. may be missing some fine but important detail of the molecular interactions, the results obtained from this model may be different from the ones that a full molecular model may produce. Unfortunately, getting a phase diagram from simulations with a full molecular description of the bilayer can be a daunting task. Nevertheless, simulations using models with full molecular detail are needed if we want to understand the nature of specific interactions between cholesterol and phospholipids that are responsible for the shape of the phase diagrams from Fig. 1.

Mixtures of cholesterol with phospholipids can be studied not just in bilayers, but also in monolayers with some very interesting results. Consider, for example, the data obtained from the fluorescence work performed by McConnell's group

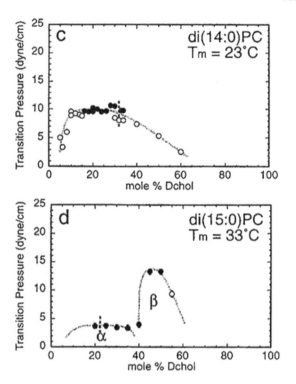

FIGURE 2   The phase diagrams for monolayers of chol/di(14:0)PC and chol/di(15:0)PC mixtures. Adopted from (Keller *et al.*, 2000).

(Keller *et al.*, 2000; McConnell and Radhakrishnan, 2003; McConnell and Vrljic, 2003). McConnell and collaborators observed that for monolayers containing a binary mixture of cholesterol and low T_m, for example chol(cholesterol)/di(10:0)PC or chol/di(12:0)PC or chol/di(14:0)PC at low surface pressures, two coexisting liquid phases were present. At higher pressures, the two phases merged into one phase, and the diagram displayed an upper miscibility critical point. When the mixture contained cholesterol and a lipid with a higher T_m, for example di(15:0), two upper critical points appeared. The phase diagrams for monolayers of chol/di(14:0)PC and chol/di(15:0)PC mixtures are shown in Fig. 2. According to McConnell and collaborators, the existence of two critical points in mixtures of di(15:0)PC phospholipids with cholesterol indicates that formation of a "condensed complex" between cholesterol (C) and phospholipid (P) is taking place. This process can be described by the reaction:

$$nqC + npP \Leftrightarrow C_{nq}P_{np},$$

where q and p are stoichiometric integers and n is the measure of the size of the complex. n also shows the degree of cooperativity in the complex formation.

Using the notion of the condensed complex, we can assign one region of the phase diagram from Fig. 2, region α, as due to immiscibility of pure phospholipid and complex and the other region, β, as due to immiscibility of complex and cholesterol. What is the molecular structure of the complex? It was pointed out in the work of McConnell's group that no molecular picture of a cholesterol/phospholipid complex exists at the present time. There is also no clear understanding what are the values of q, p and n, although some indications are that $q = 1$ or 2 and $p = 2$ or 3 and $n \sim 2$–10.

While binary mixtures of cholesterol and phospholipids show that phase separation and liquid–liquid coexistence regions are possible, ternary mixtures of cholesterol with two phospholipids, one with a high T_m and one with a low T_m, demonstrate that liquid–liquid coexistence regions are indeed present in these systems. It seems that, in this case, cholesterol displays selective properties by creating liquid ordered domains—L_o, regions enriched in cholesterol and high T_m lipids. These domains, which are lipid rafts if the high T_m lipid is a sphingomyelin, "float" in the liquid disordered phase (L_d) of the rest of the membrane. Micron-scale liquid domains (rafts) and liquid–liquid phase demixing in artificial membranes were observed in giant unilamellar vesicles (GUVs) (Veatch and Keller, 2005b; Feigenson and Buboltz, 2001) when these systems contained three lipid components. One of the earlier studied ternary mixtures contained a relatively saturated lipid such as sphingomyelin (SM), cholesterol and an unsaturated lipid such as dioleoylphosphatidylcholine (di(16:1)PC or DOPC) in ratio 1:1:1. For this mixture the lipid raft domains in L_o phase were observed to be enriched by SM and cholesterol, while the liquid disordered phase outside the rafts was enriched by DOPC (Gandhavadi et al., 2002). Later, the phase diagram for the mixture containing palmitoyl (16:0SM or PSM), POPC and cholesterol was mapped out and it displayed the coexistence of L_o and L_d phases (de Almeida et al., 2003; Veatch and Keller, 2005a). The choice of the SM molecule as a saturated lipid in a mixture was perhaps dictated by the observation that natural membranes contain a large proportion of SM. Instead of SM one can choose another saturated lipid with high T_m, such as the well studied dipalmitoylphosphatidylcholine (di(16:0)PC or DPPC), and still observe the liquid–liquid coexistence region in the phase diagram of a ternary mixture of DOPC/DPPC/Chol (Veatch and Keller, 2005b). Obviously, one wants to know what kind of interactions between cholesterol and phospholipids are responsible for the liquid–liquid coexistence regions in these mixtures. Are they very specific? What is the role of the phospholipid headgroup/cholesterol headgroup interaction and phospholipid tail/cholesterol tail interaction? What is the connection between the proposed existence of complexes in binary monolayers and domains in ternary mixtures of bilayers? With respect to the last question, McConnell, using a modified theory of regular solutions, has recently suggested that the liquid–liquid coexistence region in phase diagrams of ternary mixtures and the existence of two critical points in phase diagrams for binary monolayers (see Fig. 2) are due to the same phenomenon: creation of a complex between cholesterol and high T_m lipids (Radhakrishnan and McConnell, 2005). The complexes, as it was proposed, are not stationary and may have structures that fluctuate rapidly over a range of conformations. Further, the existence of complexes is closely connected to the existence of rafts. It seems that computer simulations describing phospholipid–cholesterol interactions in full molecular detail should be able help us to understand the validity of this proposal, and be able to identify these complexes. Detailed simulations, hopefully, should also be able to provide us with the information needed to understand the nature of lipid rafts.

II. SIMULATIONS

A. Earlier Simulations Using Simple Models

As we previously mentioned, computer simulations of mixtures containing cholesterol and phospholipids employ different models that describe the molecular interactions on a different level of detail. These levels are interconnected and the chapter in this volume by Pandit et al. on the multiscale modeling of lipid bilayer membranes describes this interconnection. The goal of the simulations is to provide a description of the phenomena observed in experiments. Ideally, full molecular details embedded in a specific force field used in the simulations should be employed to accomplish this goal. But enormous computational demands, especially in the field of membrane simulations, require "coarse graining" of the details to produce a "coarse grained" force field. Simulations using a coarse-grained description can be performed over longer time periods and larger spatial domain regions. Coarse graining can be performed systematically, or coarse grained models can be chosen *ab initio* on physical grounds. Recent reviews by Muller *et al.* (2006) and Venturoli *et al.* (2006) provide comprehensive description of coarse-graining work performed on membranes. Because these recent reviews on coarse-graining of membranes are available in the literature, we will review here simulations on membranes containing mixtures of cholesterol with phospholipids that were done using molecular detailed description of phospholipids. We will restrict our description of coarse-grained work to earlier work of Mouritsen, Zuckermann and coworkers, since it was very influential for our understanding of phase diagrams in lipid mixtures (Mouritsen, 1991). It was pointed out by Mouritsen and coworkers that phase transitions in bilayers involve two distinct, but coupled order–disorder processes: one related to the translational degrees of freedom and the other one to the conformational degrees. In a two component lipid bilayer containing cholesterol as one of the components, the decoupling between these degrees of freedom takes place and a liquid ordered phase appears. Nielsen *et al.* (1999) performed Monte Carlo simulations on a specifically designed off-lattice model where lipids and cholesterol molecules were represented as hard-core particles with internal degrees of freedom that also had specific nearest-neighbor interactions. According to Nielsen *et al.*, the model used represented the minimal model needed to describe the appropriate physics of the problem. The large number of conformational states of lipids in the model was reduced to just two conformational states and cholesterol was treated as a substitutional impurity. The interactions between particles were designed to contain the features needed to produce the simple phase behavior of one component lipid bilayer and the dual role of cholesterol as a "crystal breaker" and a "chain rigidifier". It was shown that using this minimal model, a phase diagram, such as the one depicted in Fig. 1a, could be obtained. Since obtaining a phase diagram in simulations is quite a complicated task, the authors used histogram reweighting method, finite size scaling and non-Boltzmann sampling techniques. (For a description of these and many other techniques used in molecular simulations, the reader is advised to consult an excellent book on simulations by Frenkel and Smit (2002).)

In addition to phase diagrams for the investigated cholesterol/phospholipid bilayer, Mouritsen and collaborators also studied the structural properties of different phases. The calculated structure factors clearly showed that the L_o phase is a liquid phase. It was also observed that lipid chains in direct contact with cholesterol molecules tended to align with their own kind, making a "treadlike" structure. The propagation of these "threads" was rather short, involving just a few molecules. These "threads" could be considered as hints pointing towards the existence of cholesterol/phospholipid complexes. All together, the results from the work of Nielsen et al. based on a physically simple (but in no way computationally simple) model demonstrate how much information one can obtain from such a model. Still, it is not clear how well justified are the forms of interactions assumed in the model. Such a justification should be the task of detailed molecular simulations.

B. Simulations Using Detailed Molecular Models

1. Simulations of Bilayers Containing Binary Mixtures of Cholesterol and PC Lipids

Computer simulations of lipid mixtures containing cholesterol were performed using different levels of model Hamiltonians and applying both Monte Carlo (MC) and Molecular Dynamics (MD) simulation techniques. Full molecular detail simulations of lipid bilayers containing just one phospholipid component started in earnest in early nineties and, during a five year period, reasonable success was accomplished in the building of force-fields that could reproduce many properties of such bilayers (Pastor, 1994; Tobias *et al.*, 1997; Tieleman *et al.*, 1997). These simulations also provided a molecular detailed description of properties of water next to the bilayers (Berkowitz *et al.*, 2006). The first full molecular detailed simulations of bilayers containing binary mixtures

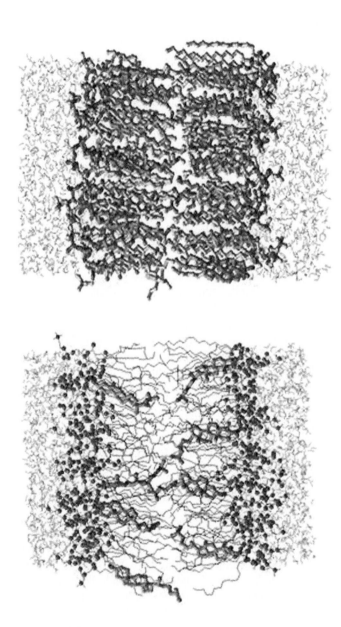

FIGURE 3 Snapshots from simulations performed on the cholesterol/DPPC mixtures. Left panel is for the mixture at low cholesterol concentration; right panel at high cholesterol concentration. Note the difference in the conformational states of phospholipids in the two panels.

of cholesterol and phospholipids provided a glimpse on the structural changes in properties of lipids next to cholesterol, but the simulations suffered from different problems, including the small number of statistically significant data due to the shortness of the production runs. An earlier simulation that produced reliable data on the binary cholesterol/phospholipid system was the simulation of Tu *et al.* (1998). In this simulation, a 1.4 ns MD run was performed at a constant-pressure on a system containing 12.5 mol% cholesterol in DPPC bilayer at temperature of 50 °C. Many of the initial simulations on the binary mixtures of cholesterol and phospholipids were performed on the mixtures of cholesterol and DPPC, since the properties of DPPC were already well established in simulations. In the simulation of Tu et al., the total size of the system was relatively small as it contained only 64 lipids in the bilayer. The results from this simulation were compared with the results obtained from a simulation on a pure DPPC performed under the same conditions (Tu *et al.*, 1995). It was observed that in the bilayer interior, cholesterol at this concentration did not significantly affect the conformations and packing of the hydrocarbon chains, only slightly reducing the free volume. The simulations revealed that cholesterol had a significant effect on lipid dynamics on the sub-nanosecond range, slowing down the C–H reorientational motion along the lengths of the hydrocarbon chains. To understand the influence of cholesterol on the neighboring phospholipids when cholesterol concentration can be low and high, Smondyrev and Berkowitz performed molecular dynamics simulations at 11 mol% and 50 mol% concentrations (Smondyrev and Berkowitz, 1999a) and compared the results with the results of their previous simulation on pure DPPC (Smondyrev and Berkowitz, 1999b). Two simulations (A and B) were performed at high cholesterol concentration that differed by the initial arrangement of cholesterol in bilayers. Since the total run time for each simulation was only 2 ns, the arrangements of lipids in bilayers did not change in any substantial way during the time of the simulations and, therefore, the results showed dependence on the initial structural arrangement. In the simulation of Smondyrev and Berkowitz, bilayers also contained only 64 lipids. In simulation A, cholesterol was initially placed to be alternating with DPPC molecules, so that nearest-neighbor interactions occurred between DPPC and cholesterol molecules. In simulation B, strips of cholesterol molecules were placed between strips of DPPC molecules and, therefore, interactions between similar molecules (DPPC:DPPC or cholesterol:cholesterol) were predominant. The structures of the bilayers obtained from the simulations snap-shots are shown in Fig. 3. From this figure it is clear that the effect of cholesterol on the chain ordering at high cholesterol concentrations is rather large and that the conformational degrees of freedom at 50 mol% cholesterol are ordered, indicating the existence of L_o phase in the system. Smondyrev and Berkowitz also observed that the area of the bilayer decreased substantially at a large cholesterol concentration, thus confirming the condensation effect that cholesterol produces in the bilayer. Cholesterol molecules also displayed a tilt, so that they could be accommodated in membranes. While the tilt was larger in bilayers with small amount of cholesterol, it decreased in bilayers with 50 mol% cholesterol since the size of the hydrophobic region in this bilayer increased. Chiu *et al.* (2002) performed a series of simulations on bilayers containing mixtures of DPPC/cholesterol at different composition ratios such as 24:1, 47:3, 11.5:1, 8:1, 7:1, 4:1, 3:1, 2:1 and 1:1. Clear onset of the condensation effect beyond 10 mol% fraction of cholesterol was observed in their simulations by analyzing the behavior of the area per molecule. When the calculated area per molecule was plotted as a function of the cholesterol fraction (in the range of 0.1 to 0.5) a linear plot emerged. This plot was fitted to a straight line given by Eq. (1)

$$A_{mol} = X_{DPPC} A_{DPPC} + X_{chol} A_{chol}, \tag{1}$$

where X_i represented the molar fraction of molecule i. The following values for areas were obtained using Eq. (1): $A_{DPPC} = 0.507$ nm^2 and $A_{chol} = 0.223$ nm^2. Although the meaning of the areas obtained from Eq. (1) was not discussed by Chiu et al., one can think of them as partial areas that had composition independent values when cholesterol composition was in the range of 10 to 50 mol%. Extraction of values for areas per lipid molecule from simulations containing lipid mixtures is not a trivial issue, and different approaches may produce different values (Edholm and Nagle, 2005). The partial areas per lipid obtained from procedure proposed by Chiu *et al.* required that simulations be performed in a range of different fractions of lipid components. The method described in Hofsass *et al.* (2003) provided the values of areas per lipid at a given composition from the simulation data performed at that composition only. Nevertheless, Hofsass *et al.* (2003) performed a series of simulations on bilayers with different cholesterol content in the range of 0–40 mol%. Their simulations were performed on larger sized patches of bilayers containing 1024 lipid molecules, which improved the statistics and also allowed them to study elastic properties of these bilayers. According to Hofsass et al., the values for areas that can be assigned to different molecules in the mixed bilayer, can be obtained from an analysis based on simple geometry using the following equations:

$$a_{PL} = \frac{2A}{(1-x)N_{lipid}} \left[1 - \frac{x N_{lipid} V_{chol}}{V - N_W V_W} \right], \tag{2}$$

$$a_{chol} = \frac{2A V_{chol}}{V - N_W N_W}, \tag{3}$$

where A is the cross section $(x-y)$ area of the simulation box, N_{lipid} is the total number of molecules in the bilayer, x is the mole fraction of cholesterol: $x = N_{chol}/N_{lipid}$, V is the volume of the simulation box, N_W is the number of water molecules in the system. The volume per Chol molecule, V_{chol}, is taken to be 0.593 nm^3 and that of water, V_W, is 0.0305 nm^3. Hofsass *et al.* observed that both areas per DPPC and cholesterol were composition dependent, although the dependence for cholesterol was very weak. The calculated area per cholesterol of

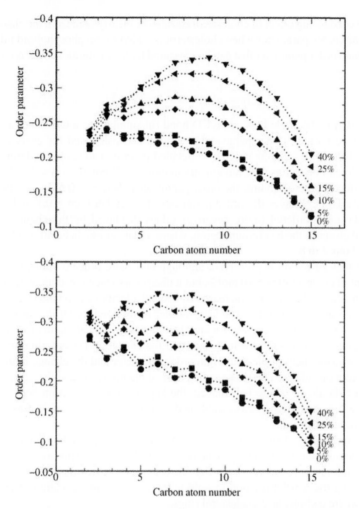

FIGURE 4   Calculated order parameters corresponding to NMR deuterium ones vs chain position for different cholesterol concentrations (top for the Sn1 chain and bottom for the Sn2 chain). Reproduced from (Hofsass *et al.*, 2003).

0.27 nm^2 obtained by Hofsass *et al.* for larger cholesterol content was somewhat larger than the one obtained by Chiu et al., but still was much smaller than the value of 0.38 nm^2 obtained from calculations performed on a cholesterol crystal (see Fig. 4). The reason cholesterol is having a smaller area in the mixed bilayer is that it is partly buried in the phospholipid bilayer, as considered in the "umbrella" model (Huang and Feigenson, 1999) and as also was pointed out by Smondyrev and Berkowitz (1999a) in their discussion on the area reduction in mixed bilayer containing cholesterol.

The condensing effect of cholesterol can also be clearly seen from the change in the chain order parameter when cholesterol is added to the phospholipid bilayer. The chain order parameter that can be measured by the deuterium NMR is defined as

$$S_{CD} = \frac{1}{2}(3 < \cos^2 \theta_{CD} > -1),\qquad(4)$$

where θ_{CD} is the angle between a CD bond (in simulations a CH bond) and the membrane normal. Order parameter measured and calculated for every carbon atom of the DPPC chains provides a detailed picture of the chain ordering. For the hydrocarbon chains with random orientation of CD bonds, $S_{CD} = 0$, while for the perfectly ordered chains, the order parameter is $S_{CD} = -0.5$. In simulations, for pure DPPC bilayer the order parameter changes between -0.1 and -0.22 with the more ordered (more negative) values displayed by the chain carbons located closer to the headgroup region, where a plateau-like behavior is observed for carbons 2 to 8.

The plateau region is preserved, although at a more ordered value, when the amount of cholesterol is ~10 mol%, but it disappears when the amount of cholesterol increases. The order is increased by around a factor of two when cholesterol fraction is around 40 mol%. The largest order is observed for DPPC carbons that face the ring system of cholesterol and the order is decreased rapidly towards the tail region of phospholipids. It is clear that the change in the ordering observed by the measurement of the deuterium order parameter and the change in the area per lipid molecule are due to the same physical phenomenon and therefore, these changes should be connected. Indeed, the approximate connection between the average S_{CD} and A_{DPPC} was established and the simulations of Hofsass et al. support this assertion.

Another means of studying the ordering of the tails in phospholipid molecules is to measure the fraction of gauche defects in these tails. This was also calculated and, again, it was observed that cholesterol increases the tail ordering by decreasing the gauche fraction in lipid tails, especially in the region of carbons 4–10 where carbons face cholesterol rings.

All the simulations on DPPC/cholesterol bilayers we mentioned above clearly demonstrated the condensing effect of cholesterol and a strong increase in tail ordering at larger mol fraction of cholesterol. Simulations can also provide molecular details related to specific interactions such as hydrogen bonding in bilayers and how it is affected when the bilayer composition is changed. Thus, Smondyrev and Berkowitz determined the average number and the specific atoms of phospholipids that formed hydrogen bonds with cholesterol (Smondyrev and Berkowitz, 1999a). Since, traditionally, these are considered to be O–H type bonds, for the DPPC/cholesterol/water system, it was observed that the hydroxyl hydrogen of cholesterol was engaged in hydrogen bonding with oxygens in carbonyl groups and one of the oxygens of phosphate group. It was also observed that cholesterol makes hydrogen bonds with water molecules. Chiu et al. found that when the cholesterol fraction in the bilayer is below 50 mol%, cholesterol displays a clear preference for hydrogen bonding with carbonyl oxygens over the phosphate oxygens (Chiu et al., 2002). The patterns of direct and indirect (through water) hydrogen bonding between cholesterol and PC molecules was investigated in detail in works of Rog and Pasenkiewicz-Gierula (2001) and Pasenkiewicz-Gierula et al. (2000), where results from 14 ns MD simulation of dimyristoylphosphatidylcholine (DMPC) and cholesterol mixture containing 56 DMPC and 16 cholesterol (22 mol%) were presented. Chiu et al. observed in their simulations that each cholesterol molecule is hydrogen bonded to at least one phospholipid. They proposed that this bonding is responsible for the creation of a two molecule (cholesterol and DPPC) complex that serves as a "building block" for the creation of an oligomeric structure of complexes as was proposed by McConnell and co-workers. Pandit et al. moved this idea even further (Pandit et al., 2004a). By noticing that complexes of cholesterol and phospholipids can contain 1 cholesterol and 2 phospholipid molecules, they suggested that cholesterol can hydrogen bond to DPPC lipids not just through a cholesterol headgroup hydrogen, but also through its oxygen, making a hydrogen bond to the methyl group of tetramethylammonia of the DPPC choline. The existence of hydrogen bonding of the O–CH$_3$ type is well accepted concept in chemistry (Gu et al., 1999). The most probable number of hydrogen bonds per cholesterol in simulations of Pandit et al. turned out to be two when the bilayer contained a mixture of cholesterol and DPPC. When the phospholipid in the bilayer was DLPC (12:0, 12:0 PC), which contains shorter hydrocarbon chains, the most probable number of hydrogen bonds per cholesterol was just one. Pandit et al. speculated that with two hydrogen bonds per cholesterol, one can create a linear network of hydrogen bonded molecules Ph(phospholipid)–chol–Ph–chol… when Ph is a DPPC molecule. Such a network has a smaller chance to exist when Ph is a DLPC molecule. Therefore, if one assumes that presence of a hydrogen bonded network is correlated with the existence of oligomeric complexes, one finds that such complexes are more probable to exist in bilayers containing mixtures of cholesterol and DPPC compared to bilayers with mixtures of cholesterol and DLPC. According to Pandit et al., the reason that cholesterol is most likely engaged in two hydrogen bonds with DPPC and only one bond with DLPC is that the cholesterol axis had a larger tilt in DLPC to accommodate cholesterol in the hydrophobic region of this membrane. Due to the larger tilt, cholesterol was not in a favorable position to engage in two hydrogen bonds with neighboring phospholipids. The phase diagrams obtained by McConnell and coworkers for cholesterol/DPPC and cholesterol/DLPC monolayers, and their interpretation of these diagrams, indicated that complexes exist for DPPC/cholesterol system and do not exist for DLPC/cholesterol case. Therefore, the interpretation of the results obtained from the simulations on DPPC/cholesterol and DLPC/cholesterol bilayers presented by Pandit et al. was consistent with the results observed in the experiment. In spite of this agreement, one should be careful when comparing results obtained from bilayers and monolayers, since these may display profound differences in their structures.

The cholesterol molecule has a smooth α-face and a rough β-face with two protruding methyl groups. This kind of a special molecular structure is believed to be important in shaping the interaction of cholesterol with lipid molecules containing both saturated and unsaturated chains. To study this situation Pitman et al. performed a simulation wherein the binary bilayer contained a 1:3 mixture of cholesterol and stearoyl-docosahexanenoyl—phosphocholine (SDPC) lipids (Pitman et al., 2004). One of the SDPC chains is strongly unsaturated. Pitman et al. observed that cholesterol had a preference for a solvation by a saturated fatty acid chain, especially the smooth side of the cholesterol. This interaction bias created an inhomogeneous environment around cholesterol that could play an important role in the formation of microdomains. The picture that emerged from the simulation of Pitman et al. is consistent with the model of Mitchell and Litman (1998), who proposed that formation of microdomains is solely due to the acyl chain-cholesterol interactions.

2. Simulations of Binary Mixtures Containing Cholesterol and SM Lipids

As we have seen, the definition of lipid rafts requires the presence of cholesterol and sphingomyelin molecules in the bilayers. Therefore, the emphasis of simulations shifted to simulations of mixtures containing these two lipids. Prior to simulations on such mixtures, simulations on pure sphingomyelin bilayers were performed. Thus, Mombelli et al. reported results from three 8 ns simulations of a palmitoylsphingomyelin (16:0 SM or PSM) bilayer in water using different force fields and found that independently of the force fields, the major physico-chemical parameters of the simulated bilayer agreed with experimental data (Mombelli et al., 2003). They studied the hydrogen-bonded network of the PSM bilayer and determined that an extensive network of such bonds exists in the PSM bilayer and that the sphingosine OH group was mainly involved in the intramolecular hydrogen bonding, while the amide NH moiety was participating in intermolecular hydrogen bonding. This conclusion was confirmed later in the simulations of Niemela et al. (2004), who performed a systematic comparison between properties of PSM and DPPC bilayers. It was emphasized by Niemela et al. that dynamical properties of PSM were slowed down and the ordering of PSM was increased and that these dynamical changes were due to the presence of the hydrogen-bonded network in the PSM bilayer.

Naturally, the next step in our quest for understanding the nature of lipid rafts was to study binary mixtures of SM and cholesterol and, further, ternary mixtures of SM, cholesterol and unsaturated phosphatidylcholine, like POPC. Thus, Khelashvili and Scott simulated a bilayer containing 266 stearoyl SM (18:0 SM or SSM) and 122 cholesterol molecules at two different temperatures, 20 and 50 °C (Khelashvili and Scott, 2004). These two temperatures bracket the main transition in the pure SSM, but due to the presence of cholesterol this transition is eliminated and the results of the simulations were quite similar, despite the 30 °C difference. Comparison between the results from simulations performed on SSM bilayers with cholesterol and pure SSM bilayers showed that SSM lipid chains were more ordered when cholesterol was present in the system. Little tendency was observed for cholesterol to form hydrogen bonds with SM. A more detailed study of cholesterol–sphingomyelin hydrogen bonding pattern and the difference between cholesterol–PC and cholesterol–SM pair interaction was presented by Rog and Pasenkiewicz-Gierula (2006).

Although simulations have shown that there is a difference in the interaction of cholesterol with SM compared to the interaction with PC, the issue of a specificity of cholesterol-SM interaction still remained unclear. While some experimental studies indicated the specific character of cholesterol–SM interactions (Silvius, 2003), other studies found evidence for a lack of such interactions (Holopainen *et al.*, 2004). It was suspected that the specific nature of the interaction between cholesterol and SM is due to the direct hydrogen bonding between cholesterol and SM, but in a recent simulation of Aittoniemi *et al.* (2007) it was shown that this is not the case. The total interaction between SM and cholesterol is not determined by the hydrogen bonding only; it is more subtle and comprises of contributions from electrostatic, van der Waals and hydrophobic effects. If it is not just the hydrogen bond that determines the interaction between sphingomyelin (or other phospholipid) and cholesterol, one has to study the total energy of the cholesterol–phospholipid interaction to understand when and why the cholesterol–phospholipid complexes may be created. Therefore, we recently studied the energy of interaction between POPC and cholesterol and compared it to the interaction energy of stearoyl SSM with cholesterol. Since POPC and SSM differ in the structure of both their tails and their headgroups we also studied the interaction of oleoyl SM (18:1 SM or OSM) with cholesterol (Zhang *et al.*, 2007). OSM is a molecule that is somewhat intermediary between POPC and SSM: it has the same headgroup as SSM and tails similar to tails in POPC.

Can one use the energetic criteria to distinguish between cholesterol/phospholipid complex from the non-complex?

To answer this question we calculated the distribution for the lowest nearest neighbor interaction energies between cholesterol and our phospholipid molecules. The distributions for Chol/SSM and Chol/OSM interactions look very similar, and the distribution for the Chol/POPC interaction did not differ much from them either (see Fig. 5). We also calculated the distribution of lowest energies for the triplet of molecules containing one cholesterol and two (but same type) phospholipids, and observed that all three distributions were very similar (also shown in Fig. 5). Moreover, the energy distributions did not display any pronounced bimodal structure that could be used to identify which of the cholesterol–phospholipid 1:1 pair or 1:2 triplet were engaged in a creation of a complex. The distributions also did not indicate, from the energetic point of view, that cholesterol should prefer SM over POPC. That means that if cholesterol

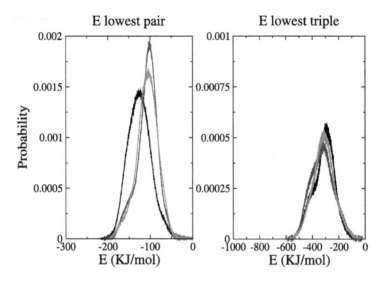

FIGURE 5 Distributions of lowest energy of interaction between cholesterol and phospholipids (left) and energies of a triplet of molecules consisting of cholesterol and two (same type) phospholipids (right). Red line is for the SSM cholesterol interactions, green for cholesterol OSM interactions and black for POPC cholesterol interactions.

molecules are engaged in a complex creation with the SM molecules, judging by the energy distributions cholesterol is also engaged in a complex creation with the POPC molecules, and this contradicts the assumption made by McConnell and collaborators. Therefore, a simple criterion based on the idea that one could explain complex formation between cholesterol and phospholipid by looking at the corresponding interaction energy is not applicable. One needs to look for another criterion to prove or disprove complex existence from the simulations. Our calculations also showed that, on average, the POPC/cholesterol interaction was slightly stronger than the SM/cholesterol interaction. We observed that among three bilayers containing pure phospholipids (POPC, SSM and OSM) the larger change upon addition of cholesterol occurred in the POPC bilayer. Specifically we observed a large change in the tail ordering of this phospholipid. Therefore we concluded that the preference of cholesterol for a specific phospholipid is due to a *delicate* balance between energy and entropy. The delicate balance indicates a possibility that in ternary mixtures of sphingomyelin/phosphatidylcholine/Chol, the domain structure may exist when sphingomyelin has both saturated chains like in the SSM molecule, while the domain structure may disappear when the SSM is replaced by the OSM molecule. Indeed, Epand and Epand noted that this is the case for the ternary mixture of OSM/SOPC/Chol (Epand and Epand, 2004). Also, that a delicate balance between energy and entropy is indeed present in nature was concluded by Frazier *et al.* (2007). They studied a bilayer containing a ternary mixture of bovine SM/POPC/Chol by using Fluorescence Resonance Energy Transfer experimental technique and also performed Monte Carlo simulations on a simple lattice. Frazier *et al.* determined that one obtains phase separations and domains in their system when Flory–Huggins like parameters that describe the difference between the free energies of interactions between components in the bilayers have values of order of just 300 cal/mol.

To add to the complexity of biological membranes, the identity of lipids in the leaflets of the bilayers that one finds in cell membranes is different. Thus, for example, lipids such as SM and PC can be often found in the outer leaflet of the plasma membrane, while two other typical lipids: phosphatidylserine (PS) and phosphatidylethanolamine (PE) are found in the inner leaflet of the membrane (Devaux and Morris, 2004). Nearly all simulations on bilayers containing lipid mixtures that were reported in the literature were performed on symmetric bilayers containing the same composition of both leaflets; however Bhide *et al.* (2007) recently reported results from a molecular dynamics (MD) simulation performed on an asymmetric bilayer containing a mixture of cholesterol and SSM in one leaflet and SOPS (stearoyl-oleoyl-phosphatidylserine) and cholesterol in the other leaflet. The raft forming ratio of concentrations of cholesterol to SM (1:2) was chosen for the one leaflet and the same ratio between cholesterol and PS was chosen for the other leaflet. Since the interaction between PS and cholesterol is favorable, and the PS molecules are condensed in the bilayer due to interlipid hydrogen bonding interactions and condensing effect due to counterions (Pandit *et al.*, 2003), the expectation was that the phase of the mixture containing SOPS and cholesterol was also the L_o phase. Thus Bhide et al. simulated a patch of an asymmetric bilayer where both monolayers were in the L_o phase. They also compared the results from the simulations on the asymmetric bilayer with the results from simulations on two symmetric bilayers containing binary mixtures of cholesterol and SSM and cholesterol and POPS respectively. From their simulations, Bhide et al. concluded that properties of the monolayers in the leaflets did not change much when going from a symmetric to an asymmetric bilayer, indicating that cross-leaflet interactions such as interdigitation were unimportant for the bilayers containing these specific leaflets. The work of Bhide et al. also indicates that PS phospholipids may be engaged in the creation of lipid raft domains in the inner leaflet of biological membranes.

3. Simulations of Ternary Mixtures Containing Cholesterol, SM and PC

Recently, simulations were performed on systems containing lipid membranes with three components, the same components that are able to create lipid rafts in model membranes: cholesterol, sphingomyelin and unsaturated phosphatidylcholine. One of the major problems with the simulations of bilayers with a ternary composition is that, due to inhomogeneous structure of these bilayers, it is hard to find proper starting configurations. This problem may be not that severe in simulations with coarse-grained potentials. In such simulations, one can start with a random distribution of lipids and observe how the system phase separates, since with a coarse grained potential one can perform simulations on a larger sized patches of membranes and over longer simulation times. The problem can be rather severe when a detailed microscopic description is used. In latter case, due to the slow diffusion of lipid molecules, the configurations change slowly and one may need to simulate the system containing a large number of particles over microseconds of time in order to establish an equilibrium distribution, which is impractical. To avoid this problem, it is possible to prepare the system in a special arrangement that may be closer to the equilibrium state. Therefore, simulating a mixture containing 1424 DOPC, 266 molecules of SSM and 122 molecules of cholesterol, Pandit et al. (2004c) prepared a patch that contained all the SSM and cholesterol molecules. This patch was imbedded in the rest of membrane containing all the DOPC molecules. This way, the condition that the L_o phase of the SM–cholesterol mixture is imbedded in the L_d phase containing the majority of unsaturated PC molecules was reproduced. This solution of the initial problem is not perfect: in experiment, the L_o domain has much larger dimensions, and it is not clear if one can reproduce the correct physics of the inhomogeneous membrane by scaling down the patch size. In any case, it was observed that the SM–cholesterol domain was stable and remained stable for the length of the analyzed simulation run (8.5 ns). It was observed that the liquid-ordered patch was 0.45 nm thicker than the rest of the membrane. To understand what is happening during the initial stages of the L_o domain formation, Pandit et al. (2004b) also simulated a ternary mixture of DOPC/SSM/cholesterol in a proportion of 1:1:1 (100 molecules of each lipid) starting from a random initial distribution of lipids. The simulation run was performed for 200 ns and it was concluded that on the initial stage of possible phase separation between the L_o and L_d domains, cholesterol is located on the domain border, thus reducing the line tension due to the presence of the interface between the liquid phases. This can be also understood on the microscopic level. As was already mentioned, the structure of a cholesterol molecule displays two faces: a smooth one and a rough one. According to Pandit et al., due to steric constrains, cholesterol orients its smooth face to SM molecule and shows its rough face to DOPC, therefore acting as a two-dimensional surfactant and therefore reducing the line tension. This argument is quite similar to the argument invoked by Pitman et al. (2004) that we discussed above.

Tension and other mechanical quantities play an important role in the stability of the membranes. It was recently proposed that it is not just the tension, which is an integral over the local pressure, but the whole distribution of the local pressure in the bilayer that is determining the functioning of membranes, including the functioning of proteins in membranes (Cantor, 1997). Niemela et al. (2007) recently studied large sized patches (1024 lipid molecules) of ternary mixtures of POPC/PSM/cholesterol by performing molecular dynamics runs for relatively long (for systems of this size) time periods of 100 ns. Two raft-like concentrations were studied in two systems A and B at concentrations having a 1:1:1 ratio and

2:1:1 ratio of POPC:PSM:Chol, respectively, and compared the results to the results from a simulation on system C at a non-raft concentration mixture of 62:1:1, where only few PSM and cholesterol molecules were present. It was observed that elastic properties and the local pressure profiles of mixtures A and B differed substantially from properties measured in the mixture C. Full implications of this difference for the functioning of peptides and proteins in membranes is still not clear, although experiments (Allende *et al.*, 2005) indicate that the transfer of peptide from the L_d domain to the L_o domain produces a change of 4–8 kT, thus substantially influencing the partitioning of peptides in membranes.

As we can see, while the issue of equilibration of membranes containing ternary mixtures is still not completely resolved, simulation of these mixtures has provided us with interesting observations and suggestions.

III. CONCLUDING REMARKS

Cholesterol plays an important role in the creation of liquid ordered domains in biological membranes. Experiments show that such domains can be created in synthetic membranes containing just three components: cholesterol and two phospholipids. What is the molecular detailed background that is responsible for the creation of these domains? Detailed molecular simulations of membranes containing cholesterol already produced a large number of observations and data, but since we want to extract at least a semi-quantitative, and, if possible, quantitative understanding of the processes occurring in these membranes, we may need even more data. For example, although a proposed idea of complexes that exists between cholesterol and certain lipids is becoming more popular (Zhang *et al.*, 2007), molecular simulations still did not confirm the existence of these complexes or show how to identify them. The quantitative information obtained from detailed simulations can be used to get, in a systematic way, coarse grained potentials for larger scale simulations, but it is also possible to use this information to extract parameters needed for phenomenological theories that explain domain formations in three component membranes (Komura *et al.*, 2004). Since these parameters may reflect the delicate balance of energy and entropy, the extraction of their values from detailed simulations may be a computationally difficult task, given the accuracy of force fields we use today. Altogether, the author feels that the task of understanding why cholesterol creates domains in membranes, even in model membranes, is still not accomplished. More work, including work in both simulation and theory, is needed.

Acknowledgements
This work was supported by the National Science Foundation under grant number MCB-0615469. The author would like to thank C. Davis for careful reading of the manuscript. It is the author's pleasure to acknowledge his collaboration on the described subject with A. Smondyrev, S. Pandit, D. Bostick, S. Bhide and Z. Zhang.

References

Aittoniemi, J., Niemela, P.S., Hyvonen, M.T., Karttunen, M., Vattulainen, I. (2007). Insight into the putative specific interactions between cholesterol, sphingomyelin, and palmitoyl-oleoyl phosphatidylcholine. *Biophys. J.* **92**, 1125–1137.

Allende, D., Simon, S.A., McIntosh, T.J. (2005). Melittin-induced bilayer leakage depends on lipid material properties: Evidence for toroidal pores. *Biophys. J.* **88**, 1828–1837.

Bagatolli, L.A. (2003). Direct observation of lipid domains in free standing bilayers: From simple to complex lipid mixtures. *Chem. Phys. Lipids* **122**, 137–145.

Bagatolli, L.A., Gratton, E. (2001). Direct observation of lipid domains in free-standing bilayers using two-photon excitation fluorescence microscopy. *J. Fluoresc.* **11**, 141–160.

Baird, B., Sheets, E.D., Holowka, D. (1999). How does the plasma membrane participate in cellular signaling by receptors for immunoglobulin E? *Biophys. Chem.* **82**, 109–119.

Berkowitz, M.L., Bostick, D.L., Pandit, S. (2006). Aqueous solutions next to phospholipid membrane surfaces: Insights from simulations. *Chem. Rev.* **106**, 1527–1539.

Bhide, S.Y., Zhang, Z.C., Berkowitz, M.L. (2007). Molecular dynamics simulations of SOPS and sphingomyelin bilayers containing cholesterol. *Biophys. J.* **92**, 1284–1295.

Brown, D.A., London, E. (1998a). Functions of lipid rafts in biological membranes. *Annu. Rev. Cell Dev. Biol.* **14**, 111–136.

Brown, D.A., London, E. (1998b). Structure and origin of ordered lipid domains in biological membranes. *J. Membr. Biol.* **164**, 103–114.

Burgos, P., Yuan, C.B., Viriot, M.L., Johnston, L.J. (2003). Two-color near-field fluorescence microscopy studies of microdomains ("Rafts") in model membranes. *Langmuir* **19**, 8002–8009.

Burns, A.R. (2003). Domain structure in model membrane bilayers investigated by simultaneous atomic force microscopy and fluorescence imaging. *Langmuir* **19**, 8358–8363.

Cantor, R.S. (1997). Lateral pressures in cell membranes: A mechanism for modulation of protein function. *J. Phys. Chem. B* **101**, 1723–1725.

Chiu, S.W., Jakobsson, E., Mashl, R.J., Scott, H.L. (2002). Cholesterol-induced modifications in lipid bilayers: A simulation study. *Biophys. J.* **83**, 1842–1853.

de Almeida, R.F.M., Fedorov, A., Prieto, M. (2003). Sphingomyelin/phosphatidylcholine/cholesterol phase diagram: Boundaries and composition of lipid rafts. *Biophys. J.* **85**, 2406–2416.

Devaux, P.F., Morris, R. (2004). Transmembrane asymmetry and lateral domains in biological membranes. *Traffic* **5**, 241–246.

Draber, P., Draberova, L., Kovarova, M., Halova, I., Tolar, P., Cerna, H., Boubelik, M. (2001). Lipid rafts and their role in signal transduction—Mast cells as a model. *Trends Glycosc. Glycotechnol.* **13**, 261–279.

Edholm, O., Nagle, J.F. (2005). Areas of molecules in membranes consisting of mixtures. *Biophys. J.* **89**, 1827–1832.

Epand, R.M., Epand, R.F. (2004). Non-raft forming sphingomyelin–cholesterol mixtures. *Chem. Phys. Lipids* **132**, 37–46.

Feigenson, G.W., Buboltz, J.T. (2001). Ternary phase diagram of dipalmitoyl-PC/dilauroyl-PC/cholesterol: Nanoscopic domain formation driven by cholesterol. *Biophys. J.* **80**, 2775–2788.

Frazier, M.L., Wright, J.R., Pokorny, A., Almeida, P.F.F. (2007). Investigation of domain formation in sphingomyelin/cholesterol/POPC mixtures by fluorescence resonance energy transfer and Monte Carlo simulations. *Biophys. J.* **92**, 2422–2433.

Frenkel, D., Smit, B. (2002). "Understanding Molecular Simulations. From Algorithms to Applications". Academic Press, San Diego.

Gandhavadi, M., Allende, D., Vidal, A., Simon, S.A., McIntosh, T.J. (2002). Structure, composition, and peptide binding properties of detergent soluble bilayers and detergent, resistant rafts. *Biophys. J.* **82**, 1469–1482.

Gaus, K., Gratton, E., Kable, E.P.W., Jones, A.S., Gelissen, I., Kritharides, L., Jessup, W. (2003). Visualizing lipid structure and raft domains in living cells with two-photon microscopy. *Proc. Natl. Acad. Sci. USA* **100**, 15554–15559.

Giocondi, M.C., Milhiet, P.E., Dosset, P., Le Grimellec, C. (2004). Use of cyclodextrin for AFM monitoring of model raft formation. *Biophys. J.* **86**, 861–869.

Gu, Y., Kar, T., Scheiner, S. (1999). Fundamental properties of the CH···O interaction: Is it a true hydrogen bond? *J. Am. Chem. Soc.* **121**, 9411–9422.

Hancock, J.F. (2006). Lipid rafts: Contentious only from simplistic standpoints. *Nat. Rev. Mol. Cell Biol.* **7**, 456–462.

Hofsass, C., Lindahl, E., Edholm, O. (2003). Molecular dynamics simulations of phospholipid bilayers with cholesterol. *Biophys. J.* **84**, 2192–2206.

Holopainen, J.M., Metso, A.J., Mattila, J.P., Jutila, A., Kinnunen, P.K.J. (2004). Evidence for the lack of a specific interaction between cholesterol and sphingomyelin. *Biophys. J.* **86**, 1510–1520.

Huang, J.Y., Feigenson, G.W. (1999). A microscopic interaction model of maximum solubility of cholesterol in lipid bilayers. *Biophys. J.* **76**, 2142–2157.

Ianoul, A., Burgos, P., Lu, Z., Taylor, R.S., Johnston, L.J. (2003). Phase separation in supported phospholipid bilayers visualized by near-field scanning optical microscopy in aqueous solution. *Langmuir* **19**, 9246–9254.

Ipsen, J.H., Karlstrom, G., Mouritsen, O.G., Wennerstrom, H., Zuckermann, M.J. (1987). Phase-equilibria in the phosphatidylcholine–cholesterol system. *Biochim. Biophys. Acta* **905**, 162–172.

Kahya, N., Scherfeld, D., Bacia, K., Schwille, P. (2004). Lipid domain formation and dynamics in giant unilamellar vesicles explored by fluorescence correlation spectroscopy. *J. Struct. Biol.* **147**, 77–89.

Keller, S.L., Radhakrishnan, A., McConnell, H.M. (2000). Saturated phospholipids with high melting temperatures form complexes with cholesterol in monolayers. *J. Phys. Chem. B* **104**, 7522–7527.

Khelashvili, G.A., Scott, H.L. (2004). Combined Monte Carlo and molecular dynamics simulation of hydrated 18:0 sphingomyelin–cholesterol lipid bilayers. *J. Chem. Phys.* **120**, 9841–9847.

Komura, S., Shirotori, H., Olmsted, P.D., Andelman, D. (2004). Lateral phase separation in mixtures of lipids and cholesterol. *Europhys. Lett.* **67**, 321–327.

Mayor, S., Rao, M. (2004). Rafts: Scale-dependent, active lipid organization at the cell surface. *Traffic* **5**, 231–240.

McConnell, H.M., Radhakrishnan, A. (2003). Condensed complexes of cholesterol and phospholipids. *Biochim. Biophys. Acta* **1610**, 159–173.

McConnell, H.M., Vrljic, M. (2003). Liquid–liquid immiscibility in membranes. *Ann. Rev. Biophys. Biomol. Struct.* **32**, 469–492.

Mitchell, D.C., Litman, B.J. (1998). Effect of cholesterol on molecular order and dynamics in highly polyunsaturated phospholipid bilayers. *Biophys. J.* **75**, 896–908.

Mombelli, E., Morris, R., Taylor, W., Fraternali, F. (2003). Hydrogen-bonding propensities of sphingomyelin in solution and in a bilayer assembly: A molecular dynamics study. *Biophys. J.* **84**, 1507–1517.

Mouritsen, O.G. (1991). Theoretical-models of phospholipid phase-transitions. *Chem. Phys. Lipids* **57**, 179–194.

Muller, M., Katsov, K., Schick, M. (2006). Biological and synthetic membranes: What can be learned from a coarse-grained description? *Phys. Rep. Rev. Sect. Phys. Lett.* **434**, 113–176.

Nicolini, C., Thiyagarajan, P., Winter, R. (2004). Small-scale composition fluctuations and microdomain formation in lipid raft models as revealed by small-angle neutron scattering. *Phys. Chem. Chem. Phys.* **6**, 5531–5534.

Nielsen, M., Miao, L., Ipsen, J.H., Zuckermann, M.J., Mouritsen, O.G. (1999). Off-lattice model for the phase behavior of lipid–cholesterol bilayers. *Phys. Rev. E* **59**, 5790–5803.

Niemela, P., Hyvonen, M.T., Vattulainen, I. (2004). Structure and dynamics of sphingomyelin bilayer: Insight gained through systematic comparison to phosphatidylcholine. *Biophys. J.* **87**, 2976–2989.

Niemela, P.S., Ollila, S., Hyvonen, M.T., Karttunen, M., Vattulainen, I. (2007). Assessing the nature of lipid raft membranes. *PLoS Comput. Biol.* **3**, 304–312.

Ohvo-Rekila, H., Ramstedt, B., Leppimaki, P., Slotte, J.P. (2002). Cholesterol interactions with phospholipids in membranes. *Prog. Lipid Res.* **41**, 66–97.

Pandit, S.A., Bostick, D., Berkowitz, M.L. (2003). Mixed bilayer containing dipalmitoylphosphatidyl-choline and dipalmitoylphosphatidylserine: Lipid complexation, ion binding, and electrostatics. *Biophys. J.* **85**, 3120–3131.

Pandit, S.A., Bostick, D., Berkowitz, M.L. (2004a). Complexation of phosphatidylcholine lipids with cholesterol. *Biophys. J.* **86**, 1345–1356.

Pandit, S.A., Jakobsson, E., Scott, H.L. (2004b). Simulation of the early stages of nano-domain forma-tion in mixed bilayers of sphingomyelin, cholesterol, and dioleoylphosphatidylcholine. *Biophys. J.* **87**, 3312–3322.

Pandit, S.A., Vasudevan, S., Chiu, S.W., Mashl, R.J., Jakobsson, E., Scott, H.L. (2004c). Sphingo-myelin–cholesterol domains in phospholipid membranes: Atomistic simulation. *Biophys. J.* **87**, 1092–1100.

Pasenkiewicz-Gierula, M., Rog, T., Kitamura, K., Kusumi, A. (2000). Cholesterol effects on the phos-phatidylcholine bilayer polar region: A molecular simulation study. *Biophys. J.* **78**, 1376–1389.

Pastor, R.W. (1994). Molecular-dynamics and Monte-Carlo simulations of lipid bilayers. *Curr. Opin. Struct. Biol.* **4**, 486–492.

Pike, L.J. (2006). Rafts defined: A report on the keystone symposium on lipid rafts and cell function. *J. Lipid Res.* **47**, 1597–1598.

Pitman, M.C., Suits, F., MacKerell, A.D., Feller, S.E. (2004). Molecular-level organization of sat-urated and polyunsaturated fatty acids in a phosphatidylcholine bilayer containing cholesterol. *Biochemistry* **43**, 15318–15328.

Radhakrishnan, A., McConnell, H. (2005). Condensed complexes in vesicles containing cholesterol and phospholipids. *Proc. Natl. Acad. Sci. USA* **102**, 12662–12666.

Rog, T., Pasenkiewicz-Gierula, M. (2001). Cholesterol effects on the phosphatidylcholine bilayer non-polar region: A molecular simulation study. *Biophys. J.* **81**, 2190–2202.

Rog, T., Pasenkiewicz-Gierula, M. (2006). Cholesterol-sphingomyelin interactions: A molecular dy-namics simulation study. *Biophys. J.* **91**, 3756–3767.

Silvius, J.R. (2003). Role of cholesterol in lipid raft formation: Lessons from lipid model systems. *Biochim. Biophys. Acta Biomembranes* **1610**, 174–183.

Simons, K., Ehehalt, R. (2002). Cholesterol, lipid rafts, and disease. *J. Clin. Invest.* **110**, 597–603.

Simons, K., Ikonen, E. (1997). Functional rafts in cell membranes. *Nature* **387**, 569–572.

Simons, K., Ikonen, E. (2000). How cells handle cholesterol. *Science* **290**, 1721–1726.

Simons, K., Toomre, D. (2000). Lipid rafts and signal transduction. *Nat. Rev. Mol. Cell Biol.* **1**, 31–39.

Simons, K., Vaz, W.L.C. (2004). Model systems, lipid rafts, and cell membranes. *Annu. Rev. Biophys. Biomol. Struct.* **33**, 269–295.

Smondyrev, A.M., Berkowitz, M.L. (1999a). Structure of dipalmitoylphosphatidylcholine/cholesterol bilayer at low and high cholesterol concentrations: Molecular dynamics simulation. *Biophys. J.* **77**, 2075–2089.

Smondyrev, A.M., Berkowitz, M.L. (1999b). United atom force field for phospholipid membranes: Constant pressure molecular dynamics simulation of dipalmitoylphosphatidylcholine/water sys-tem. *J. Comput. Chem.* **20**, 531–545.

Subczynski, W.K., Kusumi, A. (2003). Dynamics of raft molecules in the cell and artificial mem-branes: Approaches by pulse EPR spin labeling and single molecule optical microscopy. *Biochim. Biophys. Acta Biomembranes* **1610**, 231–243.

Tieleman, D.P., Marrink, S.J., Berendsen, H.J.C. (1997). A computer perspective of membranes: Molecular dynamics studies of lipid bilayer systems. *Biochim. Biophys. Acta Rev. Biomem-branes* **1331**, 235–270.

Tobias, D.J., Tu, K.C., Klein, M.L. (1997). Atomic-scale molecular dynamics simulations of lipid membranes. *Curr. Opin. Colloid Interface Sci.* **2**, 15–26.

Tu, K., Tobias, D.J., Klein, M.L. (1995). Constant pressure and temperature molecular dynamics sim-ulation of a fully hydrated liquid crystal phase dipalmitoylphosphatidylcholine bilayer. *Biophys. J.* **69**, 2558–2562.

Tu, K.C., Klein, M.L., Tobias, D.J. (1998). Constant-pressure molecular dynamics investigation of cholesterol effects in a dipalmitoylphosphatidylcholine bilayer. *Biophys. J.* **75**, 2147–2156.

Veatch, S.L., Keller, S.L. (2005a). Miscibility phase diagrams of giant vesicles containing sphingomyelin. *Phys. Rev. Lett.* **94**, 148101.

Veatch, S.L., Keller, S.L. (2005b). Seeing spots: Complex phase behavior in simple membranes. *Biochim. Biophys. Acta Mol. Cell Res.* **1746**, 172–185.

Venturoli, M., Sperotto, M.M., Kranenburg, M., Smit, B. (2006). Mesoscopic models of biological membranes. *Phys. Rep. Rev. Sect. Phys. Lett.* **437**, 1–54.

Yuan, C.B., Furlong, J., Burgos, P., Johnston, L.J. (2002). The size of lipid rafts: An atomic force microscopy study of ganglioside GM1 domains in sphingomyelin/DOPC/cholesterol membranes. *Biophys. J.* **82**, 2526–2535.

Zhang, Z., Bhide, S.Y., Berkowitz, M.L. (2007). Molecular dynamics simulations of bilayers containing mixtures of sphingomyelin with cholesterol and phosphatidylcholine with cholesterol. *J. Phys. Chem. B* **111**, 12888–12897.

Zhang, J.B., Cao, H.H., Regen, S.L. (2007). Cholesterol–phospholipid complexation in fluid bilayers as evidenced by nearest-neighbor recognition measurements. *Langmuir* **23**, 405–407.

CHAPTER 10

Atomistic and Mean Field Simulations of Lateral Organization in Membranes

Sagar A. Pandit[*]**, See-wing Chiu**[†]**, Eric Jakobsson**[†] **and H. Larry Scott**[‡]

[*]Department of Physics, University of South Florida, Tampa, FL 47907, USA
[†]Department of Molecular and Integrative Physiology, Department of Biochemistry,
UIUC programs in Biophysics, Neuroscience, and Bioengineering,
National Center for Supercomputing Applications, and Beckman Institute, University of Illinois,
Urbana, IL 61801, USA
[‡]Department of Biological, Chemical and Physical Sciences, Illinois Institute of Technology,
3101 S. Dearborn, Chicago, IL 60616, USA

I. INTRODUCTION

In the popular conceptual view of biomembranes, the Fluid Mosaic model, membrane proteins, sterols (such as cholesterol), and other biologically essential molecules reside in or on or penetrate an underlying lipid bilayer. In the years since its inception in 1972, the Fluid Mosaic cartoon has evolved, as a consequence of many detailed experiments, into a highly dynamic and complex structure. Within the plane of the lipid bilayer there is rapid lateral diffusion that promotes and is promoted by dynamical changes in structure that occur at a localized level, on a nanometer or sub-nanometer scale. Since a typ-

Current Topics in Membranes, Volume 60
Copyright © 2008, Elsevier Inc. All rights reserved
1063-5823/08 $35.00
DOI: 10.1016/S1063-5823(08)00010-0

ical membrane is made up of up perhaps a dozen different lipids, and also contains sterol and proteins, to dissect the underlying physical interactions is a formidable problem. One important example of biologically important structure emerging from the complex interactions within a heterogeneous lipid bilayer are lipid rafts. Rafts are nanometer sized regions that can spontaneously form and dissipate in membrane outer leaflets, and that are enriched in cholesterol and sphingolipids (Simons and Ikonen, 1997; Brown and London, 1998; Reitveld and Simons, 1998; Pralle *et al.*, 2000; Jacobson and Dietrich, 1999; Dietrich *et al.*, 2002). Rafts have been identified as important membrane structural components in signal transduction (Manes *et al.*, 1999; Aman and Ravichandran, 2001; Xavier *et al.*, 1998; Kawabuchi *et al.*, 2000) protein transport (Rozelle *et al.*, 2000; Cheong *et al.*, 1999; Viola *et al.*, 1999), and sorting of membrane components (Manie *et al.*, 2000; Harder *et al.*, 1998; Sönnichsen *et al.*, 2000). There is evidence for rafts functioning as sites for the binding and transport into the cell of pathogens and toxins, including the human immunodeficiency virus 1 (HIV-1) and the prion protein PrP^{sc} (Fantini *et al.*, 2002).

Lipid rafts are found in caveolae, nanoscopic flask-shaped invaginations common in a large family of cells. Caveolae are principal locations for endocytosis. Typical sizes of caveolae are 100–200 nm. The curvature radii are quite large: 30–50 nm for the neck and 10–15 nm for the central bag (Westermann *et al.*, 1999). Lipids found in caveolae generally consist primarily of sphingolipids, cholesterol, and gangliosides. In entothelial cells and fybroblasts, caveolae are coated with the protein caveolin which may affect the membrane curvature. GPI proteins tend to anchor in caveolae (Anderson, 1998). A recent volume reviews many aspects of the structure and function of caveolae, and their association with lipid rafts (Fielding, 2006). There is increasing evidence that nanoscopic lipid rafts in association with gangliosides, both in caveolae and outside caveolae, play roles in the biological activities of cells. In a recent review, Fantini *et al.* (2002) argue that lipid rafts in caveolae are preferential sites for the fusion of HIV-1 with a host cell, for the formation of the pathological prion protein PrP^{Sc}, and for the location of amyloid peptides linked to Alzheimers disease. Gellerman *et al.* (2005) have identified raft lipids as components of human extracellular amyloid fibrils associated with several diseases, suggesting a common mechanism for the formation of these deposits. Molander-Melin *et al.* (2005) have analyzed detergent-resistant membranes (DRMs) from the brains of Alzheimers diseased patients. they find an, in addition to raft lipids, an increased concentration of gangliosides G_{M1} and G_{M2}, compared to non-diseased tissues. The diseased tissues were also depleted in cholesterol. They suggest that G_{M1} and G_{M2} may disrupt membrane raft structure, with pathological consequences. Maguy *et al.* (2006) have reviewed the roles of lipid raft domains internal and external to caveolae in cardriac ion channel properties.

The high complexity of biological membranes precludes any easy characterization of the structure and composition of rafts, if and when they form. Consequently model membranes consisting of binary and ternary mixtures of lipids are useful to study in order to reveal the underlying mechanisms that lead to raft formation and control their stability. Keller and co-workers, and Feigenson and co-workers (Feigenson and Buboltz, 2001; Veatch and Keller, 2005) have shown through fluorescence imaging that in simple ternary mixtures of lipids complex and large domains can form under the right thermodynamic conditions. The domains are often micrometer sized, much larger rafts or domains in biological membranes. But their formation is likely driven by the same interactions that drive the assembly of rafts.

Simulation is a tool that can examine the interactions involved in mixed bilayers atomic detail, and thereby provide atomic level support for experiments. With accurate intermolecular force fields, atomistic simulations will produce trajectories that evolve the bilayer in time (Molecular Dynamics, or MD) or in configuration space (Monte Carlo, or MC). The trajectories provide a detailed data set of atomic coordinates and (in the case of MD) velocities that are a direct consequence of the force fields. At present MD trajectories of large lipid bilayers (250–1000 lipids plus waters of hydration) have been run for up to 100 ns (Lindahl and Edholm, 2000; Niemelä et al., 2007; Pandit et al., 2004b, 2004c) but this is still well short of the timescale needed for the simulated bilayer to undergo any large scale lateral reorganization. However, the interactions that drive large scale lateral reorganization can be observed on an interatomic scale early in the simulations, through analysis of radial distribution functions and other correlation functions. This type of analysis has led to important conceptual and quantitative insights into the driving forces behind lipid lateral organization in bilayers. The pictures that emerge are complex, and sometimes surprising.

In this Chapter will provide a synopsis and overview of MD simulations of mixed lipid bilayers. We will then show how the data from MD simulations can be incorporated into a larger scale, coarse grained model that predicts lateral organization and phase properties over times and distances well beyond the reach of MD. One of the most critical issues, the determination of the simulation force field parameters, will be discussed at length, describing new and highly accurate set of parameters our groups have developed. This section will complement the Chapter by Pastor, MacKerell, and Klauda. Their effort is based on the inclusion of all hydrogen atoms on the lipids as well as the water, whereas our effort uses a "united atom" approach for hydrogens on most lipid molecules (exceptions will be noted below). In Section IV we describe the MD simulation of lipid bilayers of mixed composition. In final Section V we describe how the MD data from the mixed bilayer simulations can be are used as input to construct a mean field-based coarse grained model that can serve as a predictive platform for the lateral organization of the membranes. This section complements the Chapter by Ayton and Voth, who describe a different scheme for reaching longer length and time scales from MD simulations.

II. MOLECULAR DYNAMICS SIMULATION OF HYDRATED LIPID BILAYERS

In molecular dynamics, snap shots of the system are generated by starting with initial positions and velocities of all the atoms in the system and integrating Newton's equations of motion. Typically the classical motion of atoms in the system are propagated according to forces obtained from a potential energy function of the form of Eq. (1):

$$V_{\text{tot}} = \sum_{\text{bonds}} K_b(r - r_0)^2 + \sum_{\text{angles}} K_\theta(\theta - \theta_0)^2 + \sum_{\text{impropers}} K_\Phi(\Phi - \Phi_0)^2$$

$$+ \sum_{\text{dihedrals}} K_\phi\big[1 + \cos(n\phi - \phi_0)\big]$$

$$+ \sum_{\text{non-bonded pairs}} \left\{ \frac{q_i q_j}{r_{ij}} + \left[\frac{C_{ij}^{(12)}}{r_{ij}^{12}} - \frac{C_{ij}^{(6)}}{r_{ij}^6} \right] \right\}. \tag{1}$$

The sum in Eq. (1) runs over (in order) bonds, bond angles, improper and proper dihedrals, and all pairs of atoms that are on different molecules or are separated by more than four interatomic bonds on a molecule, and ϕ is the angle between the planes of atoms 1–2 and atoms 3–4 of any group of 4 consecutively bonded atoms on a chain. The dihedral function in Eq. (1) represents the energy of a connected set of four consecutive atoms on a molecule. The parameters in each term in the sum must be determined as described in the next section. The negative gradient of this potential provides the force used in the solution of Newton's equations. For a system of N atom one has $6N - 3$ coupled equations for position and momentum variables, subtracting the center of mass coordinates. The equations are solved numerically using an algorithm such as that of Verlet (Frenkel and Smit, 2002).

In working with Eq. (1) several critical implicit assumptions are made. First of all, it is assumed that the dynamical time scales for electronic degrees of freedom are several orders of magnitude faster than the motion of the individual atomic nuclei. Then it is possible to approximate the effects of electronic motions by averages when deriving bonded and non-bonded interactions. Second, it is assumed that all the non-bonded interactions are represented by pairwise interactions rather than more complicated and computationally expensive many-body interactions. Third, classical expressions for intra and intermolecular interactions are used. In the final analysis the parameters in Eq. (1) must be tuned correctly to optimize the accuracy of the classical force expressions, as we discuss below.

Straightforward integration of the coupled, nonlinear atomic equations of motion, using the Verlet or an equivalent algorithm, gives rise to a simulation of a lipid bilayer under conditions of a fixed number of particles (N), a fixed volume (V), and a fixed energy (E). An NVE simulation therefore produce samples from the micro-canonical ensemble. To produce samples from different ensembles, the system can be coupled to a heat bath at constant temperature (NVT) ensemble and also can be coupled to pistons at constant temperature and pressure (NPT) ensemble. Such coupling are achieved by extending the phase space of the hydrated lipid bilayer to incorporate degrees of freedom corresponding to the scaling variables used to scale velocities and the simulation cell. Pressure and temperature coupling schemes are discussed in several textbooks, such as Frenkel and Smit (2002), and will not be elaborated on here.

There are a number of code packages available, some in the public domain, that can be employed to run an MD simulation of a lipid bilayer. Open source software includes GROMACS (http://www.gromacs.org) (Berendsen *et al.*, 1995; Lindahl *et al.*, 2001; van der Spoel *et al.*, 2001) and LAMMPS (http://lammps.sandia.gov/). Other popular MD simulation software includes NAMD (http://www.ks.uiuc.edu/Research/namd/) and the commercial software package CHARMM (http://www.charmm.org/). All software packages rely on Message Passing Interface (MPI) based parallelism so they can be easily installed and used on wide variety of parallel computers ranging from SMP to small and large scale clusters. Not all these codes are equally scalable on parallel computers, however. In general GROMACS and NAMD perform better in parallel environment than CHARMM. The choice of GROMACS and NAMD depends on the system size, the model used for electrostatics and personal preference. LAMMPS is a relatively new product that appears to have excellent scalability, although at this writing LAMMPS has not yet been tested in a lipid bilayer simulation.

Probably the single most important part of any simulation is the set of parameters that are used in the potential energy function (Eq. (1)). This set is collectively referred to as the "force field" for the simulation. Force fields vary according to the details of the basic underlying atoms in the simulation. In Section III we describe several procedures that are employed for the determination of a force field, based on the force field for use with the GROMACS code package. The chapter by MacKerell in this volume describes a complementary calculation of a force field based on the CHARMM code package.

III. GENERAL ISSUES IN FORCE FELD DEVELOPMENT FOR BIOMOLECULAR SIMULATIONS

Atomistically detailed molecular dynamics simulations are based on the approximation that molecules are collections of charged polarizable balls (atoms) connected by springs (bonds). This approximation permits the motions of molecules to be simulated by classical Newtonian mechanics, with great savings in computing time. In fact, of course, a molecule is comprised of atomic nuclei embedded in a cloud of electronic charge. The art/science of molecular dynamics force field construction is the assignment of charge, size, and polarizability to the atoms; and stretching, twisting, and bending coefficients to the springs, in such a way that the springs and balls behave as much as possible like real molecules.

The theoretical science component of force field construction is quantum chemistry. This involves solving the Shrödinger equations for the electronic distribution around a set of atomic nuclei as accurately as possible for various possible configurations of the molecule. Completely accurate solutions are not possible for a complex molecule, so the science/art of quantum chemistry lies in making such approximations that the accuracy is optimum for a given expenditure of compute power. Because the compute power necessary for a given accuracy increases rapidly as the size of the molecule, sometimes the optimum accuracy for a large molecule is achieved by doing the quantum chemistry for smaller molecules that are part of the larger molecule. Two aspects of force fields emerge from quantum chemistry: (1) the partial charges on the individual atoms, and (2) the properties of the bonds: energetic cost of stretch and torsion. Force fields may also be inferred from experiment. Indeed inference from experiment will trump computation from theory when the two are in conflict. This is because the theoretical results, as mentioned above, are only approximate, and fidelity to experimental reality is the final standard. In particular, the effective size of atoms, manifest in the force fields as the short range van der Waals repulsion, is typically largely determined by experimental data such as overall specific volume. In other cases the partial charges, initially determined by quantum chemistry, may be slightly modified to improve the fit to experimental data. For example, slight adjustments to partial charges of water molecules that were initially determined by quantum chemistry can improve the fit to important experimental properties such as self-diffusion coefficient and dielectric constant (Berendsen *et al.*, 1987).

For biomolecular simulations, the key to force field development that leads to added value for our understandings of these large and complex molecules, is that force fields that are derived from microscopic physics on the atomic scale can, through simulations, lead to enhanced understanding of the emergent properties of the larger systems.

A. United Atom vs. All Atom Force Fields for Membrane Molecules

Several groups have developed force fields for use in membrane simulations in the context of different molecular dynamics simulation packages. Since force fields are readily transferable between simulation packages, there has been a large amount of cross-fertilization among the force fields used in such packages as CHARMM, AMBER, and GROMACS. However there are two distinct classes of force fields, *united atom* and *All Atom*. In the united atom force fields, non polar hydrocarbon groups are represented as single carbon atoms. The charge on the hydrogen is taken into account only implicitly, by its effect on the long range attractive component of the van der Waals force. In the all atom force fields, each hydrogen atom is represented explicitly, as its own charge center. Seminal papers representing the two lines of development are (MacKerell *et al.*, 1998) for all atom force fields and (Smondyrev and Berowitz, 1999).

There is no disagreement or significant variation of use between different research groups with respect to the necessity for explicit polar hydrogens (where the magnitude of the partial charge on the hydrogen is a least a couple of tenths of an electronic charge) or for cyclic hydrocarbon groups. Explicit polar hydrogens are essential for capturing hydrogen bonding interactions. Explicit hydrogens are essential for cyclic groups, even if the partial charges are quite small, in order to provide for the important pi-electron effect, in which ions or positively charged groups are attracted to the center of ringed carbon structures. The definitive demonstration that explicit hydrogens are necessary for cyclic groups comes from simulations of liquid benzene, in which it is found that accurate representation of properties require either explicit hydrogens or an additional radial quadrupolar term (M. Clark, personal communication). One of the major implications of this result for membrane simulations is that ring structures in cholesterol and other molecules with cyclic hydrocarbon groups must be represented by explicit hydrogens. On the other hand, it seems that united atom models for liquid linear hydrocarbons describe the volumetric and thermodynamic properties very well (Chiu *et al.*, 1995). Thus it seems justified to use united atoms for hydrophobic interiors of bilayer membranes, which are comprised of linear hydrocarbons. A conceptual issue with respect to this is that the dielectric constant of liquid hydrocarbon is approximately 2 rather than 1, implying some non-zero charge separation. Since the charge on each carbon group in a unit atom model is zero, the non-zero charge is implicitly represented in the attractive component of the van der Waals force (whose physical basis is dipole–dipole interactions) rather than explicitly represented by charged hydrogens and carbons.

B. Some History of United Atom Force Fields for Membrane Molecules

For initial simulations on lipid bilayer membranes Chiu et al. performed quantum calculations to determine the partial charges on the phosphatidylcholine head groups (Chiu *et al.*, 1995). They used these partial charges in conjunction with what were at the time default values for van der Waals parameters in the GROMOS (subsequently evolved to GROMACS) molecular dynamics package. These partial charges have since been used in many other simulations of PC membranes, either exactly or with minor variations. However, Berger *et al.* (1997) pointed out that the standard GROMOS hydrocarbon parameters we had used resulted in the volume of the hydrocarbon interior being too small. They developed new parameters based on fitting volumetric and thermodynamic data for liquid pentadecane, as an analogue for the 16-carbon chains of the dipalmitoylphosphatidylcholine (DPPC) phospholipid molecule and showed that this modification, combined with our head group charges, provided a very good fit to experimental data on the DPPC bilayers.

A further stage in the evolution of the force fields came from our realization that fitting hydrocarbon parameters to a single length hydrocarbon chain left the parameters underdetermined. This is because there are multiple combinations of methyl and methylene specific volumes that will result in the correct specific volume for a linear hydrocarbon of any specific length, such as pentadecane. Consequently we developed hydrocarbon force fields that provided an excellent fit for volume and heat of vaporization over a wide range of hydrocarbon lengths, Chiu et al. (1999, 2003), and have used these with slight refinement for all our subsequent bilayer simulations.

Petrache *et al.* (1997) then presented an additional factor useful in validation of membrane force fields, namely the comparison of specific volume of the lipids in membrane simulations with volumes inferred from X-ray diffraction data of multilamellar preparations. Armen *et al.* (1998) refined the analysis by breaking it down into components of lipids. From this perspective we found that in simulations using the above improved hydrocarbon chain parameters, with the right specific volumes for the hydrocarbons and the right partial charges for the head groups, the specific volumes of the lipids were still too small. At this point we were still using the van der Waals parameters and torsion angles for the head groups from the GROMOS96 43A1 parameter set, but had re-parameterized the hydrocarbon tails and the head group partial charges. The reparameterization of the head group torsion angle and van der Waals parameters, which we describe briefly below and will be described in detail in a manuscript in preparation, has led to creation of our 43A1-S-3 Membrane Force Field, which provides excellent correspondence with experiment for a range of simulation conditions, including correct specific volumes. The reparameterization was based on ab initio calculations and MD simulations of three smaller model compounds that each represent part of the head group atom types; ethyltrimethylammonium to represent the choline group, dimethyl phosphate to represent the phosphate group, and methyl acetate to represent the ester group. Specifically the ab initio calculations were used to reparameterize the torsion angles. Then MD simulations at constant pressure were performed to reparameterize the van der Waals parameters. The single change that perhaps had the most effect on the specific volumes was an increase in the van der Waals radius of the oxygens, inferred from specific volumes of methyl acetate as seen in molecular dynamics simulations and compared to experiment.

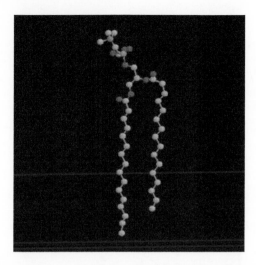

FIGURE 1   Stick model of DPPC with CH_2 and CH_3 united atoms (gray) oxygen (red), phosphorous (orange) and nitrogen (blue). (For interpretation of the references to color in this figure legend, the reader is referred to the web version of this book.)

C. Head Group Torsion Angle Determination for 43A1-S-3 Membrane Force Field

The PC head group consists of choline and phosphate fragments as well as two ester groups which link the head group and the hydrocarbon tails, as illustrated in Fig. 1.

Figure 2 illustrates the molecular fragments used for modeling the dihedrals for these fragments. Partial charges assigned to the atoms of these fragments and the dihedrals in interest are also displayed in the figure. Here we describe the model compounds and outline how torsion angle parameters were obtained.

1. Choline Group: Ethyltrimethylammonium

Torsional parameters for the dihedral at the choline group were fitted to the corresponding ab initio torsional profile for $[(CH_3)_3NCH_2CH_3]^+$. They were in turn accepted in modeling the torsional parameters for the adjacent dihedral $N–CH_2–CH_2–O$ of the choline fragment Fig. 1.

2. Phosphate Group: Dimethyl Phosphate

The dihedrals of the phosphodiester group, $CH_2–O–P–O$ and $C–O–P–O$ were modeled using dimethyl phosphate (DMP) as the model system. The potential energy surface of DMP as a function of the two P–O single bonds is quite complex, with local minima at the gauche–gauche, trans–gauche, and trans–trans conformations. To simplify the situation, we kept one P–O torsion in the trans conformation

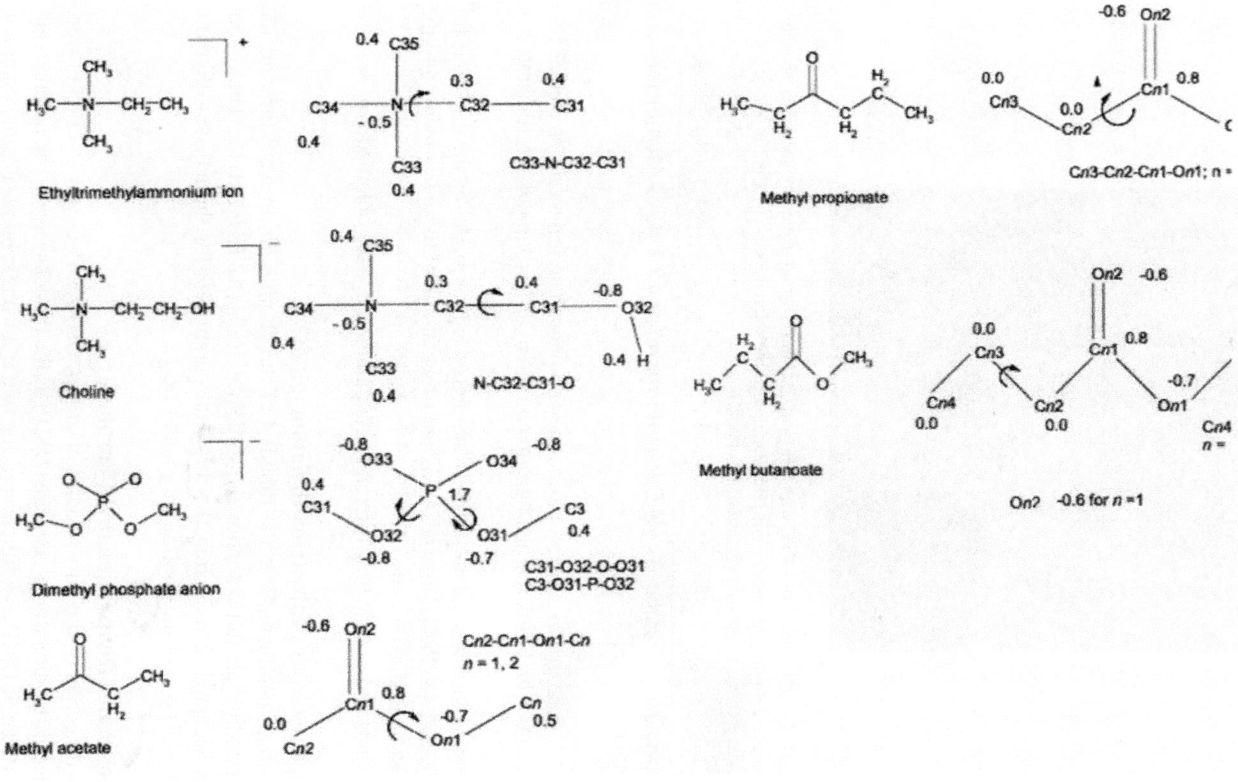

FIGURE 2 Molecular fragments used in simulations and in quantum calculations for the determination of united atom force field parameters.

initially and calculated the potential energy surface. Using the torsional parameters X–P–OA–X from the GROMOS96 43A1 parameter set, the molecular mechanics torsional profile matched the ab initio torsional profile.

3. Ester Group Region: Methyl Acetate

Linking the head group and the hydrocarbon tails are two ester groups in PC lipid (Fig 1). Methyl acetate CH_3COOCH_3 was used to model the methyl acetate fragments (C_{12}–$C_{11}(O_{12})$–O–C_1/C_{22}–$C_{21}(O_{22})$–O_{21}–C_2) in optimizing the torsional parameters for the dihedrals C_{12}–C_{11}–O_{11}–C_1 and C_{22}–C_{21}–O_{21}–C_2. These optimized torsional parameters were then accepted in parameterizing the torsional parameters of the dihedrals C_{n3}–C_{n2}–C_{n1}–O_{n1} ($n = 1, 2$), _7/_8, using methyl propionate as a model system. Then the torsional parameters for the dihedrals C_{n4}–C_{n3}–C_{n2}–C_{n1} ($n = 1, 2$), of the methylbutanoate fragment were parameterized to reproduce the corresponding ab initio torsional profiles. Finally, the torsional parameters for the dihedrals, $C_{(n+1)}$–C_n–O_{n1}–C_{n1} ($n = 1, 2$), of the ethyl acetate fragment were optimized to fit the quantum mechanically calculated.

D. Validation of the 43A1-S-3 Membrane Force Field

As always, the ultimate proof of the quality of the forcefield lies in comparison with experimental data. We have recently extended simulations of DPPC, DOPC and POPC. Figure 3 shows X-ray form factors calculated from simulation, and compared with experimental data from the group of Nagle et al. (2000a, 2000b), Greenwood et al. (2006), Petrache et al. (1997). As can be seen, the agreement is quite good.

Structural properties of DPPC, palmitoyl-oleoyl phosphatidylcholine (POPC) and di-oleoyl phosphatidylcholine (DOPC) bilayers can be calculated for comparison with experiment. Table I shows various structural properties of the bilayers calculated in simulations. The total volume per lipid (V_l) is computed by summing up the headgroup + gycerol backbone volume (V_{HG}) and the hydrocarbon volume (V_c). These volumes are computed by the method proposed by Petrache et al. (1997). In this method the system is divided into n_s slabs and the partial specific volumes of the atom types (v_i) in the system are obtained by imposing conservation of total volume for each slice element through minimization of the function

$$F(v_i) = \sum_{z_j}^{n_s} \left(1 - \sum_{i=1}^{3} n_i(z_j)v_i\right)^2 \qquad (2)$$

over v_i, where $n_i(z_j)$ is the number density of ith type of molecule in the slice at z_j. The computed volumes of DPPC, POPC and DOPC headgroup, hydrocarbon core, and total lipid volumes agree well with the experimental volumes (Kucerka et al., 2005). In all the cases the difference is less than the volume

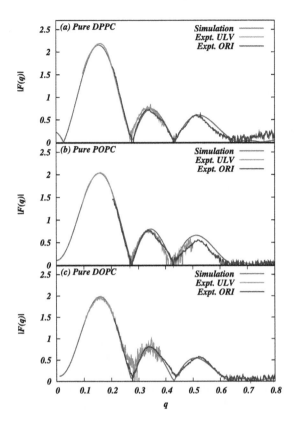

FIGURE 3 Comparison of experimental and simulation X-ray form factors for DPPC, POPC, and DOPC. Experimental data from (Greenwood *et al.*, 2006; Petrache *et al.*, 1997). Reprinted from Pandit *et al.* (2008) with permission.

of one water molecule. The area per lipid in the table is

$$A_l = \frac{V_c}{D_c}, \tag{3}$$

where D_c is the thickness of the hydrocarbon core (Nagle and Tristram-Nagle, 2000a, 2000b). In the simulation the hydrocarbon thickness, D_c, is calculated by determining the half value of the probability distribution of the hydrocarbon core (Nagle and Tristram-Nagle, 2000a, 2000b). The hydrocarbon thickness determined by this method is in good agreement with the experimental values (see Table I).

In summary, the 43A1-S-3 force field when used to simulate three different phospholipids, yielded results for properties of those lipids that are in excellent agreement with experiment. All calculated parameters for the 43A1-S-3 membrane force field are available by download from the web site: http://www.nanoconductor.org. A full description of the 43A1-S-3 forcefield will

TABLE I

Structural Quantities for Hydrated DPPC, POPC, and DOPC Bilayers

	DPPC MD	DPPC Expt	POPC MD	POPC Expt	DOPC MD	DOPC Expt
Temperature (K)	323	323	303	303	303	303
V_l (Å^3)	1212.8	1228.5	1241.8	1256	1269.7	1303
V_c (Å^3)	869.0	895.6[a]	911.4	924.2	949.3	971
V_{HG} (Å^3)	343.8	332.9[a]	330.4	331	320.4	331
V_{CH_2} (Å^3)	27.1	–	27.1	27.6	27.0	27.6
V_{CH_3} (Å^3)	54.2	–	54.2	53.6	54.0	53.6
$V_{CH=CH}$ (Å^3)	–	–	43.4	44.2	43.2	44.2
$2D_c$ (Å)	27.0	27.9	27.4	27.1	27.6	26.8
D_{HH} (Å)	36.7	37.8	36.6	37.0	37.2	36.7
A_l (Å^2)	64.3	64.2	66.5	68.3	68.8	72.5
A_l^g (Å^2)	61.9	–	63.0	–	64.0	–

[a] V_c is calculated as $A_l \times D_c$ from reported values and $V_{HG} = V_l - V_c$. Experimental data are from (Kucerka et al., 2005).

be published elsewhere. We can now proceed with confidence to the modeling of more complex lipid bilayers of mixed composition.

IV. SIMULATION OF HETEROGENEOUS LIPID MEMBRANES

The first and simplest type of heterogeneous membrane that has been studied by simulations is also one with considerable biological importance: bilayers composed of a single species of lipid and a varying concentration cholesterol. Lipid–cholesterol mixtures have been studied by MC and MD simulation for over twenty years by several groups including (Scott, 1993; Tu et al., 1998; Pandit et al., 2004a; Edholm and Nagle, 2005; Chiu et al., 2001). The work on lipid-cholesterol bilayer simulations prior to 2002 has been reviewed by (Scott, 2002). One comprehensive set of DPPC–cholesterol simulations was run in 2002 by (Chiu et al., 2002) This suite of calculations used Configurational-bias Monte Carlo and Molecular Dynamics simulations for bilayers of dipalmitoylphosphatidylcholine (DPPC) and cholesterol for DPPC:cholesterol ratios of 24:1, 47:3, 23:2, 8:1, 7:1, 4:1, 3:1, 2;1, and 1:1, using 5 nanosecond (ns) MD runs and interspersed Configurational Bias Monte Carlo to ensure equilibration. As Fig. 4 shows, for cholesterol concentrations above 12.5% the area per molecule of the heterogeneous membrane varied linearly with cholesterol fraction. From the intercept of the area/molecule line it was found that DPPC has an average cross-sectional area per lipid of 52 square angstroms. The area per cholesterol, found

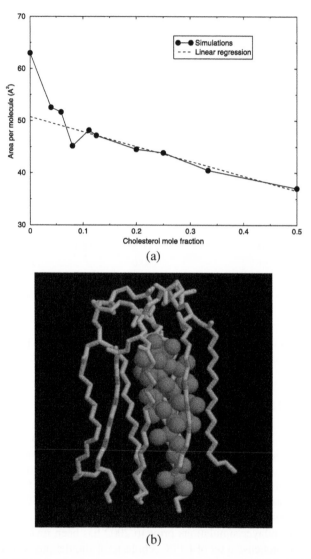

(a)

(b)

FIGURE 4 (a) Plots of molecular areas versus cholesterol concentration for DPPC–cholesterol simulations. Reprinted from Chiu *et al.* (2002) with permission. (b) Snapshot of one cholesterol molecule interacting with neighboring DPPC molecules, illustrating "wetting" of cholesterol by DPPC chains.

from the slope of the line, is surprisingly small, ≈ 24 Å^2 per molecule. The low area of cholesterol is a consequence of the ability of the flexible DPPC chains to back closely around the cholesterol rings. Figure 4 shows a snapshot that illustrates the "wetting" of cholesterol by the DPPC chains.

Analysis of the lateral distribution of cholesterol molecules in the bilayer reveals a tendency for small subunits of one lipid plus one cholesterol, hydrogen bonded together, to act as one composite particle, and perhaps to aggregate with other composites. The conclusions drawn from simulations were consistent with experimentally observed effects of cholesterol, including the condensation effect of cholesterol in phospholipid monolayers and the tendency of cholesterol-rich domains to form in cholesterol-lipid bilayers. This work was extended to mixtures of cholesterol with POPC and DOPC. Pandit *et al.* (2008) using the 43A1-S-3 force fields for lipids described in Section III.

Figure 5 shows partial molecular volumes and areas calculated from the simulations for DPPC, DOPC, and POPC as functions of cholesterol concentration.

The method for extracting the partial molecular areas proposed by Edholm and Nagle (2005) was applied, based on the partial specific area relation:

$$a(x) = \frac{A_{\text{box}}}{(N_{\text{l}} + N_{\text{ch}})} = (1 - x)A_{\text{l}} + xA_{\text{ch}}.$$

A plot of $a(x)/(1-x)$ as a function of $x/(1-x)$ for all simulated concentrations then yields A_{ch}, the cholesterol partial specific area.

The partial specific volume was calculated using the method described above (Petrache *et al.*, 1997). Unlike the volume curves (Fig. 5a), the area curves (Fig. 5b) are not linear. In fact for cholesterol concentrations cholesterol is negative. Negative partial molecular areas are interpreted by Edholm and Nagle (2005) as a manifestation of the condensation effect of cholesterol on surrounding lipids. The curves in Fig. 5b are almost linear above 20% cholesterol concentrations. The partial molecular areas of cholesterol and lipids are listed in Table I. The computed cholesterol partial molecular area is largest in saturated DPPC and smallest in POPC where one chain is saturated and the other chain is unsaturated. Again with both unsaturated chains, like in DOPC, the area per cholesterol is higher. This implies the important new insight that in POPC bilayers, cholesterol admits different packing structure as compared to the mixtures with fully saturated DPPC and unsaturated DOPC.

For a more quantitative analysis of sub-molecular interactions, it is useful to consider radial distribution functions between the C9–C10 double bond group on the Sn-2 chains of POPC and DOPC, and the CH_3 group on the methylated (β) face of cholesterol. Figure 6 shows these rdfs for POPC and DOPC molecules. They observed that there are two peaks in the rdf. The first peak is around ~4.5 Å from the methyl group and the second peak is around ~7 Å from the methyl group. The two peaks correspond to two closest positions for CH=CH group to the nearest β- and α-faces respectively. In Fig. 6, in the case of DOPC both the peaks are approximately same height whereas for POPC the first peak is taller than the second peak. This shows that there is a higher correlation of POPC CH=CH groups with the β-face of cholesterol, compared to DOPC.

Sphingomyelin (SM) is an important component of the outer leaflet of mammalian cell membranes, yet much less is known about SM structure in bilayers

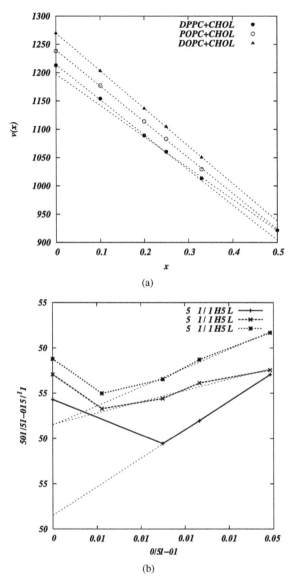

FIGURE 5 (a) Plots of partial specific volumes (a) and areas (b) versus cholesterol concentration. Reprinted from Pandit *et al.* (2008) with permission.

than is known about phospholipids. Do begin to address this issue, Chiu and co-workers constructed a fully hydrated bilayer of (18:0) SM. The size of this system exceeded any lipid bilayer simulation reported to date: 1600 SM molecules and

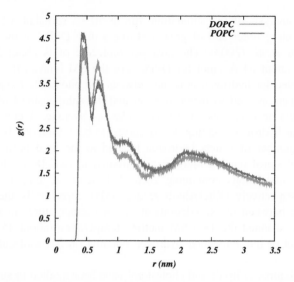

FIGURE 6 Plots of radial distribution functions between the C_9–C_{10} double bond group on the Sn-2 chains of POPC and DOPC, and the CH_3 group on the methylated (β) face of cholesterol. Reprinted from Pandit *et al.* (2008) with permission.

FIGURE 7 Snapshot of a large bilayer of 18:0 sphingomyelin after 20 ns of MD simulation.

51,200 water molecules. Figure 7 shows a snapshot of this bilayer after 20 ns of MD simulation

Structural analysis of the simulation, run at 50 °C, revealed a bilayer with significantly more order and of greater stiffness than DPPC at the same temperature Chiu *et al.* (2003). The area per molecule of is about 52 Å²/mol compared to about 64 Å²/mol for DPPC. For the SM bilayer, there is significant intramolecular hydrogen bonding between the phosphate oxygen and the amide hydrogen, as well as between water and the sphingosine hydroxyl moiety, that occurs on over 75% of the molecules, on average. A consequence of intramolecular hydrogen bonding was observed to be a reduction by one molecule of the number of waters hydrogen bonded to the lipid headgroups. The effect of cholesterol on SM bilayers was studied in two 18:0 sphingomyelin lipid bilayer simulations containing 20% cholesterol, at temperatures of 50 and 20 °C, respectively (Khelashvili *et al.*, 2005). Structurally there was little difference between the simulations at the two temperatures in spite of the fact that they spanned the 18:0 SM melting temperature (about 45 °C). About the same level of intramolecular hydrogen bonding on SM molecules was observed.

Ternary mixtures of lipids and cholesterol have been studied in simulation by Pandit et al. (2004b, 2004c), and most recently by Niemelă *et al.* (2007). In two sets of simulations, Pandit *et al.* examined the structure of lipid bilayers containing mixtures of 18:0 SM, DOPC, and cholesterol. In the first case, they constructed a large bilayer consisting of 1424 molecules of DOPC, 266 molecules of 18:0 SM, 122 molecules of cholesterol, and 62,561 water molecules. The simulations were run for 100 ns each. This study provides pressure profiles and order profiles that show the rigidity of the bilayers, possibly a consequence of the use of POPC instead of DOPC as the phospholipid. The lateral distribution of chain order is indicative of the beginnings of a lateral organization process. These profiles are similar to ones that can be constructed by Self Consistent Mean Field modeling as we describe below. [Niemelă *et al.* ran large scale simulations of lipid bilayers consisting of 1024 lipids in 1:1:1 and 2:1:1 POPC:16:0SM:cholesterol.]

A set of ternary mixture simulations of smaller bilayers (Pandit *et al.*, 2004a) consisted of 100 DOPC, 100 SM, and 100 cholesterol molecules plus 9600 waters. These simulations were started from random distributions of DOPC, SM, and cholesterol. The simulations were run for 250 ns with the goal of determining the interactions, structural, and dynamical parameters that drive the formation of domains. As a control, a simulation of a binary system consisting of 100 SM plus 100 DOPC, with no cholesterol. Figure 8 shows before and after snapshots of the ternary system, and a 200 ns snapshot of the binary system. We have performed the same series of structural calculations on the random mixtures as we did on the single-domain simulation described in the preceding paragraph. One of the important findings of this work, revealed in quantitative analysis of the lateral molecular area correlations, is that, on the 200 ns timescale, the presence of cholesterol enhances the formation of segregated domains rich in either SM or DOPC.

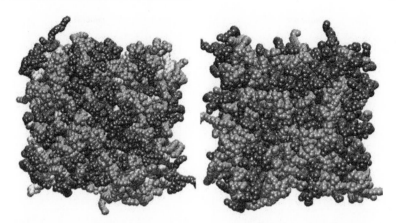

FIGURE 8 (Left) Snapshots of ternary DOPC–SM–cholesterol left: initial states (after 4 ns of equilibration). (Right) After 200 ns; far right: binary system after 200 ns. Color code is: Black: DOPC, orange: SM, yellow: cholesterol. Reprinted from Pandit *et al.* (2004a) with permission.

After a more detailed analysis of cholesterol–SM and cholesterol–DOPC interactions and radial distribution functions, it was found that the rough methylated face of cholesterol is preferentially oriented towards DOPC molecules, while the smooth face is preferentially orientated towards SM molecules. This is consistent with the findings for DOPC and POPC–cholesterol bilayers described above. For the 250 ns runs, a sufficient statistical data base of conformations was present to allow the construction of two-dimensional radial distribution functions in both distance in the membrane plane, and the plane polar angle ϕ:

$$g(r, \varphi) = \frac{N(r, \varphi)}{2\pi r \rho \delta r \delta \varphi},$$

Figure 9 shows two-dimensional radial distribution function between cholesterol and SM, with the orientation of the cholesterol taken into account. Darker color indicates stronger correlation as reflected through larger $g(r, \phi)$. Cholesterol is represented by the **T** figure, with the horizontal appendage as the methylated rough face. The color distribution shows a clear preference of SM for the smooth face of cholesterol. By default a similar calculation for DOPC–cholesterol shows a slight preponderance of DOPC nearer to the rough face of cholesterol molecules.

In summary, simulations of heterogeneous membranes have provided a number of important insights into interactions between lipids and cholesterol:

- Simulations have observed the change from fluid to liquid ordered structure in DPPC as cholesterol concentration increases.
- Simulations have shown that the partial specific area of cholesterol in DPPC is less in all cases than the area of cholesterol in crystals of pure cholesterol. POPC bilayers minimize the partial specific area of cholesterol.

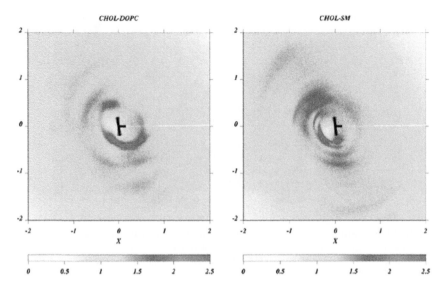

FIGURE 9 (Left) Density plot of $g(r, \phi)$ on the XY plane of the cholesterol body coordinate system at 200 ns. A top view of the cholesterol molecule reference frame is schematically shown as black T. The radial distribution function was calculated between O atoms of cholesterol and the backbone CH_1 moiety of SM (and also DOPC). Left: DOPC–cholesterol; right: SM–cholesterol. Reprinted from Pandit *et al.* (2004b) with permission.

- The asymmetric structure of cholesterol plays an important role in lipid–cholesterol interactions, with the smooth face favoring contact with saturated lipid chains strongly over unsaturated chains.
- A consequencs of the above is that cholesterol seems to prefer to lie at an interface between lipids with saturated chains and lipids with monounsaturated chains, or, in the case of POPC, seems to invite an asymmetric packing that minimizes the cholesterol partial specific area.

The above are atomistic interactions that are observed in simulations of at most a few hundred ns. To verify the role of the interactions in the lateral organization of membranes, it is necessary to extend the simulation predictions to significantly larger scales. We describe such an effort in the following section.

V. APPLICATION: A MEAN FIELD BASED LIPID BILAYER MODEL BASED ON MDSIMULATIONS

One way of achieve a multi-millisecond time scale simulations is to replace atomic level MD with a coarse-grained model that uses Langevin Dynamics (LD) for the time propagation. The LD algorithm requires a free energy functional that drives, along with a stochastic component, the evolution of the model under simulation. The free energy function in general will depend on local variables that are related to thermodynamic properties. In turn, the free energy has its origin in the atomistic properties of the system under simulation. If the molecular properties used in the model free energy functional that drives the LD approach have their basis in atomistic simulations then this approach represents an extention of MD to larger scales. An MD-based LD methodology was the basis for an effort to model mixed lipid bilayers to mixtures of DPPC and cholesterol Khelashvili *et al.* (2005). It was then extended to ternary mixtures using interaction functions and coupling constants calculated from the MD simulations, as described by us in Pandit *et al.* (2007). In this section we describe the model, the underlying assumptions, the interplay with MD simulations, and the results of this effort.

The Langevin modeling as applied to a lipid bilayer is based on a Ginzburg–Landau approach to the study of coexisting phases Lubensky and Chaikin (1995). The Langevin approach calculates temporal and spatial evolution of an order parameter, that characterizes the phase transition, s, and a free energy functional $F(s, t)$. The order parameter s can be, for example, the difference in densities of liquid and vapor phases, or the spontaneous magnetization per particle in simple systems. For application to a lipid bilayer the order parameter is defined as the average chain order parameter field:

$$S(\vec{r}) = \frac{1}{N} \sum_n S_n(\vec{r}), \tag{4}$$

where $S_n(\vec{r})$ is the C–H order parameter at carbon n for the chain at position $\vec{r}$, and N is the total number of carbons for which S_n is calculated in the chains of the lipids. Furthermore, the system contains discrete objects (in our case, cholesterol molecules), each of which has an orientation θ and a position $\vec{r}$. A Langevin dynamics algorithm drives the cholesterol degrees of freedom in the model according to Peng *et al.* (2000):

$$\frac{d\vec{r}_i}{dt} = -M \frac{dF}{d\vec{r}_i} + \vec{\eta}_i, \tag{5}$$

$$\frac{d\theta_i}{dt} = -M' \frac{dF}{d\theta_i} + \zeta_i, \tag{6}$$

where M and M' are mobilities, and η_i and ζ_i are thermal fluctuations, which are derived from an appropriate Gaussian distribution.

In the adaptation of this methodology to a lipid bilayer a leaflet of the bilayer is mapped onto a large (at least $\sim 100 \times 100$) two dimensional lattice that contains lipid chains plus cholesterols. The lipid chains occupy points on the lattice, while the cholesterols are modeled as small hard rods that move continuously over the plane. The key to modeling the properties of this model system is the

Free Energy functional. In typical modeling work Balazs *et al.* (2000), Peng *et al.* (2000) a generic Ginzburg–Landau free energy functional is used, to which are added interaction functions between the rods and the surrounding fluid. In order to specifically target this approach to lipid bilayers, However for a lipid bilayer a better choice is the mean field functional proposed by Marčelja (1974). The basic hypothesis of the Marčelja approach is that the interaction energy between two neighboring lipid chains may be approximated as being proportional to the product of their respective average order parameters times a coupling constant, or equivalently,

$$H = - \sum_{\langle ij \rangle} V_0 S_i S_j - \sum_{\alpha} \sum_{\alpha_i} V_{lc} s_{\alpha_i} - \sum_{\langle ij \rangle} V_{cc}, \tag{7}$$

where V_0 is a lipid–lipid coupling constant obtained from MD tuned to achieve correct phase transition temperature, second sum is over all the cholesterols and over all the chains that are in the neighborhood of cholesterol. The determination of V_{lc} is method on data from MD simulations of DPPC–cholesterol mixtures (Chiu *et al.*, 2002; Khelashvili *et al.*, 2005). The V_{cc} term corresponds to repulsion between two neighboring cholesterols, and can be modeled as a simple excluded volume contribution. The bilinear form of Eq. (7) has its origin in the Maier–Saupe Theory of liquid crystals (Maier and Saupe, 1959), and is amenable to a mean-field decomposition.

In order to calculate the free energy within mean field theory it is necessary to solve a series of coupled self-consistent equations for the local order parameters, and this requires a partition function sum over all lipid chain conformational states. Marčelja accomplished this by generating the states, within a rotational-isomeric model. A better solution to the configuration sum problem is to use libraries of chain configurations generated in MD simulations of large hydrated lipid bilayers for the self-consistent MFT (SCMFT) calculation. It is important to note that, in contrast to Marčelja's original model the order parameter in this model is a local rather than a global variable. That is, it is necessary to solve the self-consistent equations at each site at each time step. This means that the order parameter field re-equilibrates after cholesterols move through the system according to Eqs. (5) and (6).

Once the parameters for the model are determined, and a chain conformation library is assembled, the SCMFT method proceeds as follows:

1. Initialize by setting simulation lattice size and cholesterol content. Randomly distribute cholesterol rods over lattice.
2. Solve self-consistent equations with cholesterols fixed positions. Calculate free energy function at all points on simulation lattice.
3. Move cholesterols one time step using Langevin equation with free energy from step 2.
4. Go to step 2, repeat for desired total number of time steps.

Using this approach have carried out SCMFT calculation for a pure DPPC bilayer. With a single phenomenological parameter, the lipid–lipid coupling constant, it was possible to reproduce the main DPPC phase transition at 42 °C for a system of 10000 lipid chains on a 100×100 lattice. At the phase transition temperature, the system has coexisting domains of ordered and fluid phases, as measured by the relative values of the chain order parameters at the different sites on the lattice. The interaction energies between DPPC and cholesterol were calculated from MD simulation data. The dimensionless mobilities used in Eqs. (5) and (6) are related to the experimental diffusion constant D by: $M_r = D/k_bT$, $M_\theta \approx (10 - 100) \times M_r$, and $\Gamma \approx 10M_r$. The dimensionless simulation timestep $\Delta\tau$ is related to the real-time step length Δt by $\Delta\tau = \frac{D}{a^2}\Delta t \approx 10^7 \Delta t$, where a is the lattice constant, set in preliminary work at 0.65 nm. This allows for nanosecond or greater real time step sizes with reasonable time step for the coarse grained simulation. For higher cholesterol concentration, smaller timesteps are needed but the self-consistent equations converge faster in this case, compensating for the shorter timestep.

MD simulations of SM:DOPC:cholesterol mixtures described above have shown that cholesterol has preference for saturated chain lipid towards its α-face (Pandit *et al.*, 2004a). This result was input into the SCMFT model by changing the form of the interaction between the jth cholesterol and ith lipid chain to:

$$V_{j,i}^{lc} = V_{lc}^c\left(1 - \Delta\sin(\theta_{j,i})\right)s_i, \tag{8}$$

where the coupling constant V_{lc}^c is still the same V_{lc} used in the isotropic cholesterol model, and the trigonometric term accounts for the increased ordering towards α-face and reduced ordering towards the β-face of cholesterol. The parameter Δ is a measure of the amount of anisotropy in the structure of the cholesterol molecule.

Using the asymmetric lipid–cholesterol coupling interaction, SCMFT simulations for DPPC + cholesterol bilayers at 5, 10, 15, 20, 25, 30, and 35% cholesterol concentration and over a wide range of temperatures from 303 to 328 K were performed. Fig. 10 is a plot of the average order parameter as a function of temperature for various cholesterol concentrations within this model. In the absence of chol, the model clearly exhibits a first order phase transition at 315 K. The order parameters at temperatures below and above phase transition are similar to the values expected in gel and liquid crystalline phases of DPPC, respectively. A closer examination of Fig. 10 reveals that with increasing cholesterol concentration, the phase transition diminishes and completely vanishes for concentrations above 15%. At concentrations above 25%, the system has uniform order across all the temperatures investigated here, representing the liquid ordered (β) phase. Similar results are reported by the NMR and DSC experiments of Huang *et al.* (1993). Figures reftransition show a comparison of average order parameters obtained using our model, and the NMR quadrupolar shift reported by Huang *et al.*

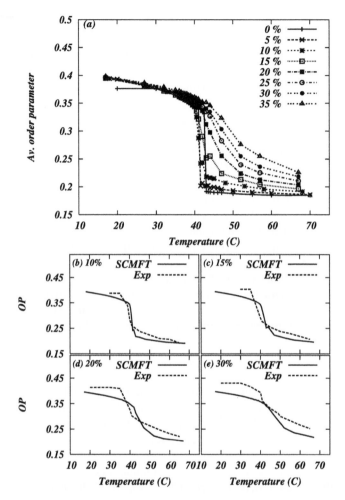

FIGURE 10   (a) Average order parameter as a function of temperature for various chol concentrations. Comparison of the average order parameter in the model with the order parameter of the sixth carbon atom reported by Huang *et al.* (1993) for (b) 10% chol, (c) 15% chol, (d) 20% chol, and (e) 30% chol. Reprinted from (Pandit *et al.*, 2007) with permission.

In their study, Huang *et al.* report the shift for sixth carbon atoms in DPPC hydrocarbon chains. Since the order parameter of the sixth atom is on the plateau region of the order parameter profile, we expect the average order parameter over the entire chain to be lower than the order parameter of the sixth atom. As Figs. 10b–10e show, this is indeed the case. The model average order parameters show remarkable agreement with the NMR data for various temperatures and cholesterol concentrations.

In order to quantify the existence of regions within the model field of different levels of chain order, a binning procedure was used to obtain the distribution of chain order over the lattice. It was found that for certain temperature and cholesterol concentration values, that the systems had coexisting regions of different order, as revealed by a bimodality in the distribution. This property to detect different regions of that correspond to localized high order and lo order nanodomains.

A plot that traces out the boundaries between regions of bimodal order and homogeneous regions is shown in Fig. 11 and is in excellent agreement with the phase diagram proposed by Vist and Davis (1990). In particular, both model and experiment exhibit two coexisting regions of different chain order at the same cholesterol concentrations and temperatures as the Vist–Davis diagram. At temperatures below 315 K, within the two dimensional field of chain order, there are separate regions of gel-like order and regions of intermediate order, between gel-like and fluid-like states. This is the same distribution of chain order identified by Vist and Davis as the *gel + β* region Vist and Davis (1990), where *β* signifies the "liquid ordered" region in which lipid chains are highly ordered but maintain liquid-like diffusion constants. At temperatures above 315 K, simulations predict separate regions of intermediate and low levels of fluid-like order, similar to the $L_\alpha + \beta$ region of Vist and Davis (1990). The model could not resolve the three phase region near the lipid melting temperature and low chol concentration observed by Vist and Davis. The agreement between our model diagram and the Vist–Davis diagram is especially important, since no parameters in the model were set to force it to match the experimental plot.

In summary, in spite of the rather severe degree of coarse graining involved in the SCMFT model, we are able to retain much of the relevant detail of the system. The model accurately predicts the phase behavior of the DPPC–cholesterol system, a level of modeling that is far beyond conventional MD simulation.

It is possible to extended this model to DOPC–SM mixtures. This is done by constructing a new composition field on the lattice, defined at each site $\vec{r}$, by:

$$\psi(\vec{r}) = \phi_{X_1}(\vec{r}) - \phi_{X_2}(\vec{r}), \tag{9}$$

where ϕ_x represents the number fraction of species x, at the point $\vec{r}$ and ψ is the local difference in concentration. We construct a Free Energy functional from a quadratic Hamiltonian, which includes the interactions of X_1–X_1, X_1–X_2, and X_2–X_2 along with interactions of X_1 and X_2 with cholesterol, all using inputs from the simulations we describe in Khelashvili *et al.* (2005). The functional contains coupling terms between different molecules, which are calculated from simulations, and are used to predict the lateral organization of membranes on the scale of the TDGL method. Since the new field ψ is a conservative field, it is evolved using Cahn–Hilliard equation

$$\frac{d\psi(\vec{r})}{dt} = -\Gamma \nabla^2 \frac{dF}{d\psi}.$$

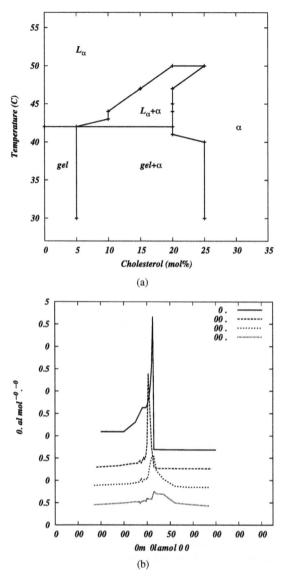

(a)

(b)

FIGURE 11   (a) Computed phase diagram from the model, (b) plots of specific heat as a function of temperature for several cholesterol concentrations, reprinted from (Pandit *et al.*, 2007) with permission.

These major issues have been resolved, and we can now simulate SM/DOPC mixtures to 50 *microseconds* or more. The next step in this modeling is to develop a robust method to estimate the coupling parameters from MD simulations of mix-

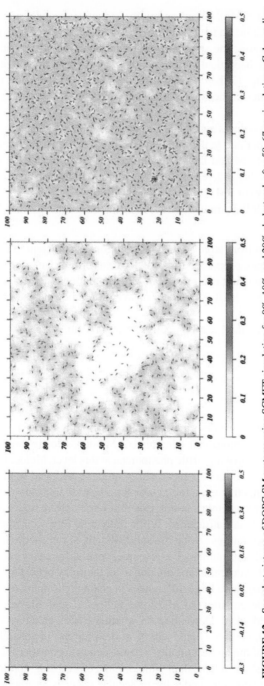

FIGURE 12 Snapshot pictures of DOPC:SM systems using SCMFT simulation for 0%, 10%, and 20% cholesterol, after 58–67 ns simulations. Color coding represents distribution of the chain order field, and small segments represent cholesterol molecules.

tures Pandit *et al.* (2004a). Based on rough estimates of model parameters, two systems, 3:1 DOPC:SM and 1:1 DOPC:SM, have been modeled thus far. We have observed in both cases, the spontaneous formation of domains, and we believe this is the first such theoretical observation for a model based on direct input from MD simulations. Fig. 12, below shows 'before-and-after' snapshots of the distribution of chain order in the two systems consisting of 10000 chains, for 58 and 67 μs, respectively.

In Fig. 12, chain order is quantified by color, with blue representing the highest degree of chain order and yellow the lowest. Initial states were formed by a random distribution of SM and DOPC. Both systems clearly show the evolution of small clusters of highly ordered chains representing domains consisting primarily (but not exclusively) of SM.

In summary, SCMFT simulations of heterogeneous membranes have shown that:

- Accurate input from atomistic MD simulations is sufficient to reproduce experimental phase properties for DPPC–cholesterol mixtures.
- The SCMFT method can propagate MD simulation data to time and length scales that are at least 2–3 orders of magnitude beyond the reach of brute force MD.
- The SCMFT method can be applied to ternary mixtures of two lipids plus cholesterol.

VI. SUMMARY

In this Chapter we have described modeling of lipid bilayers starting from the atomistic level, where molecular dynamics or Monte Carlo are the method capable of providing details of atomistic phenomena at this level. Molecular dynamics, must be applied with caution to successfully model a hydrated lipid bilayer, due to complex and competing interatomic interactions. Individual atoms may experience large forces from chemical bonds, dihedral forces, steric forces, electrostatic forces, and van der Waals attractions. While individual forces may be quite large, there are often cancellations so that the subtle details of the energy landscape are highly complex and non-intuitve. For example, small differences in each dihedral along a single lipid hydrocarbon chain, due to an interplay between various terms in Eq. (1) will propagate to large differences in chain conformations.

In general, the development of MD force fields for lipid bilayers is a process that cannot be overlooked. Depending on whether hydrogen atoms are to be included or excluded, the force fields will be quite different. In general, both types of force fields are of the highest quality when tested against experimental smaller "model compounds" in independent simulations before being applied to a lipid simulation. In this vein, the model for molecular water is an important consideration. We described in detail the determination of force field parameters in the United Atom Approximation in this Chapter.

While atomistic simulations are of great interest and provide many new insights, they are limited by the relatively slow rate of diffusion of lipids in a bilayer to describing the bilayer over time scales where individual molecules do not move very far laterally. However, atomistic simulations provide clues for larger scale thermodynamic properties of bilayers in the local interactions. We described efforts to use input from MD simulations to construct models based on self consistent mean field theory that can describe larger scale lateral organizational, development, over time and across the membrane. Self Consistent Mean Field Theory, based on models with parameters calculated from atomistic simulations has, as we illustrated in this chapter, great promise for understanding the properties of lipid bilayers of complex composition. The long range goal of this type of modeling is to begin to understand properties of membranes of biological composition.

Acknowledgements

Author S.A.P. thanks Prof. Ananth Grama for financial support under NSF grant number DMR0427540. E.J. and H.L.S. acknowledge support under NIH grant UIUC/NIH 2006-139-1.

References

Aman, M.J., Ravichandran, K.S. (2001). A requirement for lipid rafts in B cell receptor induced Ca^{2+} flux. *Curr. Biol.* **10** (7), 393–396.

Anderson, R.W.G. (1998). The caveolae membrane system. *Annu. Rev. Biochem.* **69**, 199–225.

Armen, R.S., Uitto, O.D., Feller, S.E. (1998). Phospholipid component volumes: Determination from bilayer structure calculations. *Biophys. J.* **75** (2), 734–744.

Balazs, A., Ginsburg, V.V., Qiu, F., Peng, G., Jasnow, D. (2000). Multi-scale model for binary mixtures containing nanoscopic particles. *J. Phys. Chem. B* **104**, 3411–3422.

Berendsen, H.J.C., Grigera, J.R., Straatsma, T.P. (1987). The missing term in effective pair potentials. *J. Phys. Chem.* **91**, 6269.

Berendsen, H.J.C., van der Spoel, D., van Drunen, R. (1995). GROMACS: A message-passing parallel molecular dynamics implementation. *Comput. Phys. Commun.* **91**, 43–56.

Berger, O., Edholm, O., Jähnig, F. (1997). Molecular dynamics simulations of a fluid bilayer of dipalmitoylphosphatidylcholine at full hydration, constant pressure, and constant temperature. *Biophys. J.* **72**, 2002–2013.

Brown, D.A., London, E. (1998). Functions of lipid rafts in biological membranes. *Annu. Rev. Cell Dev. Biol.* **14**, 111–136.

Cheong, K.H., Zachetti, D., Schneeberger, E., Simons, K. (1999). VIP17/MAL, a lipid raft-associated protein, is involved in apical transport in MDCK cells. *Proc. Natl. Acad. Sci. USA* **96**, 6241–6248.

Chiu, S.-W., Clark, M., Subramaniam, S., Scott, H.L., Jakobsson, E. (1995). Incorporation of surface tension into molecular dynamics simulation of an interface. A fluid phase lipid bilayer membrane. *Biophys. J.* **69**, 1230–1245.

Chiu, S.-W., Clark, M.M., Jakobsson, E., Subramaniam, S., Scott, H.L. (1999). Optimization of hydrocarbon chain interaction parameters: Application to the simulation of fluid phase lipid bilayers. *J. Phys. Chem. B* **103**, 6323–6327.

Chiu, S.-W., Jakobsson, E., Scott, H.L. (2001). Combined Monte Carlo and molecular dynamics simulation of hydrated dipalmitoyl-phosphatidylcholine-cholesterol lipid bilayers. *J. Chem. Phys.* **114**, 5435–5443.

Chiu, S.-W., Jakobsson, E., Mashl, R.J., Scott, H.L. (2002). Cholesterol-induced modifications in lipid bilayers: A simulation study. *Biophys. J.* **83** (4), 1842–1853.

Chiu, S.-W., Vasudevan, S., Jakobsson, E., Mashl, R.J., Scott, H.L. (2003). Structure of sphingomyelin bilayers: A simulation study. *Biophys. J.* **85** (6), 3624–3635.

Dietrich, C., Yang, B., Fuiwara, T., Kusumi, A., Jacobson, K. (2002). Relationship of lipid rafts to transient confinement zones detected by single particle tracking. *Biophys. J.* **82**, 244–284.

Edholm, O., Nagle, J.F. (2005). Areas of molecules in membranes consisting of mixtures. *Biophys. J.* **89** (3), 1827–1832.

Fantini, J., Garmy, N., Mahfoud, R., Yahi, N. (2002). Lipid rafts: Structure, function, and role in HIV, Alzheimers, and prion diseases. *Exp. Rev. Mol. Med. December* **20**, 1–21.

Feigenson, G., Buboltz, T. (2001). Ternary phase diagram of dipalmitoyl PC/dilauroyl PC/cholesterol: Nanoscopic domain formation driven by cholesterol dispersions. *Biophys. J.* **80**, 2775–2788.

Fielding, C.J. (Ed.) (2006). "Lipid Rafts and Caveolae". Wiley–VCH, Weinheim.

Frenkel, D., Smit, B. (2002). "Understanding Molecular Simulation from Algorithm to Applications", vol. 1. Computational Science Series. Academic Press, New York.

Gellerman, G., Appel, T., Tannert, A., Radestock, A., Hortschansky, P., Schroekh, V., Leisner, C.T., Lutkepohl, T., Shtrasburg, S., Rocken, C., Pras, M., Linke, R., Diekmann, S., Fandrich, M. (2005). Raft lipids as common components of human extracellular amyloid fibrils. *Proc. Natl. Acad. Sci. USA.*

Greenwood, A.I., Tristram-Nagle, S., Nagle, J.F. (2006). Partial molecular volumes of lipids and cholesterol. *Chem. Phys. Lipids* **143**, 1–10.

Harder, T., Scheiffele, P., Verkade, P., Simons, K. (1998). Lipid domain structure of the plasma membrane revealed by patching of membrane components. *J. Cell Biol.* **141**, 929–942.

Huang, T.-H., Less, C.W.B., Das Gupta, S.K., Blume, A., Griffin, R.G. (1993). A ^{13}C and ^{2}H nuclear magnetic resonance study of phosphatidylcholine/cholesterol interactions: Characterization of liquid–gel phases. *Biochemistry* **32** (48), 13277–13287.

Jacobson, K., Dietrich, C. (1999). Looking at lipid rafts? *Trend. Cell Biol.* **9**, 87–91.

Kawabuchi, M., Satomi, Y., Takao, T., Shimonishi, Y., Nada, S., Nagai, K., Tarakhovsky, A., Okada, M. (2000). Transmembrane phosphoprotein Cbp regulates the activities of scr-family tyrosine kinases. *Nature* **404**, 999–1002.

Khelashvili, G.A., Pandit, S.A., Scott, H.L. (2005). Self-consistent mean field model based on molecular dynamics: Application to lipid–cholesterol bilayers. *J. Chem. Phys.* **123**, 034910.

Kucerka, N., Tristram-Nagle, S., Nagle, J.F. (2005). Structure of fully hydrated fluid phase lipid bilayers with monounsaturated chains. *J. Membr. Biol.* **208**, 193–202.

Lindahl, E., Edholm, O. (2000). Mesoscopic undulations and thickness fluctuations in lipid bilayers from molecular dynamics simulations. *Biophys. J.* **79** (1), 426–433.

Lindahl, E., Hess, B., van der Spoel, D. (2001). GROMACS 3.0, A package for molecular simulation and trajectory analysis. *J. Mol. Mod.* **7**, 306–317.

Lubensky, T., Chaikin, P. (1995). "Principles of Condensed Matter Physics". Cambridge University Press, Cambridge.

MacKerell Jr., A.D., Bashford, D., Bellott, M., Dunbrack Jr., R.L., Evanseck, J.D., Field, M.J., Fischer, S., Gao, J., Guo, H., Ha, S., Joseph-McCarthy, D., Kuchnir, L., Kuczera, K., Lau, F.T.K., Mattos, C., Michnick, S., Ngo, T., Nguyen, D.T., Prodhom, B., Reiher III, W.E., Roux, B., Schlenkrich, M., Smith, J.C., Stote, R., Straub, J., Watanabe, M., Wiorkiewicz-Kuczera, J., Yin, D., Karplus, M. (1998). All-atom empirical potential for molecular modeling and dynamics studies of proteins. *J. Phys. Chem. B* **102**, 3586–3616.

Maguy, A., Hebert, T., Nattel, S. (2006). Involvement of lipid rafts and caveolae in cardiac ion channel function. *Cardiovasc. Res.* **69**, 798–807.

Maier, W., Saupe, A. (1959). *Z. Naturforsch. A* **14**, 882–889.

Manes, S., Mira, E., Gomez-Moulton, C., Lacalle, R.A., Keller, P., Labrador, J.P., Martinez, A.C. (1999). Membrane raft microdomains mediate front-rear polarity in migrating cells. *EMBO J.* **18**, 6211–6220.

Manie, S.N., Debreyne, S., Vincent, S., Gerlier, D. (2000). Measles virus structural components are enriched into lipid raft microdomains: A potential cellular location for virus assembly. *J. Virol.* **74**, 305–311.

Marčelja, S. (1974). Chain ordering in liquid crystals II. The structure of bilayer membranes. *Biochim. Biophys. Acta* **367**, 156–176.

Molander-Melin, M., Blennow, K., Bogdanovic, N., Dellheden, B., Mansson, J.E., Fredman, P. (2005). Structural membrane alterations in Alzheimer brains found to be associated with regional disease development; increased density of gangliosides GM1 and GM2 and loss of cholesterol in detergent-resistant membrane domains. *J. Neurochem.* **92** (1), 171–182.

Nagle, J.F., Tristram-Nagle, S. (2000a). Lipid bilayer structure. *Curr. Opin. Struct. Biol.* **10**, 474–480.

Nagle, J.F., Tristram-Nagle, S. (2000b). Structure of lipid bilayers. *Biochim. Biophys. Acta* **1469**, 159–195.

Niemelä, P.S., Ollila, S., Hyvönen, M., Kartunen, M., Vattulainen, I. (2007). Assessing the nature of lipid raft membranes. *PLoS Comput. Biol.* **3** (2), 0304–0312.

Pandit, S.A., Bostick, D.L., Berkowitz, M.L. (2004a). Complexation of phosphatidylcholine lipids with cholesterol. *Biophys. J.* **86** (3), 1345–1356.

Pandit, S.A., Jakobsson, E., Scott, H.L. (2004b). Simulation of the early stages of nano-domain formation in mixed bilayers of sphingomyelin, cholesterol, and dioleylphosphatidylcholine. *Biophys. J.* **87** (5), 3312–3322.

Pandit, S.A., Vasudevan, S., Chiu, S.-W., Mashl, R.J., Jakobsson, E., Scott, H.L. (2004c). Sphingomyelin–cholesterol domains in phospholipid membranes: Atomistic simulation. *Biophys. J.* **87** (2), 1092–1100.

Pandit, S., Khelashvili, G., Jakobsson, E., Grama, A., Scott, H.L. (2007). Lateral organization in lipid–cholesterol mixed bilayers. *Biophys. J.* **92**, 440–447.

Pandit, S., Chiu, S.-W., Jakobsson, E., Grama, A., Scott, H.L. (2008) The effect of chain saturation on cholesterol solubility in lipid bilayers: A simulation study. *Biophys. J.*, in press.

Peng, G., Qiu, F., Ginzburg, V.V., Jasnow, D., Bakazs, A. (2000). Forming supramolecular networks from nanoscale rods in binary, phase-separating mixtures. *Science* **288**, 1802–1804.

Petrache, H.I., Feller, S.E., Nagle, J.F. (1997). Determination of component volumes of lipid bilayers from simulations. *Biophys. J.* **70**, 2237–2242.

Pralle, A., Keller, P., Florin, E.L., Simone, K., Horber, J.H.K. (2000). Sphingolipid–cholesterol rafts diffuse as small entities in the plasma membrane of mammalian cells. *J. Cell. Biol.* **148**, 997–1007.

Reitveld, A., Simons, K. (1998). The differential miscibility of lipids as the basis for the formation of functional membrane rafts. *Biochim. Biophys. Acta* **1376**, 467–479.

Rozelle, A.L., Machesky, L.M., Yamamoto, M., Driessens, M.H., Insall, R.H., Roth, M.G., Luby-Phelps, K., Marriott, G., Hall, A., Yin, H.L. (2000). Phosphatidylinositol 4,5-biphosphate induces actin-based movement of raft-enriched vesicles through WASP-Arp2/3. *Curr. Biol.* **6**, 311–320.

Scott, H.L. (1993). Lipid–cholesterol phase diagrams: Theoretical and numerical aspects. In: Finegold, L. (Ed.), "Cholesterol in Model Membranes". CRC Press, Boca Raton, pp. 197–222.

Scott, H.L. (2002). Modeling the lipid component of membranes. *Curr. Opin. Struct. Biol.* **12** (4), 495–502.

Simons, K., Ikonen, E. (1997). Functional rafts in cell membranes. *Nature* **387**, 569–572.

Smondyrev, A.M., Berowitz, M.L. (1999). United atom force field for phospholipid membranes: Constant pressure molecular dynamics simulation of diaplmitoylphosphatidylcholine/water system. *J. Comput. Chem.* **20**, 531–545.

Sönnichsen, B., de Renzis, S., Nielsen, E., Reitdorf, J., Zerial, M. (2000). Distinct membrane domains on endosomes in the recycling pathway visualized by multicolor imaging of Rab4, Rab5, and Rab1. *J. Cell Biol.* **149**, 901–914.

Tu, K., Klein, M.L., Tobias, D.J. (1998). Constant-pressure molecular dynamics investigation of cholesterol effects in a dipalmitoylphosphatidylcholine bilayer. *Biophys. J.* **75**, 2147–2156.

van der Spoel, D., van Buuren, A.R., Apol, E., Meulenhoff, P.J., Tieleman, D.P., Sijbers, A.L.T.M., Hess, B., Feenstra, K.A., Lindahl, E., van Drunen, R., Berendsen, H.J.C. (2001). GROMACS User Manual Version 3.0. Nijenborgh 4, 9747 AG Grongen, The Netherlands.

Veatch, S.L., Keller, S.L. (2005). Seeing spots: Complex phase behavior in simple membranes. *Biochim. Biophys. Acta* **1746**, 4428–4436.

Viola, A., Schroeder, S., Sakakibara, Y., Lanzavecchia, A. (1999). T Lymphocyte costimulation mediated by reorganization of membrane microdomains. *Science* **283**, 680–682.

Vist, M.R., Davis, J.H. (1990). Phase equilibria of cholesterol/dipalmitoylphosphatidylcholine mixtures: ^{2}H nuclear magnetic resonance and differential scanning calorimetry. *Biochemistry* **29** (2), 451–464.

Westermann, M., Leutbecher, H., Meyer, H.W. (1999). Membrane structure of caveolae and isolated caveolin-rich vesicles. *Histochem. Cell Biol.* **111**, 71–81.

Xavier, R., Brennan, T., Li, Q., McCormack, C., Seed, B. (1998). Membrane compartmentation is required for efficient T cell activation. *Immunity* **6**, 723–732.

CHAPTER 11

Molecular Modeling of the Structural Properties and Formation of High-Density Lipoprotein Particles

Amy Y. Shih[*,†], **Peter L. Freddolino**[*,†], **Anton Arkhipov**[†,‡],
Stephen G. Sligar[*,†,§] **and Klaus Schulten**[*,†,‡]

[*]Center for Biophysics and Computational Biology, University of Illinois at Urbana-Champaign, Urbana, IL 61801, USA
[†]Beckman Institute for Advanced Science and Technology, University of Illinois at Urbana-Champaign, Urbana, IL 61801, USA
[‡]Department of Physics, University of Illinois at Urbana-Champaign, Urbana, IL 61801, USA
[§]Department of Biochemistry, University of Illinois at Urbana-Champaign, Urbana, IL 61801, USA

I. INTRODUCTION

Lipids such as fatty acids, triglycerides, phospholipids, and cholesterol are naturally occurring molecules that can be found in higher order mammals. They play a functionally diverse role, for example, from serving as structural components of cellular membranes to providing metabolic energy reserves. Lipids are amphi-

Current Topics in Membranes, Volume 60
Copyright © 2008, Elsevier Inc. All rights reserved

pathic particles consisting of a hydrophobic and a hydrophilic domain and are not soluble in aqueous solutions. How then can these insoluble lipid molecules be transported through aqueous media such as blood? Nature has provided a solution in the form of lipoproteins, protein-lipid particles that are used to transport lipids. In humans, such lipoprotein particles are involved in the transport of lipids and cholesterol in the blood. They are divided into various classes based on their density, size, and protein composition, ranging from very low-density lipoproteins to high-density. The two most well known of these human blood lipoproteins are the high-density lipoprotein (HDL) and the low-density lipoprotein (LDL), often referred to as the "good" and the "bad" cholesterol, respectively.

Cholesterol is a lipid molecule often thought of as being harmful to humans. However, cholesterol is a naturally occurring lipid that is produced by the body and is used in maintaining the structural rigidity of membranes and as the precursor to many steroidal hormones, such as testosterone and estrogen. Cholesterol can also be obtained from the diet. The source of cholesterol's bad reputation is due to the role it plays in the formation of plaques on artery walls (a condition known as atherosclerosis) resulting in the development of coronary heart disease, often manifesting itself as a heart attack. Coronary heart disease, in fact, is the single largest cause of death in the United States. A variety of major risk factors, which increase the risk of cardiovascular disease, have been identified, including smoking, high blood pressure, physical inactivity, obesity, diabetes, and high blood cholesterol. Thus, cholesterol, lipoproteins, and a variety of enzymes, which interact with lipoproteins, have been under intense investigation for many decades. So then, what is the difference between the "good" and the "bad" cholesterol? The "good" cholesterol, HDL, is responsible for transporting cholesterol from peripheral cells to the liver where it can be processed, degraded, and excreted or used by steroidal tissues for hormone biosynthesis. The "bad" cholesterol, LDL, is responsible for transporting lipids to cells. On the one hand, low levels of HDL are a known risk factor for coronary heart disease while on the other hand, high levels of LDL are a risk factor (Nofer *et al.*, 2005). This review will focus on the structural properties and formation of HDL.

The main pathway by which excess cholesterol is removed from peripheral tissues and transported to the liver for degradation is known as reverse cholesterol transport, reviewed by Wang and Briggs (2004), of which HDL is a major component. A simplified schematic of the reverse cholesterol transport pathway is shown in Fig. 1. The primary protein component of human HDL, apolipoprotein A-I (apo A-I), is synthesized in the liver and intestines and excreted into the blood, initially forming lipid-free/poor particles. The efflux of lipids and cholesterol from peripheral tissues to apo A-I, either mediated by the transmembrane transport protein ATP-binding cassette transporter A-I (ABCA1) or through passive diffusion, results in the formation of nascent discoidal HDL particles. The continued absorption of lipids and cholesterol and the conversion of cholesterol to esterified cholesterol by the enzyme lecithin cholesterol acyl transferase (LCAT)

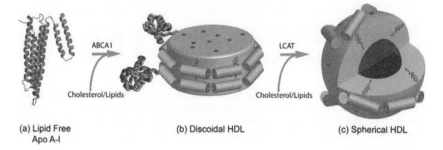

(a) Lipid Free
Apo A-I

(b) Discoidal HDL

(c) Spherical HDL

FIGURE 1 Simplified schematic representation of the reverse cholesterol transport pathway which describes the production, transformation, and degradation of cholesterol of which HDL is a major component. Apo A–I, the primary protein component of HDL, is synthesized in the liver and intestines and excreted into the blood as lipid-free particles (a). The efflux from peripheral tissues and the aggregation of cholesterol and lipids to lipid-free apo A–I, either mediated by the transmembrane transport protein ABCA1 or through passive diffusion, results in the formation of nascent discoidal HDL particles (b). The conversion of cholesterol into cholesterol esters by LCAT and the continued uptake of lipids ultimately result in the formation of a spherical HDL particle (c), the primary circulating form of HDL.

results in the formation of spherical lipoprotein particles. These mature spherical HDL particles continue to grow in size by accumulating additional lipid and cholesterol. Mature spherical HDL can exchange lipids with other lipoproteins particles facilitated by cholesterol ester transfer protein. Eventually the mature spherical HDL makes its way to the liver where membrane proteins called scavenger receptors bind to the HDL particles, internalizing the cholesterol esters and recycle the apolipoproteins. Cholesterol is eventually removed from the body either as bile acids (such as cholate) or as free cholesterol in the bile. The apo A-I is catabolized in the kidneys, facilitated by the endocytic receptor cubulin.

Apo A-I is a 243-residue amphipathic protein consisting of an N-terminal globular domain and a primarily helical C-terminal lipid binding domain (Segrest *et al.*, 1994). The lipid binding domain is characterized as having eight 22-mer and two 11-mer amphipathic α-helical repeats punctuated by the presence of prolines and glycines (Boguski *et al.*, 1986; Nolte and Atkinson, 1992). As the major protein component of HDL, apo A-I's major role is to aid in the transport of cholesterol through the blood. Additionally, portions of apo A-I act as a ligand to several of the proteins involved in the reverse cholesterol transport pathway. The tertiary structure of apo A-I has been difficult to characterize due to its inherent flexibility (Brouillette *et al.*, 2001; Marcel and Kiss, 2003) required for its biological function, namely interactions with other proteins and components of HDL involved in the reverse cholesterol transport pathway. In fact, lipid-free apo A-I has been characterized as having properties of a molten globule.

No high resolution structures have been solved of apo A-I in any of its lipid associated forms, whether as nascent discoidal or mature spherical HDL par-

a) Double-Belt Model   b) Helical-Hairpin Model   c) Picket-Fence Model

FIGURE 2 A schematic representation of the three models describing the orientation of apo A-I in discoidal HDL particles: the (a) double-belt, (b) helical-hairpin, and (c) picket-fence models. Each discoidal HDL particle has two amphipathic apo A-I proteins (black) wrapped around a lipid bilayer effectively shielding the hydrophobic lipid tail groups from the aqueous environment making the particles soluble.

ticles. Therefore, knowledge about the structure of these forms of HDL has come from lower resolution techniques such as infrared spectroscopy, fluorescence spectroscopy, mass spectroscopy, and small angle X-ray scattering. There is by far a greater wealth of experimental structural data for the discoidal form of HDL, even though spherical HDL is the primary circulating form of HDL. Nascent discoidal HDL are heterogeneous particles which can exist in a variety of sizes (7–12 nm) and exhibit varying composition (Jonas *et al.*, 1989; Jonas, 1986). There is evidence of particles with two, three, or four apo A-I proteins per HDL (Jonas *et al.*, 1989; Jonas, 1986). The most widely studied of these particles are those with two apo A-I per HDL, for which three models exist for the orientation of apo A-I around the lipid bilayer (Fig. 2): the double-belt (Segrest *et al.*, 1999; Segrest, 1977; Shih *et al.*, 2005), helical-hairpin (Brouillette and Anantharamaiah, 1995; Rogers *et al.*, 1998), and picket-fence (Jonas *et al.*, 1989; Phillips *et al.*, 1997) models. In both the double-belt and helical hairpin models, the apo A-I strands are oriented perpendicular to the lipid acyl chains. The double-belt model features the two apo A-I proteins wrapped around the circumference of a discoidal lipid bilayer core in a belt-like fashion, whereas in the helical-hairpin model each protein strand forms a single hairpin turn, resulting in intrahelical protein interactions as opposed to the interhelical interactions found in the double-belt model. The picket-fence model features the protein helices oriented parallel to the lipid acyl chains. Evidence exists for each of these models, although the majority of recent experimental evidence rules out the picket-fence model, and strongly suggests a double-belt model for discoidal HDL (Borhani *et al.*, 1997; Koppaka *et al.*, 1999; Panagotopulos *et al.*, 2001; Li *et al.*, 2000; Tricerri *et al.*, 2001; Silva *et al.*, 2005; Gorshkova *et al.*, 2006; Li *et al.*, 2006; Shih et al., 2007a, 2007b).

Very little is known about the orientation or structure of apo A-I proteins in mature spherical HDL. When the enzyme LCAT esterifies cholesterol in discoidal HDL, the cholesterol esters form a hydrophobic core around which a continuous monolayer of lipids forms. The apo A-I proteins are thought to float along the surface of this spherical particle (Mishra *et al.*, 1994), although their orientation is unknown. It is generally thought that the apo A-I conformation in spherical

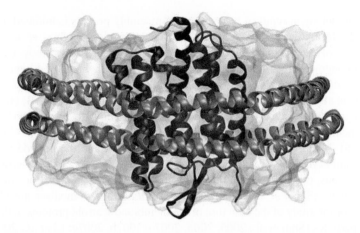

FIGURE 3 Nanodisc with a bacteriorhodopsin monomer (Shih *et al.*, 2005). The two membrane scaffold proteins are (shown in gray) arranged in an anti-parallel double-belt-like fashion surrounding a lipid bilayer (transparent gray). The bacteriorhodopsin (in black) is embedded within the lipid bilayer.

HDL differs from discoidal HDL, but to what extent is also unknown. One FRET study (Li *et al.*, 2002) has suggested that apo A-I in spherical HDL exists in a different arrangement than those of discoidal HDL, although additional distance constraints are needed before any clear conclusions can be drawn. Given the lack of experimental data on spherical HDL, the vast majority of theoretical studies, whether molecular modeling or simulations, have been focused on the discoidal form of HDL, which is also the focus of this review.

Given the heterogeneous nature of native HDL particles, reconstituted HDL (rHDL) systems, in which the ratio of components, assembly procedure, and final particle size can be precisely controlled, have been extensively used to characterize the structure of HDL (Jonas, 1986; Jonas *et al.*, 1991, 1990; Wald et al., 1990a, 1990b; Davidson *et al.*, 1995; Bayburt *et al.*, 2002; Denisov *et al.*, 2004; Shaw *et al.*, 2004; Denisov *et al.*, 2005; Shih *et al.*, 2005; Catte *et al.*, 2006; Shih et al., 2007a, 2007b, 2007c). One such rHDL system, which has been optimized to produce discoidal particles of discrete size and composition are nanodiscs (Bayburt *et al.*, 2002; Denisov et al., 2004, 2005). Nanodiscs are nanometer-sized discoidal protein-lipid particles, which are being developed as a soluble platform for embedding membrane proteins and transporting therapeutics (Sligar, 2003; Lu *et al.*, 2006; Nath *et al.*, 2007). The protein component of nanodiscs was originally engineered to contain the 200-residue C-terminal lipid binding domain of human apo A-I. These proteins, termed membrane scaffold proteins, were expressed in *E. coli* and purified before reconstitution with lipids. An optimized ratio of scaffold protein, lipids, and the detergent cholate are used in the initial reconstitution mixture and upon the removal of the cholate, either by dialysis or through the use of detergent removal beads, the self-assembly process is initiated, resulting in the formation of homogeneous discoidal particles (Bayburt *et al.*, 2002; Denisov *et al.*, 2004). Nanodiscs have been successfully used to study a variety of membrane proteins, including cytochrome P450s (Bayburt and Sligar, 2002; Civjan *et al.*, 2003; Baas *et al.*, 2004; Duan *et al.*, 2004; Davydov *et al.*, 2005; Denisov et al., 2006, 2007; Das *et al.*, 2007), bacteriorhodopsin (Fig. 3) (Bayburt and Sligar, 2003; Bayburt *et al.*, 2006), rhodopsin (Bayburt *et al.*, 2007), bacterial chemoreceptors (Boldog *et al.*, 2006), β_2-adnergenic receptors (Leitz *et al.*, 2006), blood clotting factors (Shaw *et al.*, 2007), and the peptide translocon (Alami *et al.*, 2007). Additionally, since nanodiscs are nanometer-sized and homogeneous, they make ideal platforms for studying phase transitions of finite lipid bilayers (Shaw *et al.*, 2004; Denisov *et al.*, 2005). Nanodiscs are also of interest for the study of the structure of the membrane scaffold proteins in the discoidal particles (Shih et al., 2005, 2006, 2007a, 2007b, 2007c; Li *et al.*, 2006) as nanodiscs made with the original scaffold proteins are truncated discoidal rHDL particles. Given that nanodiscs are homogeneous and well characterized experimentally (Bayburt *et al.*, 2002; Denisov *et al.*, 2004, 2005; Shaw *et al.*, 2004), their use in theoretical modeling and simulations of discoidal HDL particles offer exciting opportunities.

Even though HDL has been studied for decades, and implications of HDL for coronary heart disease are well documented, the structure of apo A-I and the transitions from lipid-free/poor to nascent discoidal to mature spherical HDL particles are not well characterized (Davidson and Silva, 2005). The large structural changes in HDL particles along with the plasticity of the apo A-I protein have been difficult to study by traditional experimental high-resolution structural techniques such as X-ray crystallography and NMR. Any structure arising from such techniques would be static by nature and would not reveal in a dynamical sense the assembly and transformations of HDL particles. Theoretical simulations, such as molecular dynamics, on the other hand, allow one to image such dynamic assembly and transformations of HDL. In the following section, the methods used to simulate lipoprotein particles, namely all-atom and coarse-grained molecular dynamics, will be reviewed along with a particularly useful low-resolution experimental technique that can possibly provide some insight into solution state transitions of HDL particles, namely solution small-angle X-ray scattering. This will then be followed by a review of recent progress towards understanding the structure and assembly of discoidal HDL and idealized nanodisc particles.

II. METHODS

Molecular dynamics is a computational method that allows one to simulate, using Newton's equations of motion, the dynamics of atomic biomolecular systems. Although all-atom molecular dynamics has been successfully used to study a range of proteins, lipids, and protein-lipid complexes (Roux et al., 2004; Gumbart et al., 2005; Bikiel et al., 2006; Tieleman, 2006) it is currently limited to the nanosecond time scale. All-atom molecular dynamics has already been used to study the structure of both the picket-fence (Phillips et al., 1997) and double-belt (Klon et al., 2002a, 2002b; Li et al., 2004; Shih et al., 2005; Catte et al., 2006) model of discoidal HDL particles. However due to the time-scale limitations of all-atom molecular dynamics, the more complicated problem of how these lipoprotein particles assemble and change from discoidal to spherical cannot be answered. Over the past several years, our group has developed a coarse-grained description of protein-lipid systems (Shih et al., 2006, 2007b) to specifically address this problem. Our coarse-grained protein-lipid model is based on work done by Marrink and coworkers in developing a coarse-grained lipid forcefield (Marrink et al., 2004, 2005; Marrink and Mark, 2003a, 2003b, 2004; Marrink, 2004). Coarse-grained molecular dynamics simulations have allowed us to self-assemble truncated discoidal HDL particles (Shih et al., 2007a, 2007b) and to disassemble nanodiscs with cholate (Shih et al., 2007c). Additionally, reverse coarse-graining methods have been developed which allows one to convert a simulated coarse-grained model into an all-atom structure (Shih et al., 2007c).

In order to the test the feasibility of the simulated coarse-grained models, comparisons to experimental structural data are needed. For this purpose, low-resolution structural data was obtained in the form of small-angle X-ray scattering measurements. Small-angle X-ray scattering is one of the most promising experimental methods for investigating protein-lipid complexes and, in particular, lipoproteins (Sachs et al., 2003; Denisov et al., 2004; Shih et al., 2005; Zagrovic et al., 2005; Denisov et al., 2005) especially when complemented by computational modeling (Shih et al., 2005, 2007a, 2007c).

A. Coarse-Grained Protein-Lipid Model

At the time we were first proposing to study the assembly of nanodiscs and lipoproteins, no coarse-grained protein-lipid model existed which would allow for the study of nanodiscs and lipoproteins. Therefore in constructing an appropriate coarse-grained model (by extension of an existing coarse-grained lipid model) several factors were considered including the ease of implementation, expected computational speed-up, and the ease of constructing a protein model. The Marrink coarse-grained lipid-water model (Marrink et al., 2004) was initially chosen due to the compatibility of its forcefield (to the CHARMM type forcefield implemented in NAMD) as well as its simple design, requiring only a minimal set of coarse-grained bead types and a discrete set of non-bonded interactions.

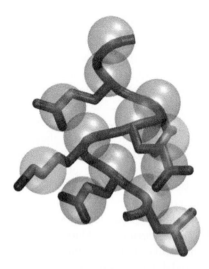

FIGURE 4 Coarse-grained protein model. The coarse-grained beads (transparent spheres) are mapped on an all-atom structure (backbone is shown as a tube and side chains are shown without hydrogens).

Using the Marrink coarse-grained lipid model as a basis, a coarse-grained protein-lipid model (Shih et al., 2006, 2007b) was developed for simulation of nanodiscs and lipoproteins. In this model, coarse-grained beads are assigned effective interaction potentials, and the time evolution of a coarse-grained system is described by Newton's equations of motion. Harmonic potentials are used to maintain bonds and angles between the bonded coarse-grained beads. Groups of atoms are represented by coarse-grained beads, with an average correspondence of four heavy (non-hydrogen) atoms per bead. Four water molecules are represented by a single coarse-grained bead; an ion together with its first hydration shell (six water molecules) is also represented by one coarse-grained bead. Coarse-grained lipids are modeled in a similar manner as Marrink's coarse-grained model (Marrink *et al.*, 2004); for example a DPPC lipid is represented by twelve coarse-grained beads; one each for the choline and phosphate, two for the glycerol esters, and eight beads for the lipid tail chain (4 beads in each tail).

The coarse-grained model for proteins (Fig. 4) was created by assigning a set of two coarse-grained beads, a backbone bead and a side-chain bead, to each amino acid residue (except for glycine, which only has a backbone bead and no side-chain bead). All the coarse-grained backbone beads are of identical type while the side chain beads are defined based on the general properties of the corresponding amino acid. The non-bonded interactions used the same finite set of interaction energies as in the original coarse-grained lipid model (i.e., attractive, semiattractive, intermediate, semirepulsive, and repulsive interaction energies) (Marrink *et al.*, 2004). The equilibrium bond and angle values for the coarse-grained protein beads were extracted from all-atom simulations of preformed nanodiscs (Shih *et al.*, 2005).

Since the proteins in HDL and nanodiscs are known to be primarily helical (Jonas *et al.*, 1990; Sparks *et al.*, 1995), it was important that the coarse-grained protein model be able to properly reproduce an α-helical secondary structure. Therefore, a dihedral force was applied to the protein backbone beads that maintain a helical topology (Shih *et al.*, 2007b). The dihedral angle and force constant used in the coarse-grained simulations is derived from all-atom simulations using Boltzmann inversion (Shih *et al.*, 2007b). In the Boltzmann inversion procedure, applied here only in the trivial one-dimensional case, one assumes that the dynamics of the system is governed by a potential $V(x)$, where x represents the degrees of freedom under consideration, i.e., a dihedral angle in this case. $V(x)$ is chosen to reproduce a sampled thermal equilibrium probability distribution $p(x)$ of configurations over x, $p(x) = Z^{-1}e^{-V(x)/(k_B T)}$, where k_B is the Boltzmann constant, T is the temperature, and Z is the partition function. Then, by running an all-atom simulation at constant temperature, one can compute the probability distribution $p(x)$ of configurations over x. The inversion of this probability distribution according to $V(x) = -k_B T \ln[p(x)] + \text{const}$ provides the shape of the effective potential. Coupling to degrees of freedom other than x is neglected.

Coarse-grained molecular dynamics simulations using this protein-lipid model resulted in up to a three orders of magnitude increase in computational speed compared with comparable all-atom simulations (Shih *et al.*, 2006) and has been successfully used to study the assembly (Shih et al., 2006, 2007a, 2007b) of truncated rHDL particles as well as disassembly (Shih *et al.*, 2007c) of nanodiscs.

B. Reverse Coarse-Graining

The coarse-grained simulations allow for much longer timescales to be simulated, but the downside is the loss of atomic level detail. Thus, a reverse coarse-graining method was developed in order to obtain all-atom structures from the coarse-grained models for testing (Shih *et al.*, 2007c). The structures resulting from the coarse-grained simulations can be recast into all-atom structures by mapping the center of mass of the group of atoms represented by a single coarse-grained bead to the bead's location, using the CGTools plugin in VMD (Humphrey *et al.*, 1996). The resulting all-atom system then needs to be minimized and equilibrated with the center of mass of each atom group restrained in order to generate the final all-atom structure. This type of reverse coarse-graining was first used to generate all-atom structures of "disassembled" nanodiscs following the addition of cholate in order to calculate small-angle X-ray scattering curves using the program CRYSOL (Svergun *et al.*, 1995) which takes an all-atom structure as input (Shih *et al.*, 2007c).

C. Small-Angle X-Ray Scattering

Small-angle X-ray scattering is a low-resolution experimental method, which does not provide a detailed atomic structure. However, it is a widely used technique that can provide important structural information, especially on the size and shape of particles in solution under native conditions. Small-angle X-ray scattering is particular suited to lipoprotein particles, since it is sensitive to changes in lipid phase and distribution. Methods of generating small-angle X-ray scattering curves from coarse-grained models have been developed. Whether by calculating directly from a coarse-grained model (Shih *et al.*, 2007a) or by first reverse-coarse graining the system and then calculating the small-angle X-ray scattering curves from an all-atom structure (using programs such as CRYSOL) (Shih *et al.*, 2007c), the calculated small-angle X-ray scattering curves can be compared with experimentally measured small-angle X-ray scattering in order to test simulations.

III. STRUCTURE OF DISCOIDAL HDL

Due to the lack of a high-resolution crystal structure of any of the lipid-bound states of HDL, there was, and still is, considerable debate as to the arrangement of the apo A-I proteins around the lipids. For discoidal HDL particles, three models have been proposed, the double-belt, helical-hairpin, and picket-fence models (Fig. 2), with modeling and molecular dynamics performed on the double-belt and picket-fence models (Segrest *et al.*, 1992, 1999, 2000; Phillips *et al.*, 1997; Sheldahl and Harvey, 1999; Klon et al., 2000, 2002a, 2002b; Li *et al.*, 2004; Shih *et al.*, 2005; Catte *et al.*, 2006; Shih et al., 2007a, 2007b, 2007c). Although the majority of recent experimental evidence rules out the picket-fence model, and strongly suggests a double-belt model for discoidal HDL particles (Koppaka *et al.*, 1999; Panagotopulos *et al.*, 2001; Li *et al.*, 2000, 2006; Tricerri *et al.*, 2001; Silva *et al.*, 2005; Gorshkova *et al.*, 2006), doubt as to the actual structure of discoidal HDL particles still exists.

A. Picket-Fence Model

In 1997, an atomic-level structure for a picket-fence model truncated discoidal HDL particle was proposed (Phillips *et al.*, 1997) based on experimental data, sequence analysis, and molecular modeling. The picket-fence model had gained support in the 1980s and early 1990s because of particle geometry arguments (Jonas *et al.*, 1989) and an internal reflection infrared spectroscopy study of dried lipoprotein films (Wald *et al.*, 1990a). State-of-the-art (at the time) molecular dynamics simulations were performed on head-to-head and head-to-tail picket-fence systems of up to 50,000 atoms. Several hundreds of picoseconds of simulated

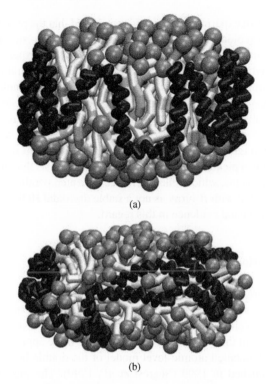

(a)

(b)

FIGURE 5 Coarse-grained simulation of a head-to-tail picket-fence discoidal HDL particle (Shih *et al.*, 2007b). The minimized coarse-grained picket-fence system (a) was simulated for 1 μs. During the course of the simulation, many of the protein helices rearranged themselves to be perpendicular to the lipid acyl chains (b). The two truncated apo A-I proteins are shown in black and the 160 DPPC lipids are shown with the lipid head groups in dark gray and the lipid tails in light gray.

annealing followed by energy minimizations were used to generate the atomic-level structures (Phillips *et al.*, 1997). The picket-fence model heavily relied on the infrared spectroscopic studies of dried lipoprotein films (Wald *et al.*, 1990a), which suggested that the helices of apo A-I are oriented perpendicular to the plane of the bilayer. However, subsequent methodologically updated polarized attenuated total internal reflection infrared spectroscopy of discoidal HDL on supported lipid monolayers (allowing measurements under more native conditions) contradicted the earlier studies, suggesting instead that the helices of apo A-I are oriented parallel to the plane of the bilayer (Koppaka *et al.*, 1999). Since then, other pieces of experimental evidence have suggested that the double-belt model is the more likely candidate for discoidal HDL particles (Borhani *et al.*, 1997; Panagotopulos *et al.*, 2001; Li *et al.*, 2000; Tricerri *et al.*, 2001; Silva *et al.*, 2005; Gorshkova *et al.*, 2006; Li *et al.*, 2006).

Even though most experimental evidence suggested that the picket-fence model is not viable for discoidal HDL particles, we recently chose to simulate a head-to-tail picket-fence model (Phillips *et al.*, 1997) discoidal lipoprotein particle using the coarse-grained model (Fig. 5) in order to see if any structural rearrangement would occur on timescales not reachable in the earlier all-atom molecular dynamics studies (Shih *et al.*, 2007b). The proteins in the picket-fence simulations were found to be unstable and upon simulation immediately started rearranging such that, after 1 μs of simulation, most of the protein helices were aligned perpendicular to the lipid tail groups (Fig. 5b), more in a double-belt or helical-hairpin like fashion than in a picket-fence like fashion. This suggests that the picket-fence arrangement of proteins, which relies on helices oriented parallel to the lipid acyl chains and connected with β-turns, is not a stable discoidal HDL structure, correlating with experimental evidence in this regard.

B. Double-Belt Model

The Borhani crystal structure of a 43-residue N-terminal truncated lipid-free apo A-I published in 1997 (Borhani *et al.*, 1997) showed a belt-shaped complex of four apo A-I proteins, and although this structure was solved in the absence of lipids, it was thought that the obtained structure was more applicable to discoidal HDL particles. A detailed atomic-level model of the double-belt discoidal HDL particle was published in 1999 (Segrest *et al.*, 1999). The model proposed an anti-parallel alignment of the two apo A-I strands in the discoidal particles with a particular orientation which optimized the intermolecular salt-bridging. This was followed by molecular belt models for two mutant forms of apo A-I, the Paris and Milano strains (Klon *et al.*, 2000).

Using these double-belt models for discoidal HDL particles, 1 ns molecular dynamics simulations were performed to investigate the mechanism of rotation of the apo A-I belts (Klon *et al.*, 2002b). The authors proposed that rotation requires the stacking of proline kinks in the two apo A-I strands. For further rotation to occur the proline kinks must straighten out and the two apo A-I monomers must rotate to prevent the proline residues from returning to their kinked state. Therefore the proline kinking introduces a kinetic barrier to the rotation of the two apo A-I strands relative to each other. The authors proposed that salt-bridging between the two protein strands provides the mechanism for rotation rearrangements on a 10-ns timescale and that the proline kinking and the presence of the two 11-mer helices select for a preferred rotational alignment.

Secondary structure predictions had originally suggested that the lipid binding domain of apo A-I contains the 200 C-terminal residues (Segrest *et al.*, 1994, 1992). However, experimental studies of nanodiscs self-assembled with membrane scaffold proteins consisting of the 200 C-terminal residues (termed MSP1) as well as nanodiscs made from membrane scaffold proteins with an

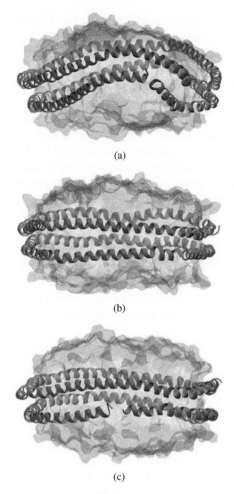

(a)

(b)

(c)

FIGURE 6   MSP1, MSP1 $\Delta(1-11)$ and MSP1 $\Delta(1-22)$ nanodiscs following all-atom molecular dynamics simulation (Shih *et al.*, 2005). The two membrane scaffold proteins are (shown in dark gray) arranged in a belt-like fashion around 160 DPPC lipids (in transparent gray).

additional 11 or 22 N-terminal residues truncated (named MSP1 $\Delta(1-11)$ and MSP1 $\Delta(1-22)$, respectively) resulted in identically sized particles (Denisov *et al.*, 2004). This suggested, for the first time, that the first 22 residues were not involved in formation of the belts surrounding the lipid-bilayer in 9.6 nm nanodisc particles. All-atom molecular dynamics simulations were used to simulate the MSP1, MSP1 $\Delta(1-11)$ and MSP1 $\Delta(1-22)$ nanodiscs in order to elucidate the actual number of residues involved in lipid binding (Shih *et al.*, 2005). Results of these all-atom molecular dynamics simulations revealed that the MSP1 nanodiscs exhibit a severe out-of-plane deformation of both scaffold protein and lipid bilayer (Fig. 6a), whereas simulations of MSP1 $\Delta(1–11)$ and MSP1 $\Delta(1–22)$ nanodiscs resulted in discoidal structures with little deformation of the lipid bilayer or scaffold proteins (Figs. 6b and 6c). As a result of the out-of-plane deformation of the MSP1 nanodisc the two scaffold proteins strands exhibit kinking at the prolines, which did not align well with each other, unlike in the MSP1 $\Delta(1–11)$ and MSP1 $\Delta(1–22)$ nanodiscs where the two protein strands as well as the proline residues align vertically over each other with minimal misalignment (Shih *et al.*, 2005). Since the alignment of MSP1, MSP1 $\Delta(1–11)$ and MSP1 $\Delta(1–22)$ scaffold proteins are very similar, the only differences stemming from the overlapping truncated regions, the out-of-plane deformation of the nanodiscs appears to be due to the overall size and most importantly the packing density of the lipids. Additional analysis, and comparison with experimental measurements (Shih *et al.*, 2005), of surface area per lipid and small-angle X-ray scattering calculated using the program CRYSOL (Svergun *et al.*, 1995), also suggested that the MSP1 nanodisc structure seen in the simulations is not optimally packed with lipids. Therefore, it was concluded that the first 11 to 22 residues of the originally predicted 200 residue lipid binding domain of apo A-I are not needed in order to completely shield the hydrophobic lipid tails with amphipathic protein.

Subsequently, coarse-grained simulations were performed on preformed MSP1, MSP1 $\Delta(1–11)$ and MSP1 $\Delta(1–22)$ nanodiscs (Shih *et al.*, 2007b). None of the resulting nanodiscs showed an out-of-plane deformation whereas all simulated nanodiscs had diameters and thicknesses comparable with experimental results (Shih *et al.*, 2007b; Denisov *et al.*, 2005). However, clear differences in the "overlap" region of the scaffold proteins were observed: MSP1 nanodiscs (Fig. 7a) exhibited a 6 residue overlap of the terminal ends of one protein strand and a 12 residue overlap in the other protein strand, whereas the MSP1 $\Delta(1–22)$ nanodisc (Fig. 7c) had a 12 residue gap and MSP1 $\Delta(1–11)$ nanodiscs (Fig. 7b) had neither a gap nor an overlap. This correlates well with the previous all-atom molecular dynamics simulation predictions (Shih *et al.*, 2005) and experimental (Denisov *et al.*, 2004) results suggesting that the first 11 to 22 N-terminal residues are not needed for the formation of nanodiscs of this size and composition. These results also demonstrate that much of the plasticity of nanodiscs resulting from lipid packing is centered at the ends of the membrane scaffold protein sequences, similar findings were also recently reported for discoidal HDL particles (Bhat *et al.*, 2007). Although these residues are not needed for formation of nanodiscs, they are obviously present in native HDL particles. Recent mass spectroscopy experiments on full-length apo A-I discoidal rHDL made with POPC lipids have revealed that these additional residues, and in fact the entire N-terminal domain, may bind with lipids by overlapping of the protein helices, similar to what is seen in Fig. 7a (Bhat *et al.*, 2007).

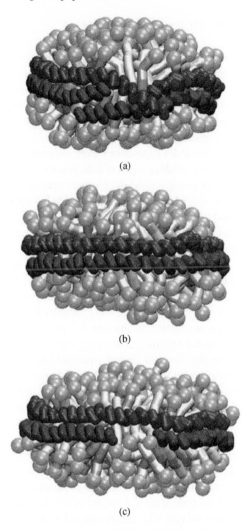

(a)

(b)

(c)

FIGURE 7   Preformed coarse-grained (a) MSP1, (b) MSP1 Δ(1–11), and (c) MSP1 Δ(1–22) nanodiscs (double-belt model) after 500 ns of dynamics (Shih *et al.*, 2007b). The two scaffold proteins, shown in black and dark gray, are wrapped around the DPPC lipids, lighter gray, in a belt-like fashion.

IV. FORMATION OF DISCOIDAL HDL

Once the detailed structure of discoidal HDL particles was known and factors important for its stability, salt-bridging and lipid-binding domains, were studied, investigations turned to the question of how these particles form.

A. How Do Discoidal HDL Particles Assemble?

One way of obtaining information on how lipids aggregate with apo A-I is to take preformed discoidal HDL particles and remove lipids, as was done by Catte *et al.* (2006). Using all-atom molecular dynamics the authors in (Catte *et al.*, 2006) progressively shrunk the HDL particle size by stepwise removal of lipids (10 in each step) either in a randomly selected location or from the center of the disk. Between each lipid removal step between 0.5 to 1 ns of simulation was performed. Numerous intermediate structures were obtained in this fashion and it was observed that removing lipids resulted in a decrease in the particle size, while the conformations became less discoidal double-belt structures, instead becoming increasingly nonplanar, twisted, ellipsoidal, and double helical structures.

The final delipidated structure containing just two apo A-I proteins revealed the formation of a bent saddle shaped protein complex (Catte *et al.*, 2006). It was proposed that this bent saddle shaped protein forms a flexible lipid pocket which captures lipids one at a time forming a "R2-0 particle" with a diameter of 78 Å (observed in discoidal HDL reconstituted with POPC only). The continued lipidation of the particles results in the formation of the "R2-1" and "R2-2" (particles with diameters of 98 Å and 106 Å, respectively) particles that are planar and contain the double-belt like orientation of the apo A-I proteins. Comparisons were made between an intermediate particle containing a ratio of 120 POPC lipids and two apo A-I proteins to the Borhani crystal structure (Borhani *et al.*, 1997). In these delipidated simulations, it was assumed that convergence of a given particle was independent of the convergence of prior intermediate species, since very little structural change occurred after the first 0.5 to 1 ns of simulation. However, in the three simulations that were run for longer periods of time (5–7 ns each), the protein RMSD after 1 ns continued to increase (although at a rate slower then the first nanosecond), and never plateaued, indicating that on the timescale simulated the intermediate structures were not truly equilibrated. It is unclear what effect this has on any of the structures obtained. Recent experimental evidence has suggested that the "R2-0" particle with a diameter of 78 Å, which in the all-atom simulations has a twisted bilayer with saddle shaped proteins, rather maintains a planar geometry with the proteins overlapping with each other to form the smaller particles (Bhat *et al.*, 2007).

We chose to tackle the problem of discoidal HDL assembly in another fashion. Using the developed coarse-grained protein-lipid model (see Section II), several ⩾10 μs simulations were performed starting from initially random placement of protein and lipid. For these simulations, a ratio of 160 DPPC lipids to two truncated apo A-I proteins was used (i.e., the ratio of lipids to proteins that results in the formation of homogeneous nanodisc particles of 9.6 nm in diameter). As an illustrative example, one of these simulations, shown in Fig. 8, will be described in detail below and used to illustrate the factors that were found to be important in lipoprotein assembly. The coarse-grained assembly simulation (Fig. 8) shows the

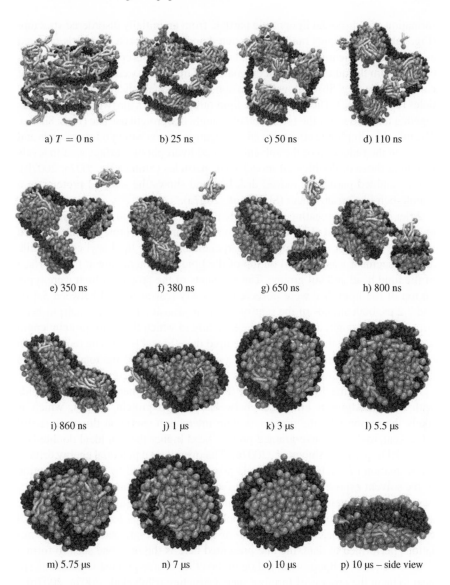

a) $T = 0$ ns   b) 25 ns   c) 50 ns   d) 110 ns

e) 350 ns   f) 380 ns   g) 650 ns   h) 800 ns

i) 860 ns   j) 1 μs   k) 3 μs   l) 5.5 μs

m) 5.75 μs   n) 7 μs   o) 10 μs   p) 10 μs – side view

FIGURE 8 Snapshots from a 10 μs simulation of the assembly of discoidal HDL particles with the lipids (shown in gray) initially randomly scattered and the two truncated apo A-I proteins (shown in black) oriented as half-circles (Shih et al., 2007a, 2007b). The lipid particles quickly aggregate and form micelles (b–d). The fusion of these micelles draws the protein strands together and by 800 ns (h) all the lipids are associated with the apo A-I proteins in the form of micelles. At 860 ns (i) the two lipid micelles fuse forming a single particle (j). The apo A-I proteins continue to slowly rearrange themselves over the surface of the lipid bilayer (k–n), eventually forming a double-belt model nanodisc (o and p).

formation of a discoidal lipoprotein particle from an initially disordered structure (Fig. 8a) with the lipids randomly scattered and the two protein strands oriented as half circles. The initial aggregation is driven by the hydrophobic effect. This is evident from the quick formation of small lipid micelles (occurring within tens of nanoseconds) (Fig. 8b), followed by fusion of these small micelles into larger particles (Figs. 8c–8g). The fusion of the lipid micelles draws the two protein strands together and allows for the formation of a single lipoprotein particle (Figs. 8h–8j). The initial hydrophobic effect-driven aggregation occurs within microseconds and results in the reduction of the solvent exposed hydrophobic surface area to levels matching those of double-belt model HDL particles (Shih et al., 2007a, 2007b). The assembled particle is indeed disk-like and shows the scaffold proteins in a double-belt like arrangement (Shih *et al.*, 2007a).

After initially aggregating into a single particle, the two apo A-I protein strands assumed with a bend (Fig. 8j) so that one portion of the proteins was oriented along the perimeter of the lipid disc perpendicular to the lipid acyl chains while another segment was oriented on top of the lipid disc interacting at the interface of the lipid head and tail groups. The segments of the proteins which were lying on top of the lipid disc were observed to slowly rearrange, and between 5.4 µs and 7.2 µs, both strands of the scaffold proteins transitioned from a hairpin-bend structure to a straight double belt-like structure in which the entire protein length was oriented along the perimeter of the lipid disc (Figs. 8l–8n). The reorientation of the scaffold proteins is reflected in a steady decrease of the total interaction energy between the two protein strands (Shih et al., 2007a, 2007b).

At the end of the 10 µs simulation (Figs. 8o and 8p) there were still 9 lipids with their headgroups embedded between the two protein strands, which is likely the reason why the protein–protein interaction energy at this time is still ~0.2 kcal/mol per coarse-grained protein bead higher then an ideal double-belt model HDL particle (Shih *et al.*, 2007a). The overall total potential energy driving the formation of the HDL particles is primarily dominated by the minimization of the solvent exposed hydrophobic surface area while the interaction energy between the two protein strands, although important for formation of protein tertiary structure (i.e., the formation of the double-belts) accounts for little of the total energy driving particle formation. Out of several coarse-grained assembly simulations performed to date, the one presented here is the only one seen to form a double-belt model lipoprotein, although the other coarse-grained simulations appear to be in the process of forming such a structure (Shih et al., 2007a, 2007b).

Little is known about the transformation of discoidal HDL to the functionally more relevant spherical particles. It has been proposed that spherical HDL particles contain a hydrophobic core composed of cholesterol esters surrounded by a lipid/cholesterol monolayer and with the apo A-I proteins arranged on the periphery of the spherical particle floating at the interface of the lipid head and tail groups (Davidson and Silva, 2005). The assembly simulations (Fig. 8) have shown that apo A-I, in fact, can float on the surface of a lipid bilayer in which the

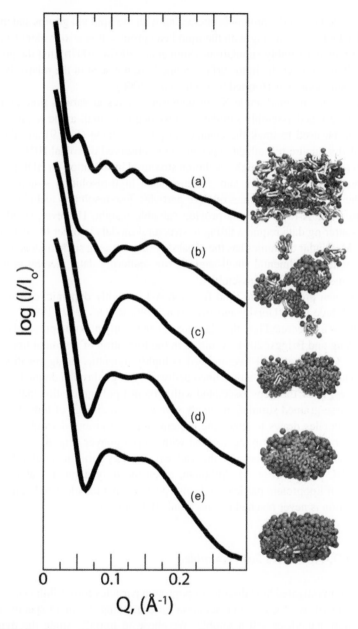

FIGURE 9 Changes in small-angle X-ray scattering curves computed over the course of the coarse-grained assembly simulation shown in Fig. 8. Small-angle X-ray scattering curves at various time points and corresponding simulation snapshots (to the right) are shown for (a) 0 ns, (b) 150 ns, (c) 850 ns, (d) 4 μs, and (e) 10 μs. The final curve (e) is comparable with experimentally measured small-angle X-ray scattering curves of nanodiscs (Shih *et al.*, 2007a).

hydrophobic face of the protein is interacting with the lipid tail groups and the hydrophilic face is interacting with the lipid head groups. It is also evident from the coarse-grained assembly simulations (Shih et al., 2007a, 2007b) that the proteins can easily move over the lipid surface, something not seen in the relatively short all-atom simulations performed by Catte *et al.* (2006).

Calculations of small-angle X-ray scattering curves at various time points in the coarse-grained assembly simulations revealed that small-angle X-ray scattering could be used to track the changes in particle shape (Fig. 9) and the final assembled structure resembled experimentally observed discoidal HDL particles (Shih *et al.*, 2007a). Since high-resolution structural information on HDL particles has been difficult to obtain and, because such high-resolution structures preclude the study of the dynamics of HDL particles, low-resolution techniques such as small-angle X-ray scattering provide valuable insight. However, small-angle X-ray scattering data requires fitting to structural models in order to interpret the results. Molecular dynamics has the capability of producing such models for comparison with experimental small-angle X-ray scattering data, thereby providing valuable insight into HDL assembly.

It has been proposed that lipid-free apo A-I is highly dynamic, existing in a molten globule state (Gursky and Atkinson, 1996; Rogers *et al.*, 1998). The apo A-I proteins are secreted into the plasma and can readily pickup lipids and cholesterol (Wang and Briggs, 2004) resulting in the formation of discoidal lipoprotein particles. Given that lipid-free apo A-I is highly dynamic, its aggregation with lipids is unlikely to follow any defined pathway or to go through defined intermediate states. Thus, the steps associated with protein-lipid aggregation that are seen in the coarse-grained simulations likely give a realistic image of how the fusion of lipid micelles draws together the protein strands forming discoidal particles with the proteins randomly aggregated with a lipid bilayer (Fig. 8). The protein strands can then freely rearrange themselves until they form the double-belt discoidal HDL particle (Fig. 8). Although, given the truly dynamic nature of the formation of lipoprotein particles, it is also possible that the idealized double-belt configuration is never reached for discoidal HDL *in vivo*.

B. How Do Nanodiscs Self-Assemble?

Having investigated how discoidal lipoprotein particles form (Shih *et al.*, 2006, 2007a, 2007b), we focused our attention on examining the more specific problem of how nanodiscs self-assemble. We chose to initially study the dynamic nanodisc assembly process of cholate removal resulting in protein-lipid aggregation and discoidal particle formation by actually studying the reverse process, the disassembly of nanodiscs. For this purpose, specific quantities of cholate were added to preformed nanodiscs particles and the resulting structural changes were

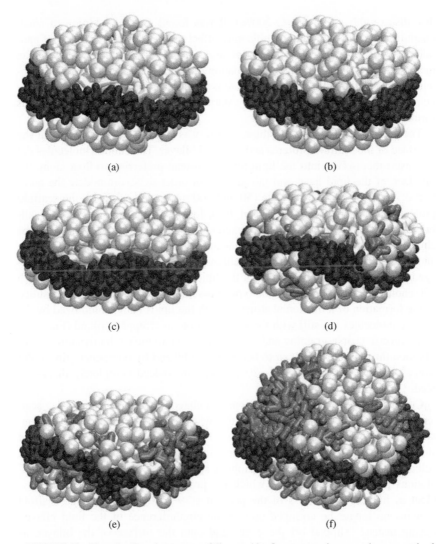

(a) (b)

(c) (d)

(e) (f)

FIGURE 10   Nanodisc in various stages of disassembly. One to two microsecond coarse-grained molecular dynamics simulations were performed in which various concentrations of cholate shown in dark gray (b–f) are added to a preformed nanodisc (a) with two membrane scaffold proteins (in black) wrapped around a 160 DPPC lipid bilayer (in light gray) (Shih *et al.*, 2007c).

quantified both experimentally, using small-angle X-ray scattering, and computationally, using coarse-grained molecular dynamics (Shih *et al.*, 2007c). All-atom (Marrink and Mark, 2002) and coarse-grained (Marrink, 2004) simulations have previously been used to study mixed micelles systems modeling human bile (i.e., cholate). These studies formed the foundation for our investigations on nanodisc disassembly (Shih *et al.*, 2007c). It was found that the addition of small quantities of cholate did not significantly affect the overall size or shape of nanodiscs (Figs. 10b and 10c). Addition of larger quantities of cholate resulted in an increase in disc diameter and a perturbation of the lipid bilayer surrounding proteins (Figs. 10d and 10e). When cholate is added at even higher concentrations, i.e., those used for nanodisc assembly starting conditions, a spherical particle is formed (Fig. 10f).

The addition of cholate into the nanodisc was seen to occur primarily at the interface of the scaffold proteins and the lipid bilayer. The cholate molecules did not insert themselves into the lipid bilayer, instead preferring to float along the interface of the lipid head and tail groups. In order to accommodate the addition of larger amounts of cholate (50 and 100 molecules per disc), the scaffold proteins had to open (Figs. 10d and 10e), losing their ideal inter-helical salt-bridging and producing a larger diameter nanodisc. The incorporation of the cholates into the nanodisc causes significant fluctuations of the nanodisc particle and especially the scaffold proteins, although the scaffold proteins never managed to slip off the circumference of the lipid bilayer (Fig. 11). Nanodiscs with 320 cholates (corresponding to the typical assembly starting conditions) resulted in the formation of a liposomal structure with the lipids being solubilized by the cholate molecules but still with the scaffold proteins wrapped around (Fig. 10f). Upon reverse coarse-graining and subsequent all-atom molecular dynamics equilibration this particle was seen to become destabilized by interpenetrating water suggesting that nanodiscs with 320 cholates are indeed completely disassembled.

Comparisons of distance distributions calculated from small-angle X-ray scattering results (Shih *et al.*, 2007c) of simulated and experimental nanodiscs containing the same cholate concentrations (Fig. 12) revealed that the simulated structures at lower cholate concentrations compare well with experimental results. However at higher concentrations it appears that the coarse-grained simulations produced a particle in which the proteins are too stable. This is evident as judged by the loss of the peak at $\sim$79 Å, that suggests the scaffold proteins no longer be arranged around the circumference of the lipid bilayer, having instead "slipped off the edge" and onto the surface of the bilayer, as seen in the increase in the peak between 50 and 70 Å. The arrangement of the scaffold proteins on the surface of the lipid bilayer was observed in previous coarse-grained simulations of nanodisc assembly (Shih *et al.*, 2007b). The discrepancy between observation and computation could be due to a number of factors, including insufficient sampling in the simulations or errors in the coarse-grained forcefield. In fact, recently, an improved coarse-grained lipid model, MARTINI, has been released (Marrink *et al.*, 2007), which solves some of the problems such as spontaneous curvature being too negative and water tending to freeze too readily with the previous forcefield (Marrink and Mark, 2004; de Joannis *et al.*, 2006; Baron et al., 2006a, 2006b, 2007).

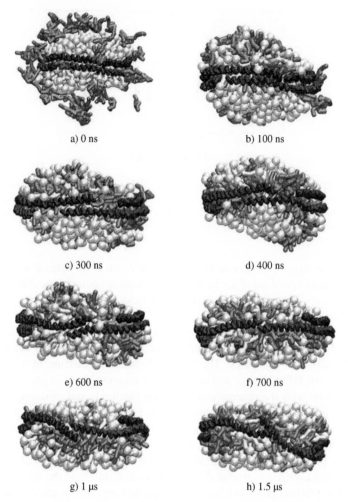

a) 0 ns

b) 100 ns

c) 300 ns

d) 400 ns

e) 600 ns

f) 700 ns

g) 1 μs

h) 1.5 μs

FIGURE 11 Snapshots of a simulation in which 100 cholate molecules (in dark grey) are added to a preformed nanodisc (membrane scaffold proteins are shown in black, and lipid bilayer in light gray) (Shih *et al.*, 2007c).

Experimental studies of native and reconstituted discoidal HDL have revealed that they can form particles with a range of diameters (Swaney, 1980; Jonas et al., 1989, 1990; Bergeron *et al.*, 1995; Rogers *et al.*, 1997). This size heterogeneity has made it difficult to study the structure of HDL particles, thus our preference for using the homogeneous nanodisc particles. However, the nanodisc disassembly simulations revealed how the proteins strands can easily open and accommodate additional lipids resulting in the formation of larger diameter particles (Figs. 10d and 10e). Although the alignment of the scaffold proteins to each

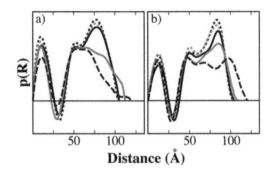

FIGURE 12   Distance distributions calculated from small-angle X-ray scattering curves (Shih *et al.*, 2007c) of simulated nanodiscs (a) and experimentally measured nanodiscs (b) with the addition of 0 (black dashed line), 5 (gray dashed line), 10 (black solid line), 50 (gray solid line), and 100 (black long dashed line) cholate molecules per particle.

other in these larger diameter particles does not result in an optimal inter-helical salt-bridging and portions of the lipid tail groups are not completely shielded from the aqueous environment, such lipoprotein particles appear to be relatively stable in the coarse-grained simulations.

Having performed simulations of lipoprotein assembly and nanodisc disassembly with cholate, one can begin to understand the factors that are important in the actual formation of nanodiscs. The initial self-assembly mixture of 320 cholate molecules, 160 DPPC lipids, and two scaffold proteins likely exists in a heterogeneous form, with the detergent solubilizing the lipids and proteins. Removal of cholate initializes a self-assembly process, causing the lipids and proteins to start aggregating, likely into micelles or liposomes. These aggregates may have a core of lipids and cholate with the scaffold proteins oriented in a random fashion around the perimeter. Once a critical concentration of cholate has been removed, the lipids will undergo a structural transformation resulting in the formation of a bilayer. The scaffold proteins likely take longer to rearrange themselves into their typical double-belt confirmation. Upon the removal of the majority of the cholate molecules a discoidal particle arises in which a discoidal lipid bilayer is encircled by the membrane scaffold proteins (i.e., nanodiscs).

V. CONCLUSIONS

Molecular modeling and relatively short all-atom molecular dynamics simulations have been used for the past decade to generate detailed atomic structures of discoidal HDL particles such as the picket-fence and double-belt models. However, it has not been until recently, with the increased computational abilities of molecular dynamics programs such as NAMD (Phillips *et al.*, 2005), that the true dynamic properties of HDL particles have been examined. It is through this increased sampling, starting with the all-atom simulations ranging to tens of nanoseconds of discoidal HDL particles (Shih *et al.*, 2005) and the removal of lipids (Catte *et al.*, 2006), followed by the dramatically increased sampling provided for by coarse-grained simulations of multiple microseconds (Shih et al., 2006, 2007a, 2007b) that light has been shed on how these amphipathic helical apo A-I proteins collect lipids and form discoidal HDL particles. The next step of the overall research program will be to examine the discoidal to spherical transformation of HDL particles, and the resulting structure of the spherical particles. For nanodiscs, the factors important to self-assembly have been determined (Shih *et al.*, 2007c) through the use of combined computational and experimental data. In the future, the incorporation of membrane proteins into nanodiscs as well as the further engineering of the scaffold proteins in nanodiscs will be investigated.

Acknowledgements

This work is supported by grants RO1-GM067887, P41-RR05969, and RO1-GM33775 from the National Institutes of Health to K.S. and S.G.S. The authors acknowledge supercomputer time provided by the National Science Foundation grant MCA93S028. Portions of this work were performed at the DuPont-Northwestern-Dow Collaborative Access Team (DND-CAT) Synchrotron Research Center located at Sector 5 of the Advanced Photon Source. DND-CAT is supported by the E.I. DuPont de Nemours and Co., the Dow Chemical Company, the U.S. National Science Foundation through Grant DMR-9304725, and the State of Illinois through the Department of Commerce and the Board of Higher Education Grant IBHE HECA NWU 96. Use of the Advanced Photon Source was supported by the U. S. Department of Energy, Office of Science, Office of Basic Energy Sciences, under Contract No. W-31-109-Eng-38.

References

Alami, M., Dalal, K., Lelj-Garolla, B., Sligar, S.G., Duong, F. (2007). Nanodiscs unravel the interaction between the SecYEG channel and its cytosolic partner SecA. *EMBO J.* **26**, 1995–2004.

Baas, B.J., Denisov, I.G., Sligar, S.G. (2004). Homotropic cooperativity of monomeric cytochrome P450 3A4 in a nanoscale native bilayer environment. *Arch. Biochem. Biophys.* **430**, 218–228.

Baron, R., de Vries, A.H., Hünenberger, P.H., van Gunsteren, W.F. (2006a). Comparison of atomic-level and coarse-grained models for liquid hydrocarbons from molecular dynamics configurational entropy estimates. *J. Phys. Chem. B* **110**, 8464–8473.

Baron, R., de Vries, A.H., Hüunenberger, P.H., van Gunsteren, W.F. (2006b). Configurational entropies of lipids in pure and mixed bilayers from atomic-level and coarse-grained molecular dynamics simulations. *J. Phys. Chem. B* **110**, 15602–15614.

Baron, R., Trzesniak, D., de Vries, A.H., Elsener, A., Marrink, S.J., van Gunsteren, W.F. (2007). Comparison of thermodynamic properties of coarse-grained and atomic-level simulation models. *ChemPhysChem* **8**, 452–461.

Bayburt, T.H., Sligar, S.G. (2002). Single-molecule height measurements on microsomal cytochrome P450 in nanometer-scale phospholipid bilayer disks. *Proc. Natl. Acad. Sci. USA* **99**, 6725–6730.

Bayburt, T.H., Sligar, S.G. (2003). Self-assembly of single integral membrane proteins into soluble nanoscale phospholipid bilayers. *Protein Sci.* **12**, 2476–2481.

Bayburt, T.H., Grinkova, Y.V., Sligar, S.G. (2002). Self-assembly of discoidal phospholipids bilayer nanoparticles with membrane scaffold proteins. *Nano Lett.* **2**, 853–856.

Bayburt, T.H., Grinkova, Y.V., Sligar, S.G. (2006). Assembly of single bacteriorhodopsin trimers in bilayer nanodiscs. *Arch. Biochem. Biophys.* **450**, 215–222.

Bayburt, T.H., Leitz, A.J., Xie, G., Oprian, D.D., Sligar, S.G. (2007). Transducin activation by nanoscale lipid bilayers containing one and two rhodopsins. *J. Biol. Chem.* **282**, 14875–14881.

Bergeron, J., Frank, P.G., Scales, D., Meng, Q.-H., Castro, G., Marcel, Y.L. (1995). Apolipoprotein A-I conformation in reconstituted discoidal lipoproteins varying in phospholipids and cholesterol content. *J. Biol. Chem.* **270**, 27429–27438.

Bhat, S., Sorci-Thomas, M.G., Tuladhar, R., Samuel, M.P., Thomas, M.J. (2007). Conformational adaptation of apolipoprotein A-I to discretely sized phospholipid complexes. *Biochemistry* **46**, 7811–7821.

Bikiel, D.E., Boechi, L., Capece, L., Crespo, A., De Biase, P.M., Di Lella, S., Gonzalez Lebrero, M.C., Marti, M.A., Nadra, A.D., Perissinotti, L.L., Scherlis, D.A., Estrin, D.A. (2006). Modeling heme proteins using atomistic simulations. *Phys. Chem. Chem. Phys.* **8**, 5611–5628.

Boguski, M.S., Freeman, M., Elshourbagy, N.A., Taylor, J.M., Gordon, J.I. (1986). On computer-assisted analysis of biological sequences: Proline punctuation, consensus sequences, and apolipoprotein repeats. *J. Lipid Res.* **27**, 1011–1034.

Boldog, T., Grimme, S., Li, M., Sligar, S.G., Hazelbauer, G.L. (2006). Nanodiscs separate chemoreceptor oligomeric states and reveal their signaling properties. *Proc. Natl. Acad. Sci. USA* **103**, 11509–11514.

Borhani, D.W., Rogers, D.P., Engler, J.A., Brouillette, C.G. (1997). Crystal structure of truncated human apolipoprotein A-I suggests a lipid-bound conformation. *Proc. Natl. Acad. Sci. USA* **94**, 12291–12296.

Brouillette, C.G., Anantharamaiah, G.M. (1995). Structural models of human apolipoprotein A-I. *Biochim. Biophys. Acta* **1256**, 103–129.

Brouillette, C.G., Anantharamaiah, G.M., Engler, J.A., Borhani, D.W. (2001). Structural models of human apolipoprotein A-I: A critical analysis and review. *Biochim. Biophys. Acta* **1531**, 4–46.

Catte, A., Patterson, J.C., Jones, M.K., Jerome, W.G., Bashtovyy, D., Su, Z., Gu, F., Chen, J., Aliste, M.P., Harvey, S.C., Li, L., Weinstein, G., Segrest, J.P. (2006). Novel changes in discoidal high density lipoprotein morphology: A molecular dynamics study. *Biophys. J.* **90**, 4345–4360.

Civjan, N.R., Bayburt, T.H., Schuler, M.A., Sligar, S.G. (2003). Direct solubilization of heterologously expressed membrane proteins by incorporation into nanoscale lipid bilayers. *Biotechniques* **35**, 556–560, 562–563.

Das, A., Grinkova, Y.V., Sligar, S.G. (2007). Redox potential control by drug binding to Cytochrome P450 3A4. *J. Am. Chem. Soc.* **129**, 13778–13779.

Davidson, W.S., Gillotte, K.L., Lund-Katz, S., Johnson, W.J., Rothblat, G.H., Phillips, M.C. (1995). The effect of high density lipoprotein phospholipid acyl chain composition on the effux of cellular free cholesterol. *J. Biol. Chem.* **270**, 5882–5890.

Davidson, W.S., Silva, R.A.G.D. (2005). Apolipoprotein structural organization in high density lipoproteins: Belts, bundles, hinges and hairpins. *Curr. Opin. Lipidol.* **16**, 295–300.

Davydov, D.R., Fernando, H., Baas, B.J., Sligar, S.G., Halpert, J.R. (2005). Kinetics of dithionite-dependent reduction of cytochrome P450 3A4: Heterogeneity of the enzyme caused by its oligomerization. *Biochemistry* **44**, 13902–13913.

de Joannis, J., Jiang, F.Y., Kindt, J. (2006). Coarse-grained model simulations of mixed lipid systems: Composition and line tension of a stabilized bilayer edge. *Langmuir* **22**, 998–1005.

Denisov, I.G., Grinkova, Y.V., Lazarides, A.A., Sligar, S.G. (2004). Directed self-assembly of monodisperse phospholipid bilayer nanodiscs with controlled size. *J. Am. Chem. Soc.* **126**, 3477–3487.

Denisov, I.G., McLean, M.A., Shaw, A.W., Grinkova, Y.V., Sligar, S.G. (2005). Thermotropic phase transition in soluble nanoscale lipid bilayers. *J. Phys. Chem. B* **109**, 15580–15588.

Denisov, I.G., Grinkova, Y.V., Baas, B.J., Sligar, S.G. (2006). The ferrous-dioxygen intermediate in human cytochrome P450 3A4: Substrate dependence of formation and decay kinetics. *J. Biol. Chem.* **281**, 23313–23318.

Denisov, I.G., Baas, B.J., Grinkova, Y.V., Sligar, S.G. (2007). Cooperativity in P450 CYP3A4: Linkages in substrate binding, spin state, uncoupling and product formation. *J. Biol. Chem.* **282**, 7066–7076.

Duan, H., Civjan, N.R., Sligar, S.G., Schuler, M.A. (2004). Co-incorporation of heterologously expressed Arabidopsis cytochrome P450 and P450 reductase into soluble nanoscale lipid bilayers. *Arch. Biochem. Biophys.* **424**, 141–153.

Gorshkova, I.N., Liu, T., Kan, H.Y., Chroni, A., Zannis, V.I., Atkinson, D. (2006). Structure and stability of apolipoprotein A-I in solution and in discoidal high-density lipoprotein probed by double charge ablation and deletion mutation. *Biochemistry* **45**, 1242–1254.

Gumbart, J., Wang, Y., Aksimentiev, A., Tajkhorshid, E., Schulten, K. (2005). Molecular dynamics simulations of proteins in lipid bilayers. *Curr. Opin. Struct. Biol.* **15**, 423–431.

Gursky, O., Atkinson, D. (1996). Thermal unfolding of human high-density apolipoprotein A-1: Implications for a lipid-free molten globular state. *Proc. Natl. Acad. Sci. USA* **93**, 2991–2995.

Humphrey, W., Dalke, A., Schulten, K. (1996). VMD—Visual Molecular Dynamics. *J. Mol. Graphics* **14**, 33–38.

Jonas, A. (1986). Reconstitution of high-density lipoproteins. *Methods Enzymol.* **128**, 553–582.

Jonas, A., Kézdy, K.E., Wald, J.H. (1989). Defined apolipoprotein A-I conformations in reconstituted high density lipoprotein discs. *J. Biol. Chem.* **264**, 4818–4824.

Jonas, A., von Eckardstein, A., Kézdy, K.E., Steinmetz, A., Assmann, G. (1991). Structural and functional properties of reconstituted high density lipoprotein discs prepared with six apolipoprotein A-I variants. *J. Lipid Res.* **32**, 97–106.

Jonas, A., Wald, J.H., Toohill, K.L.H., Krul, E.S., Kézdy, K.E. (1990). Apolipoprotein A-I structure and lipid properties in homogeneous reconstituted spherical and discoidal high density lipoproteins. *J. Biol. Chem.* **265**, 22123–22129.

Klon, A.E., Jones, M.K., Segrest, J.P., Harvey, S.C. (2000). Molecular belt models for the apolipoprotein A-I Paris and Milano mutations. *Biophys. J.* **79**, 1679–1685.

Klon, A.E., Segrest, J.P., Harvey, S.C. (2002a). Comparative models for human apolipoprotein A-I bound to lipid in discoidal high-density lipoprotein particles. *Biochemistry* **41**, 10895–10905.

Klon, A.E., Segrest, J.P., Harvey, S.C. (2002b). Molecular dynamics simulations on discoidal HDL particles suggest a mechanism for rotation in the apo A-I belt model. *J. Mol. Biol.* **324**, 703–721.

Koppaka, V., Silvestro, L., Engler, J.A., Brouillette, C.G., Axelsen, P.H. (1999). The structure of human lipoprotein A-I. Evidence for the "belt" model. *J. Biol. Chem.* **274**, 14541–14544.

Leitz, A.J., Bayburt, T.H., Barnakov, A.N., Springer, B.A., Sligar, S.G. (2006). Functional reconstitution of Beta2-adrenergic receptors utilizing self-assembling nanodisc technology. *Biotechniques* **40**, 601–612.

Li, H., Lyles, D.S., Thomas, M.J., Pan, W., Sorci-Thomas, M.G. (2000). Structural determination of lipid-bound ApoA-I using fluorescence resonance energy transfer. *J. Biol. Chem.* **275**, 37048–37054.

Li, H., Lyles, D.S., Pan, W., Alexander, E., Thomas, M.J., Sorci-Thomas, M.G. (2002). ApoA-I structure on discs and spheres. Variable helix registry and conformational states. *J. Biol. Chem.* **277**, 39093–39101.

Li, L., Chen, J., Mishra, V.K., Kurtz, J.A., Cao, D., Klon, A.E., Harvey, S.C., Anantharamaiah, G., Segrest, J.P. (2004). Double belt structure of discoidal high density lipoproteins: Molecular basis for size heterogeneity. *J. Mol. Biol.* **343**, 1293–1311.

Li, Y., Kijac, A.Z., Sligar, S.G., Rienstra, C.M. (2006). Structural analysis of nanoscale self-assembled discoidal lipid bilayers by solid-state NMR spectroscopy. *Biophys. J.* **91**, 3819–3828.

Lu, D., Aksimentiev, A., Shih, A.Y., Cruz-Chu, E., Freddolino, P.L., Arkhipov, A., Schulten, K. (2006). The role of molecular modeling in bionanotechnology. *Phys. Biol.* **3**, S40–S53.

Marcel, Y.L., Kiss, R.S. (2003). Structure-function relationships of apolipoprotein A-I: A flexible protein with dynamic lipid associations. *Curr. Opin. Lipidol.* **14**, 151–157.

Marrink, S.J. (2004). Molecular dynamics simulation of cholesterol nucleation in mixed micelles modelling human bile. In: "Proceedings of Falk Symposium", vol. **139**. Kluwer Academic, Dordrecht, pp. 98–105.

Marrink, S.J., Mark, A.E. (2002). Molecular dynamics simulations of mixed micelles modeling human bile. *Biochemistry* **41**, 5375–5382.

Marrink, S.J., Mark, A.E. (2003a). The mechanism of vesicle fusion as revealed by molecular dynamics simulations. *J. Am. Chem. Soc.* **125**, 11144–11145.

Marrink, S.J., Mark, A.E. (2003b). Molecular dynamics simulation of the formation, structure, and dynamics of small phospholipid vesicles. *J. Am. Chem. Soc.* **125**, 15233–15242.

Marrink, S.J., Mark, A.E. (2004). Molecular view of hexagonal phase formation in phospholipid membranes. *Biophys. J.* **87**, 3894–3900.

Marrink, S.J., de Vries, A.H., Mark, A.E. (2004). Coarse grained model for semiquantitative lipid simulations. *J. Phys. Chem. B* **108**, 750–760.

Marrink, S.J., Risselada, J., Mark, A.E. (2005). Simulation of gel phase formation and melting in lipid bilayers using a coarse grained model. *Chem. Phys. Lipids* **135** (2), 223–244.

Marrink, S.J., Risselada, J.H., Yefimov, S., Tieleman, D.P., de Vries, A.H. (2007). The MARTINI force field: Coarse grained model for biomolecular simulations. *J. Phys. Chem. B* **111**, 7812–7824.

Mishra, V.K., Palgunachari, M.N., Segrest, J.P., Anantharamaiah, G.M. (1994). Interactions of synthetic peptide analogs of the class A amphipathic helix with lipids. Evidence for the snorkel hypothesis. *J. Biol. Chem.* **269**, 7185–7191.

Nath, A., Atkins, W.M., Sligar, S.G. (2007). Applications of phospholipid bilayer nanodiscs in the study of membrane and membrane proteins. *Biochemistry* **46**, 2059–2069.

Nofer, J.R., Walter, M., Assmann, G. (2005). Current understanding of the role of high-density lipoproteins in artherosclerosis and senescene. *Expert. Rev. Cardiovasc. Ther.* **3**, 1071–1086.

Nolte, R.T., Atkinson, D. (1992). Conformational analysis of apolipoprotein A-I and E-3 based on primary sequence and circular dichroism. *Biophys. J.* **63**, 1221–1239.

Panagotopulos, S.E., Horace, E.M., Maiorano, J.N., Davidson, W.S. (2001). Apolipoprotein A-I adopts a belt-like orientation in reconstituted high density lipoproteins. *J. Biol. Chem.* **276**, 42965–42970.

Phillips, J.C., Wriggers, W., Li, Z., Jonas, A., Schulten, K. (1997). Predicting the structure of apolipoprotein A-I in reconstituted high density lipoprotein disks. *Biophys. J.* **73**, 2337–2346.

Phillips, J.C., Braun, R., Wang, W., Gumbart, J., Tajkhorshid, E., Villa, E., Chipot, C., Skeel, R.D., Kale, L., Schulten, K. (2005). Scalable molecular dynamics with NAMD. *J. Comput. Chem.* **26**, 1781–1802.

Rogers, D.P., Brouillette, C.G., Engler, J.A., Tendian, S.W., Roberts, L., Mishra, V.K., Anantharamaiah, G.M., Lund-Katz, S., Phillips, M.C., Ray, M.J. (1997). Truncation of the amino terminus of human apolipoprotein A-I substantially alters only the lipid-free conformation. *Biochemistry* **36**, 288–300.

Rogers, D.P., Roberts, L.M., Lebowitz, J., Engler, J.A., Brouillette, C.G. (1998). Structural analysis of apolipoprotein A-I: Effects of amino- and carboxy-terminal deletions on the lipid-free structure. *Biochemistry* **37**, 945–955.

Roux, B., Allen, T., Berneche, S., Im, W. (2004). Theoretical and computational models of biological ion channels. *Quart. Rev. Biophys.* **37** (1), 15–103.

Sachs, J.N., Petrache, H.I., Woolf, T.B. (2003). Interpretation of small angle X-ray measurements guided by molecular dynamics simulations of lipid bilayers. *Chem. Phys. Lipids* **126**, 211–223.

Segrest, J.P. (1977). Amphipathic helixes and plasma lipoproteins: Thermodynamic and geometric considerations. *Chem. Phys. Lipids* **18**, 7–22.

Segrest, J.P., Jones, M.K., De Loof, H., Brouillette, C.G., Venkatachalapathi, Y.V., Anantharamaiah, G.M. (1992). The amphipathic helix in the exchangeable apolipoproteins: A review of secondary structure and function. *J. Lipid Res.* **33**, 141–166.

Segrest, J.P., Garber, D.W., Brouillette, C.G., Harvey, S.C., Anantharamaiah, G.M. (1994). The amphipathic alpha helix: A multifunctional structural motif in plasma apolipoproteins. *Adv. Protein Chem.* **45**, 303–369.

Segrest, J.P., Jones, M.K., Klon, A.E., Sheldahl, C.J., Hellinger, M., De Loof, H., Harvey, S.C. (1999). A detailed molecular belt model for apolipoprotein A-I in discoidal high density lipoprotein. *J. Biol. Chem.* **274**, 31755–31758.

Segrest, J.P., Harvey, S.C., Zannis, V. (2000). Detailed molecular model of apolipoprotein A-I on the surface of high-density lipoproteins and its functional implications. *Trends Cardiovasc. Med.* **10**, 246–252.

Shaw, A.W., McLean, M.A., Sligar, S.G. (2004). Phospholipid phase transitions in homogeneous nanometer scale bilayer discs. *FEBS Lett.* **556**, 260–264.

Shaw, A.W., Pureza, V.S., Sligar, S.G., Morrissey, J.H. (2007). The local phospholipid environment modulates the activation of blood clotting. *J. Biol. Chem.* **282**, 6556–6563.

Sheldahl, C.J., Harvey, S.C. (1999). Molecular dynamics on a model for nascent discoidal high-density lipoprotein: Role of salt-bridges. *Biophys. J.* **76**, 1190–1198.

Shih, A.Y., Denisov, I.G., Phillips, J.C., Sligar, S.G., Schulten, K. (2005). Molecular dynamics simulations of discoidal bilayers assembled from truncated human lipoproteins. *Biophys. J.* **88**, 548–556.

Shih, A.Y., Arkhipov, A., Freddolino, P.L., Schulten, K. (2006). Coarse grained protein-lipid model with application to lipoprotein particles. *J. Phys. Chem. B* **110**, 3674–3684.

Shih, A.Y., Arkhipov, A., Freddolino, P.L., Sligar, S.G., Schulten, K. (2007a). Assembly of lipids and proteins into lipoprotein particles. *J. Phys. Chem. B* **111**, 11095–11104.

Shih, A.Y., Freddolino, P.L., Arkhipov, A., Schulten, K. (2007b). Assembly of lipoprotein particles revealed by coarse-grained molecular dynamics simulations. *J. Struct. Biol.* **157**, 579–592.

Shih, A.Y., Freddolino, P.L., Sligar, S.G., Schulten, K. (2007c). Disassembly of nanodiscs with cholate. *Nano Lett.* **7**, 1692–1696.

Silva, R.A.G.D., Hilliard, G.M., Li, L., Segrest, J.P., Davidson, W.S. (2005). A mass spectrometric determination of the conformation of dimeric apolipoprotein A-I in discoidal high density lipoproteins. *Biochemistry* **44**, 8600–8607.

Sligar, S.G. (2003). Finding a single-molecule solution for membrane proteins. *Biochem. Biophys. Res. Commun.* **312**, 115–119.

Sparks, D.L., Davidson, W.S., Lund-Katz, S., Phillips, M.C. (1995). Effects of the neutral lipid content of high density lipoprotein on apolipoprotein A-I structure and particle stability. *J. Biol. Chem.* **270**, 26910–26917.

Svergun, D.I., Barberato, C., Koch, M.H.J. (1995). Crysol—a program to evaluate X-ray solution scattering of biological macromolecules from atomic coordinates. *J. Appl. Crystallogr.* **28**, 768–773.

Swaney, J.B. (1980). Properties of lipid–apolipoprotein association products. Complexes of human apo AI and binary phospholipid mixtures. *J. Biol. Chem.* **255**, 8798–8803.

Tieleman, D.P. (2006). Computer simulations of transport through membranes: Passive diffusion, pores, channels and transporters. *Clin. Exp. Pharmacol. Physiol.* **33**, 893–903.

Tricerri, M.A., Behling Agree, A.K., Sanchez, S.A., Bronski, J., Jonas, A. (2001). Arrangement of apolipoprotein A-I in reconstituted high-density lipoprotein disks: An alternative model based on fluorescence resonance energy transfer experiments. *Biochemistry* **40**, 5065–5074.

Wald, J.H., Coormaghtigh, E., De Meutter, J., Ruysschaert, J.M., Jonas, A. (1990a). Investigation of the lipid domains and apolipoprotein orientation in reconstituted high density lipoproteins by fluorescence and IR methods. *J. Biol. Chem.* **265**, 20044–20050.

Wald, J.H., Krul, E.S., Jonas, A. (1990b). Structure of apolipoprotein A-I in three homogeneous, reconstituted high density lipoprotein particles. *J. Biol. Chem.* **265**, 20037–20043.

Wang, M., Briggs, M.R. (2004). HDL: The metabolism, function, and therapeutic importance. *Chem. Rev.* **104**, 119–137.

Zagrovic, B., Jayachandran, G., Millett, I.S., Doniach, S., Pande, V.S. (2005). How large is an alpha-helix? Studies of the radii of gyration of helical peptides by small-angle X-ray scattering and molecular dynamics. *J. Mol. Biol.* **353**, 232–241.

CHAPTER 12

Gas Conduction of Lipid Bilayers and Membrane Channels

Yi Wang, Y. Zenmei Ohkubo and Emad Tajkhorshid

Center for Biophysics and Computational Biology, Department of Biochemistry, and
Beckman Institute University of Illinois at Urbana-Champaign, Urbana, IL 61801, USA

Abstract

Exchange of gas molecules across biological membranes constitutes one of the most fundamental phenomena in biology of aerobic organisms. The primary mechanism for gas conduction across the cellular membrane is deemed to be free diffusion of the species across lipid bilayers, however, the involvement of a number of membrane channels in the process has also been suggested. In this chapter we summarize the results of recent molecular dynamics simulations investigating the mechanism and pathways through which O_2, CO_2, and other biologically relevant gas species are exchanged between the two sides of the membrane. Different

Current Topics in Membranes, Volume 60   1063-5823/08 $35.00
DOI: 10.1016/S1063-5823(08)00012-4

computational methodologies are employed, including explicit gas diffusion simulations where multiple copies of the gas species of interest are explicitly included in the simulation, under either equilibrium or biased chemical potential conditions, and implicit ligand sampling where the distribution of small, neutral ligands (gas molecules) inside membrane and/or a protein are deduced from the simulation of the ligand-free system. The results of simulations of pure lipid bilayers indicate that although the lipid bilayers are permeable by the gas species investigated, there is a significant barrier against gas permeation in the head-group layer. The barrier appears to be, at least partly, due to a tighter structure of water in the head-group region and can be markedly affected by changes in the head-group composition of the bilayer. In addition to lipid bilayers, several membrane channels were also investigated. Interestingly, almost all studied systems provide one or more pathways for gas conduction. Some of the identified gas conduction pathways are along the symmetry axis of oligomeric membrane channels, which might suggest the first functional implication for oligomerization of these proteins inside the membrane.

I. INTRODUCTION

Transport of materials across biological membranes is a fundamental process in all living cells. Cellular membranes are primarily composed of lipid molecules, whose hydrophobic tails introduce a significant energetic barrier against water soluble molecules, which are the most abundant species around living cells. Cells have developed specialized mechanisms that facilitate the transmembrane diffusion of polar, and particularly charged molecules (ions) in a highly selective manner. Two important classes of specialized proteins for this purpose are membrane channels and membrane transporters, the latter requiring energy for their function. A large portion of the cell's energy and genetic material is, in fact, spent to support such mechanisms to ensure effective and adequate exchange of water soluble molecules across the membrane in a selective manner.

In contrast to polar molecules, small, uncharged molecules, such as alcohols, gases, and other hydrophobic species, can in principle traverse pure lipid bilayers and, thus, would not require any additional effort from the cell to cross the membrane. This is the case even for the prototypical polar molecule, water. Water molecules can permeate a wide variety of pure lipid bilayers, and, thus, water diffusion through lipid molecules constitutes a major component of transmembrane water exchange in many cells. In fact, before the discovery of specialized water channels, lipid-mediated water diffusion was deemed to be the only mechanism by which water diffuses in and out of the cell. Not surprisingly, its small size and its overwhelming presence in biological fluids (55 M concentration) both contribute to this attribute.

The discovery of water-conducting properties for a family of membrane proteins, aquaporins (AQPs), and their identification and major physiological roles in a large number of cells and organs, considerably affected the prevailing view of water transport across cellular membranes. Apparently, water diffusion through lipid molecules is not adequate under certain conditions and for certain organs. For example, in order to concentrate urine, our kidneys rely on a transcellular water pathway, meaning that water needs to cross the cellular membrane twice. This pathway is used for reabsorption of hundreds of liters of water every day. Processes like this cannot rely on the apparently slow diffusion of water across lipid bilayers, and that is why several members of the AQP family have been made available at high concentrations in membranes of these kidney cells. In contrast to lipid bilayers, channel proteins provide a more controllable mechanism, and, therefore, protein-based pathways provide the additional element of physiological control to the process.

Exchange of gas molecules across cellular membranes presents a very similar problem to water. Evidently, most physiologically relevant gas molecules, e.g., O_2 and CO_2, can readily permeate ordinary pure lipid bilayers, and, thus, there does not seem to be a need for membrane channels to be involved in exchange of gas molecules across the cellular membrane. However, the involvement of proteins in this process has been suggested and indeed demonstrated for several proteins, and, interestingly AQPs are among the primary suspects in this regard. The abundance of AQPs in certain organs where a convincing physiological role for increased water permeability is not apparent, and demonstration of the composition-dependence of gas permeability of lipid bilayers are suggestive that protein-mediated gas conduction could play a role and be of physiological relevance.

In this chapter we will present the results of some of our recent studies investigating in detail the mechanism of permeation of a number of gas molecules through various lipid bilayers and membrane channels by classical molecular dynamics (MD) simulations. Due to the fast diffusion of small molecules, the problem is an ideal one for MD simulations, as satisfactory sampling of the process can be achieved by currently accessible time scales to MD simulations of system sizes reported in this chapter. As will be described later, we will be combining standard MD simulations with a number of novel computational methodologies in order to improve the sampling even further. Through employing various types of lipids and conditions we hope to shed light on some of the structural and energetic details that may help understand under what conditions protein-mediated gas diffusion might play a physiologically relevant role.

Below, after a brief description of major computational methodologies employed, we will first discuss the results of our simulation studies on gas permeation through pure lipid bilayer systems, and then continue with presenting simulations in which gas conduction through several membrane channels will be demonstrated.

II. COMPUTATIONAL METHODOLOGY

In this section we describe major computational methodologies employed for simulation and analysis of molecular systems reported in this chapter. The emphasis will be primarily on novel methodologies that were combined with MD to improve the sampling. The simulation systems and general conditions used for MD simulations will be described only briefly. We refer the reader to the papers cited herein for more details.

In general two types of molecular systems were employed for the simulations, pure lipid bilayers and membrane-embedded models of membrane channels. System sizes employed are on the order of 100,000 atoms, and between 5 and 50 ns of simulations have been conducted for each system. Relevant specifics of each simulation system will be described individually along with the results.

A. Molecular Dynamics Simulations

For all simulations, the program NAMD 2.6 (Phillips *et al.*, 2005) and the CHARMM27 parameter set (Schlenkrich *et al.*, 1996; MacKerell *et al.*, 1998) were used. Assuming periodic boundary conditions, the Particle Mesh Ewald (PME) method (Darden *et al.*, 1993) with a grid density of at least $1/\text{Å}^3$ was employed for computation of long-range electrostatic forces. All simulations were performed with time steps of 1 fs, 2 fs, and 4 fs for bonded, non-bonded and PME calculation, respectively. For constant-pressure simulations, a Langevin piston (Feller *et al.*, 1995) was employed to maintain the pressure at 1 atm. Langevin dynamics was used to keep the temperature constant at 300 K or 310 K. During induced pressure simulations, gas molecules were not coupled to the temperature bath, and counter forces were applied to C_α atoms of the protein and/or heavy atoms of the lipid bilayer to prevent net translocation of the system along the direction of external force. The total counter force on the membrane was equal to the total external force applied on gas molecules. In the protein-membrane systems, the counter force was evenly distributed between the protein and the lipids according to the ratio of their areas in the membrane.

B. Explicit and Implicit Sampling

Explicit gas diffusion simulations were performed on AQP-embedded membranes as well as pure lipid bilayers. For the AQP1 tetramer embedded in a POPE bilayer, 100 copies of either CO_2 or O_2 molecules are placed in the bulk water at the beginning of the simulations (Wang and Tajkhorshid, 2007). Control simulations were performed on pure POPE bilayers without AQP1. The explicit gas diffusion simulation of nitric oxide (NO) permeation across AQP4 started with a bigger lipid patch than the AQP1 system. To account for the larger system, a larger number (150) of NO molecules were placed in the bulk water at the beginning of the simulation.

Implicit ligand sampling is a method to infer the three-dimensional potential of mean force (3D PMF) of a weakly interacting ligand, purely based on an equilibrium simulation of the ligand-free system. This method overcomes the problem of poor sampling caused by the slow diffusion of explicit ligands exploring a protein and has proven useful in finding gas migration pathways in proteins (Cohen *et al.*, 2005, 2006; Wang and Tajkhorshid, 2007). Multiple conformations of an imaginary ligand molecule are placed at every point of a grid, on top of an equilibrium trajectory of a simulated protein system. At each time step, and for each ligand position and orientation, the protein-ligand interactions are computed. These interaction energies are then used to reconstruct the PMF of placing the ligand at any point inside the protein using a perturbative approach, as described in detail in (Cohen *et al.*, 2006).

In order to compute the PMFs, equilibrium simulations of pure lipid bilayers and membrane-embedded proteins in the absence of gas molecules were performed for 1 ns and 5 ns, respectively (Wang and Tajkhorshid, 2007). For each of the systems, a 3D PMF map at 1 Å resolution was computed, using the `volmap` feature of the VMD software (Humphrey *et al.*, 1996). For the lipid bilayer and bulk water regions, the projected PMFs along the z-axis, $F(z)$, were then calculated from the 3D PMF map, using the formula:

$$F(z) = -RT \ln \sum_{x,y=0}^{L_x,L_y} \frac{e^{-F(x,y,z)/RT}}{L_x L_y}, \tag{1}$$

where $F(x, y, z)$ is the local 3D PMF computed by implicit ligand sampling, and L_x and L_y are the dimensions of the PMF map in units of Å. In the case of AQP1 and AQP4, the calculation is the same as in Eq. (1), except that the summations were performed over restricted cross-sectional areas, e.g., the central pore only, of the total PMF map in order to isolate the targeted gas channels. The resulting PMF was then shifted by $-RT \ln(L_x L_y / A_0)$, to account for the entire AQP tetramer, where $L_x L_y$ is the area of the summation, and A_0 is roughly the area occupied by an AQP tetramer ($A_0 = 50$ nm^2). This is done so that the absolute values of the PMF can be compared with those from explicit gas diffusion simulations.

C. Induced Pressure Simulations

After the first 10 ns of equilibrium simulation in each system, we applied a force of 20 pN on the center of mass of every gas molecule along the $-z$ direction in a 10 ns "induced pressure" simulation (Wang *et al.*, 2007). Previous simulations (Zhu *et al.*, 2002, 2004) had shown that such an external force can generate an effect equivalent to a concentration gradient across the membrane, biasing the diffusion of gas molecules in one direction. Using the force applied in our simulations, a pressure difference of $\sim$20 MPa was generated across the membrane (Wang and Tajkhorshid, 2007).

D. Analysis

1. Free Energy Profile Calculation
The free energy profile associated with gas permeation is calculated from both implicit ligand sampling, as described above, and explicit gas diffusion simulations, using the relation (Marrink and Berendsen, 1994):

$$\Delta G(z) = -RT \ln \frac{C(z)}{C_{\text{bulk}}}, \qquad (2)$$

where R is the gas constant, T is the temperature, C_{bulk} is the concentration of gas molecules in bulk water, and $C(z)$ is the concentration of gas molecules in each 1 Å thick layer parallel to the xy plane along the z-axis (membrane normal). For AQP-embedded membranes, $C(z)$ is the gas concentration inside the central pore, except in the bulk water region, where gas concentration in the entire layer was calculated and then normalized according to the area of an AQP tetramer (50 nm^2). This normalization ensures that our calculated PMF corresponds to a definite density of AQP (1 AQP per 50 nm^2 of bilayer). As a result, by comparing the PMFs for the protein and lipids, we can directly determine whether replacing a patch of lipids by the protein will increase or decrease the barrier against transmembrane gas permeation (Wang and Tajkhorshid, 2007).

2. Order Parameter
The lipid order parameter S_{CD} is calculated as

$$S_{\text{CD}} = \left\langle \frac{3}{2} \cos^2 \theta - \frac{1}{2} \right\rangle, \qquad (3)$$

where θ is the angle between each C–H bond and the membrane normal, and $\langle \ldots \rangle$ stands for the ensemble average.

III. LIPID-MEDIATED GAS CONDUCTION

In this section, we report a number of MD simulations of various lipid bilayer systems used to investigate gas permeation across biological membranes. Both implicit and explicit sampling methods described in Computational Methodology were employed. This section will start with a zwitterionic (electroneutral) lipid (POPE) as a prototypical lipid bilayer widely used in simulation of membranes and membrane proteins. As described below, the results of POPE simulations revealed that the main origin of the barrier against gas diffusion across lipid bilayers is the head group region and most likely the structure and dynamics of water play an important role in determining the barrier. In order to investigate this effect in more detail, two different types of charged lipids were then simulated: DOPS and cardiolipin (CL). As these systems present rather new simulation systems, we will begin our discussion for DOPS and CL with a brief description of their structural and dynamical properties before addressing their gas permeability.

A. Energetics of Gas Permeation Across Pure Lipid Bilayers

A typical free energy profile associated with gas permeation through POPE is shown in Fig. 1. In general, the bilayer has a rather "flat" free energy profile for O_2, with the largest energy barrier being 0.4 kcal/mol relative to the bulk water (Wang *et al.*, 2007). The PMF for CO_2 permeation (data not shown) reveals a similar energy barrier (Wang *et al.*, 2007). Such a profile allows gas molecules to readily diffuse through the lipid bilayer, a behavior observed in our simulations and reflected in the number of permeation events during the explicit gas diffusion simulations; on the average one permeation event per nanosecond is observed for both CO_2 and O_2. An energy well of -1.0 kcal/mol for CO_2 and -1.3 to -1.5 kcal/mol for O_2 is located at the bilayer center, while a small barrier of 0.1 to 0.4 kcal/mol for CO_2 and 0.1 to 0.6 kcal/mol for O_2 has to be overcome at

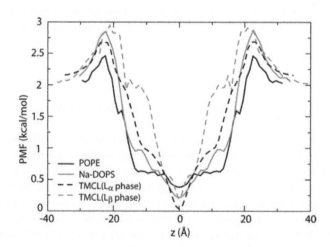

FIGURE 1 Gas permeability of pure lipid bilayers. Free energy profiles of O_2 permeation across different lipid bilayers. The PMFs for O_2 permeation in POPE (black solid line), DOPS with Na^+ ions (gray solid line), the L_α phase TMCL (black dashed line) and the L_β phase TMCL (gray dashed line) are shown.

the head group region where gas molecules cross the lipid–water interface (Wang and Tajkhorshid, 2007). It is noteworthy that the head group region is also the region in which gas molecules exhibit the least mobility (Wang and Tajkhorshid, 2007). Inside the bilayer, traversal (along the membrane normal) gas diffusion is almost always faster than the lateral diffusion, however, both the traversal and lateral diffusions are significantly slower in the head group region. We have also calculated the available free volume from a 100 ps equilibration of the POPE bilayer with no gas molecules. It is found that the diffusion coefficients correlate closely with the free volume inside the bilayer in all regions, suggesting that the gas diffusion rate across a lipid bilayer is proportional to the available free space in the bilayer.

Besides the free energy profile, one can also assess the gas permeability of a membrane by counting the number of permeation events in a given period of time. However, as only one permeation per nanosecond is observed for POPE-CO_2 and POPE-O_2 during equilibrium simulations, better statistics are required to meaningfully represent the permeability of the POPE bilayer. This was achieved by induced pressure simulations, in which the diffusion of gas molecules was biased in one direction through application of external force, in order to increase the total number of permeation events across the membrane (Wang and Tajkhorshid, 2007). Through this approach, the number of gas permeation events was increased by approximately an order of magnitude, allowing us to compare the gas permeability of different types of membranes directly (Wang and Tajkhorshid, 2007). During induced pressure simulations, 203 CO_2 and 258 O_2 permeation events were observed for pure POPE bilayers in 10 ns, corresponding to approximately 20 to 25 permeation events per nanosecond. These values are compared with the AQP1-embedded membrane to illustrate the difference in gas permeability of pure lipid bilayers and the membrane protein AQP1.

B. Electrostatic Effects in Gas Permeability of Lipid Bilayers

The simulations on POPE clearly indicated that the main barrier against gas permeation in lipid bilayers is in the head group region. Furthermore, analysis of water dynamics in the head group region also revealed that the structure and dynamics of water in this region might play a role in the barrier. In order to investigate the effect of the head group composition on the permeation rate, we decided to simulate two negatively charged lipid bilayer and see how the energetics of permeation might be affeced. Charged head groups are expected to exhibit stronger electrostatic effects which would result in a different behavior for water and thus are expected to create a different barrier against gas diffusion.

The simulated DOPS bilayer used to investigate gas permeability of bilayers consists of 288 DOPS molecules solvated in 14,851 water molecules. The system was simulated with two different counterions; the first system included 288 Na^+ ions (Na-DOPS) (Ohkubo and Tajkhorshid, 2008), whereas in the second simulations (Ca-DOPS) 144 Ca^{2+} ions were used to neutralize the system. The bilayer was equilibrated for 10 ns under NP_nTA conditions. The area of the bilayer was fixed at 97×97 $Å^2$, corresponding to an average area per lipid, A_L, of 65.3 $Å^2$, i.e., the experimentally measured value for DOPS (Petrache *et al.*, 2004). The currently used lipid parameters of the CHARMM force-field (MacKerell *et al.*, 1998) show a tendency of shrinking lipid bilayers, thus, resulting in lower A_L values, especially after long simulation times (Feller and Pastor, 1999; Jensen *et al.*, 2004; Sonne *et al.*, 2007). During the initial 50 ps of equilibration of Na-DOPS, the z dimension of the simulation box dropped from 100 Å to an asymptotic value of 84 Å. The P–P distance of the two leaflets, D_{P-P} (or D_{HH} (Nagle and Tristram-Nagle, 2000)), also converged quickly to 39.4 Å, which is very close to the experimentally determined value of 38.4 Å (Petrache *et al.*, 2004). The Ca-DOPS system was equilibrated in a similarly manner ($D_{P-P} = 39.3$ Å).

The atomic density profiles of the two systems along the membrane normal averaged over the last 5 ns of the equilibration are shown in Fig. 2. The Na^+ and Ca^{2+} counterions were initially placed uniformly in the bulk water. These counterions diffuse during the equilibration and quickly accumulate at the surface of the membrane. At the end of the 10 ns equilibration, they are mostly within or very close to the lipid head group region, interacting with both carboxy and phosphate groups of the DOPS molecules (Figs. 2a and 2b). The peak (± 22 Å) and inner-most (± 15 Å) locations of these different counterions match, respectively. In Na-DOPS, at the end of the equilibration, 14% of Na^+ ions are still in the bulk water, whereas in Ca-DOPS, almost all Ca^{2+} ions are positioned close to the membrane surface at the end of the simulation. Also evident from the atomic density is a significant penetration of water molecules into the head group region all the way into the ester groups of DOPS. The adequate hydration of the head group region of the lipid bilayer is also discernible from the bilayer structure at the end of the equilibration (Figs. 2c and 2d).

Examination of the S_{CD} order parameters (van der Ploeg and Berendsen, 1982; Heller *et al.*, 1993) was used to ensure the fluidity of the DOPS bilayer (Fig. 2e). The calculated order parameters for the DOPS acyl chains are around 0.2 or less, depending on the positions of carbons. The low values of S_{CD} implies that the bilayer has retained its liquid phase during the simulation. These properties of the simulated DOPS bilayer indicate altogether that the membrane has been well equilibrated and is in a good condition for calculation of gas permeability.

The free energy profile associated with permeation of O_2 across Na-DOPS is presented in Fig. 1. As depicted, the energy barrier of the head group region is significantly higher in DOPS than POPE. The observed barrier in POPE (0.4 kcal/mol) has almost doubled due to introduction of charged head groups in DOPS (0.8 kcal/mol). The additional charge and the accumulation of coun-

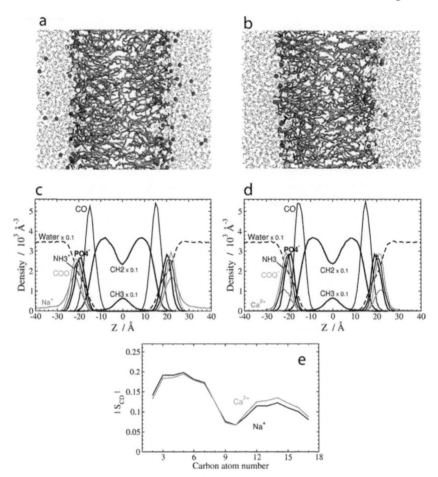

FIGURE 2 DOPS simulation results. The results of the two simulation systems of DOPS neutralized with (a, c, e) Na[+] ions (Na-DOPS) and (b, d, e) Ca[2+] ions (Ca-DOPS) are shown. (a, b) Snapshots of the systems after 10 ns of equilibration (30-Å-thick slices along the membrane normal). Water molecules are shown in wires. The carbon atoms in DOPS are shown in gray sticks. The moieties in the head groups of DOPS are in dark gray. Counterions, shown in dark spheres, accumulate quickly at the surface of the membrane in both simulation systems. (c, d) The density profiles along the membrane normal as a function of the distance from the DOPS membrane center (Z). Deep penetration of both Na[+] and Ca[2+] ions are discernible. (e) Calculated order parameter $|S_{CD}|$ of the carbons in the DOPS tails in Na-DOPS and Ca-DOPS systems.

terions in the head group region of DOPS can significantly affect the structure and dynamics of water molecules in this region, and, thus, create a less flexible hydration shell, which in turn affects the barrier against permeation of small molecules.

C. Gas Conduction Across Mitochondrial Lipid Bilayers

Cardiolipin (CL), or diphosphatidylglycerol, is an anonic phospholipid found extensively in energy-transducing membranes in bacteria and mitochondria (Hoch, 1992; Mileykovskaya *et al.*, 2004). Every CL molecule has two head groups which carry two negative charges, and four fatty acyl tails. CL has been found to be crucial to function of many protein complexes involved in oxidative phosphorylation and photophosphorylation (Mileykovskaya *et al.*, 2004). For instance, CL is required for full electron transport activity of mammalian cytochrome c oxidase (Robinson, 1993; McMillin and Dowhan, 2002). Another example is the mitochondrial ADP/ATP carrier, which is responsible for exchange of ADP and ATP across the inner mitochondrial membrane. The activity of this carrier is reduced to a 20% level when CL is depleted from the membrane (Jiang *et al.*, 2000).

As each CL molecule has two negative charges, one may expect the CL bilayer to be less permeable to hydrophobic gas molecules. To test this hypothesis, we performed altogether 60 ns of MD simulations on a bilayer composed of tetramyristol cardiolipin (TMCL). TMCL has four 14-carbon, saturated fatty acyl chains and is found almost exclusively in bacterial membranes (Mileykovskaya *et al.*, 2004). At 35 °C TMCL is in a gel (L_β) phase, and at 55 °C the bilayer is found to be in a liquid-crystalline phase (L_α) (Lewis *et al.*, 2007). Based on recent experimental findings (Lewis *et al.*, 2007), we built an L_β phase and an L_α phase TMCL bilayer, as shown in Fig. 3. Every monolayer in the two systems is comprised of 100 TMCL molecules. The area per lipid is 83.2 $Å^2$ in the gel phase TMCL, and 103.3 $Å^2$ in the liquid-crystalline phase TMCL. The thickness of the bilayer, defined as the distance between phosphorus atoms of the two monolayers, is 43.5 Å and 37.8 Å in the L_β and L_α TMCL, respectively. To balance the charge of the lipids, 400 Na^+ ions were added to the bulk water at the beginning of the simulation. Similar with the DOPS system, as the simulation proceeds, most of the counterions bind to the head groups of the TMCL bilayers.

After a series of initial minimization and partially constrained MD cycles, each of the two TMCL bilayers are simulated for 30 ns under NP_nTA conditions. The temperature is kept constant at 55 °C for L_α TMCL, and 35 °C for L_β TMCL. A constant pressure is maintained only along the membrane normal while the area of the membrane is fixed. Comparison of the calculated order parameters of the two CL systems (Fig. 3d) clearly indicates the significantly higher order of lipid molecules in the gel phase. Using the last 2 ns of these simulations, we calculated the permeability of the two TMCL bilayers to O_2 using the implicit ligand sampling method. As shown in Fig. 1, the gel phase TMCL bilayer exhibits a 0.9 kcal/mol energy barrier to O_2 permeation at the head group region. Among all the simulated lipid bilayers, the barrier of the L_β phase TMCL is the largest and is more than twice of the energy barrier presented by POPE. The L_α phase TMCL

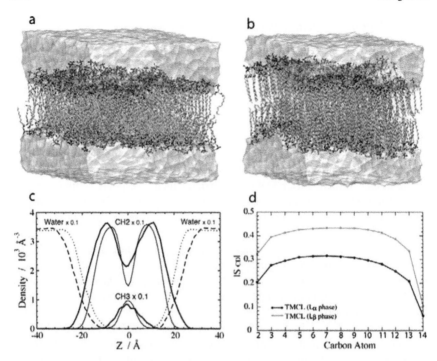

FIGURE 3   TMCL simulation results. (a) The L_α phase TMCL and (b) the L_β phase TMCL after 30 ns of equilibration. Water molecules are shown as transparent boxes. The carbon atoms in TMCL are shown in gray sticks, and the oxygen atoms from the carbonyl groups are colored in black. (c) The density profile of TMCL. The L_α phase TMCL is shown in thick lines (solid or dashed), and the L_β phase TMCL is shown in thin lines (solid or dotted). (d) Calculated order parameter $|S_{CD}|$ for TMCL.

has a slightly lower energy barrier, which may be explained by the less rigidity of the head groups and a larger area per lipid compared with the L_β phase TMCL. However, compared with POPE, which has a 0.4 kcal/mol energy barrier, the liquid-crystalline phase TMCL still presents a 50% higher barrier. These results, together with the PMF obtained for DOPS, suggest that the major energy barrier of a lipid bilayer against gas permeation is determined by the charge of the head group, most likely through affecting the flexibility and hydration of the head group region.

Interestingly, compared with other lipids, the gel phase TMCL has a distinct free energy profile in the lipid tail region. As shown in Fig. 3, the PMF of gas permeation in the gel phase TMCL is elevated by about 1.0 kcal/mol in the lipid tail region (5 Å < |z| < 16 Å), compared with POPE, DOPS and the liquid-crystalline phase TMCL. This difference can be easily explained by the density profile of the lipid bilayers. As shown in Fig. 3c, the density of the CH_2 groups is consistently higher in the gel phase TMCL than the liquid-crystalline phase

TMCL. This results in a smaller free volume in the lipid tail region in the gel phase TMCL, which causes an elevated free energy profile for O_2 permeation. However, it should be noted that the lipid tail region of the gel phase TMCL is still more favorable by O_2 molecules than bulk water. In fact, the tail regions in all of the simulated lipid bilayers have an energy well that attracts the hydrophobic oxygen, even though this energy well is smaller in magnitude for the gel phase TMCL.

IV. PROTEIN-MEDIATED GAS CONDUCTION

In this section we discuss the results of our simulation of gas permeation across a number of membrane channels. The membrane channels to be discussed are primarily AQP water channels. After a brief introduction of the AQP family and relevant structural and functional properties, we will describe in detail gas permeation of two members of the family, AQP1 and AQP4, studied through computational methodologies. We will conclude this section with a brief report of some of our recent preliminary results implicating other oligomeric membrane channels in conduction of gas molecules.

A. Aquaporin Water Channels

AQPs are a family of membrane channels (Preston *et al.*, 1992; Agre *et al.*, 1998; Borgnia *et al.*, 1999; Heymann and Engel, 1999; Hohmann *et al.*, 2001; Fujiyoshi *et al.*, 2002; de Groot and Grubmüller, 2005; Agre, 2006) that, through modulating the water permeability of biological membranes, play a significant role in water homeostasis in living cells (Deen and van Os, 1998; Li and Verkman, 2001; Agre and Kozono, 2003; Agre *et al.*, 2004; King and Agre, 2004). Due to their structural simplicity and the approachable timescale of their basic function, i.e., water conduction, AQPs have been investigated extensively by MD simulations (Zhu *et al.*, 2004; de Groot and Grubmüller, 2001, 2005; Tajkhorshid *et al.*, 2002, 2003, 2005a, 2005b; Zhu *et al.*, 2001; Jensen *et al.*, 2001, 2003; Wang *et al.*, 2005; Grayson *et al.*, 2003; Törnroth-Horsefield *et al.*, 2006). These studies have significantly contributed to our understanding of the molecular basis of their function and selectivity. A particularly intriguing property of AQPs is their ability to block protons (Pohl *et al.*, 2001; Saparov *et al.*, 2005) while allowing water to cross, a problem that has attracted marked attention from theoretical studies (de Groot and Grubmüller, 2005; Tajkhorshid *et al.*, 2002; Jensen *et al.*, 2003; de Groot *et al.*, 2003; Burykin and Warshel, 2003, 2004; Chakrabarti *et al.*, 2004; Ilan *et al.*, 2004).

AQPs present the richest family of membrane channels with regard to the number of known high resolution structures, a fact that makes them a unique, ideal family for investigation of the structural and physical basis of these functional aspects among membrane channels. The structures of two bacterial AQPs (GlpF, a glycerol channel (Tajkhorshid *et al.*, 2002; Fu *et al.*, 2000), and AqpZ (Savage *et al.*, 2003), an orthodox water channel), and five mammalian AQPs (human AQP1 (Murata *et al.*, 2000), bovine AQP1 (Sui *et al.*, 2001), rat AQP4 (Hiroaki *et al.*, 2005), as well as bovine and sheep AQP0 at two different pH conditions (Gonen *et al.*, 2004; Harries *et al.*, 2004; Gonen *et al.*, 2005)) are available at full atomic resolution. Recently, two new AQP structures have been reported, the structure of AqpM, an archaeal H_2S channel (Lee *et al.*, 2005), and the structures of a plant AQP (spinach PIP2) in both closed and open gated states (Törnroth-Horsefield *et al.*, 2006). All members of the family form homotetramers in the membrane in which four functionally independent pores provide highly selective, yet efficient pathways for water permeation across the low dielectric barrier of lipid bilayers. The function of a fifth pore, the "central pore", which is located at the fourfold symmetry axis of the four monomers, remains unclear.

Although generally known for endowing a high transmembrane permeability to water (Preston and Agre, 1991; Agre, 2004), involvement of AQPs in the permeation of small molecules other than water has been reported (Borgnia *et al.*, 1999; Hohmann *et al.*, 2001; Agre and Kozono, 2003; Agre *et al.*, 2004). AQPs have been suggested to participate in gas conduction across cellular membranes (Uehlein *et al.*, 2003; Nakhoul *et al.*, 1998; Cooper and Boron, 1998; Prasad *et al.*, 1998; Cooper *et al.*, 2002; Terashima and Ono, 2002; Blank and Ehmke, 2003; Hanba *et al.*, 2004; Endeward *et al.*, 2006; Herrera *et al.*, 2006). For instance, a recent study on tobacco aquaporin NtAQP1 has suggested that NtAQP1 not only facilitates CO_2 permeation when expressed in *Xenopus* oocytes, but also has a significant function in photosynthesis and in stomatal opening (Uehlein *et al.*, 2003). Another study on *Xenopus* oocytes has shown that the CO_2 permeability is increased by about 40% when AQP1 is expressed (Nakhoul *et al.*, 1998; Cooper and Boron, 1998). AQP1 reconstituted in proteoliposomes has been found to increase both water and CO_2 permeabilities markedly (Prasad *et al.*, 1998). However, as lipid membranes are often gas-permeable, characterizing the gas permeability of AQPs is a challenging experimental problem (Yang *et al.*, 2000; Fang *et al.*, 2002; Verkman, 2002). Moreover, the pathway through which gas molecules might permeate AQPs cannot be identified directly through presently available experimental methods.

B. Conduction of O_2 and CO_2 by AQP1

To shed light on the gas permeability of AQP1 and on its physiological relevance, we performed a total of $\sim$80 ns MD simulations on a membrane-embedded AQP1 tetramer (Wang *et al.*, 2007). Using two different methodologies, namely explicit gas diffusion simulation and implicit ligand sampling (Cohen *et al.*,

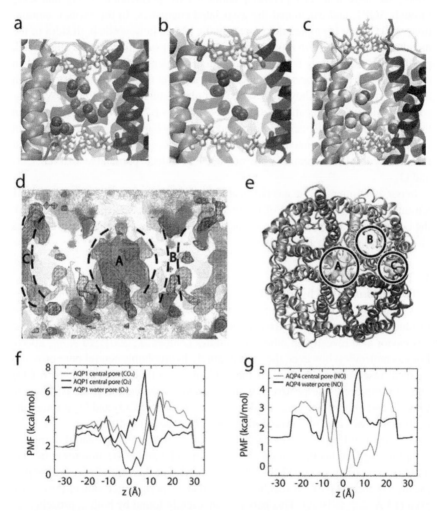

FIGURE 4 Gas permeability of AQP1 and AQP4. The results of explicit gas diffusion simulations are shown in panels a, b, and c. 8 CO_2 (a) and 5 O_2 (b) entered the central pore of AQP1 after 30 ns and 26 ns explicit simulations, respectively. 7 NO molecules (c) were found inside the AQP4 central pore after 50 ns of explicit gas diffusion simulation. (d) The agreement between explicit and implicit simulation results. Areas of high O_2 occupancy predicted by implicit sampling (0.6 kcal/mol energy isosurface shown in black wireframe) are compared with O_2 locations revealed by explicit simulations (gray lines). (e) Top view of the AQP1 tetramer, along with the cylinders used to compute projected PMFs for each of AQP1's various pores: (A) the central pore, (B) the monomeric water pores, and (C) the side pores. (f) PMFs for CO_2 and O_2 permeation through AQP1 central pore and water pores. (g) PMFs for NO permeation through AQP4 central pore and water pores.

2006), we have analyzed potential pathways for permeation of O_2 and CO_2 through AQP1, and calculated the associated energetics. In the explicit simulations, 100 CO_2 or O_2 molecules were initially placed in water on both sides of the membrane and allowed to move freely. The implicit sampling, on the other hand, predicts the PMF purely based on simulations of membrane-embedded AQP1 in the absence of explicit gas molecules. Although the two methods work through disparate mechanisms, they have yielded similar results for AQP1. The PMFs associated with gas permeation revealed a -0.4 to -1.7 kcal/mol energy well in the middle of the AQP1 central pore, and an energy barrier of 3.6 to 4.6 kcal/mol in the periplasmic entrance (Fig. 4f) (Wang *et al.*, 2007). Our results suggest that the hydrophobic central pore of AQP1 is indeed permeable to both CO_2 and O_2.

In our assessment of the gas permeability of AQP1, we have divided the AQP1 tetramer into various regions of interest which we describe separately (Fig. 4e): the central pore, the monomeric water pores, and a side pore that we propose as an additional possible gas pathway.

1. The Central Pore

Two quadruplets of hydrophobic residues, Val52 and Leu172, form the outermost doorways of the AQP1 central pore. Water molecules can hardly pass these residues and are kept away from the inside of the pore during our simulations, a behavior also revealed in other MD studies (Zhu *et al.*, 2001; Yu *et al.*, 2006). For gas molecules, however, the vacant, purely hydrophobic central pore provides an ideal reservoir. At the end of the explicit gas diffusion simulations, 8 CO_2 and 5 O_2 molecules, respectively, were found inside the central pore, as shown in Figs. 4a and 4b. Furthermore, a full permeation event of CO_2 and O_2 through the central pore was also observed during each corresponding simulation.

As shown in Fig. 4f, PMFs calculated from both the explicit and the implicit sampling suggest that the central pore is the most favorable location for both CO_2 and O_2. According to our PMFs, the main barrier of the central pore to gas permeation is about 3.6 to 4.6 kcal/mol, located at the periplasmic side of the central pore ($13 \text{ Å} \leqslant z \leqslant 19 \text{ Å}$). This barrier, consistently found by both approaches, as well as by an independent study (Hub and de Groot, 2006), surprisingly, does not correspond to a region that is sterically blocked directly by the protein. Specifically, this barrier is located *above* the region of maximum protein contraction where the four hydrophobic residues Val52 reside; rather, it corresponds to a region that is occupied primarily by water. A calculation of the local occupancy of water reveals that the barrier corresponds to a dense layer of water molecules surrounded by four aspartate (Asp50) residues. With a higher density than the bulk water, this water layer reduces the chance of gas molecules to access the central pore. We predict that mutation of these aspartate residues to neutral residues, e.g., alanines or asparagines, would weaken the electrostatic effects and results in a less dense water structure. The central pore of AQP1 might then become better gas-conductive.

2. The Water Pores

Compared with the central pore, the four monomeric water pores in AQP1 were found to be less permeable to gas molecules (Wang *et al.*, 2007). In the explicit simulations, although both CO_2 and O_2 could enter the two vestibules of water pores and reach the NPA motifs, no gas molecule passed through the selectivity filter (SF), i.e., the narrowest part of an AQP channel. The single file of water molecules at the SF was almost always kept intact during the simulations. This may be explained by the steric barrier imposed by the protein, as well as the strong tendency of water molecules to keep their hydrogen-bonded structure; as the SF has a diameter of only ~ 2 Å, gas molecules could hardly pass this region without perturbing the water file, which is held strongly by hydrogen bonds between water molecules and with the protein. Using the implicit ligand sampling method, we measured the PMF for O_2 permeation across the four water pores counted as one. As shown in Fig. 4f, the highest energy barrier is located at the SF and is found to be 5.7 kcal/mol with respect to the water solution.

The central pore also appears to be more gas-permeable than the water pores in our induced pressure simulations. During these simulations, the single file of water was largely disturbed when CO_2 penetrated the water pores. In contrast, no obvious disturbance to the protein was observed when CO_2 permeated the central pore. During the induced pressure simulations, four CO_2 and six O_2 permeated the single central pore, while six CO_2 and one O_2 passed through the four water pores. Given that there are four water pores, the ratio of gas permeation between the central pore and a water pore is $\sim 3:1$ for CO_2 and 24:1 for O_2. The difference between CO_2 and O_2 suggests that the water pores of AQP1 are less favorable to the more hydrophobic O_2 molecules. Therefore, while CO_2 might take advantage of both the central pore and water pores to permeate AQP1, O_2 is primarily conducted through the central pore.

3. The Side Pores

An interesting and somewhat unexpected result of our implicit sampling is the identification of a favorable gas channel spanning AQP1, located between each pair of AQP1 monomers near the protein-lipid interface. These "side pores", which to our knowledge have not been considered before, have periplasmic and cytoplasmic exits which are completely enclosed inside the AQP1 tetramer. The shape of the side pores can be seen on either sides of Fig. 4d. The O_2 PMF for the four side pores suggests that it offers low entry and exit barriers to O_2 in AQP1 along with a flat energy profile inside the pore, making it a strong candidate for a favorable gas pathway across AQP1. Furthermore, because the middle of these four pores are directly connected to the center of the lipid bilayer, the side pores can increase the rates of O_2 entry into and exit from the bilayer, further facilitating permeation across the membrane.

Our simulations suggest that aside from its conventional role as the water channel, AQP1 might provide a pathway for small, neutral gas molecules across the membrane. However, compared with a pure POPE bilayer with a similar area, AQP1 is less gas-permeable to both CO_2 and O_2. This is also revealed by the induced pressure simulations: the AQP1-embedded membranes have approximately 8 permeation events per nanosecond in both CO_2 and O_2 simulations, ~60% smaller than the number of pure POPE bilayers. Therefore, gas conduction through AQP1 may be of physiological importance only in membranes with low intrinsic gas permeability or where a major fraction of the surface area of the membrane is occupied by AQPs.

C. Conduction of Nitric Oxide by AQP4

AQP4 is the predominant water channel in the mammalian brain (Amiry-Moghaddam *et al.*, 2003; Hiroaki *et al.*, 2006). Concentrated in the perivascular and subpial membrane domains of astrocytes, AQP4 is an important drug target for the treatment of cerebral edema (Amiry-Moghaddam *et al.*, 2003; Hiroaki *et al.*, 2006). Although both AQP1 and AQP4 are water channels, the two have been shown to possess distinctive biological features. AQP4 forms orthogonal arrays in intact membranes and its water permeation cannot be inhibited by mercurials (Hiroaki *et al.*, 2006). The importance of AQP4 as the brain water channel and its tendency to form ordered arrays inspired us to examine its gas permeability, in particular, its permeability to the gaseous neurotransmitter nitric oxide (NO) (Ignarro, 1999, 2002; Murad, 2004). NO, also known as "endothelium-derived relaxing factor" (EDRF), can trigger the relaxation of vascular smooth muscles and therefore increase the blood flow (Ignarro, 1999, 2002; Murad, 2004). To investigate the permeability of AQP4 to NO, we used both the explicit and implicit sampling methods described before.

As mentioned in the Computational Methodology, 150 NO molecules are placed in bulk water at the beginning of the AQP4 explicit gas diffusion simulations. After 50 ns of equilibration, we observed the entrance of seven NO molecules into the central pore of AQP4, as shown in Fig. 4c. The PMF of NO permeation calculated using implicit ligand sampling reveals a 2.5 kcal/mol energy barrier in the central pore of AQP4, and a 3.5 kcal/mol barrier in the monomeric water pores (Fig. 4g). These results clearly suggest that the AQP4 central pore can provide a pathway for NO permeation across the membrane. The water pores, on the other hand, are less favored by NO, which is similar to the case of AQP1. Likewise, the steric barrier imposed by the protein and the hydrogen-bonded structure of water molecules are most likely the cause of a higher barrier in water pores.

The structure of AQP4 is highly homologous to AQP1. However, a significant difference between the two AQPs is in the periplasmic side of the central pore: AQP4 has an almost completely closed pore at this region, while the central pore of AQP1 is readily accessible to water at this location. This structural difference turns out to be of high relevance to the difference in the gas permeability of the two AQPs as revealed by our simulations. As shown in Fig. 4g, AQP4 has a maximum of 2.5 kcal/mol energy barrier to NO permeation, while AQP1 has a 3.6 to 4.6 kcal/mol energy barrier to O_2 permeation. A close examination of Fig. 4 reveals that the PMF of NO in bulk water is 0.5 kcal/mol smaller than the PMF of O_2 in water. This difference agrees well with experimental measurements of the solvation energy of the two gas molecules (Scharlin *et al.*, 1998). However, even after taking this difference into account, the energy barrier in AQP4 is still smaller than AQP1.

We have shown that the main barrier against gas permeation in AQP1 is caused by a dense layer of water molecules anchored at the mouth of the central pore by four aspartate residues. In contrast to AQP1, AQP4 has an almost completely closed central pore at the periplasmic side. Hydrophobic residues block the mouth of the pore and prevent water molecules from entering. Surprisingly, these hydrophobic residues do not form a barrier against NO. As NO is less polar and smaller than a water molecule, it can readily penetrate the mesh formed by the hydrophobic residues and enter the central pore of AQP4. Therefore, the "closed" central pore of AQP4 is in fact open for hydrophobic gas molecules, and compared with AQP1, the more hydrophobic environment in AQP4 central pore is more favorable to gas molecules.

D. Gas Conduction in Other Membrane Channels

Involvement of other membrane channels in gas permeation across biological membranes can also be envisioned. The results of our simulations studies along with structural examination of other membrane proteins, in particular membrane channels, indicate that gas pathways are likely very common among membrane proteins. In this regard, a particularly interesting group of membrane proteins are those that oligomerize inside the membrane. In fact, for the most part, the significance and reason behind oligomerization of these proteins have not been understood. The results of our AQP1 and AQP4 simulations suggesting the involvement of the central, four-fold axis of the channel as the main pathway for conduction of small gas molecules provides for the first time a possible relevant biological reason for oligomerization of these proteins.

Along the idea of involvement of oligomerization pores in gas transport, we also examined another membrane channel, the ammonia/ammonium channel AmtB (Khademi *et al.*, 2004). Our preliminary results show that the free energy profile for O_2 permeation has an about 3 kcal/mol energy barrier in the central pore of AmtB, suggesting that the three fold symmetry axis of AmtB can indeed play a similar role to the four-fold symmetry axis (the central pore) in AQPs. Very similar structural and gas permeability properties are very likely to exist for other membrane proteins that oligomerize inside the membrane.

V. CONCLUDING REMARKS

Conduction of gas molecules across the cellular membrane is one of the most fundamental transport processes in biology. While pure lipid bilayers provide a "convenient" pathway for gas permeation across the membrane, the role of membrane proteins in this process cannot be neglected. Proteins are among major constituents of biological membranes and can take up to 75% of the total area of the membrane. As such, even small single-channel conduction rates for a membrane protein of high abundance can result in a significant effect on the total gas permeability of the membrane.

In this chapter we presented the results of our recent MD simulations investigating the mechanism, dynamics, and energetics associated with permeation of different gas species across pure lipid bilayers as well as through several membrane channels. Two different computational techniques were employed to calculate the free energy profiles associated with gas permeation of these systems and yielded in consistent results. The results clearly show that the effective barrier against gas diffusion in a lipid bilayer is due to the composition of the head-group region, and, thus, can be significantly altered by the lipid composition of the bilayer. We propose that the level of hydration and the structure and dynamics of water in the interfacial region are key factors in determining the height of the barrier in this region. Interestingly, the phase of the lipid bilayer can also affect the energetics of gas conduction, however, this effect is limited to the lipid tail region, and does not seem to affect the height of the main barrier.

We also presented several simulations in which gas permeation across membrane channels was clearly demonstrated. Our simulations suggest that apart from their conventional role as water channels, AQPs provide pathways for small, neutral gas molecules across the membrane. However, compared with a patch of pure bilayer with a similar area, AQPs are found to be less gas-permeable. It is noteworthy that the calculated barrier against gas permeation in AQP1, revealed by both of our approaches, is not directly caused by the protein, but rather by a dense layer of water molecules "anchored" at the periplasmic mouth of the central pore by four charged residues.

Both the explicit gas diffusion simulations and the implicit ligand sampling method suggest that the monomeric water pores of AQPs are less gas-permeable than the central pore, likely due to the strong hydrogen bonds between the protein and water molecules inside the water pores. A somewhat unexpected discovery is the identification of "side pores" located in between the AQP monomers, which may facilitate gas conduction across the membrane. However, due to the fluctuation of the protein, these side channels might not be permanently open. It is therefore of interest and importance to further investigate the role of these side pores in gas conduction mediated by AQPs.

These results provide a clear demonstration for involvement of the members of the AQP family, and probably other membrane channels, in gas conduction. The question remains as to whether protein-mediated gas conduction is physiologically relevant. Given the sensitivity of gas permeability of lipid bilayers to lipid composition, as demonstrated here, it is possible that certain cellular membranes exhibit a low intrinsic gas permeability. In such membranes, or where a major fraction of the surface area of the membrane is occupied by membrane proteins, protein-mediated gas conduction can indeed play a major role in the exchange of gas molecules across the cellular membrane and thus be of physiological significance.

Acknowledgements

This work was supported by grants from NIH (R01-GM067887 and P41-RR05969). The authors acknowledge computer time at TERAGRID resources (grant number MCA06N060), specially on the BigRed cluster at Indiana University, as well as computer time from the DoD High Performance Computing Modernization Program at the Arctic Region Supercomputing Center, University of Alaska at Fairbanks.

References

Agre, P. (2004). Aquaporin water channels (Nobel lecture). *Angew. Chem. Int. Ed. Engl.* **43**, 4278–4290.

Agre, P. (2006). The aquaporin water channels. *Proc. Am. Thor. Soc.* **3**, 5–13.

Agre, P., Kozono, D. (2003). Aquaporin water channels: Molecular mechanisms for human diseases. *FEBS Lett.* **555**, 72–78.

Agre, P., Bonhivers, M., Borgnia, M.J. (1998). The aquaporins, blueprints for cellular plumbing systems. *J. Biol. Chem.* **273**, 14659–14662.

Agre, P., Nielsen, S., Ottersen, O.P. (2004). Towards a molecular understanding of water homeostasis in the brain. *Neuroscience* **129**, 849–850.

Amiry-Moghaddam, M., Otsuka, T., Hurn, P., Traystman, R., Haug, F., Froehner, S., Adams, M., Neely, J., Agre, P., Ottersen, O., Bhardwaj, A. (2003). An alpha-syntrophin-dependent pool of aqp4 in astroglial end-feet confers bidirectional water flow between blood and brain. *Proc. Natl. Acad. Sci. USA* **100**, 2106–2111.

Blank, M., Ehmke, H. (2003). Aquaporin-1 and HCO_3^-–Cl^- transporter-mediated transport of CO_2 across the human erythrocyte membrane. *J. Physiol.* **550**, 419–429.

Borgnia, M., Nielsen, S., Engel, A., Agre, P. (1999). Cellular and molecular biology of the aquaporin water channels. *Annu. Rev. Biochem.* **68**, 425–458.

Burykin, A., Warshel, A. (2003). What really prevents proton transport through aquaporin? Charge self-energy versus proton wire proposals. *Biophys. J.* **85**, 3696–3706.

Burykin, A., Warshel, A. (2004). On the origin of the electrostatic barrier for proton transport in aquaporin. *FEBS Lett.* **570**, 41–46.

Chakrabarti, N., Tajkhorshid, E., Roux, B., Pomès, R. (2004). Molecular basis of proton blockage in aquaporins. *Structure* **12**, 65–74.

Cohen, J., Kim, K., King, P., Seibert, M., Schulten, K. (2005). Finding gas diffusion pathways in proteins: Application to O_2 and H_2 transport in CpI [FeFe]-hydrogenase and the role of packing defects. *Structure* **13**, 1321–1329.

Cohen, J., Arkhipov, A., Braun, R., Schulten, K. (2006). Imaging the migration pathways for O_2, CO, NO, and Xe inside myoglobin. *Biophys. J.* **91**, 1844–1857.

Cooper, G., Boron, W. (1998). Effect of PCMBS on CO_2 permeability of *Xenopus* Oocytes expressing aquaporin 1 or its C189S mutant. *Am. J. Physiol.* **275**, C1481–C1486.

Cooper, G., Zhou, Y., Bouyer, P., Grichtchenko, I., Boron, W. (2002). Transport of volatile solutes through AQP1. *J. Physiol.* **542**, 17–29.

Darden, T., York, D., Pedersen, L. (1993). Particle mesh Ewald. An $N \cdot \log(N)$ method for Ewald sums in large systems. *J. Chem. Phys.* **98**, 10089–10092.

de Groot, B.L., Grubmüller, H. (2001). Water permeation across biological membranes: Mechanism and dynamics of aquaporin-1 and GlpF. *Science* **294**, 2353–2357.

de Groot, B.L., Grubmüller, H. (2005). The dynamics and energetics of water permeation and proton exclusion in aquaporins. *Curr. Opin. Struct. Biol.* **15**, 1–8.

de Groot, B.L., Frigato, T., Helms, V., Grubmüller, H. (2003). The mechanism of proton exclusion in the aquaporin-1 water channel. *J. Mol. Biol.* **333**, 279–293.

Deen, P.M.T., van Os, C.H. (1998). Epithelial aquaporins. *Curr. Opin. Cell Biol.* **10**, 435–442.

Endeward, V., Musa-Aziz, R., Cooper, G., Chen, L., Pelletier, M., Virkki, L., Supuran, C., King, L., Boron, W., Gros, G. (2006). Evidence that Aquaporin 1 is a major pathway for CO_2 transport across the human erythrocyte membrane. *FASEB J.* **20**, 1974–1981.

Fang, X., Yang, B., Matthay, M., Verkman, A. (2002). Evidence against aquaporin-1-dependent CO_2 permeability in lung and kidney. *J. Physiol.* **542**, 63–69.

Feller, S.E., Pastor, R.W. (1999). Constant surface tension simulations of lipid bilayers: The sensitivity of surface areas and compressibilities. *J. Chem. Phys.* **111**, 1281–1287.

Feller, S.E., Zhang, Y.H., Pastor, R.W., Brooks, B.R. (1995). Constant pressure molecular dynamics simulation—the Langevin piston method. *J. Chem. Phys.* **103**, 4613–4621.

Fu, D., Libson, A., Miercke, L.J.W., Weitzman, C., Nollert, P., Krucinski, J., Stroud, R.M. (2000). Structure of a glycerol conducting channel and the basis for its selectivity. *Science* **290**, 481–486.

Fujiyoshi, Y., Mitsuoka, K., de Groot, B.L., Philippsen, A., Grubmüller, H., Agre, P., Engel, A. (2002). Structure and function of water channels. *Curr. Opin. Struct. Biol.* **12**, 509–515.

Gonen, T., Cheng, Y., Sliz, P., Hiroaki, Y., Fujiyoshi, Y., Harrison, S.C., Walz, T. (2005). Lipid–protein interactions in double-layered two-dimensional AQP0 crystals. *Nature* **438**, 633–638.

Gonen, T., Sliz, P., Kistler, J., Cheng, Y., Walz, T. (2004). Aquaporin-0 membrane junctions reveal the structure of a closed water pore. *Nature* **429**, 193–197.

Grayson, P., Tajkhorshid, E., Schulten, K. (2003). Mechanisms of selectivity in channels and enzymes studied with interactive molecular dynamics. *Biophys. J.* **85**, 36–48.

Hanba, Y.T., Shibasaka, M., Hayashi, Y., Hayakawa, T., Kasamo, K., Terashima, I., Katsuhara, M. (2004). Overexpression of the barley aquaporin HvPIP2;1 increases internal CO_2 conductance and CO_2 assimilation in the leaves of transgenic rice plants. *Plant Cell Physiol.* **45**, 521–529.

Harries, W.E.C., Akhavan, D., Miercke, L.J.W., Khademi, S., Stroud, R. (2004). The channel architecture of aquaporin 0 at a 2.2-angstrom resolution. *Proc. Natl. Acad. Sci. USA* **101**, 14045–14050.

Heller, H., Schaefer, M., Schulten, K. (1993). Molecular dynamics simulation of a bilayer of 200 lipids in the gel and in the liquid crystal-phases. *J. Phys. Chem.* **97**, 8343–8360.

Herrera, M., Hong, N.J., Garvin, J.L. (2006). Aquaporin-1 transports NO across cell membranes. *Hypertension* **48**, 157–164.

Heymann, J.B., Engel, A. (1999). Aquaporins: Phylogeny, structure, and physiology of water channels. *News Physiol. Sci.* **14**, 187–193.

Hiroaki, Y., Tani, K., Kamegawa, A., Gyobu, N., Nishikawa, K., Suzuki, H., Walz, T., Sasaki, S., Mitsuoka, K., Kimura, K., Mizoguchi, A., Fujiyoshi, Y. (2005). Implications of the Aquaporin-4 structure on array formation and cell adhesion. *J. Mol. Biol.* **355**, 628–639.

Hiroaki, Y., Tani, K., Kamegawa, A., Gyobu, N., Nishikawa, K., Suzuki, H., Walz, T., Sasaki, S., Mitsuoka, K., Kimura, K., Mizoguchi, A., Fujiyoshi, Y. (2006). Implications of the aquaporin-4 structure on array formation and cell adhesion. *J. Mol. Biol.* **355**, 628–639.

Hoch, F.L. (1992). Cardiolipins and biomembrane function. *Biochim. Biophys. Acta* **1113**, 71–133.

Hohmann, S., Nielsen, S., Agre, P. (2001). "Aquaporins". Academic Press, San Diego.

Hub, J.S., de Groot, B.L. (2006). Does CO_2 permeate through Aquaporin-1? *Biophys. J.* **91**, 842–848.

Humphrey, W., Dalke, A., Schulten, K. (1996). VMD—Visual Molecular Dynamics. *J. Mol. Graphics* **14**, 33–38.

Ignarro, L.J. (1999). Nitric oxide: A unique endogenous signaling molecule in vascular biology. *Biosci. Rep.* **19**, 51–71.

Ignarro, L.J. (2002). Nitric oxide as a unique signaling molecule in the vascular system: A historical overview. *J. Physiol. Pharmacol.* **53**, 503–514.

Ilan, B., Tajkhorshid, E., Schulten, K., Voth, G.A. (2004). The mechanism of proton exclusion in aquaporin channels. *Proteins: Struct., Funct., Bioinf.* **55**, 223–228.

Jensen, M.Ø., Mouritsen, O., Peters, G.H. (2004). Simulations of a membrane-anchored peptide: Structure, dynamics, and influence on bilayer properties. *Biophys. J.* **86**, 3556–3575.

Jensen, M.Ø., Tajkhorshid, E., Schulten, K. (2001). The mechanism of glycerol conduction in aquaglyceroporins. *Structure* **9**, 1083–1093.

Jensen, M.Ø., Tajkhorshid, E., Schulten, K. (2003). Electrostatic tuning of permeation and selectivity in aquaporin water channels. *Biophys. J.* **85**, 2884–2899.

Jiang, F., Ryan, M.T., Schlame, M., Zhao, M., Gu, Z., Klingenberg, M., Pfanner, N., Greenberg, M.L. (2000). Absence of cardiolipin in the crd1 null mutant results in decreased mitochondrial membrane potential and reduced mitochondrial function. *J. Biol. Chem.* **275**, 22387–22394.

Khademi, S., Remis, J.O. III, Robles-Colmenares, Y., Miercke, L., Stroud, R. (2004). Mechanism of ammonia transport by Amt/MEP/Rh: Structure of AmtB at 1.35 Å. *Science* **305**, 1587–1594.

King, L., Agre, P. (2004). From structure to disease: The evolving tale of aquaporin biology. *Nat. Rev. Mol. Cell Biol.* **5**, 687–698.

Lee, J.K., Kozono, D., Remis, J., Kitagawa, Y., Agre, P., Stroud, R. (2005). Structural basis for conductance by the archaeal aquaporin AqpM at 1.68 Å. *Proc. Natl. Acad. Sci. USA* **102**, 18932–18937.

Lewis, R.N., Zweytick, D., Pabst, G., Lohner, K., McElhaney, R.N. (2007). Calorimetric, X-ray diffraction, and spectroscopic studies of the thermotropic phase behavior and organization of tetramyristoyl cardiolipin membranes. *Biophys. J.* **92**, 3166–3167.

Li, J., Verkman, A.S. (2001). Impaired hearing in mice lacking aquaporin-4 water channels. *J. Biol. Chem.* **276**, 31233–31237.

MacKerell Jr., A.D., Bashford, D., Bellott, M., *et al.* (1998). All-atom empirical potential for molecular modeling and dynamics studies of proteins. *J. Phys. Chem. B* **102**, 3586–3616.

Marrink, S., Berendsen, H. (1994). Simulation of water transport through a lipid membrane. *J. Phys. Chem.* **98**, 4155–4168.

McMillin, J.B., Dowhan, W. (2002). Cardiolipin and apoptosis. *Biochim. Biophys. Acta* **1585**, 97–107.

Mileykovskaya, E., Zhang, M., Dowhan, W. (2004). Cardiolipin in energy transducing membranes. *Biochemistry (Moscow)* **70**, 154–158.

Murad, F. (2004). Discovery of some of the biological effects of nitric oxide and its role in cell signaling. *Biosci. Rep.* **24**, 452–474.

Murata, K., Mitsuoka, K., Hirai, T., Walz, T., Agre, P., Heymann, J.B., Engel, A., Fujiyoshi, Y. (2000). Structural determinants of water permeation through aquaporin-1. *Nature* **407**, 599–605.

Nagle, J.F., Tristram-Nagle, S. (2000). Structure of lipid bilayers. *Biochim. Biophys. Acta* **1469**, 159–195.

Nakhoul, N., Davis, B., Romero, M., Boron, W. (1998). Effect of expressing the water channel aquaporin-1 on the CO_2 permeability of *Xenopus* oocytes. *Am. J. Physiol.* **274**, C543–C548.

Ohkubo, Y.Z., Tajkhorshid, E. (2008). Distinct structural and adhesive roles of Ca^{2+} in membrane binding of blood coagulation factors. *Structure* **16**, 72–81.

Petrache, H.I., Tristram-Nagle, S., Gawrisch, K., Harries, D., Parsegian, V.A., Nagle, J.F. (2004). Structure and fluctuations of charged phosphatidylserine bilayers in the absence of salt. *Biophys. J.* **86**, 1574–1586.

Phillips, J.C., Braun, R., Wang, W., Gumbart, J., Tajkhorshid, E., Villa, E., Chipot, C., Skeel, R.D., Kale, L., Schulten, K. (2005). Scalable molecular dynamics with NAMD. *J. Comput. Chem.* **26**, 1781–1802.

Pohl, P., Saparov, S.M., Borgnia, M.J., Agre, P. (2001). Highly selective water channel activity measured by voltage clamp: Analysis of planar lipid bilayers reconstituted with purified AqpZ. *Proc. Natl. Acad. Sci. USA* **98**, 9624–9629.

Prasad, G.V.T., Coury, L., Finn, F., Zeidel, M. (1998). Reconstituted aquaporin 1 water channels transport CO_2 across membranes. *J. Biol. Chem.* **273**, 33123–33126.

Preston, G.M., Agre, P. (1991). Isolation of the cDNA for erythrocyte integral membrane-protein of 28-kD—member of an ancient channel family. *Proc. Natl. Acad. Sci. USA* **88**, 11110–11114.

Preston, G.M., Carroll, T.P., Guggino, W.B., Agre, P. (1992). Appearance of water channels in Xenopus oocytes expressing red cell CHIP28 protein. *Science* **256**, 385–387.

Robinson, N.C. (1993). Functional binding of cardiolipin to cytochrome c oxidase. *J. Bioenerg. Biomembr.* **25**, 153–163.

Saparov, S., Tsunoda, S., Pohl, P. (2005). Proton exclusion by an aquaglyceroprotein: A voltage clamp study. *Biol. Cell* **97**, 545–550.

Savage, D.F., Egea, P.F., Robles-Colmenares, Y., O'Connell III, J.D., Stroud, R.M. (2003). Architecture and selectivity in aquaporins: 2.5 Å X-ray structure of aquaporin Z. *PLoS Biology* **1**, E72.

Scharlin, P., Battino, R., Silla, E., Tuñón, I., Pascual-Ahuir, J.L. (1998). Solubility of gases in water: Correlation between solubility and the number of water molecules in the first solvation shell. *Pure Appl. Chem.* **70**, 1895–1904.

Schlenkrich, M., Brickmann, J., MacKerell Jr., A.D., Karplus, M. (1996). Empirical potential energy function for phospholipids: Criteria for parameter optimization and applications. In: Merz, K.M., Roux, B. (Eds.), "Biological Membranes: A Molecular Perspective from Computation and Experiment". Birkhauser, Boston, pp. 31–81.

Sonne, J., Jensen, M.Ø., Hansen, F.Y., Hemmingsen, L., Peters, G.H. (2007). Reparameterization of all-Atom dipalmitoylphosphatidylcholine lipid parameters enables simulation of fluid bilayers at zero tension. *Biophys. J.* **92**, 4157–4167.

Sui, H., Han, B.-G., Lee, J.K., Walian, P., Jap, B.K. (2001). Structural basis of water-specific transport through the AQP1 water channel. *Nature* **414**, 872–878.

Tajkhorshid, E., Aksimentiev, A., Balabin, I., Gao, M., Isralewitz, B., Phillips, J.C., Zhu, F., Schulten, K. (2003). Large scale simulation of protein mechanics and function. In: Richards, F.M., Eisenberg, D.S., Kuriyan, J. (Eds.), "Advances in Protein Chemistry", vol. 66. Elsevier Academic Press, New York, pp. 195–247.

Tajkhorshid, E., Nollert, P., Jensen, M.Ø., Miercke, L.J.W., O'Connell, J., Stroud, R.M., Schulten, K. (2002). Control of the selectivity of the aquaporin water channel family by global orientational tuning. *Science* **296**, 525–530.

Tajkhorshid, E., Cohen, J., Aksimentiev, A., Sotomayor, M., Schulten, K. (2005a). Towards understanding membrane channels. In: Martinac, B., Kubalski, A. (Eds.), "Bacterial ion channels and their eukaryotic homologues". ASM Press, Washington, DC, pp. 153–190.

Tajkhorshid, E., Zhu, F., Schulten, K. (2005b). Kinetic theory and simulation of single-channel water transport. In: Yip, S. (Ed.), Handbook of Materials Modeling, vol. I, Methods and Models. Springer, Netherlands, pp. 1797–1822.

Terashima, I., Ono, K. (2002). Effects of $HgCl_2$ on CO_2 dependence of leaf photosynthesis: Evidence indicating involvement of aquaporins in CO_2 diffusion across the plasma membrane. *Plant Cell Physiol.* **43**, 70–78.

Törnroth-Horsefield, S., Wang, Y., Hedfalk, K., Johanson, U., Karlsson, M., Tajkhorshid, E., Neutze, R., Kjellbom, P. (2006). Structural mechanism of plant aquaporin gating. *Nature* **439**, 688–694.

Uehlein, N., Lovisolo, C., Siefritz, F., Kaldenhoff, R. (2003). The tobacco aquaporin NtAQP1 is a membrane CO_2 pore with physiological functions. *Nature* **425**, 734–737.

van der Ploeg, P., Berendsen, H.J.C. (1982). Molecular dynamics simulation of a bilayer membrane. *J. Chem. Phys.* **76**, 3271–3276.

Verkman, A. (2002). Does aquaporin-1 pass gas? An opposing view. *J. Physiol.* **542**, 31.

Wang, Y., Tajkhorshid, E. (2007). Molecular mechanisms of conduction and selectivity in aquaporin water channels. *J. Nutr.* **137**, 1509S–1515S.

Wang, Y., Schulten, K., Tajkhorshid, E. (2005). What makes an aquaporin a glycerol channel: A comparative study of AqpZ and GlpF. *Structure* **13**, 1107–1118.

Wang, Y., Cohen, J., Boron, W.F., Schulten, K., Tajkhorshid, E. (2007). Exploring gas permeability of cellular membranes and membrane channels with molecular dynamics. *J. Struct. Biol.* **157**, 534–544.

Yang, B., Fukuda, N., van Hoek, A., Matthay, M.A., Ma, T., Verkman, A.S. (2000). Carbon dioxide permeability of aquaporin-1 measured in erythrocytes and lung of aquaporin-1 null mice and in reconstituted proteoliposomes. *J. Biol. Chem.* **275**, 2682–2692.

Yu, J., Yool, A.J., Schulten, K., Tajkhorshid, E. (2006). Mechanism of gating and ion conductivity of a possible tetrameric pore in Aquaporin-1. *Structure* **14**, 1411–1423.

Zhu, F., Tajkhorshid, E., Schulten, K. (2001). Molecular dynamics study of aquaporin-1 water channel in a lipid bilayer. *FEBS Lett.* **504**, 212–218.

Zhu, F., Tajkhorshid, E., Schulten, K. (2002). Pressure-induced water transport in membrane channels studied by molecular dynamics. *Biophys. J.* **83**, 154–160.

Zhu, F., Tajkhorshid, E., Schulten, K. (2004). Theory and simulation of water permeation in aquaporin-1. *Biophys. J.* **86**, 50–57.

CHAPTER 13

A Brief Introduction to Voltage-Gated K$^+$ Channels

Benoît Roux

The University of Chicago, Institute of Molecular Pediatric Sciences,
Gordon Center for Integrative Science, 929 East 57th Street, room W323B, Chicago, IL 60637, USA

I. INTRODUCTION

Voltage-gated K$^+$ (Kv) channels are transmembrane proteins that control and regulate the flow of K$^+$ ions across cell membranes (Sigworth, 1993; Yellen, 1998; Hille, 2001; Bezanilla et al., 2003). In response to changes in the membrane potential, they undergo conformation changes, thereby allowing or blocking the conduction of ions. Structurally, Kv channels are formed by four subunits surrounding a central aqueous pore for K$^+$ permeation. Each subunit comprises six transmembrane segments (Tempel et al., 1987), S1–S6, the first four transmembrane segments, S1–S4, constituting the voltage sensor domain, while the last two transmembrane segments, S5 and S6, from each of the four subunits join around a common axis to form a selective ion-conducting pore (Bezanilla, 2000). Positively charged residues along S4, mostly arginines, appear to be mainly responsible for the coupling to the membrane voltage (Papazian et al., 1991; Aggarwal and MacKinnon, 1996; Seoh et al., 1996). Upon membrane depolarization, the voltage sensor in each subunit undergoes voltage-dependent transition from a resting to an activated state, resulting in a conformation that allows the

1063-5823/08 $35.00
DOI: 10.1016/S1063-5823(08)00013-6

opening of the (Bezanilla *et al.*, 1994; Zagotta *et al.*, 1994). Once all of the four subunits are in the activated state, opening of the pore gate occurs cooperatively via a concerted transition that is weakly voltage-dependent (Ledwell and Aldrich, 1999). After their opening, the channels start to enter into a non-conductive state via a process that is called slow inactivation.

To understand how Kv channels function, requires knowledge of the structure of the channel in its various states to atomic resolution, and knowledge of how the different parts of the protein change their conformation during voltage-gating. Ultimately, one would like to be able to "visualize", atom-by-atom, how the protein moves as a function of time in response to the membrane potential. Little by little, progress is being made toward this goal. In this brief chapter, I will try to provide a view of the recent progress that has been made in recent years. In the first section, I review the current state of knowledge about the structure of Kv channels.

II. STRUCTURE OF Kv CHANNELS AND GATING MECHANISMS

A. *Structural Information*

X-ray crystallography of the KvAP channel offered the first view of the molecular architecture of a voltage-activated K^+ channel at atomic resolution (Jiang et al., 2003a, 2003b). Two independent X-ray structures were obtained. They are shown in Fig. 1. The full tetrameric channel was crystallized in complex with a monoclonal Fab antibody fragment bound to the S3–S4 loop. In addition, the isolated voltage sensor, consisting of S1 to S4, was also crystallized, again with a monoclonal Fab antibody bound to the S3–S4 loop. The conformations of the

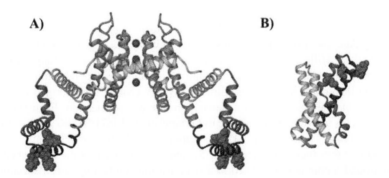

FIGURE 1 Crystallographic structures of the KvAP channel (S1, white; S2, yellow; S3, red; S4, blue; S5–S6 pore, green). (A) Structure of the full length channel is with the four arginine residue in space filling (magenta) is displayed in a transmembrane-like orientation (only two subunits are shown for the sake of clarify). (B) Structure of the isolated voltage sensor S1–S4. (For interpretation of the references to color in this figure legend, the reader is referred to the web version of this book.)

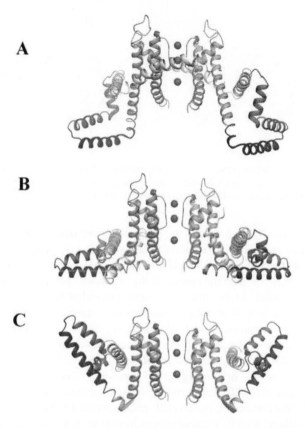

FIGURE 2 Structural models of the KvAP channel from (Jiang et al., 2003a, 2003b). (A) X-ray structure of the full-length KvAP channel viewed from the side with the intracellular solution below. (B) Identical view with the X-ray structure of the isolated voltage sensor (S1–S4) docked according to the position of S2 in the X-ray structure of the full-length channel (obtained by superposing the backbone of S2 from the two X-ray structures as described in (Jiang et al., 2003a, 2003b). (C) Representation of the modeled open state structure of KvAP based on the paddle depth and orientation as described in (Jiang et al., 2003a, 2003b).

voltage-sensor in the two X-ray structures are quite different (Fig. 1). Recognizing that the structure of the full channel was not entirely consistent with the expected architecture of the functional channel based on numerous experimental results, the authors concluded that the X-ray structure was probably distorted. Occurrence of such a non-native conformation in protein crystallography is unusual, but in this case may have been caused by the combined effect of the detergent and the interaction of the Fab fragment. To circumvent the problems with the distorted X-ray structure, the authors used the structure of the isolated voltage sensor and accessibility data from trapping experiments of biotinylated channels by avidin to

A)   **B)**

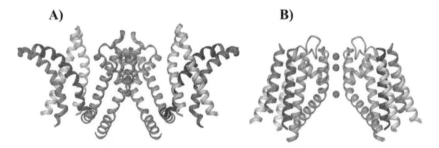

FIGURE 3 Comparison of the crystallographic structure of the Kv1.2 channel (Long *et al.*, 2005) with the approximate model of the Shaker channel in the open state from (Laine *et al.*, 2003) build from the X-ray structure of the MthK channel (Jiang *et al.*, 2002).

construct models of the channel in the open and closed states (Jiang et al., 2003a, 2003b). Those models were generated as shown in Fig. 2, by superimposing the S2 helix from the full length channel and the isolated voltage sensor. This modeling is suggestive of the basic feature of the "up-down" movement characteristic of the paddle model. At the time, the resulting model of KvAP in an open state differed extensively with an approximate model of the *Shaker* channel based on a wide range of experimental data (Laine *et al.*, 2003, 2004). The controversy about the structure of Kv channels then reached a maximum (Cohen *et al.*, 2003; Laine *et al.*, 2004). The crystal structure of Kv1.2 from rat brain provided the first atomic resolution view of a voltage-gated potassium channel in a native conformation (Long *et al.*, 2005) has considerably clarified these matters. According to the experimental conditions, the crystallographic structure of Kv1.2 should correspond to a channel with its voltage-sensors near an their activated position.

As shown in Figs. 3 and 4, the overall topological features of the X-ray structure of Kv1.2 are in excellent accord with a model of *Shaker* K^+ channel in its activated open state deduced on the basis of a wide range of structural, functional and biophysical experiments. Namely, the voltage sensor is formed by a bundle of four anti-parallel transmembrane α-helices, S1–S4, each with their N- and C-terminal ends exposed alternatively to the intra and extracellular solution. Seen from the extracellular side, the S1–S4 helices of the voltage sensor are packed in a counterclockwise fashion, and the S4 helix of a subunit is making contact with the S5 helix of the adjacent subunit in the clockwise direction (Laine *et al.*, 2003). This is displayed in Fig. 4. The excellent accord between the X-ray structure and results from numerous functional and previous models deduced from a wide range of biophysical experiments considerably strengthens the consensus about voltage-gated K^+ channels. It is now apparent that the broad features of the model of the *Shaker* channel in the open state based on all available data at the time and the open-state X-ray structure of the MthK channel (Jiang *et al.*, 2002) was much closer to the X-ray structure of the Kv1.2 channel than the models presented by (Jiang et al., 2003a, 2003b), where the S4 segment was located at the

A)   B)

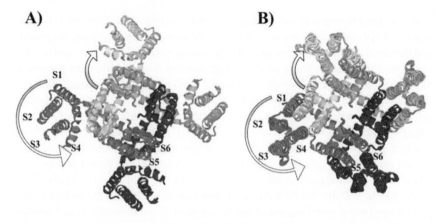

FIGURE 4   Comparison of the X-ray structure of the Kv1.2 channel seen from the extracellular side (Long *et al.*, 2005) with the approximate model of *Shaker* in the open state from (Chanda *et al.*, 2005). As previously established in (Laine *et al.*, 2003), the four transmembrane helices of the voltage sensor are packed in a counterclockwise direction, and the voltage sensor from one subunit is within proximity of the pore domain from the adjacent subunit in the clockwise direction.

periphery of the protein and the S2 was close to the pore. The good agreement also validates the general strategy for translating available experimental data into structural models. Additional efforts incorporating also the information from the X-ray structure of the KvAP channel together with available experimental data led to better models of the *Shaker* K$^+$ channel in the open and the closed state (Chanda *et al.*, 2005). Though there clearly some structural differences, the main structural features of the open state model are in are in excellent agreement with the X-ray structure of the Kv1.2 channel (Long *et al.*, 2005).

An interesting feature of the X-ray structure is the modular nature of the voltage sensor domain and its lack of extensive interactions with the pore domain. About 66% of molecular surface of the transmembrane region of each voltage-sensor S1–S4 is exposed to lipids. The existence of nearly independent voltage-sensing modules is consistent with the functional chimeras engineered by substituting the pore domain of the KcsA channel into the voltage-gated Shaker channel (Lu *et al.*, 2001) and naturally compatible with the allosteric model of channel gating developed by Aldrich and co-workers (Ledwell and Aldrich, 1999). Furthermore, the discovery of voltage-sensors-like domains with high sequence similarity to the S1–S4 helices in two unrelated membrane-associated proteins lacking any channel-like central pore domain (Murata *et al.*, 2005; Okamura *et al.*, 2006; Ramsey *et al.*, 2006) reinforced the concept of the S1–S4 helical bundle as independent functional modules. This suggests that biophysical experiments performed on isolated voltage-sensors may be able to reveal some of the universal features of gating motions, without the complicated coupling to the pore domain.

While there is now a broad agreement concerning the open (activated) conformation of Kv channels, there is no apparent consensus concerning the magnitude and the character of the motion underlying voltage gating. Three principal models of the conformational change of the voltage-sensing domain have been proposed: the sliding helical screw model (Catterall, 1988), the paddle model (Jiang et al., 2003a, 2003b), and the transporter model (Chanda et al., 2005). The helical screw model is most closely associated with the most traditional view of voltage-gating (Catterall, 1988). It pictures the S4 helix as being completely enclosed inside a gating pore formed by the rest of the protein in which it could slide up and down during gating. The paddle model describes the gating process as a whole body translocation of the S3–S4 helix-turn-helix through the lipids (Jiang et al., 2003a, 2003b). While both the sliding helical screw and the paddle model posit a substantial translocation of the S4 segment, the transporter model invokes smaller movements (Chanda et al., 2005). This model pictures the coupling to the transmembrane voltage through small movements of the charged residues of the voltage sensor across a transmembrane electric field focused by high dielectric aqueous crevices.

B. Gating Movements, Large or Small?

Arguments about "large" versus "small" movements within the voltage sensing domain now appear to be at the heart of the controversy concerning the mechanism of voltage-gating. To make matters worse, there exist experimental data supporting both large and small movements and this certainly contributes to create confusion. Luminescence resonance energy transfer (LRET) experiments for a fluorophore attached to various parts of the voltage–sensor (Cha et al., 1999; Posson et al., 2005), fluorescence transfer to the hydrophobic anion dypicrylamine (DPA) localized at the lipid–aqueous intersurface (Chanda et al., 2005), and kinetic analysis of inhibitory toxin binding to the voltage sensor (Phillips et al., 2005), all seem to be consistent with small (1–2 Å) movements upon channel gating. In contrast, trapping experiments of biotinylated channels by avidin seem to indicate rather large (15–20 Å) movements of the S3–S4 helices (Jiang et al., 2003a, 2003b; Ruta et al., 2005). In many ways, the controversy about the magnitude and character of the movement underlying voltage gating reflects the uncertainty of the various experimental methods and the limited information about the conformation of the resting (closed) state.

The difficulty in interpreting the various results in terms of a well-defined movement is illustrated schematically in Fig. 5 for three types of experiments, LRET (Posson et al., 2005), hannatoxin inhibition (Phillips et al., 2005), and biotin–avidin trapping experiments (Ruta et al., 2005). For example, LRET experiments use the rate of fluorescence energy transfer as a spectroscopic ruler by estimating the donor–acceptor distance according to Förster theory (Posson et al.,

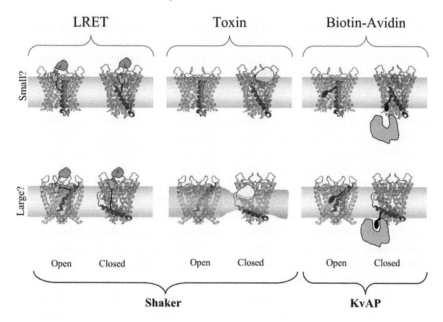

FIGURE 5 Schematic illustration of the possible interpretation as "large" or "small" movements from three methods commonly used to determine the magnitude of the movement upon channel gating. The experiments include LRET (Posson *et al.*, 2005), toxin inhibition (Phillips *et al.*, 2005), and biotin–avidin trapping experiments (Ruta *et al.*, 2005).

2005). First, a cysteine is substituted at a specific position in the channel, and then linked to a thiol-reactive terbium chelate to act as the donor. An acceptor is then placed elsewhere in the channel or in a toxin binding bound to the channel. While the energy transfer rate may accurately report the donor–acceptor distance, it is important to keep in mind that the latter is attached to the backbone of the channel via a flexible molecular linker that is 5–10 Å long. As illustrated in Fig. 5, a small observed change in donor–acceptor could be consistent with both a small or relatively large movement within the voltage sensor. In addition, if there are large conformational fluctuations in the donor–acceptor distance, the short values tend to weight more because the Förster energy transfer varies like the inverse 6th power of the donor–acceptor distance (Selvin, 2002). Channel inhibition of *Shaker* channels by hannatoxin from tarantula also provide information about the movement of the voltage sensor upon channel activation (Phillips *et al.*, 2005). This toxin inhibit the movement of the opening of Kv channels by binding to the S3–S4 region of the voltage sensor, keeping it near its resting position. Hana-toxin forms a strong and stable complex with the voltage sensors and its affinity is reduced by voltage-sensor activation, explaining why the toxin stabilizes the resting conformation. The toxin is also known to partition near the membrane–solution interface, which puts a limit on the amount of motion. Consequently, the

S3–S4 region cannot move than the outer half of the bilayer during activation. However, the exact magnitude of the movement depends on assumptions about the local shape and the amount of distortion of the membrane–solution interface in the neighborhood of the voltage sensor (Fig. 5). Finally, the magnitude of the movement has been estimated with trapping experiments of biotinylated KvAP channels by avidin (Jiang et al., 2003a, 2003b; Ruta *et al.*, 2005). In those experiments, a biotin is tethered to the voltage sensor via a cysteine using linkers of different lengths and the closed and open conformations are trapped irreversibly by avidin binding to the biotin from the intra or extracellular solution. Those experiments indicate that the S1 and S2 helices do not move significantly during gating. But movements on the order of 15–20 Å were inferred for the S3–S4 region of the voltage sensor. In interpreting those results, it is important to keep in mind that the binding of biotin with avidin is essentially irreversible. This means that any encounter of biotin with avidin, even through a rare thermal fluctuations, could lead to binding and, thereby, channel inhibition. From this point of view, biotin–avidin trapping experiments provide an upper bound to the magnitude of typical movements that can occur.

Despite the apparent discrepancies, the situation is progressively becoming clearer as more experimental data is accumulated about the closed state. The model of *Shaker* in the closed state assumes an inward vertical movement of ~4 Å at the level of the first arginine in the S4 helix (Chanda *et al.*, 2005). A refined model of the closed state based on the structure of the Kv1.2 channel using the Rosetta method suggests a similar inward movement of S4 (Yarov-Yarovoy *et al.*, 2006). Such inward displacement is compatible with the LRET measurements, but appears much shorter than the 15–20 Å extracted from the biotin–avidin experiments. However, a careful analysis reveals that the discrepancy with the biotin–avidin experiments is more limited than expected (see also Yarov-Yarovoy *et al.*, 2006). This is illustrated in Fig. 6 using the approximate models from (Chanda *et al.*, 2005), where residues are color-coded according to the accessibility deduced from the biotin–avidin experiment. The main discrepancies, indicated by the colored arrows, are on the order of 5–8 Å. While this is not perfect accord, the biotin–avidin results could be rationalized if the voltage-sensor was undergoing inward–outward thermal fluctuations on the order 3–4 Å. In fact, this seems likely given its high flexibility as indicated by electron paramagnetic resonance (EPR) (Cuello *et al.*, 2004). This means that the "gap" between the extreme views with "small" and "large" movements is not as large as commonly thought.

Recently, the approximate model of *Shaker* in the closed state from (Chanda *et al.*, 2005) was used as a template to guide a search for possible cross-links between S4 and other parts of the voltage sensor (Campos *et al.*, 2007). In qualitative agreement with the model, experiments with substituted cysteine residues in *Shaker* showed that, at hyperpolarized potentials, both I241C (in S1) and I287C (in S2) can spontaneously form disulfide and metal bridges with R362C, the

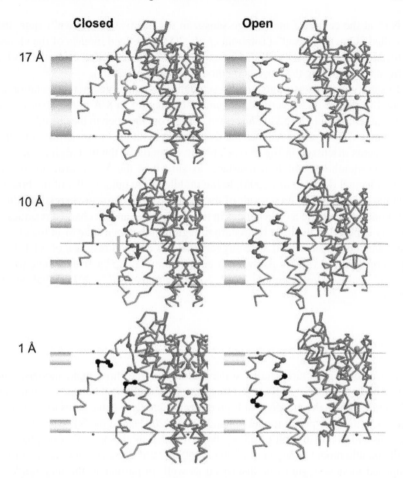

FIGURE 6   Comparison of the approximate open and closed models of *Shaker* from (Chanda *et al.*, 2005) to the biotin–avidin trapping experimental data (Ruta *et al.*, 2005). The color code indicates the residues accessible only from the extracellular side (red), the residues accessible only from the intracellular side (blue), the residues accessible only from both side (yellow), and the residues that are inaccessible from either side (black). The grey slab area indicate the accessible depth with the different length of biotin linkers (17, 10 and 1 Å) and the colored arrows indicate the discrepancies between the models and the biotin–avidin trapping data. (For interpretation of the references to color in this figure legend, the reader is referred to the web version of this book.)

position of the first charge-carrying residue in S4. These results constrain unambiguously the closed-state positions of the S4 segment with respect to the S1 and S2 segments, which are known to undergo little or no transmembrane (vertical) movement during gating. Remarkably, these results are entirely compatible with a series of mutations yielding a favorable pathway for the rapid conduction of Na^+ at the center of the voltage sensor in the inactivated ("down") state, the so-called "Omega current" (Tombola *et al.*, 2007). Refined models of the closed states in good accord with all these results have been generated using the Kv1.2 X-ray structure using the Rosetta method (Pathak *et al.*, 2007).

To satisfy these constraints, the S4 segment must undergo an axial rotation of about 180 degree and a vertical movement of about 6 to 8 Å at the level of the first charged residue (R362 in *Shaker*) in going from the open to the closed state of the channel. This further reduces the discrepancy with the biotin–avidin trapping experiments discussed in Fig. 6 (see above). Essentially, the first arginine moves from the membrane–solution interface, as shown in the X-ray structure of the Kv1.2 channel (Long *et al.*, 2005), to the middle of the outer leaflet of the bilayer (Campos *et al.*, 2007). As a result of the rotation, the positively charged residues of S4 point toward the S1–S3 helices in the closed state and no charged residues is directly exposed to the lipid hydrocarbon region in the open state or in the closed state. Models of the closed state suggest that the charged residues of S4 follow an energetically favorable pathway in going from the "up" activated state to the "down" resting state (Tombola *et al.*, 2007; Campos *et al.*, 2007; Pathak *et al.*, 2007).

C. Common Grounds

The emerging picture retains some features of the three main models. The whole body movement of S4 is reminiscent of the traditional helical screw model (Catterall, 1988), the presence of high dielectric crevices contributes to a focused field as in the transporter model (Chanda *et al.*, 2005), and the S4 segment is not moving within a hydrophilic "gating-pore" formed by the protein as emphasized in the paddle model (Jiang et al., 2003a, 2003b). But some specific details of the proposed models begin to be disproved as well. In particular, the magnitude of the movement compatible with the data is 7–8 Å, which is clearly larger than 1–2 Å of the "small movement" camp (Cha *et al.*, 1999), but yet smaller than the 15–20 Å of the "large movement" camp (Jiang et al., 2003a, 2003b; Ruta *et al.*, 2005). The S4 helix is not predominantly surrounded by lipids as initially pictured in the paddle model. In addition, the concept of arginines residues acting as hydrophobic cations that project directly into the nonpolar hydrocarbon core of the membrane (Jiang et al., 2003a, 2003b; Long *et al.*, 2005) does not seem relevant to channel gating, mostly like due to the large free energy for dragging the side chain from water to the membrane interior (Dorairaj and Allen, 2007). This is further supported by molecular dynamics simulations of the Kv1.2 channel in the open state, which shows that the hydration of the voltage sensor is extensive and that the arginine along the S4 segment are not directly exposed directly to the hydrocarbon region of the membrane (Treptow and Tarek, 2006; Jogini and Roux, 2007).

III. FREE ENERGY AND MEMBRANE VOLTAGE

The transmembrane potential affects the configurational equilibrium of membrane-bound proteins, and provide the driving force for activating voltage-gated channels. Assuming only two conformational states, open and closed for the sake of simplicity, the probability of the open state can be expressed in terms of the total free energy as (Sigworth, 1993),

$$P_o = \frac{\exp[-G_o^{tot}/k_B T]}{\exp[-G_o^{tot}/k_B T] + \exp[-G_c^{tot}/k_B T]}.$$

The total free energy of the protein in the open state can be written as the sum of two contributions. The first contribution represents the intrinsic energy of the protein, which is independent of the membrane potential. The second contribution represents the coupling between the protein and the transmembrane potential V_{mp}. This coupling takes the form of an effective state-dependent charge Q multiplying the transmembrane potential V_{mp}. The total free energy of the open state is thus,

$$G_o^{tot} = G_o + Q_o V_{mp}$$

and likewise for the closed state. A simple re-arrangement allows to re-write P_o as,

$$P_o = \frac{\exp[\Delta Q(V_{mp} - V_{1/2})/k_B T]}{1 + \exp[\Delta Q(V_{mp} - V_{1/2})/k_B T]},$$

where $\Delta Q = (Q_c - Q_o)$ and $V_{1/2} = (G_o - G_c)/\Delta Q$. The quantity $V_{1/2}$ corresponds to the voltage at which one half of the population of proteins are in the open state, and one half are in the closed state. It is directly related to the relative free energy of these two states in the absence of an applied voltage. The quantity ΔQ is the "gating charge" (Sigworth, 1993). Kv channels display a large gating charge, on the order of about 12 to 14 elementary charge (Schoppa et al., 1992). In simple terms, channel opening corresponds to the outward translocation of a large positive charge ΔQ. This charge is responsible for the strong coupling of the conformation of the channel to the transmembrane voltage.

It is important to try to reconcile the observed gating charge to the structure. In the simplest case of a perfectly planar membrane, the atomic charges of an intrinsic protein are expected to interact directly with the (constant) transmembrane electric field. In such a case, the gating charge ΔQ would simply be related to the displacement of each atomic charge of the protein in the direction perpendicular to the membrane surface. However, as illustrated in Fig. 7 the situation is more complicated if the surface of the protein is irregular and there are high dielectric aqueous crevices. For instance, there appears to be an aqueous channel-like region at the center of the voltage sensor. Such aqueous pore has been inferred by mutations in S4 that enabled the rapid conduction of ions in the inactivated ("down") state (Starace and Bezanilla, 2004; Tombola et al., 2005, 2007; Campos

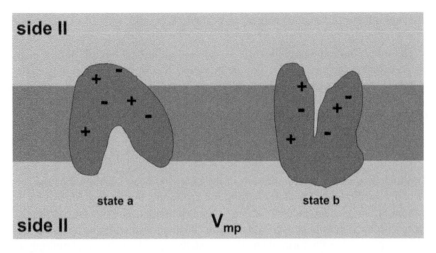

FIGURE 7 Schematic representation of a protein with embedded charges with irregular boundaries with the high dielectric solvent. In such complex geometry, the naïve concept of a constant transmembrane field is not valid.

et al., 2007). For example, conduction of H^+ through the voltage sensor is not normally observed in the case of the wild-type channel, but substitution of R371 by a histidine in *Shaker* (corresponding to R303 in Kv1.2) is sufficient to allow H^+ conduction. The presence of water molecules within the voltage sensor is qualitatively consistent with the concept of high dielectric aqueous crevices suggested previously by Islas and Sigworth (2001) and by Bezanilla and co-workers (Starace and Bezanilla, 2004). As illustrated by continuum electrostatic computations using simple geometries (Islas and Sigworth, 2001; Chanda *et al.*, 2005; Grabe *et al.*, 2004), a fundamental consequence of such high dielectric regions is to focus the transmembrane potential. To think rationally about those effects, it is necessary to have a theory of voltage gating that can be applied at the atomic level.

An effective theoretical formulation of the membrane potential and its influence on the configurational free energy of an intrinsic protein of arbitrary shape can be derived using macroscopic continuum electrostatics (Roux and Simonson, 1999). Following a Green's function decomposition of the total free energy, we obtained the modified Poisson–Boltzmann (PB) equation to account for the effect of the transmembrane potential (Roux, 1997),

$$\nabla \cdot \left[\varepsilon(\mathbf{r})\nabla\phi_{\mathrm{mp}}(\mathbf{r})\right] - \bar{\kappa}^2(\mathbf{r})\left[\phi_{\mathrm{mp}}(r) - \Theta(\mathbf{r})\right] = 0,$$

where $\varepsilon(\mathbf{r})$ and $\kappa(\mathbf{r})$ are the space-dependent dielectric coefficient and Debye–Hückel ionic screening factor, respectively, and $\Theta(\mathbf{r})$ is a Heaviside step function equal to 0 on the intracellular side of the membrane and equal to 1 on the extracellular side. While the atomic charges of the protein do not appear in the

CLOSED

OPEN

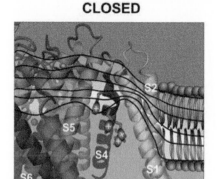
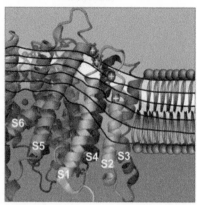

FIGURE 8 The PB-V equation (Roux, 1997) was solved numerically with a finite-difference relaxation algorithm for the approximate model of the Shaker channel in the closed and open states (Chanda *et al.*, 2005). In the calculations, the channel is represented explicitly with atomic details while the membrane is represented as a planar slab of dielectric 2, and the ions and water molecules by a continuum with dielectric constant of 80 and a salt concentration of 150 mM. The transmembrane field is focused and is most intense toward the extracellular half of the membrane.

PB-V equation, its molecular shape is taken into account in constructing the functions $\varepsilon(\mathbf{r})$ and $\kappa(\mathbf{r})$. The gating charge of the channel is then calculated from the ϕ_m calculated from the PB-V equation for the both the open and closed states,

$$\Delta Q = \sum_i q_i \left[\phi_{mp}^{(o)}(\mathbf{r}_i) \right] - \sum_i q_i \left[\phi_{mp}^{(c)}(\mathbf{r}_i) \right],$$

where q_i are the protein charges and $\mathbf{r}_i$ is their position. Mathematically, $\phi_{mp}(x, y, z)$ varies between 0 (on the extracellular side) to 1 (on the intracellular side) and corresponds to the dimensionless fraction of the applied transmembrane potential V_{mp} at the point $\mathbf{r}$. To illustrate how the PB-V theory works, we have calculated the gating charge for the approximate models of the open and closed states of the *Shaker* channel from (Chanda *et al.*, 2005). The result of the calculated transmembrane field is shown in Fig. 8. As expected, it is observed that the transmembrane field across the voltage sensor is focused. Summing over all partial charges of the *Shaker* channel yielded a gating charge of about 12 to 14 elementary charges, in good accord with experiments (Schoppa *et al.*, 1992). As shown in Fig. 9 the dominant contributions comes from arginine residues in the S4, in accord with experiment (Aggarwal and MacKinnon, 1996; Seoh *et al.*, 1996). While this calculation is based on approximate structural models (Chanda *et al.*, 2005), it helps illustrate that the observed gating charge

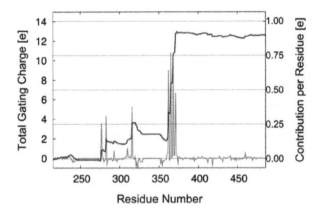

FIGURE 9 Computation of the gating charge from the approximate models of (Chanda *et al.*, 2005) using the PB-V equation (Roux, 1997). The cumulative gating charge (left axis) is shown in blue and the contributions from the individual residues (right axis) is shown in red. (For interpretation of the references to color in this figure legend, the reader is referred to the web version of this book.)

can, in principle, be reproduced without a full translocation of S4 across the bilayer.

IV. CONCLUSION

The progress in understanding the mechanism of voltage-gated channels in recent years has been remarkable. In many ways, there remains much to be done. Even the available X-ray structure of the Kv1.2 channel, which represents a remarkable achievement, is only of moderate resolution and several important regions of the protein are poorly defined (Long *et al.*, 2005). There is currently no atomic resolution structure of the closed state. All we have are approximate models generated from a wide range of data (Chanda *et al.*, 2005; Yarov-Yarovoy et al., 2006; Campos *et al.*, 2007; Pathak *et al.*, 2007; Tombola *et al.*, 2007). While there is little doubt that those models are qualitatively correct, they do not provide the sufficient atomic resolution that is required for detailed quantitative computations. Furthermore, there are probably a number of intermediate states, which will need to be characterized to reach a complete understanding of the mechanism (Bezanilla *et al.*, 1994; Zagotta *et al.*, 1994; Ledwell and Aldrich, 1999). More efforts will also be needed to better characterize the membrane environment of the protein. For example, recent experimental results indicated that the channel does not function well in membranes that formed by lipids that lack the phospho-di-ester group (Schmidt *et al.*, 2006). In the years to come, it can be anticipated that structural modeling and molecular dynamics simulations based on detailed models will continue to contribute to the progress in this exciting field.

Acknowledgements

This work was supported by grants form the National Institute of Health GM62342.

References

Aggarwal, S., MacKinnon, R. (1996). Contribution of the S4 segment to gating charge in the Shaker K$^+$ channel. *Neuron* **16**, 1169–1177.

Bezanilla, F. (2000). The voltage sensor in voltage-dependent ion channels. *Physiol. Rev.* **80**, 555–592.

Bezanilla, F., Perozo, E., Stefani, E. (1994). Gating of Shaker K$^+$ channels: II. The components of gating currents and a model of channel activation. *Biophys. J.* **66**, 1011–1021.

Bezanilla, F., Ruta, V., Chen, J., Lee, A., MacKinnon, R. (2003). The principle of gating charge movement in a voltage-dependent K$^+$ channel. *Nature* **423**, 42–48.

Campos, F.V., Chanda, B., Roux, B., Bezanilla, F. (2007). Two atomic constraints unambiguously position the S4 segment relative to S1 and S2 segments in the closed state of Shaker K channel. *Proc. Natl. Acad. Sci. USA* **104** (19), 7904–7909.

Catterall, W.A. (1988). Structure and function of voltage-sensitive ion channels. *Science* **242**, 50–61.

Cha, A., Snyder, G., Selvin, P., Bezanilla, F. (1999). Atomic scale movement of the voltage-sensing region in a potassium channel measured via spectroscopy. *Nature* **402**, 809–813.

Chanda, B., Asamoah, O.K., Blunck, R., Roux, B., Bezanilla, F. (2005). Gating charge displacement in voltage-gated ion channels involves limited transmembrane movement. *Nature* **436**, 852–856.

Cohen, B.E., Grabe, M., Jan, L.Y. (2003). Answers and questions from the KvAP structures. *Neuron* **39**, 395–400.

Cuello, L., Cortes, D., Perozo, E. (2004). Molecular architecture of the KvAP voltage-dependent K$^+$ channel in a lipid bilayer. *Science* **306**, 491–495.

Dorairaj, S.J., Allen, T.W. (2007). On the thermodynamic stability of a charged arginine side chain in a transmembrane helix. *Proc. Natl. Acad. Sci. USA* **104** (12), 4943–4948.

Grabe, M., Lecar, H., Jan, Y.N., Jan, L.Y. (2004). A quantitative assessment of models for voltage-dependent gating of ion channels. *Proc. Natl. Acad. Sci. USA* **101**, 17640–17645.

Hille, B. (2001). "Ion Channels of Excitable Membranes", 3rd ed. Sinauer, Sunderland, MA.

Islas, L.D., Sigworth, F.J. (2001). Electrostatics and the gating pore of Shaker potassium channels. *J. Gen. Physiol.* **117**, 69–89.

Jiang, Y., Lee, A., Chen, J., Cadene, M., Chait, B.T., MacKinnon, R. (2002). Crystal structure and mechanism of a calcium-gated potassium channel. *Nature* **417**, 515–522.

Jiang, Y., Lee, A., Chen, J., Ruta, V., Cadene, M., Chait, B., MacKinnon, R. (2003a). X-ray structure of a voltage-dependent K$^+$ channel. *Nature* **423**, 33–41.

Jiang, Y., Ruta, V., Chen, J., Lee, A., MacKinnon, R. (2003b). The principle of gating charge movement in a voltage-dependent K$^+$ channel. *Nature* **423**, 42–48.

Jogini, V., Roux, B. (2007). Dynamics of the Kv1.2 voltage-gated K$^+$ channel in a membrane environment. *Biophys. J.* **93** (9), 3070–3082.

Laine, M., Lin, M.C.A., Bannister, J.P.A., Silverman, W.R., Mock, A.F., Roux, B., Papazian, D.M. (2003). Atomic proximity between S4 segment and pore domain in Shaker potassium channels. *Neuron* **39**, 467–481.

Laine, M., Papazian, D.M., Roux, B. (2004). Critical assessment of a proposed model of Shaker. *FEBS Lett.* **564**, 257–263.

Ledwell, J.L., Aldrich, R.W. (1999). Mutations in the S4 region isolate the final voltage-dependent cooperative step in potassium channel activation. *J. Gen. Physiol.* **113**, 389–414.

Long, S.B., Campbell, E.B., MacKinnon, R. (2005). Crystal structure of a mammalian voltage-dependent Shaker family K$^+$ channel. *Science* **309**, 897–903.

Lu, Z., Klem, A., Ramu, Y. (2001). Ion conduction pore is conserved among potassium channels. *Nature* **413**, 809–813.

Murata, T., Yamato, I., Kakinuma, Y., Leslie, A.G.W., Walker, J.E. (2005). Structure of the rotor of the V-type Na-ATPase from enterococcus hirae. *Science* **308**, 654–659.

Okamura, Y., Murata, Y., Iwasaki, H., Sasaki, M. (2006). [New role of voltage sensor: Voltage-regulated phosphatase recently identified from ascidian genome]. *Tanpakushitsu Kakusan Koso* **51**, 18–26.

Papazian, D.M., Timpe, L.C., Jan, Y.N., Jan, L.Y. (1991). Alteration of voltage-dependence of Shaker potassium channel by mutations in the S4 sequence. *Nature* **349**, 305–310.

Pathak, M.M., Yarov-Yarovoy, V., Agarwal, G., Roux, B., Barth, P., Kohout, S., Tombola, F., Isacoff, E.Y. (2007). Closing in on the resting state of the Shaker K(+) channel. *Neuron* **56** (1), 124–140.

Phillips, L.R., Milescu, M., Li-Smerin, Y., Mindell, J.A., Kim, J.I., Swartz, K.J. (2005). Voltage-sensor activation with a tarantula toxin as cargo. *Nature* **436**, 857–860.

Posson, D.J., Ge, P., Miller, C., Bezanilla, F., Selvin, P.R. (2005). Small vertical movement of a K^+ channel voltage sensor measured with luminescence energy transfer. *Nature* **436**, 848–851.

Ramsey, I.S., Moran, M.M., Chong, J.A., Clapham, D.E. (2006). A voltage-gated proton-selective channel lacking the pore domain. *Nature* **440**, 1213–1216.

Roux, B. (1997). The influence of the membrane potential on the free energy of an intrinsic protein. *Biophys. J.* **73**, 2980–2989.

Roux, B., Simonson, T. (1999). Implicit solvent models. *Biophys. Chem.* **78**, 1–20.

Ruta, V., Chen, J., MacKinnon, R. (2005). Calibrated measurement of gating-charge arginine displacement in the KvAP voltage-dependent K^+ channel. *Cell.* **123**, 463–475.

Schmidt, D., Jiang, Q.X., MacKinnon, R. (2006). Phospholipids and the origin of cationic gating charges in voltage sensors. *Nature* **444**, 775–779.

Schoppa, N.E., McCormack, K., Tanouye, M.A., Sigworth, F.J. (1992). The size of gating charge in wild-type and mutant Shaker potassium channels. *Science* **255**, 1712–1715.

Selvin, P.R. (2002). Principles and biophysical applications of lanthanide-based probes. *Ann. Rev. Biophys. Biomol. Struct.* **31**, 275–302.

Seoh, S., Sigg, D., Papazian, D., Bezanilla, F. (1996). Voltage-sensing residues in the S2 and S4 segments of the Shaker K^+ channel. *Neuron* **16**, 1159–1167.

Sigworth, F.J. (1993). Voltage gating of ion channels. *Quat. Rev. Biophys.* **27**, 1–40.

Starace, D., Bezanilla, F. (2004). A proton pore in a potassium channel voltage sensor reveals a focused electric field. *Nature* **427**, 548–553.

Tempel, B.L., Papazian, D.M., Schwarz, T.L., Jan, Y.N., Jan, L.Y. (1987). Sequence of a probable potassium channel component encoded at Shaker locus of Drosophila. *Science* **237**, 770–775.

Tombola, F., Pathak, M.M., Gorostiza, P., Isacoff, E.Y. (2007). The twisted ion-permeation pathway of a resting voltage-sensing domain. *Nature* **445**, 546–549.

Tombola, F., Pathak, M.M., Isacoff, E.Y. (2005). Voltage-sensing arginines in a potassium channel permeate and occlude cation-selective pores. *Neuron* **45**, 379–388.

Treptow, W., Tarek, M. (2006). Environment of the gating charges in the Kv1.2 Shaker potassium channel. *Biophys. J.* **90**, L64–L66.

Yarov-Yarovoy, V., Baker, D., Catterall, W.A. (2006). Voltage sensor conformations in the open and closed states in ROSETTA structural models of K(+) channels. *Proc. Natl. Acad. Sci. USA* **103**, 7292–7297.

Yellen, G. (1998). The moving parts of voltage-gated ion channels. *Quant. Rev. Biophys.* **31**, 239–295.

Zagotta, W.N., Hoshi, T., Aldrich, R.W. (1994). Shaker potassium channel gating. III: Evaluation of kinetic models for activation. *J. Gen. Physiol.* **103**, 321–362.

CHAPTER 14

Computational Models for Electrified Interfaces

Elizabeth J. Denning and Thomas B. Woolf

Johns Hopkins University, School of Medicine, Departments of Physiology and of Biophysics and Biophysical Chemistry, Baltimore, Maryland, USA

I. INTRODUCTION

Ion channels are one of the most studied membrane proteins and understanding the behavior of ions and water near the pores of ion channels is essential for a molecular theory of function (Hille, 1992). One channel family of particular interest, especially due to recent X-ray structures, is the voltage-dependent potassium channel family (e.g. Doyle *et al.*, 1998; Long *et al.*, 2005a, 2005b). These channels conduct potassium ions across the cell membrane and play a key role in maintaining ionic balance across the membrane. The channel undergoes a structural change to facilitate the passage of potassium ions across the cellular membrane that depends on the transmembrane voltage, a process called 'gating'. The voltage-gating motion is coupled to differences in electrical potential across the membrane, initiating a transition between the open and closed states of the channel (e.g. Roux, 1997, 2002). The channel movement and the connection of movement to the transmembrane electrical potential are not fully understood at

Current Topics in Membranes, Volume 60
1063-5823/08 $35.00
DOI: 10.1016/S1063-5823(08)00014-8

an atomistic level, despite ongoing simulation work (e.g. Hoyles *et al.*, 1998). The problem is in understanding how the electrical potential difference at the interface is connected to gating and how a change in potential leads to large-scale molecular motions in the protein (e.g. Woolf *et al.*, 2004).

The voltage-gating motion couples to differences in electrical potential across the membrane initiating a transition between the open and closed states of the channel (Hille, 1992). The first models for ion channel gating were the result of voltage clamp experiments from Hodgkin and Huxley (1952). They fit their data from a series of voltage clamp experiments to a conceptual description of simple particles involved in kinetics between two states. In the resulting electrical model, the rate constants defining conductance were time and voltage dependent. The challenge is to understand this type of result on a molecular level.

To investigate the voltage dependence of the potassium channel, we first need to understand the electrostatic potential drop over a pure membrane bilayer (Sachs *et al.*, 2004b). In the past, the membrane bilayer and its components have been viewed as an electrical equivalent circuit with properties varying with channel type, with Nernst potentials describing reversal potentials, and with a membrane capacitance (Hille, 1992). Electrical models of ion channels date from Hodgkin and Huxley (1952) and consist of an equivalent circuit showing the membrane as a capacitance and the ion channel as a time and voltage dependent conductance. Fluctuations in the membrane structure contribute to channel conductance and are affected by solvent properties (e.g., Roux 1999, 2002; Roux *et al.*, 2000). The degree to which ions and other solvent molecules contribute to the electrified potential is dependent on ion type and charge. For this chapter we will be focusing on the electrified interface of a membrane bilayer and in particular the ionic and surface charge interactions.

II. GENERAL CONCEPTS OF SURFACE CHARGE AND CO/COUNTER IONS

Solvent molecules are the focus of all interactions concerning solute molecules and their surfaces (e.g. Benjamin, 1993; Balbuena *et al.*, 1996; Michael and Benjamin, 2001; Schnell *et al.*, 2004). Solute molecules suspended in water or any liquid exhibiting a high dielectric constant are usually ionic in nature. A pure membrane environment contains dielectric heterogeneity with electrostatic forces playing a central role involving intermolecular forces and long-range interactions between molecules and the surface of the membrane bilayer (White, 1994). For this section, we will examine the general concepts required for charging a surface and the role of co/counter ions in surface charging.

If no ions were present between two similarly neutral surfaces there will be no electric field in the gap between them and thus no surface charge. Surface charging in a solvent occurs in one of two ways. One mechanism of charging a surface is by ionization or dissociation of surface groups. An example of this mechanism would be the protonation or deprotonation of the terminal group of the protein through either the carboxylic acid group ($-COOH \rightarrow -COO^- + H^+$) or the amino acid group ($-NH_3^+ \rightarrow -NH_2 + H^+$). Another mechanism for charging a surface can come about by the adsorption of ions from solution onto a surface that is uncharged or oppositely charged in relation to the nature of the ion. Near a charged surface over time there is an increase in the density of counterions (ions with opposite charge of the surface) and a decrease in co-ion (ions charged similarly to the surface) density thus charging the surface (e.g. Boda *et al.*, 1998). An example of an ion adsorption onto an uncharged surface would be an interface with air–water or hydrocarbon–water forcing the surface of the bilayer to become charged. Another example, the association of opposite charges to a surface, is when Na^+ ions within a solvent associated with the anionic lipid headgroups of a lipid bilayer. This association makes the net surface charge of the bilayer neutral. As noted by the examples mentioned above, it is seen that this mechanism is dependent on the ionic conditions of the solvent.

Since voltage-dependent channels are sensitive to the local electric field, neutralization of the surface charge is expected to cause changes in voltage-dependent properties, such as conductance and time constants. If surface charge was static, it would be expected that a particular neutralization of surface charge would cause a corresponding change in electric field, leading to a translation of all voltage-dependent properties along the voltage axis. Before we can discuss how this relates to a membrane gating potential, we must consider the fundamental models that describe the co/counterion distribution between charged surfaces in solution.

III. CONTINUUM MODELS

In the past, the membrane had been viewed as a region of low polarizability that acted as a high-energy barrier for the passage of solute molecules and to be homogeneous in nature (e.g. Parsegian, 1969). However, we recognize the bilayer cannot be homogeneous due to the distribution of co/counterions at the surface of a membrane system. Thus, we currently view the membrane to have dielectric heterogeneity and to be a fluid mosaic adorned with various proteins, carbohydrates, and ion associations (e.g. White, 1994). There are two approaches to handling this heterogeneous type of environment: continuum and molecular theories. For this section we will focus on the continuum model.

The continuum model assumes that the bulk properties of the system such as dielectric constant and density of solvent hold true on the molecular level. The model represents the entire solvent and interactions among the atoms of the solute though dielectric constants. There are several continuum solvent models available to describe a bilayer.

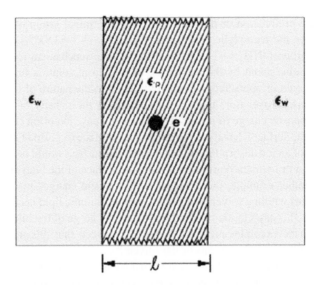

FIGURE 1 The implicit dielectric sandwich model (from Parsegian, 1969). The striped region is representative of the implicit membrane of thickness L and a low dielectric constant value of ε_p. The yellow regions are representative of the bulk solvent with a high dielectric constant of ε_w which includes water and salt contributions.

The simplest continuum model has been the implicit dielectric sandwich model (see Fig. 1). The dielectric sandwich model views the membrane as a slab of uniform thickness with a low dielectric constant and the bulk solvent modeled as a high dielectric constant on either side of the low dielectric slab. The interaction energies are modeled similarly to that of a modified Coulomb approach, specifically referred to as a Born energy (see Eq. (1)).

$$\Delta G = \frac{332q^2}{2a}\left(\frac{1}{\varepsilon_2} - \frac{1}{\varepsilon_1}\right). \tag{1}$$

The Born energy of a charge is measure of the free energy associated with the electric field around the charge. Basically, the Born energy is calculated by integrating over the free energy density of an electric field which arises from the charged state of a fixed charge within a volume divided by the dielectric permittivity of the solvent and the solvation radius of the charge.

The important aspect of the Born energy is that we are looking at the "effective" size/radius of an ion that stems from the hydration shells surrounding the ion.

This type of continuum calculation is computationally inexpensive. However, the simplicity of the calculation does not take into account the molecular details of charged surfaces such as the headgroup region. For this, we need to understand the solvent effects for Coulomb interactions using a more complex continuum algorithm or directly at the molecular level in order to better appreciate membrane surface characteristics. This means we cannot make assumptions for bulk properties and that it is necessary to examine how solvent affects the electric fields around dissolved ions.

IV. BRIDGING FROM THE CONTINUUM TO MOLECULAR CALCULATIONS

The simple continuum model has some advantages such as reduced computational time. However, it has obvious limitations in terms of understanding membrane surface properties. In order to incorporate these properties and thus enhance the success of the model, we need to review the following notion. When there is a charged surface, such as the headgroup region of a lipid bilayer, solvent molecules, especially ions, will segregate and perform one of two functions. The counterions will associate with the headgroup region to equally balance the surface or the counterions will build up in between the two surfaces and mix with the rest of the solvent molecules. This segregation of ions leads to what is commonly referred to as the electrical double bilayer. In this section, we will examine how this electrical double layer is incorporated into biological systems through various methods and bridge the gap between the simple dielectric sandwich model and molecular calculations.

A. Gouy–Chapman Theory/Poisson–Boltzmann

Gouy–Chapman theory explains the effect of a static smoothed surface charge on the membrane potential (e.g. Sharp and Honig, 1990; Netz, 1999). When no surface charge is present, the resting potential across a membrane depends on the concentration and membrane permeability of the ions. For example, the addition of a negative surface charge leads to formation of a double layer. The double layer establishes an electric field through an atmosphere of ions in rapid thermal motion of additional counterions in addition to those balancing the surface charge. The electric field results in a potential difference across the membrane which is in the same direction as the electric field in the membrane.

The overall potential difference does not change, since it must still balance the effect of the concentration and membrane permeability of the several ions. Therefore, the addition of a negative surface charge causes a decrease in the magnitude of the potential difference across the membrane surface. The usual method for obtaining experimental information about the surface charge is to neutralize it, at least partially, by the addition of polyvalent cations, and to measure voltage-dependent properties for different amounts of neutralization. The neutralization of a negative surface charge can be caused by binding, by screening, or by both since positive ions in solution tend to balance the negative surface charge. The Gouy–Chapman theory describes the static smoothed surface charge on the membrane potential, but the question remains how to begin to apply this theory to calculate the distribution of the electrostatic potential of a dynamic surface charge system.

To apply the Gouy–Chapman idea to computational calculations, the distribution of the electrostatic potential in electrolytic solute is commonly determined by computations with the nonlinear Poisson–Boltzmann (PB) equation (e.g. Sharp and Honig, 1990; Simonson, 2003). The PB equation approximates the electrostatic energy of a fixed system of solute charges within a thermally averaged salt medium, by computing the total electrostatic potential as follows:

$$\nabla \cdot \left[\varepsilon(\vec{r})\nabla\phi(\vec{r})\right] - 4\pi \sum_s q_s c_s \exp\left[-\beta q_s \phi(\vec{r})\right] + 4\pi \rho_{\text{int}}(\vec{r}) = 0, \qquad (2)$$

where s indicates the salt type, q_s is the charge of each ion, c_s is the average number density of each species, and $\beta = 1/k_B T$. The PB equation is solved for arbitrary solute charge distributions using self-consistent grid calculations.

The advantages for using the PB equation is that it accounts for a uniform surface charge density and it qualitatively predicts the energy trends. The PB model breaks down at small distances where it no longer realistically describes the ionic distribution and forces between two surfaces (e.g. Torrie and Valleau, 1980, 1982; Valleau and Torrie, 1982). The method ignores interactions that occur between dissolved ions and solvent molecules within the immediate vicinity. The method also imposes restrictions on ionic strength and surfaces charge densities in that it works well for a 1:1 electrolyte ratio, but is less effective for divalents or situations with ionic size effects. While there have been important advances in using PB for membrane systems, it is generally considered too slow for calculations with membrane proteins (e.g. Forsten et al., 1994; Im et al., 1998).

B. Generalized Born/Effective Dipoles

Although the PB method describes the distribution of the electrostatic potential in electrolyte solutions, the method is computationally expensive for membrane proteins. Recent advances in the application of the generalized Born formalism (Still et al., 1990) to membrane systems has shown that accuracy comparable to the PB equation can be achieved at a more moderate computer cost (e.g. Spassov et al., 2002; Im et al., 2003; Feig et al., 2004; Tanizaki and Feig, 2006; Ulmschneider et al., 2007). Still et al. developed the generalized Born (GB) formula so that it would incorporate interactions between atoms.

$$\Delta G_{\text{elec}} = -\frac{1}{2}\left(\frac{1}{\varepsilon_p} - \frac{1}{\varepsilon_s}\right) \sum_{\alpha\beta} \frac{q_\alpha q_\beta}{\sqrt{r_{\alpha\beta}^2 + R_\alpha^{\text{GB}} R_\beta^{\text{GB}} \exp(-r_{\alpha\beta}^2/4R_\alpha^{\text{GB}} R_\beta^{\text{GB}})}}, \qquad (3)$$

where ε_p is the dielectric constant in the interior of the solute (which normally has values between 1 and 20), ε_s describes the high dielectric/bulk solvent region,

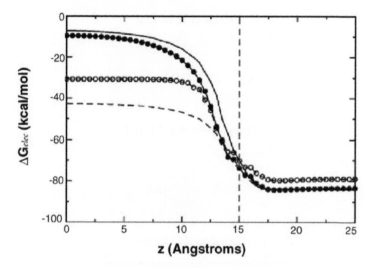

FIGURE 2   The electrostatic solvation free energy as a function of distance from the bilayer center (0 Å) in a bilayer of 30-Å thickness. The PB solutions are shown as solid and dashed lines where the internal membrane dielectric is 1 or 2 respectively. For both PB calculations, bulk water is taken as epsilon 80. The GB calculations are shown as filled or open circles and differ only in the number of radial integration points that were used. Notice the range of estimates for ion solvation in these two implicit models of the membrane bilayer (from Im *et al.*, 2003).

and R^{GB} is the effective Born radius of an atom. The GB method depends strongly on evaluating the effective Born radii properly. However, GB method has similar tradeoffs in comparison with the PB method, with a large advantage for conformational searching, but a lack of ability to comment on specific salt or headgroup effects on membrane protein function. Fig. 2 shows a comparison of both PB and GB methods for an explicit ion within an implicit bilayer of 30-angstroms thickness. The Parsegian estimate from a dielectric sandwich model (interior dielectric of 2) is about 40 kcal/mol of energy from water to the center of the bilayer. The GB/PB computation shows about 40–70 kcal/mol of energy from water to the center of the bilayer, depending on the choice of the dielectric constant for the membrane interior (values of 1 or 2 were used).

Extensions of the GB methods have focused on evaluating the Born radii within a heterogeneous dielectric environments by setting the Born radii as a function of dielectric constant (e.g. Tanizaki and Feig, 2005, 2006). Fig. 3 shows the concept behind the extension proposed by Tanizaki and Feig. The heterogeneous dielectric generalized Born model, HDGB, is the first to attempt to compensate for the GB method assumption that a sharp boundary that defines the region between high dielectric and low dielectric regions of the bilayer. This model is more computationally intensive than most GB methods in that it takes twice the amount of computer time as a standard GB method. However, the HGDB method represents

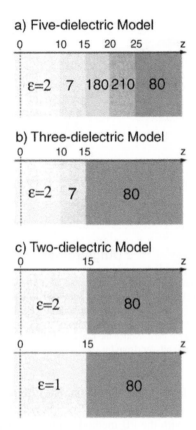

FIGURE 3 The implicit heterogeneous dielectric models (from Tanizaki and Feig, 2005). The striped regions is representative of the implicit membrane at a particular distance from the center of the implicit bilayer and at a specific dielectric constant value of ε.

the heterogeneous interfacial region of the bilayer as seen in both experiment and all-atom simulation studies. The HDGB method is limited in that it does not allow for changes in the shape or local distortions of the membrane–water interface.

An alternative continuum model, which also takes into account the heterogeneity of the interface though less well explored, is the Langevin dipole model (e.g. Grossfield *et al.*, 2000). This method can be pictured as a mesoscale model for the bilayer (e.g. Grossfield *et al.*, 2000). The approach builds from work in the Warshel lab (e.g. Warshel and Russell, 1984). The idea behind the Langevin dipole model is to vary the effects of effective dipoles as a function of membrane bilayer depth. Since the intrinsic moment of a dipole is a local property, local variations of electrostatic properties can be explored in a physically consistent manner. For the dipole model, the solvation free energy for a solute in a specific location in the membrane is estimated by calculating the response of the lattice dipoles to the solute partial charges. For this model, the implicit membrane is represented by a lattice of freely reorientable dipoles whose properties change with membrane position. The solute is represented in a fully detailed molecular form with partial charges, which is inserted into the membrane lattice. The dipoles that overlap it are displaced into the vacuum, still in their lattice configuration and as a result, interactions between these dipoles do not change and can be neglected. The energy change upon their removal is proportional to the average interaction between them and the dipoles remaining in the system. This method is advantageous because it addresses the heterogeneous interface by providing a smooth, gradual transition between the high dielectric and low dielectric regions. The electrostatic heterogeneity of the membrane results from varying the magnitude of the permanent dipoles as a function of their position in the bilayer.

To summarize this section, the Gouy–Chapmann theory, the PB method, the GB methods, and the Langevin dipole methods are efficient in terms of bridging the gap between the simple implicit dielectric sandwich model and the all-atom method. However, these models make key assumptions in that the external surface charge is uniformly distributed and that the membrane is homogeneous and cannot deform. The methods still make assumptions about the bulk properties of the solvent and omitting the important atomic details of dissolved ions. For us to more accurately comment on electrical interfaces of a complex heterogeneous bilayer, we need to look at all-atom models.

V. ALL-ATOM MODELS

Membrane characteristics have profound consequences for the biophysical processes involved in transmembrane potentials and channel function. Even with the advances of continuum models, several important characteristics of membrane systems are still not captured. These bilayer characteristics include asymmetry of lipid composition, ionic strength and concentration, water penetration of the hydrophobic core, and local ordering of water and salt near the surface of the bilayer. In this section, the discussion will revolve around recent developments and advantages of all-atom models, molecular theory calculations.

The advantages of explicit all-atom calculations are accuracy, capturing atomic details, and temporal effects (e.g. Glosli and Philpott, 1996; Yeh and Berkowitz, 2000; Tobias, 2001; Sachs and Woolf, 2003; Mukhopadhyay et al., 2004; Sachs et al., 2004a, 2004b; Kandasamy and Larson, 2006a, 2006b; Gurtovenko and Vattulainen, 2007). It is important to emphasize all-atom calculations do not have the restrictions on lipid heterogeneity, ionic strength and concentration, and surfaces charge densities that can be present in continuum models. The explicit model captures all the above-mentioned characteristics as well as both the long and short-range electrostatics and important aspects of ion hydration shells.

Ionic radii differences influence the balance of enthalpic and entropic constraints on hydration shell structure of an ion playing a significant role in membrane protein structure and function (e.g. Aroti et al., 2004; Grossfield, 2005). Continuum models only consider the primary hydration shell for the partial charge of specific atoms to reduce computational intensity. Explicit models allow solvent molecule calculations to account for the structure of the hydration shell of ions. The explicit nature of the calculations can show the hydration of every atom and multiply the hydration shells for a single atom. Ions with a small atomic radius tend to be more hydrated and have a high exchange rate with the surrounding solvent, thus they have a large hydration radius. Water molecules surrounding the ions at the surface of membranes may play a significant role in influencing headgroup conformations and in determining the strength of interactions between headgroups and ions.

The explicit solvent calculations also account for dielectric screening of interionic interactions that are dependent on solution conditions. This in turn may enable the computational reproduction of experimental results that emphasize the effects of ionic size (e.g. Harries et al., 2006, Petrache et al., 2005, 2006; Wang et al., 2006). This dielectric screening, referred to as the Debye length, describes the ionic atmosphere at a certain distance from a charged surface and only occurs when the Debye–Hückel approximation, $k_B T \gg (|z_i e \psi|)$, is valid. The magnitude of the Debye length depends solely on the properties of the solution and excludes properties of the surface such as charge or potential. The length is the distance over which the usual Coulomb $1/r$ field exponentially decays to $1/e$ of its unscreened value. This decay is due to the polarization of the surface. For example, at a distance of one Debye length away from the surplus charge, its effect on the mobile charges of the surface are no longer felt.

Typically, all-atom simulations have been modeled as a single bilayer and utilize the Ewald sum to account for the electrostatic potential. However, there is a problem with using the Ewald sum for a periodic single bilayer system because an electrified potential across the membrane cannot occur. The Ewald sum requires that a system be replicated periodically in all three dimensions but for a single bilayer the transmembrane concentration gradient simulation exhibit only two-dimensional periodicity. One solution is to simulate more than one bilayer per unit cell thus creating a periodic double bilayer system (see Fig. 4). An all-atom double bilayer system maintains the advantages given through single bilayer calculations such as multiple ion hydration shells, temporal effects, etc. but contributes the additional advantage of allowing an electrified transmembrane potential to be more accurately portrayed (Sachs et al., 2004a; Leekumjorn and Sum, 2006). An example of this type of calculation is shown in Fig. 5, where adding one salt anion/cation is sufficient for a 170 mV potential drop within the system. Figs. 4 and 5 summarize the intricate nature of a "simple" pure membrane system.

The periodic double bilayer system establishes a transmembrane concentration gradient as seen in Fig. 4. Figure 5 shows that this is a temporal effect in the

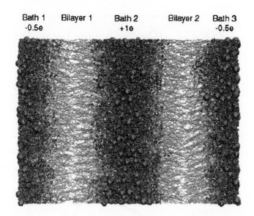

Bath 1   Bilayer 1   Bath 2   Bilayer 2   Bath 3
-0.5e   +1e   -0.5e

FIGURE 4 Snapshot from the 170 mV simulation showing the three regions of salt–water separated by the two bilayers. The left and right edges are connected by periodic boundary conditions. An excess of 1 Na^+ is in the central bath and 1 Cl^- in the two outer baths combined (see Table I). The ions are represented as oversized spheres (from Sachs *et al.*, 2004b).

TABLE I

Number of ions in the starting configuration for each of the three salt–water baths shown in Fig. 4. In each case the concentration is 1 M, with the central region having 1 excess Na^+ and each of the two periodically connected outer regions having 0.5 excess Cl^- ions on average (from Sachs *et al.*, 2004b)

Simulation	Bath 1	Bath 2	Bath 3
Neutral	60 Na^+, 60 Cl^-	120 Na^+, 120 Cl^-	60 Na^+, 60 Cl^-
Electrified	60 Na^+, 60 Cl^-	121 Na^+, 120 Cl^-	60 Na^+, 61 Cl^-

example of Fig. 4, where the Nernst potential is seen as a time and spatial average over the simulation cell. At a particular short time-interval the electrostatic force felt at a particular point in the system will depend on the ion distribution. That is, it is not the same as the long-time average and so motions under the explicit ion distribution for membrane, water and protein will depend on the details of the salt distribution. An example of the types of detail that can be influenced by electrostatics is the water dipole orientations shown in Fig. 6. Ion-exchange can be an important part of the electrified interface and the build-up of an electrical potential.

Membrane systems are complex especially their bilayer interfacial regions. To further emphasize the all-atom complexity of the interfacial zones, Fig. 7 shows the average densities of different chemical groups from the bilayer center. In particular, with different concentration and/or different anion sizes, the distribution functions change. This type of effect would not be seen in a continuum model

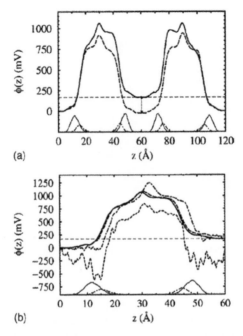

(a)

(b)

FIGURE 5 Transmembrane potential profiles for the neutral and electrified systems. The profile from the neutral long-dashed line and 170 mV simulations solid line across the entire 120 Å system shown in Fig. 4 (0 Å, the left-most point, 120 Å, the right-most point). The vertical arrow highlights the difference in potential gradient between the two simulations. A temporal build-up of the equilibrated potential profile can be seen in part b. Symmetrized half-cell profiles from the 170 mV simulation extracted from the first 1 ps dashed, 10 ps dot-dashed, 1 ns long dashed and 10 ns solid. For spatial reference, arbitrarily scaled density profiles for the lipid phosphate dotted and carbonyl groups dot-dashed from the electrified simulation are given at the bottom of both plots (from Sachs *et al.*, 2004b).

of the bilayer. Figure 8 emphasizes that residency times for ions in an all-atom model will depend on the concentration and the bilayer surface properties.

To summarize this section, charged biological bilayers are not symmetric due to the asymmetry of lipid composition and local ordering of water and salt in the double-layer (e.g. Gurtovenko, 2005; Gurtovenko and Vattulainen, 2005; Kotulska and Kubica, 2005; Song *et al.*, 2005; Cascales *et al.*, 2006). The importance of all-atom methods, to molecular theory calculations, has been highlighted. All-atom simulations of ion solvation have shown the advantages of explicit simulations for understanding the relative tradeoffs in modeling effects either directly through an all-atom approach or non-specifically through an implicit model (e.g. Hummer *et al.*, 1997; Canales and Sese, 1998; Chandra and Bagchi, 1999; Keblinski *et al.*, 2000; Stern and Feller, 2003). These types of calculations can be

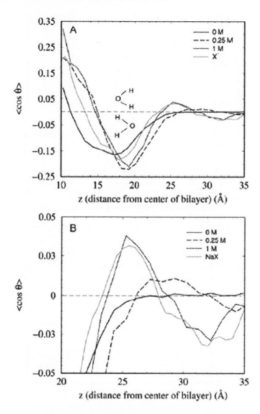

FIGURE 6   Shown are water dipole orientations as a function of distance from the bilayer center from the no-salt, 0.25 M and 1 M NaCl simulations and results from a NaX simulation. As illustrated in the figure, $\cos \theta > 0$ corresponds to water oxygens pointing inward, and $\cos \theta < 0$ corresponds to water oxygens pointing outward. (A) displays the full profile, (B) shows the oscillations originating in the headgroup region and propagating into the aqueous phase (from Sachs *et al.*, 2004b).

computationally intensive although with advances in computational power these calculations have become more attractive.

VI. LONG-TERM TRENDS

The question with all these options for computational calculations becomes: Is it better to calculate direct ion voltage effects or non-specific voltage effects? Ion and water behavior at the membrane surface have been highlighted both by experimental and computational data on lipid headgroup conformational dynamics. The degree to which ions contribute to the electrified potential is dependent on ion type and charge. Ions that have a small atomic radius tend to be more hy-

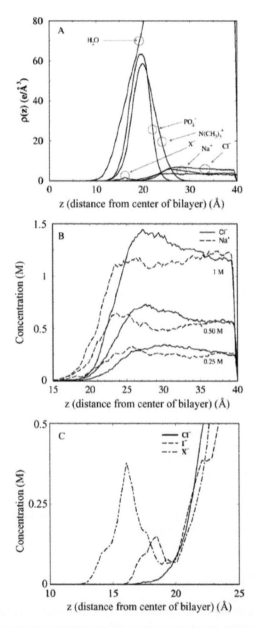

FIGURE 7 Component distributions for various groups in the simulations as a function of distance from the center of the bilayer. (A) Electron densities for water, headgroup phosphate and choline groups, and salt from the 1 M NaCl simulation, plus the profile for X^- from the NaX simulation. (B) Concentration profiles for Na^+ and Cl^- from the 0.25 M, 0.50 M, and 1 M simulations. (C) Concentration profiles for the three anion types in the interfacial regions (from Sachs *et al.*, 2004b).

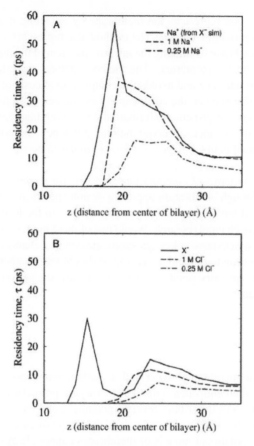

FIGURE 8 Residency times, τ, for ions are shown as a function of distance from the bilayer center, (a) for cations and (b) anions at 1 M (solid lines), 0.25 M (long-dashed lines) or from the NaX simulation (short-dashed lines) (from Sachs *et al.*, 2004b).

drated and have a higher exchange rate with the surrounding solvent, thus they have a large hydration radius. Atoms with large hydration radii include Na^+, Li^+, etc. Weakly solvated ions include Cl^-, Br^-, I^-, and $N(CH_3)^+$ and their solvent exchange rates are comparable to water-water exchange. The solvation effect of water and ions directly contributes to fluctuations of the membrane bilayer and moreover to the transmembrane electrical potential.

With the significance of solvation effects in mind, does it make sense to model implicitly with more complex algorithms or with advances in computer technology to use all-atom models? Advances such as the dipole lattice membrane model have contributed considerably to enhancing implicit solvation models (Grossfield *et al.*, 2000). The dipole lattice membrane model uses the Langevin dipoles method. This method represents the membrane by modeling it as a meso-scale lattice of freely reorienting dipoles. The method also represents dielectric heterogeneity of the membrane by making the intrinsic moment of the dipoles a function of their position in the membrane. The method captures the effects of a broad membrane–water interface and avoids the computational costs of all-atom simulations. In a similar manner, the advances in Generalized Born theories enable a much greater degree of protein conformational sampling than the all-atom models. However, with advances in computational efficiency and power, the ability to simulate all-atom double bilayer systems continues to become more attainable and appealing.

The reduction in terms of computations tradeoffs for modeling voltage effects either directly through an all-atom approach or non-specifically through implicit model have raised the membrane bilayer calculations to the level where we can begin to apply it to channel research. Would one of these models or models similar in respect to them help answer key questions surrounding channel gating? At this moment it is too soon to tell and so the future should tell whether these tools are sufficient or only an entry point for beginning to investigate ion channel gating within a membrane.

VII. SUMMARY

In the past, the membrane bilayer and its surface interactions have been pictured as an electrical equivalent circuit with varying properties such as channel type, Nernst potentials describing reversal potentials, and the membrane capacitance (e.g. Hille, 1992). To better understand the nature of an electrified potential of a bilayer, computational models of membranes cannot always be treated as a continuum solvent defined just by bulk properties due to the heterogeneity and specific solvent effects of the membrane surface. Significant advances in computational power have made it more exacting and appealing to model a membrane system explicitly in order to capture the heterogeneity of the bilayer and important solvent effects.

References

Aroti, A., Leontidis, E., Maltseva, E. (2004). Effects of Hofmeister anions on DPPC Langmuir monolayers at the air–water interface. *J. Phys. Chem. B* **108** (39), 15238–15245.

Balbuena, P.B., Johnston, K.P., Rossky, R.J. (1996). Molecular dynamics simulation of electrolyte solutions in ambient and supercritical water. 1. Ion solvation. *J. Phys. Chem.* **100**, 2706–2715.

Benjamin, I. (1993). Mechanism and dynamics of ion transfer across a liquid–liquid interface. *Science* **261**, 1558–1560.

Boda, D., Chan, K.Y., Henderson, D. (1998). Monte Carlo simulation of an ion-dipole mixture as a model of an electrical double layer. *J. Chem. Phys.* **109** (17), 7362–7371.

Canales, M., Sese, G. (1998). Generalized Langevin dynamics simulations of NaCl electrolyte solutions. *J. Chem. Phys.* **109** (14), 6004–6011.

Cascales, J.J.L., Otero, T.F., Smith, B.D. (2006). Model of an asymmetric DPPC/DPPS membrane: Effect of asymmetry on the lipid properties. A molecular dynamics simulation study. *J. Phys. Chem. B* **110** (5), 2358–2363.

Chandra, A., Bagchi, B. (1999). Ion conductance in electrolyte solutions. *J. Chem. Phys.* **110** (20), 10024–10034.

Doyle, D.A., Cabral, J.M., Pfuetzner, A., Juo, A., Gulbis, J.M., Cohen, S.L., Chait, B.T., MacKinnon, R. (1998). The structure of the potassium channel: Molecular basis of K^+ conduction and selectivity. *Science* **280**, 69–77.

Feig, M., Im, W., Brooks III, C.L. (2004). Implicit solvation based on generalized Born theory in different dielectric environments. *J. Chem. Phys.* **120** (2), 903–911.

Forsten, K.E., Kozack, R.E., Lauffenburger, D.A., Subramaniam, S. (1994). Numerical solution of the nonlinear Poisson–Boltzmann equation for a membrane–electrolyte system. *J. Phys. Chem.* **98**, 5580–5586.

Glosli, J.N., Philpott, M.R. (1996). Molecular dynamics study of interfacial electric fields. *Electrochim. Acta* **41** (14), 2145–2158.

Grossfield, A. (2005). Dependence of ion hydration on the sign of the ion's charge. *J. Chem. Phys.* **122** (2). Art. No. 024506.

Grossfield, A., Sachs, J., Woolf, T.B. (2000). Dipole lattice membrane model for protein calculations. *Proteins* **41** (2), 211–223.

Gurtovenko, A.A. (2005). Asymmetry of lipid bilayers induced by monovalent salt: Atomistic molecular-dynamics study. *J. Chem. Phys.* **122** (24). Art. No. 244902.

Gurtovenko, A.A., Vattulainen, I. (2005). Pore formation coupled to ion transport through lipid membranes as induced by transmembrane ionic charge imbalance: Atomistic molecular dynamics study. *J. Am. Chem. Soc.* **127** (50), 17570–17571.

Gurtovenko, A.A., Vattulainen, I. (2007). Ion leakage through transient water pores in protein-free lipid membranes driven by transmembrane ionic charge imbalance. *Biophys. J.* **92** (6), 1878–1890.

Harries, D., Podgornik, R., Parsegian, V.A. (2006). Ion induced lamellar–lamellar phase transition in charged surfactant systems. *J. Chem. Phys.* **124** (22). Art. No. 224702.

Hille, B. (1992). "Ionic Channels of Excitable Membranes". Sinauer Associates, Inc.

Hodgkin, A.L., Huxley, A.F. (1952). A quantitative description of membrane current and its application to conduction and excitation in nerve. *J. Physiol. (London)* **117**, 500–544.

Hoyles, M., Kuyucak, S., Chung, S.H. (1998). Computer simulation of ion conductance in membrane channels. *Phys. Rev. E* **58** (3), 3654–3661.

Hummer, G., Pratt, L.R., Garcia, A.E. (1997). Ion sizes and finite-size corrections for ionic solvation free energies. *J. Chem. Phys.* **107** (21), 9275–9277.

Im, W., Beglov, D., Roux, B. (1998). Continuum solvation model: Electrostatic forces from numerical solutions to the Poisson–Boltzmann equation. *Comput. Phys. Commun.* **111**, 59–75.

Im, W., Feig, M., Brooks III, C.L. (2003). An implicit membrane generalized Born theory for the study of structure, stability, and interactions of membrane proteins. *Biophys. J.* **85**, 2900–2918.

Kandasamy, S.K., Larson, R.G. (2006a). Cation and anion transport through hydrophilic pores in lipid bilayers. *J. Chem. Phys.* **125** (7). Art. No. 074901.

Kandasamy, S.K., Larson, R.G. (2006b). Effect of salt on the interactions of antimicrobial peptides with zwitterionic lipid bilayers. *Biochim. Biophys. Acta Biomembranes* **1758** (9), 1274–1284.

Keblinski, P., Eggebrecht, J., Wolf, D., Phillpot, S.R. (2000). Molecular dynamics study of screening in ionic fluids. *J. Chem. Phys.* **113** (1), 282–291.

Kotulska, M., Kubica, K. (2005). Structural and energetic model of the mechanisms for reduced self-diffusion in a lipid bilayer with increasing ionic strength. *Phys. Rev. E* **72** (6). Art. No. 061903.

Leekumjorn, S., Sum, A.K. (2006). Molecular study of the diffusional process of DMSO in double lipid bilayers. *Biochim. Biophys. Acta Biomembranes* **1758** (11), 1751–1758.

Long, S.B., Campbell, E.B., MacKinnon, R. (2005a). Crystal structure of a mammalian voltage-dependent Shaker family K^+ channel. *Science* **309** (5736), 897–903.

Long, S.B., Campbell, E.B., MacKinnon, R. (2005b). Voltage sensor of KV1.2: Structural basis of electromechanical coupling. *Science* **309** (5736), 903–908.

Michael, D., Benjamin, I. (2001). Molecular dynamics computer simulations of solvation dynamics at liquid/liquid interfaces. *J. Chem. Phys.* **114** (6), 2817–2824.

Mukhopadhyay, P., Monticelli, L., Tieleman, D.P. (2004). Molecular dynamics simulation of a palmitoyl–oleoyl phosphatidylserine bilayer with Na^+ counterions and NaCl. *Biophys. J.* **86** (3), 1601–1609.

Netz, R.R. (1999). Debye–Huckel theory for interfacial geometries. *Phys. Rev. E* **60** (3), 3174–3182.

Parsegian, A. (1969). Energy of an ion crossing a low dielectric membrane: Solutions to four relevant electrostatic problems. *Nature* **221**, 844–846.

Petrache, H.I., Kimchi, I., Harries, D., Parsegian, V.A. (2005). Measured depletion of ions at the biomembrane interface. *J. Am. Chem. Soc.* **127** (33), 11546–11547.

Petrache, H.I., Zemb, T., Belloni, L., Parsegian, V.A. (2006). Salt screening and specific ion adsorption determine neutral-lipid membrane interactions. *Proc. Natl. Acad. Sci.* **103** (21), 7982–7987.

Roux, B. (1997). Influence of the membrane potential on the free energy of an intrinsic protein. *Biophys. J.* **73**, 2980–2989.

Roux, B. (1999). Statistical mechanical equilibrium theory of selective ion channels. *Biophys. J.* **77**, 139–153.

Roux, B. (2002). Theoretical and computational models of ion channels. *Curr. Opin. Struct. Biol.* **12**, 182–189.

Roux, B., Berneche, S., Im, W. (2000). Ion channels, permeation, and electrostatics: Insight into the function of KcsA. *Biochemistry* **39**, 13295–13306.

Sachs, J.N., Woolf, T.B. (2003). Understanding the Hofmeister effect in interactions between chaotropic anions and lipid bilayers: Molecular dynamics simulations. *J. Am. Chem. Soc.* **125** (29), 8742–8743.

Sachs, J.N., Crozier, P.S., Woolf, T.B. (2004a). Atomistic simulations of biologically realistic transmembrane potential gradients. *J. Chem. Phys.* **121** (22), 10847–10851.

Sachs, J.N., Nanda, H., Petrache, H.I., Woolf, T.B. (2004b). Changes in phosphatidylcholine headgroup tilt and water order induced by monovalent salts: Molecular dynamics simulations. *Biophys. J.* **86** (6), 3772–3782.

Schnell, B., Schurhammer, R., Wipff, G. (2004). Distribution of hydrophobic ions and their counterions at an aqueous liquid–liquid interface: A molecular dynamics investigation. *J. Phys. Chem. B* **108** (7), 2285–2294.

Sharp, K.A., Honig, B. (1990). Electrostatic interactions in macromolecules: Theory and applications. *Annu. Rev. Biophys. Biophys. Chem.* **19**, 301–332.

Simonson, T. (2003). Electrostatics and dynamics of proteins. *Rep. Prog. Phys.* **66** (5), 737–787.

Song, Y.H., Guallar, V., Baker, N.A. (2005). Molecular dynamics simulations of salicylate effects on the micro- and mesoscopic properties of a dipalmitoylphosphatidylcholine bilayer. *Biochemistry* **44** (41), 13425–13438.

Spassov, V.Z., Yan, L., Szalma, S. (2002). Introducing an implicit membrane in generalized Born/Solvent accessibility continuum solvent models. *J. Phys. Chem. B* **106**, 8726–8738.

Stern, H.A., Feller, S.E. (2003). Calculation of the dielectric permittivity profile for a nonuniform system: Application to a lipid bilayer simulation. *J. Chem. Phys.* **118** (7), 3401–3412.

Still, W.C., Tempczyk, A., Hawley, R.C., Hendrickson, T. (1990). Semianalytical treatment of solvation for molecular mechanics and dynamics. *J. Am. Chem. Soc.* **112**, 6127–6129.

Tanizaki, S., Feig, M. (2005). A generalized Born formalism for heterogeneous dielectric environments: Application to the implicit modeling of biological membranes. *J. Chem. Phys.* **122**, 548–556.

Tanizaki, S., Feig, M. (2006). Molecular dynamics simulations of large integral membrane proteins with an implicit membrane model. *J. Phys. Chem.* **110**, 548–556.

Tobias, D.J. (2001). Electrostatics calculations: Recent methodological advances and applications to membranes. *Curr. Opin. Struct. Biol.* **11** (2), 253–261.

Torrie, G.M., Valleau, J.P. (1980). Electrical double layers. I. Monte Carlo study of a uniformly charged surface. *J. Chem. Phys.* **73** (11), 5807–5816.

Torrie, G.M., Valleau, J.P. (1982). Electrical double layers. 4. Limitations of the Gouy–Chapman Theory. *J. Phys. Chem.* **86**, 3251–3257.

Ulmschneider, M.B., Ulmschneider, J.P., Sansom, M.S.P., Di Nola, A. (2007). A generalized Born implicit-membrane representation compared to experimental insertion free energies. *Biophys. J.* **92** (7), 2338–2349.

Valleau, J.P., Torrie, G.M. (1982). The electrical double layer. III. Modified Gouy–Chapman theory with unequal ion sizes. *J. Chem. Phys.* **76** (9), 4623–4630.

Wang, L.G., Bose, P.S., Sigworth, F.J. (2006). Using cryo-EM to measure the dipole potential of a lipid membrane. *Proc. Natl. Acad. Sci.* **103** (49), 18528–18533.

Warshel, A., Russell, S.T. (1984). Calculations of electrostatic interactions in biological systems and in solutions. *Quart. Rev. Biophys.* **17** (3), 283–422.

White, S.H. (Ed.) (1994). "Membrane Protein Structure". Oxford University Press, New York.

Woolf, T.B., Zuckerman, D.M., Lu, N.D., Jang, H.B. (2004). Tools for channels: Moving towards molecular calculations of gating and permeation in ion channel biophysics. *J. Mol. Graph. Model.* **22** (5), 359–368.

Yeh, E.C., Berkowitz, M.L. (2000). Effects of the polarizability and water density constraint on the structure of water near charged surfaces: Molecular dynamics simulations. *J. Chem. Phys.* **112** (23), 10491–10495.

CHAPTER 15

Charged Protein Side Chain Movement in Lipid Bilayers Explored with Free Energy Simulation

Libo Li, Igor Vorobyov, Sudha Dorairaj and Toby W. Allen

Department of Chemistry, University of California, One Shields Avenue, Davis, CA 95616, USA

Abstract

Biological membranes exhibit a bilayer arrangement of lipid molecules forming a hydrophobic core that presents a physical barrier to all polar and charged species. This universally accepted view has been challenged by biophysical partitioning experiments that suggest small free energies to insert charged side chains into the center of a membrane. We survey theoretical, experimental and simulation approaches and report free energy simulations that reveal large barriers for charged side chain movement across membranes. In simulations of an arginine side chain attached to a transmembrane α-helix and its simple side chain analog molecule, significant penetration of water and lipid head groups into the membrane core is seen. Yet there exists differences in the shape and magnitude of the free energy profiles due to the presence of the host helix. Calculated pK_a shifts within

Current Topics in Membranes, Volume 60
1063-5823/08 $35.00
DOI: 10.1016/S1063-5823(08)00015-X

the membrane suggest that arginine will remain mostly protonated throughout the membrane if attached to a transmembrane helix. The discrepancy between simulation and a recent translocon-based assay is explained in terms of several factors, including the difficulties in obtaining a spatial interpretation from these experiments. These simulations have implications for the gating mechanisms of voltage gated ion channels, suggesting that large paddle-like movements of lipid-exposed arginines are unlikely.

I. INTRODUCTION

Biological membranes are complex dynamical structures consisting of a bilayer arrangement of phospholipid molecules with a purely hydrophobic core that provides a barrier to polar and charged molecules. At the same time these membranes provide home to a wide range of proteins that make up around one third of the human genome and which are essential for a variety of cellular functions, including pumps, channels, carriers, receptors, enzymes and energy transducers. A molecular-level interpretation of protein–lipid interactions is essential to understand the insertion, folding and function of these membrane proteins. Here we explore the driving forces of protein stability and conformational changes in membranes and undertake a review of different experimental and computational approaches that have attempted to elucidate the thermodynamics underlying charged protein side chain movement in cell membranes.

Charged protein side chains play important roles in biology, including interactions that assist in protein folding (Yang and Honig, 1992), enzyme activity (Huang and Briggs, 2002), protein-DNA interactions (Luscombe *et al.*, 2001), pH activation of proteins (Cuello *et al.*, 1998), proton transport (Mathies *et al.*, 1991) and the voltage sensitivity of ion channels (Armstrong and Bezanilla, 1973). These amino acids can also be controlling factors in membrane insertion (Ozdirekcan *et al.*, 2005), nuclear localization (Kalderon *et al.*, 1984), antimicrobial activity (Brown and Hancock, 2006) and protein-mediated membrane fusion (Han *et al.*, 2001). Charged side chains are almost always localized outside the membrane interface (Davis *et al.*, 1983; Killian and von Heijne, 2000), unless stabilized by other interactions. For example, the crystal structure of the KcsA potassium channel, solved in 1998 (Doyle *et al.*, 1998), revealed a tetramer of two transmembrane (TM) helices, shielding a pore and narrow selectivity region from the surrounding membrane (see Fig. 1A). Each TM helix exposes hydrophobic residues to the lipid core, and aromatic Trp and Phe residues to the lipid interface, which provide an anchoring role (Killian, 2003). As expected, all charged residues (red and blue in the figure, in the web version of this book) exist outside the membrane, which we understand in terms of the large dehydration energy penalties to leave the aqueous phase.

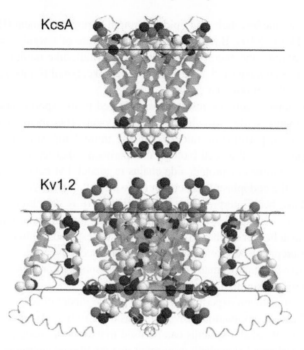

FIGURE 1   Potassium channel structures and amino acid positions. (A) KcsA potassium channel with α carbon atoms of aromatic residues shown in white, basic residues in blue and acidic residues in red. (B) Kv1.2 structure. Horizontal lines are rough estimates of the membrane region. (For interpretation of the references to color in this figure legend, the reader is referred to the web version of this book.)

An unusual situation is presented by voltage gated ion channels which rely on the sensitivity of charged side chains to changes in membrane potential. Fig. 1B shows the Kv1.2 mammalian voltage gated potassium channel crystal structure (Long *et al.*, 2005). Unlike KcsA, this channel possesses 4 additional segments (S1–S4) per monomer, called voltage sensors, which are coupled to the pore lining segments (S5–S6) via a linker to provide a gating mechanism (opening and closing of the ion conduction pathway). In this structure, charged side chains, contained in S4, exist within the membrane interface, stabilized by either salt bridges or interfacial water (Treptow and Tarek, 2006; Jogini and Roux, 2007). Yet activation of this channel requires movement of these charged residues. One proposed model of voltage gating, the so called "paddle" model, involves the passage of lipid-exposed Arginine (Arg) side chains completely across the membrane hydrophobic core upon membrane depolarization (Jiang *et al.*, 2003; Ruta *et al.*, 2005); very different to conventional models that keep charges completely (or partially (Elinder *et al.*, 2001)) sequestered from the low dielectric lipid by protein during movement (Hille, 2001; Ahern and Horn, 2004). Based on a simplified picture of a rigid continuum dielectric membrane, the energies associated with paddle-like motions of voltage sensors have been suggested to be as large as 265 kcal/mol (Grabe *et al.*, 2004) and thus should be prohibited.

However, some surprising results have emerged from experimental measurements in support of this ion channel gating model. This includes a range of measurements of partitioning into "membrane mimicking" solvents (discussed in this chapter) as well as cell biological experiments that report apparent thermodynamic stabilities of protein side chains in model TM helices, utilizing the translocon in the endoplasmic reticulum (Hessa *et al.*, 2005a, 2005b). These experiments have been interpreted to suggest that there is very little barrier for a charged residue to move across a membrane, challenging our understanding of the membrane as a barrier to charged species, with widespread implications in biology. The discrepancy between existing theory and recent experiments and models of protein activity requires that we revisit our description of protein–lipid interactions with modern simulation methods.

While the translocon experiments and the paddle model of ion channel gating require dedicated computational investigations to be fully understood, simulation studies on simple model systems can be used to extract the underlying thermodynamics for charged side chain movement in a membrane environment. In fact, a diverse range of membrane proteins lead to specific protein–lipid interactions and membrane perturbations (Jost *et al.*, 1973; Weiss *et al.*, 2003), illustrating the utility of general models of membrane proteins to reveal common mechanisms. In particular, amino acid side chain analogs and simple TM segments, as used experimentally in studies of protein–lipid interactions (Killian, 2003; Weiss *et al.*, 2003), offer a tractable route to a microscopic understanding via computer simulation. Such models have been used to calculate free energy profiles across a membrane (Dorairaj and Allen, 2007; McCallum *et al.*, 2007) to provide the key ingredients to understanding all biological processes that involve charged side chain movement in membranes. In comparison, experiments have been designed to tackle only what is amenable to experiment, including bulk solvent mimics for a membrane or complex biological systems with many uncharacterized features, often making direct interpretations challenging. While one may envision similar experiments to those free energy simulations to be reported here, using atomic force microscopy (Binning *et al.*, 1982) or laser tweezers (Block, 1992; Ashkin, 2000), such methods will either interfere with the membrane or lack the required resolution. In comparison, fully atomistic simulations can reveal not only governing thermodynamics, but also uncover the ways in which the protein is perturbing the bilayer environment and the interactions that determine the protein's stability, providing microscopic insight that is a welcome complement to experiment.

II. THEORETICAL PERSPECTIVES

Biological membranes separate two aqueous phases, effectively blocking the passage of charged and polar species without the aid of specialized proteins to catalyze permeation. The membranes can do this by formation of a bilayer structure of lipid molecules where hydrocarbon tails are segregated away from the aqueous phase to form a core, just a couple of nanometers thick, that will be highly unfavorable to polar and charged species. To enter this core, a molecule must leave an aqueous phase and give up its interactions with water molecules. The hydration energies of polar and charged molecules are often 10's of kcal/mol, which would be lost upon entry into a non-polar solvent.

The dominant contribution to the energetics for moving a polar or charged residue from water into hydrocarbon will be electrostatic. The electrostatics in bulk media can be well modeled by continuum solvation methods which treat the solvent as a uniform dielectric medium. The Born solvation free energy for a charge is given by (Born, 1920)

$$\Delta G_{\text{solvation}} = \frac{q^2}{8\pi \epsilon_0 R} \times \left(\frac{1}{\epsilon} - 1 \right), \tag{1}$$

where q is the charge on the atom, R is its Born radius, ϵ_0 is the permittivity of free space and ϵ is the relative permittivity (dielectric constant) of the solvent. The Born radii of protein atoms have previously been parameterized to reproduce hydration free energies (Nina *et al.*, 1997) i.e. solvation free energy in bulk water with dielectric constant 78.4 (Lide, 1992). For example, the hydration free energy for Arg in its protonated form is around -62 kcal/mol (Vorobyov *et al.*, 2008). If this residue moves from water to hydrocarbon (with dielectric constant around 2 (Lide, 1992)) the free energy cost will be a difference in two Born solvation free energies

$$\Delta \Delta G_{\text{solvation}}^{\text{monopole}} = \frac{q^2}{8\pi \epsilon_0 R} \times \left(\frac{1}{2} - \frac{1}{78.4} \right), \tag{2}$$

which is approximately half of the hydration free energy of Eq. (1) in magnitude, or around $+30$ kcal/mol. An equilibrium distribution spanning these two solvents at room temperature would therefore have only of the order of a single molecule per mole residing in the oily phase. We can conclude from this, presumably safely, that charged and polar species will avoid movement into the hydrophobic core of a membrane because of these large electrostatic driving forces. Through Eq. (2) we also see the sensitivity of free energies to the hydrophobicity or hydrophilicity of the chosen model solvent. E.g. changing the dielectric constant from 2 to 4, corresponding roughly, for example, a change from cyclohexane to chloroform, would halve the amount of work required to move the residue into the solvent from water.

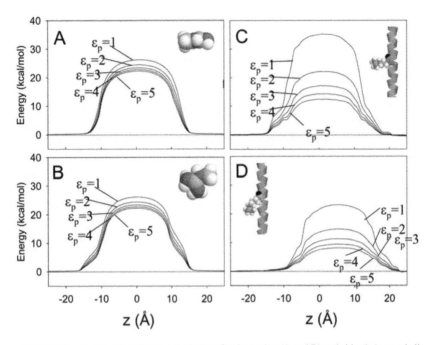

FIGURE 2 Continuum dielectric calculations for the analog (A and B) and side chain on a helix (C and D) based on solutions to Poisson's equation for a uniform membrane slab of dielectric constant 2 for protein dielectric constants of 1–5. In A the methyl-guanidinium Arg analog molecule was oriented with N atoms parallel to the xy plane (A) and in (B) it was oriented parallel to the yz plane. In (C) and (D) we show Poisson solutions for an Arg side chain attached to a TM helix in outward and downward rotameric states, respectively.

However, the approximation that the core of a membrane behaves like a bulk solvent is a fairly crude approximation. We therefore turn to a continuum membrane model. This model includes a rigid low dielectric slab representing the hydrocarbon core of the membrane, chosen to be of thickness 25 Å. Inside this slab the dielectric constant is assigned the value for bulk hydrocarbon, while outside it is assigned a value for bulk water, corresponding to the traditional picture of a membrane. Solutions to Poisson's equation have been carried out using either an isolated methyl-guanidinium ion, as an analog of an Arg side chain, or a TM α-helix containing an Arg side chain (Dorairaj and Allen, 2007). The protein was placed at various positions spanning the membrane and Poisson's equation was solved. Protein was given a dielectric constant of between 1 and 5, mapped by a 1.4-Å reentrant probe after assigning Born radii that have been optimized to reproduce aqueous phase solvation (Nina *et al.*, 1997).

Figure 2 shows solutions for the analog which has been oriented such that the three guanidine nitrogens reside in a plane parallel to the xy plane (A) or yz plane (B) while the bond to the methyl group is directed parallel to the y

axis. It can be seen that the profile depends somewhat on the orientation of the molecule, but overall the shape and magnitude of the barrier is constant. Not surprisingly, the choice of dielectric constant for the small analog molecule, between 1 and 5 here, has only a small influence on the barrier, dropping from 26 to 22 kcal/mol with increased dielectric. As a comparison we show energy profiles for the Arg side chain on the helix for two rotameric states corresponding to the lowest lying outward (C) and downward (D) directed rotamers (explained in a later section). The barrier ranges from around 10 to 30 kcal/mol depending on the choice of side chain rotamer and protein dielectric constant, comparable to previous calculations (Honig and Hubbell., 1984; Kessel and Ben-Tal, 2002). Unlike the analog case, the energy profiles vary considerably with the choice of side chain rotameric state, with the downward directed rotamer yielding a smaller barrier due to increased stabilization by the host helix. In addition to these ambiguities, continuum calculations rely on spatial assignments of dielectrics and charges, and while we have tested convergence based on grid size, the assignment of atomic radii and method for probing solvent accessibility is not unique (Allen *et al.*, 2004a). In addition, the contribution from the membrane interface is lacking from this model. This leads to a dipolar potential that can raise the barrier for a cationic species by over 10 kcal/mol (explained in a later section) and would have to be corrected for. However, as we will show, the greater error in these calculations is due to the assumption that the membrane is a rigid slab of low dielectric, when in fact it is a heterogeneous arrangement of amphiphilic molecules which will be perturbed by the strong electric fields associated with charged groups. Uniform dielectric slab membrane models are therefore likely to overestimate the dehydration barriers for charged side chains.

III. EXPERIMENTAL HYDROPHOBICITY AND MEMBRANE PARTITIONING MODELS

Membrane proteins consist of common structural elements which stabilize their 3D structures in this unique environment while carrying out biological functions. They contain hydrophobic segments, consisting usually of α-helix or β-sheet (Pauling *et al.*, 1951; Pauling and Corey, 1951) secondary structures, lined with hydrophobic residues, of a length that approximately matches the thickness of the membrane hydrocarbon region, thus avoiding the significant energy costs of exposing polar moieties to hydrocarbon or non-polar groups to water. This principle of 'hydrophobic matching' has been used to explain the insertion and folding of proteins (Killian *et al.*, 1996; Mouritsen and Bloom, 1984), protein activity (Johansson *et al.*, 1981; Andersen and Koeppe, 1992), lipid micro-domain formation and protein aggregation (Sperotto and Mouritsen, 1993; Lehtonen and Kinnnunen, 1995). Furthermore, in an attempt to overcome a mismatch of hydrophobic thickness, proteins exert their influence on the lipid bilayer through elastic distortion, bilayer curvature and compression or dilation of the lipids, with associated free energy costs (Lundbaek *et al.*, 1996). It is these energy costs that can control function if the protein changes shape during activation. It is therefore important to understand the underlying energetics associated with protein hydrophobicity when studying membrane protein function.

"There are numerous ways in which amino acid hydrophobicity can be quantified and they are not all consistent" (Gennis, 1989). A variety of scales have been developed to describe amino acid hydrophobicity based on partitioning free energies (Radzicka and Wolfenden, 1988), positioning of amino acids within globular proteins (Kyte and Doolittle., 1982), solvent accessible areas (Engelman *et al.*, 1986) and data mining of membrane protein structures (Kuhn and Leigh, 1985; Rao and Argos, 1986). While statistical analysis of protein structures can guide us to the hydrophobic nature of an amino acid, this approach differs from a true thermodynamic equilibrium distribution. The most direct thermodynamic measure of hydrophobicity would come from the extent of association of peptides with membranes. However, more amenable to experiment is an idealized approach of analyzing the partitioning of amino acids or side chain analogs into bulk "membrane-mimicking" substances (e.g. Radzicka and Wolfenden, 1988; Wimley and White, 1992; White, 2003), from which a thermodynamic estimate of the propensity for amino acids to enter the membrane core can be derived. Measurement of the distribution coefficient, $K_d = x_2/x_1$, where x_1 and x_2 represent amino acid mole fractions as estimates of the activities of amino acids in two immiscible (or partially miscible) solvents, allows apparent free energies to be calculated

$$\Delta G^{app} \approx -kT \ln(K_d). \tag{3}$$

These scales depend much on the choice of bulk solvents and offer simplified views of a membrane which, in reality, does not consist of two separated phases, but is instead a highly complex and dynamic heterogeneous dielectric environment. However, such scales can be successful in predicting the positioning of protein residues in globular and membrane proteins.

Given the nature of the hydrocarbon core of a membrane, the most logical choice for a mimicking solvent is bulk hydrocarbon (e.g. cyclohexane or octane) or a highly non-polar solvent (carbon tetrachloride or chloroform). In particular, studies have focused on the partitioning from water to cyclohexane (Radzicka and Wolfenden, 1988; Wimley and White, 1992, 1993) as a measure of hydrophobicity. As a result of the low dielectric of this reference phase (around 2 (Lide, 1992) for saturated hydrocarbon and 1 for vapor), polar residues experience large unfavorable free energy changes passing from water. However, these experiments are limited in their analysis of ionizable residues, being unable to provide free energies of partitioning for charged species because the vapor or non-polar phase should leave the residue solely in its neutral form. In fact, interpretation of results relies on this assumption (Radzicka and Wolfenden, 1988;

Wolfenden, 2007). For example, based on these partitioning experiments, the Arg side chain analog propyl-guanidinium experiences about 15 kcal/mol change in free energy moving from water (mostly protonated at pH 7) to cyclohexane (deprotonated) and around 20 kcal/mol to vapor. We know that the free energy to move the charged species from the aqueous to the non-polar phase must be much larger, being ∼60 kcal/mol to vacuum and 30 kcal/mol to cyclohexane, based on *ab initio* and empirical force field based calculations (Vorobyov *et al.*, 2008; Norberg *et al.*, 2005; Deng and Roux, 2004), yet there are no direct experiments to compare to these results (though indirectly a similar estimate can be obtained (Vorobyov *et al.*, 2008)).

Long-chain alcohols have also been used to test the hydrophobic nature of residues (Cohn and Edsall, 1943; White and Wiener, 1996), reproducing trends in amino acid affinities for soluble protein interiors or the exteriors of membrane-bound proteins. The most commonly accepted representative scale for the propensity for amino acids to reside inside a membrane is based on *n*-octanol partitioning measurements (Wimley and White, 1996). However, the free energy to move, for example, an Arg side chain into *n*-octanol is a mere 1.5 kcal/mol (White, 2003). This suggests that this "non-polar" solvent is nearly as efficient as water at stabilizing this charged side chain. i.e. The large hydration free energy of Arg is almost matched by *n*-octanol. Thus, while partitioning into bulk solvents is clearly a simplified approximation for the heterogeneous lipid bilayer environment, this low free energy of partitioning indicates that *n*-octanol cannot adequately represent the hydrophobic nature of a membrane hydrocarbon core. We understand this because octanol possesses the ability to hydrogen bond, but also because saturated wet octanol contains 20–29 mol% water (Sangster, 1997; Chen and Siepmann, 2006) and the charge will be coordinated by water molecules that are either contained in, or which have been dragged into the alcohol solvent. This water-like substance therefore gives rise to a narrow range of low partitioning free energies from water (Wolfenden, 2007).

The challenge therefore is to seek a measure of hydrophobicity directly from a membrane system. White and co-workers have studied partitioning of small polypeptides into the membrane interface to try to decompose the thermodynamics of folding and insertion in membranes (White and Wimley, 1998). These polypeptides have been designed to have little secondary structure and reside near the membrane-water interface. In the case of charged side chains, small free energy barriers have been reported (e.g. ∼1 kcal/mol for Arg). The results of this interfacial scale have been compared with previous octanol partitioning scale to reveal a fairly close correlation (Wimley and White, 1996). This may imply that octanol maybe a good mimic for a membrane interface. However, these methods do not tell us anything about the free energies to enter or cross the hydrocarbon core of a membrane: information that is needed to understand how the stability of a membrane protein is determined by the arrangement of amino acids in the three dimensional structure, and associated conformational changes in the membrane.

Recently, von Heijne, White and colleagues developed cell biological experiments that are designed to extract apparent free energies for membrane insertion of hydrophobic (H) segments on a chosen membrane protein (the Leader peptidase, Lep, from Escherichia coli) by the translocon in the endoplasmic reticulum (ER) (Hessa *et al.*, 2005a; White and von Heijne, 2005). The H-segment used experimentally for studying Arg, for example, is GGPG**AAAAALALALRLALAL AAAA**GPGG, consisting of a series of non-polar residues (bold), which has been designed to reside approximately transmembrane by placing proline residues at each end to prevent secondary structure formation (and thus make it somewhat difficult for the ends of the H-segment to be pulled into the core of the membrane). The protein also contains N-linked glycosylation sites on either side of the H-segment, to provide a means by which a transmembrane H-segment can be detected. The translocon is complexed with an enzyme called the oligosaccharyl transferase, that adds polysaccharides to these sites when exposed to the lumen of the ER membrane. If the protein with the engineered H-segment is soluble it will have both sides glycosylated in the lumen whereas if it is inserted as transmembrane (or for some other reason were to have just one site exposed to the lumen) it will have only one. The probability of insertion can then provide an equilibrium thermodynamic measure for partitioning. These translocon-based assays have yielded apparent free energies which indicate that there is very little free energy penalty ($\sim$2.5 kcal/mol) to include an Arg residue in the middle of an H-segment or to move it along its sequence (Hessa *et al.*, 2005b). One possible interpretation of these results is that charged side chains face small barriers to the physical motion across the membrane. If this is the case, it suggests that all polar and charged species may easily cross the membranes that protect cells, opposing the prevailing view of membrane biophysics which dictates that such motions must be facilitated by specialized proteins.

While these measurements represent a significant advance in biophysics, unveiling energetics associated with the sequence specific insertion by the translocon, there are, however, several complexities in this biological system that need to be considered, including the mechanisms of translocation itself and the possibility that what is measured may be free energies of a protein still strongly interacting with the translocon and not an isolated TM segment in the membrane. In fact, the correlation between these measurements and the interfacial and octanol scales suggests that what is observed corresponds to a fairly hydrophilic environment and not isolated in the membrane of the ER. The magnitudes simply do not compare to results for partitioning into alkanes, even for neutral species, despite any reported correlations (e.g. Wolfenden, 2007). This may suggest that the translocon based assays may successfully reveal the propensities for the onset of membrane incorporation, but not the actual free energies for insertion, and also not the free energies as a function of position within the membrane. Here we shall discuss several factors involved, and explore the design of these experiments for elucidating free energies of membrane insertion.

IV. FULLY ATOMISTIC COMPUTER SIMULATIONS

With improved computational resources available today, molecular dynamics (MD) simulation has become an essential tool for investigating a wide range of chemical and biological phenomena (Karplus, 2002). Modern computers and force-fields have led to simulations that not only reproduce experiment, but also make significant predictions of biological function. However, it is only in the last decade that enough computer power has been available to allow simulation of membrane systems and today there are numerous simulations of membrane proteins using explicit lipid bilayers (e.g., Woolf, 1998; Johansson and Lindahl, 2006; Treptow and Tarek, 2006; Jogini and Roux, 2007). Unlike previous membrane modeling techniques that employ continuum electrostatics, bilayer elasticity, chain packing theory, integral equation theory, coupled harmonic oscillators or coarse-grained dynamics, fully-atomistic MD simulation includes both the dynamics of the lipid membrane and the chemically accurate interactions.

MD may be used to study the time evolution of structure, dynamics, interactions and correlations of a thermally fluctuating system and visualize microscopic phenomena, or to sample phase space for the calculation of ensemble averaged equilibrium observables and the thermodynamics governing microscopic processes. Consider an ensemble average of a quantity A which is function of atomic coordinates:

$$\langle A \rangle = \frac{\int d^{3N} \mathbf{R} A(\mathbf{R}) e^{-U(\mathbf{R})/kT}}{\int d^{3N} \mathbf{R} e^{-U(\mathbf{R})/kT}}, \qquad (4)$$

where U is the potential energy as a function of N atomic coordinates, k is the Boltzmann constant and T is the temperature. A range of quantities describing the structure and dynamics of the protein and membrane can be extracted via such ensemble averaging, each adds to our understanding of the origins of microscopic behavior. However, what if one aims to observe and quantify rare events, such as protein insertion into or movement within a membrane? The obvious difficulty of all-atom MD is that the trajectory of a large number of atoms must be computed, even though only a few coordinates maybe truly relevant to a process (e.g. the depth of a partitioning amino acid). Rather than trying to observe rare events with brute force simulation, a better approach is to calculate a free energy landscape, characterized by identified relevant coordinates. One may integrate over the uninteresting degrees of freedom to reveal the average systematic force, $\langle F(z) \rangle$, that governs the thermodynamic equilibrium distribution along an order parameter, z, that is judged to be the most relevant coordinate (Zwanzig, 2001; Mori, 1965). The potential of the mean force (PMF), W, is the thermodynamic work along z (equal to the negative integral of the mean force), relative to some reference value z',

$$W(z) - W(z') = -\int_{z'}^{z} d\zeta \langle F(\zeta) \rangle. \qquad (5)$$

Using Eq. (4) this can readily be expressed in terms of configurational integrals

$$W(z) - W(z') = -kT \ln \frac{\int d^{3N-1}\mathbf{R}e^{-U(\mathbf{R},z)/kT}}{\int d^{3N-1}\mathbf{R}e^{-U(\mathbf{R},z')/kT}}, \tag{6}$$

where we have set $W(z') = 0$. This free energy profile provides a measure of effective potential that is an average over the thermal fluctuations of the other $3N - 1$ degrees of freedom. It is this function that we aim to capture for protein–lipid interactions.

PMF calculations entail simulating long enough so that all interesting values of the relevant coordinates are encountered, and an equilibrium distribution of all other coordinates are integrated over. However, in many cases it is difficult to sample all values of the relevant coordinates, simply because they correspond to the rate limiting steps and thus usually involve significant free energy barriers; not thermally accessible in a manageable simulation time-frame. The ability to control degrees of freedom on a computer enables us to efficiently sample these "slow" coordinates. To assist sampling, harmonic biasing (or 'window') potentials can be added to the system potential, U, at regular intervals spanning the order parameter in the method of Umbrella Sampling (Torrie and Valleau, 1977). Computation of the free energy profile for a molecule across a membrane requires a set of equally spaced simulations i, biased by window functions $w_i(z) = \frac{1}{2}K_i(z - z_i)^2$ that hold the molecule near positions along the z-axis. Trajectories were then combined and rigorously unbiased with the Weighted Histogram Analysis Method (Kumar *et al.*, 1992), which consists of solving the coupled equations for the optimal estimate for the unbiased equilibrium distribution.

A. Design of Model Systems

Biological membranes are truly complex, containing over 200 species of lipid, cholesterol, lysophospholipids, various amphiphilic molecules (including drugs) and proteins that can affect the structure, dynamics and mechanical properties of the bilayer (Gennis, 1989; Seeman, 1997) and which have a general influence on protein stability (Scotto and Zakim, 1985) and activity (Lundbaek *et al.*, 1997; Gandhavadi *et al.*, 2002). However, it is difficult to sample the dynamics of multicomponent membranes with such a detailed model, and thus we restrict ourselves to single component membranes. Homogeneous membranes provide a good model for a cell membrane in both experiment (Nagle and Tristram-Nagle, 2000; Gennis, 1989) and simulation (Tobias *et al.*, 1997; Pastor *et al.*, 2002). For a number of different bilayers, good agreement with experiment has been obtained, yet dipalmitoylphosphatidylcholine (DPPC) is the best understood bilayer system (Tieleman *et al.*, 1997; Feller *et al.*, 1997; Chiu *et al.*, 1999) and a suitable model of a biological membrane. Membrane patches of 48 DPPC lipid molecules (∼2 shells around the protein) were used in the simulations reported here.

We have simulated a simple model of a charged protein side chain using methyl-guanidinium in place of the Arg side chain (Li *et al.*, 2008). This is an even simpler model than the actual side chain (propyl-guanidinium), yet is expected to give almost the same results (Wolfenden *et al.*, 1981). However, any analog is going to be an oversimplification of the actual side chain attached to a TM protein segment, because of its greater ability to reorient, greater exposure to solvent and the missing influence of the nearby protein. We have therefore also simulated an Arg residue attached to a helix to provide a more realistic environment for a TM segment-bound side chain (Dorairaj and Allen, 2007). As we shall see, the effects of the host helix on the free energy profile are fairly significant, yet consistent results emerge.

Experimentally it is known that poly-Leu (Liu *et al.*, 2004) and poly-(Leu-Ala) (Killian *et al.*, 1996) form stable α-helices in a membrane bilayer. Poly-Leu was chosen as a suitable background TM helix, with 80 residues, spanning a length of $\sim$120 Å. The helix must be this long ($\sim$120 Å) so that it can be translated within $-30 \leqslant z \leqslant 30$ Å and avoid terminal-interface interaction. The C and N termini were chosen to be neutral, using acetyl and ethanolamine groups, respectively, to ensure a PMF that is independent of the host helix. The α-helical peptides were initially built in ideal conformations ($\phi = -57°$, $\psi = -47°$) (Hovmoller *et al.*, 2002), placed at a position of interest, and then a new hydrated membrane was built around it, using methods that are extensions of early techniques (Woolf and Roux, 1996).

In the case of the helix model, the system was made very long so that the protein did not pass through the periodic boundary to ensure negligible interaction between image proteins. Hexagonal periodic boundaries of xy-translation length 43.75 and average height of $\sim$190 Å were imposed. Pressure coupling was used in the z direction (parallel to the membrane normal). Lateral periodic box dimensions were based on a calculated protein cross-sectional area of $\sim$180 Å^2 and an area/lipid of 64 Å^2 (Nagle, 1993). To neutralize the net charge on the protein and provide good sampling of the ionic baths, a 0.5 M KCl solution was used, corresponding to 7,895 water molecules and 72 K$^+$ and 73 Cl$^-$ ions. In total the helix system contained 31,592 atoms. The analog system, though including the same membrane patch, was smaller due to the absence of the long helix, and totaled 2,186 water molecules, 20 K$^+$ and 21 Cl$^-$ ions, and a total of 12,852 atoms. Because much of the helix remains outside the membrane, it will tend to depart from α-helical and would require more configurational sampling that adds little to our knowledge of side chain stability within the membrane. To prevent unfolding of the hydrophobic poly-Leu helix in bulk water, we maintained an approximately α-helical conformation with dihedral constraints of 100 kcal/(mol deg^2), corresponding to an RMS fluctuation of $\sim$5° in each dihedral angle. Cylindrical constraints of 5 kcal/(mol Å^2) were also applied to the centers of mass of each residue to keep the peptide vertical. The constraints do, however, alter the bulk reference state free energy, but not the relative free energies for transmembrane positions important for studying side chain movement within the membrane. We have chosen our order parameter to be the position, parallel to the membrane normal, of the helix (or analog) center of mass (COM) relative to the membrane COM. After building membranes around the helix, placed at regular positions ranging from -30 to 30 Å in 1 Å steps, and equilibrating for 2 ns, these systems were each then simulated for another 5 ns to compute the free energy profile with umbrella sampling. Simulations began with a biasing harmonic force constants corresponding (via equipartition) to an RMS fluctuation of 1 Å. However, the helix could not be held near the membrane center and for 1.5 ns the restoring force constant was increased, corresponding to RMS fluctuations as low as 0.1 Å. The difficulty in holding the helix is already evidence that Arg is unstable inside the membrane.

The problem of studying a side chain attached to a helix is more complicated than that of a side chain analog. It is also describing a different system that is meant to provide a better representation of the situation presented by a membrane protein. Without the peptide's influence on the membrane and solvent, the energetics within the membrane is somewhat different, as observed in the above continuum calculations. But a further difficulty arises due to the structural isomers of the side chain and the long timescales of isomerizations, in particular in a membrane environment where large forces are imparted on the side chain (Dorairaj and Allen, 2007). The adiabatic energy map of Fig. 3, provides a guide to the lowest energy Arg rotamers, based on its first two dihedrals (χ_1 and χ_2). There are 8 low lying states (labeled *a–h*), with rotamer *a* the lowest in energy (by 2.0 kcal/mol), in accord with the backbone dependent rotamer libraries of Dunbrack and Karplus (Dunbrack and Karplus, 1993). This rotamer has its guanidine carbon directed downward toward the N-terminus (and toward the lower interface when inside the membrane), as shown in Fig. 3. The second lowest energy rotamer is the *e* rotamer which is directed upward and will be seen to play an important role inside the membrane as the side chain makes contact with the upper interface. Rotamer *d* is an outward directed rotamer that also plays a role in being the most common intermediate state between *a* and *e* and being directed away from the helix. Other rotamers are also directed away from the helix and are seen frequently when the side chain is at the interface of the membrane. Though this energy map is not likely to bear close resemblance to the corresponding free energy surface in a membrane environment, it does provide a guide to the accessible rotamers of Arg on an α-helix and has revealed a slight inherent preference for a structural isomer in which the guanidinium charge is directed down and close to the helix. This lowest (intrinsic) energy rotamer was therefore chosen as the initial state for all umbrella sampling simulations.

In all simulations to follow we have used the program CHARMM (Brooks *et al.*, 1983) using the PARAM27 (MacKerell *et al.*, 1998) force field with standard protein, lipid (Feller and MacKerel, 2000; Yin and Mackerell, 1998), ions (Beglov and Roux, 1994) and TIP3P (Jorgensen *et al.*, 1983) water parameters. The elec-

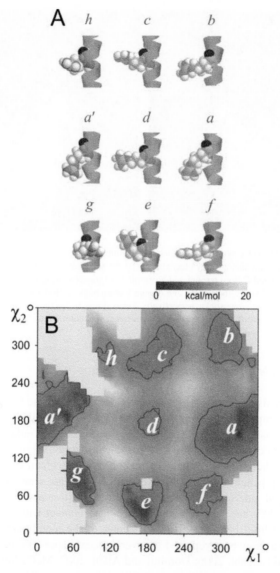

FIGURE 3   (A) Low lying energy rotamer structures for a charged Arg side chain on a TM helix. (B) Adiabatic energy map for an Arg side chain with low energy states labeled *a–h*. Adapted from (Dorairaj and Allen, 2007).

trostatics was handled by using the particle-mesh Ewald (Darden *et al.*, 1993), and bonds to H atoms were maintained with the SHAKE (Ryckaert *et al.*, 1977) algorithm. Simulations were performed under constant normal pressure and temperature conditions with the Nose-Hoover method (Feller *et al.*, 1995).

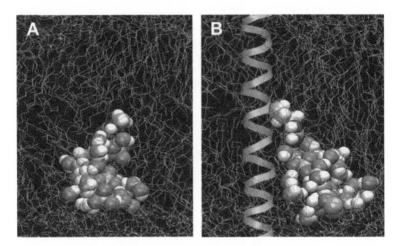

FIGURE 4   TM helix system and membrane perturbations. (A) System containing a methyl-guanidinium side chain analog (gray/blue/white). (B) An 80 residue poly-Leu α-helix (green) with neutral N (bottom) and C (top) termini and a central Arg (gray/blue/white). Each system contains a DPPC bilayer (gray hydrocarbon sticks and orange P atom balls) with 0.5 M KCl baths with water molecules shown as red/white sticks, respectively. Waters drawn inside the core of the membrane are shown as red/white balls. (For interpretation of the references to color in this figure legend, the reader is referred to the web version of this book.)

B. Charged Side Chain Solvation in a Membrane

The large fields associated with a charged side chain impose forces on the lipid bilayer that disrupt its uniform planar geometry. This is revealed in Fig. 4, showing local deformations due to the presence of the side chain as either an analog (A) or on a TM helix (B). Clearly the uniform slab structure imagined in continuum models has broken down. In both the analog and side chain cases, both water and head group penetration into the hydrocarbon core are seen. While "snorkeling" of charged side chains to the interface (Segrest *et al.*, 1990; Strandberg and Killian, 2003) has been reported before, here contact with the aqueous environment and lipid interface, from the center of the membrane, is seen. This deformation has now been reported in several recent papers (Freites *et al.*, 2005; Johansson and Lindahl, 2006; Dorairaj and Allen, 2007; McCallum *et al.*, 2007) and is becoming an accepted behavior of lipid membranes. We expect that these perturbations will incur strain energies that will influence the shape of the free energy profiles to come.

Figure 5 gives an indication of the times required to equilibrate the system undergoing membrane deformations in the presence of the charged side chain (similar observations were seen for the analog (Li *et al.*, 2008)). The figure shows two sample configurations from umbrella sampling trajectories where the side chain is in contact with either the lower or upper interface of the membrane. This

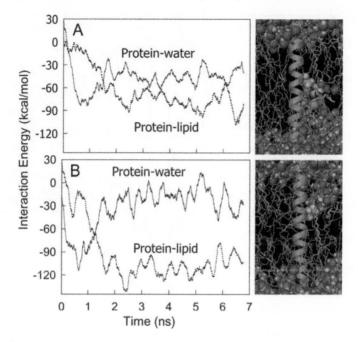

FIGURE 5 Equilibration of the TM helix system. A and B show two sample windows from umbrella sampling simulations to reveal convergence of interaction energies with water and lipid head groups. The right panels show samples corresponding configurations from the simulations. (Adapted from the supporting text of (Dorairaj and Allen, 2007).)

figure shows that Arg–lipid and Arg–water interactions require of the order of 2 ns of equilibration for these most challenging cases. In contrast, when the Arg side chain was at the interface or in bulk water, equilibration was achieved within shorter times. For all windows, the first 2 of 7 ns was excluded from analysis.

Figure 6A shows coordination numbers as a function of position of the side chain analog methyl-guanidinium for water, phosphate oxygen and carbonyl oxygen atoms (taken from Ref. (Li *et al.*, 2008)). Due to membrane rearrangement, the molecule remains coordinated by at least one lipid head group at every point throughout the membrane, even at the membrane center. The analog is also coordinated by usually five water molecules throughout the membrane. Figure 6C shows the same analysis for the Arg side chain on the helix, with similar solvation deep inside the membrane. However, increased coordination by head groups occurs at the interface for the analog due to the absence of the host helix: typically coordinated by 2 or 3 phosphates instead of 1–2 for the side chain, with increased hydration by 1 or more water molecules. Thus, we expect greater stabilization of the analog near the interface relative to the side chain. We also note that the analog is coordinated by as many as 12 water molecules in bulk water in comparison

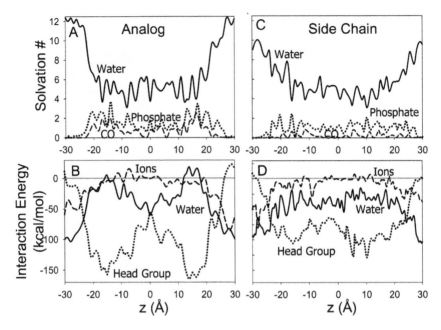

FIGURE 6   Solvation and interaction energy analysis from 1D umbrella sampling trajectories for the side chain analog (A and B) and the side chain on the helix (C and D). A and C) Mean first shell solvation numbers for Arg-water (solid), Arg-head group (dotted) and Arg-carbonyl (dashed). Solvation numbers are average number of atomic species within the first shells defined by radii 4.85, 4.55 and 5.55 Å for water, phosphate and carbonyl oxygen, respectively, relative to the guanidine carbon atom. Coordination numbers by ions are not shown as they were negligible. B and D) Average side chain interaction energies for each component: water (solid), lipid head groups (dotted) and ions (anions and cations, dashed).

to the side chain which reaches 9–10 waters in aqueous solution. This will lead to increased stabilization of the analog in the bulk as well (likely increasing the barrier to enter the membrane). We observed that counter-ions (Cl^-) played only a small role in stabilizing the charge in the membrane; seen inside the core at two locations (near -4 to -1 Å) for periods of 1–1.5 ns only for the helix simulation, for example. This is because those ions face significant dehydration energies, despite the presence of some water within the membrane due to local deformations. We have also seen that lipid carbonyl oxygen atoms do not contribute greatly to the Arg side chain stabilization within the membrane core, and were either absent from the coordination shell or spent $\leqslant 10\%$ of the time there. However, an increased involvement of carbonyls is observed for the side chain analog both in the core and interface of the membrane. We understand this because of the limited conformational flexibility of both the lipids and side chain on the helix leading to difficulties in coordinating the side chain, whereas the analog molecule may more easily reorient to be coordinated by those carbonyl groups.

Figures 6B and 6D show interaction energies for the analog and side chain, respectively. Consider first the helix + side chain system. It can be seen that lipid head groups interact very strongly with the charged side chain throughout the bilayer, with typical values of −100 kcal/mol. It is this strong interaction that provides the force to disrupt the bilayer structure and ensures that such lipid displacements will occur despite the energetic costs evident in such a perturbation. The interaction with water is also strong throughout the bilayer with values typically being −50 kcal/mol. Although these interactions are reduced at the bilayer center, relative to the interface, which will decrease protein stability there, they are clearly contributing to side chain free energy even far from the aqueous phase. While near the middle of the membrane the side chain analog reproduces similar interaction energies to the side chain itself, the interactions are quite different as the interface is approached. The variation in interaction energies across the membrane is great (varying by up to 100 kcal/mol with position) for the analog molecule in comparison to the side chain which is fairly constant in comparison. We can understand the large variation for the analog in terms of the excess solvation of the molecule at the interface relative to the center of the membrane and this is likely to lead to additional binding of that analog to the membrane interface. But so far we know nothing about thermodynamic stability and we now turn our attention to the calculation of free energies.

C. Free Energy Profile of a Charged Side Chain on a TM Helix

We begin with results for the side chain on the helix which proved to be the most challenging. Fig. 7A shows the first attempts to obtain the PMF for the poly-Leu + Arg system based on umbrella sampling trajectories for the protein's position relative to the membrane. The PMF (solid black curve, right axis) increases as Arg enters the bilayer and shows no obvious attraction to the interface, in accordance with experiment (Killian, 2003; Wimley *et al.*, 1996). It then rises sharply to a maximum of over 20 kcal/mol inside the membrane core, suggesting that movement of an Arg side chain into the membrane core will be prohibited. However, the maximum in free energy is off-center by ∼4 Å, due to the preferred downward directed side chain in the central region of the membrane, connected to the lower interface. Furthermore, the free energy with Arg in upper and lower bulk solutions differ by 8 kcal/mol, and has also led to some variations in previous analyses of interactions and solvation. We can understand this due to the greater extent of connection to the lower interface (up to $z \sim 4$ Å) which leads to a large negative force over an extended range, as also shown in Fig. 7A (gray line, left axis). We have discovered that when biased density is removed from a narrow range of just a few Å near the bilayer center this asymmetry is almost completely eliminated and the peak becomes more centered, suggesting that more effort is

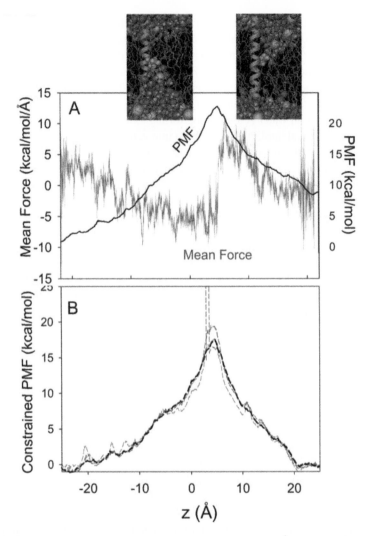

FIGURE 7 MD rotamer distributions for Arg in the: bulk ($|z| > 22$ Å), lower interface ($-22 \leqslant z < -13$ Å), upper interface ($13 < z \leqslant 22$ Å), lower core ($-13 \leqslant z < -4$ Å), central core ($-4 \leqslant z < 4$ Å) and upper core ($4 \leqslant z < 13$ Å). Each horizontal and vertical axis ranges from 0 to 360°. Adapted from (Dorairaj and Allen, 2007).

needed to sample the side chain orientations and interfacial connections with the side chain when it resides near the center of the bilayer.

The convergence of the PMF over simulation time obtained from the 61 × 7-ns simulation with z-bias is shown as a series of dashed curves in Fig. 7B which correspond to the PMF constrained to have a value of 0 as bulk reference at $z =$

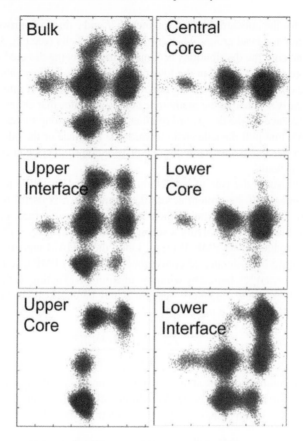

FIGURE 8 1D PMF asymmetry and convergence analysis. (A) The mean force experienced by the protein as a function of position is shown as a gray curve (left axis). Insets show sample MD snapshots explaining the direction of the force. Superimposed on this graph is the PMF that was obtained from this 1D umbrella sampling only (right axis). (B) The dashed black and gray curves show convergence of the PMF over 3–7 ns, which has been constrained to have zero bulk reference values ($z = \pm 30$ Å).

± 30 Å (forcing it to be symmetric for the purpose of convergence analysis only). The gray curves in Fig. 7B are PMFs after 3–6 ns while the black curve is the PMF after 7 ns sampling. The PMF is well converged after 7 ns (Fig. 7, black dashed curve), with average deviation of 0.08 kcal/mol from the 6-ns PMF.

What has caused this asymmetry? Fig. 8 contains several Ramachandran plots to reveal the extent of isomerizations and common rotamers of the side chain in different regions across the bilayer. When the side chain is in bulk solution all low lying rotamers are accessed with no obvious preference for an upward or downward directed rotamer. This indicates a fairly good sampling in aqueous solution within a 5 ns period. Sampling near the interfaces was also good, but there exists a preference for the outward directed rotamers b, f and c due to improved coordination by water and head groups when further from the helix (see Fig. 3A). Inside the membrane core the distributions become restricted, being directed upward in the upper core and downward in the lower core. However, in the central core region the side chain is directed mostly downward toward the lower interface. This "flipping" nearly 4 Å above the membrane center indicates that either there is an inherent preference for the downward rotamer, or that the strong forces (of several kcal/(mol Å)) that pull the side chain toward one interface or the other, create a significant barrier that makes sampling the distribution of interfacial connections exceedingly difficult.

To better illustrate the role of side chain orientational sampling in determining the free energy of the protein across the membrane it is useful to represent the trajectories in terms of a multidimensional free energy surface. Equilibrium distributions, biased in z, involving a secondary variable s, $\langle \rho(z, s) \rangle^{\text{bias}}$, may be unbiased to produce a 2D PMF $W(z, s)$. Once the WHAM equations (Kumar *et al.*, 1992) have been iterated to convergence for the 1D PMF calculation, any distribution $\langle \rho(z, s) \rangle^{\text{bias}}_{(i)}$, biased by window function $w_i(z)$ acting on the z coordinate, but involving some secondary variable s, may be unbiased via an extension of the 1D equation (Allen *et al.*, 2006a)

$$\langle \rho(z, s) \rangle = \frac{\sum_{i=1}^{N} n_i \langle \rho(z, s) \rangle^{\text{bias}}_{(i)}}{\sum_{i=1}^{N} n_i e^{-[w_i(z) - F_i]/kT}}, \qquad (7)$$

where n_i is the number of time steps in window i, and F_i are the free energy constants obtained from solution to the WHAM equations. However, this unbiasing procedure can be circumvented. Defining the function $\rho(s|z)$ by

$$\langle \rho(z, s) \rangle^{\text{bias}}_{(i)} = \langle \rho(z) \rangle^{\text{bias}}_{(i)} \cdot \rho(s|z), \qquad (8)$$

where the conditional probability $\rho(s|z)$ (probability of having a value of s, given a value of z) is normalized such that

$$\int_{-\infty}^{\infty} ds \, \rho(s|z) = 1, \qquad (9)$$

then the unbiased distributions, related to the biased distributions via Eq. (7), possess a similar relation

$$\langle \rho(z, s) \rangle = \langle \rho(z) \rangle \cdot \rho(s|z). \qquad (10)$$

The 2D PMF, $W(z, s)$, is then simply obtained from $W(z)$ by

$$W(z, s) = -kT \ln \langle \rho(z, s) \rangle = W(z) - kT \ln \langle \rho(s|z) \rangle + C, \qquad (11)$$

where C is some additive constant. The resulting 2D free energy surface as a function of helix position, z, and relative side chain (guanidine C atom) position ($s = z_C - z$) is shown in Fig. 9A. In bulk water, the side chain uniformly samples

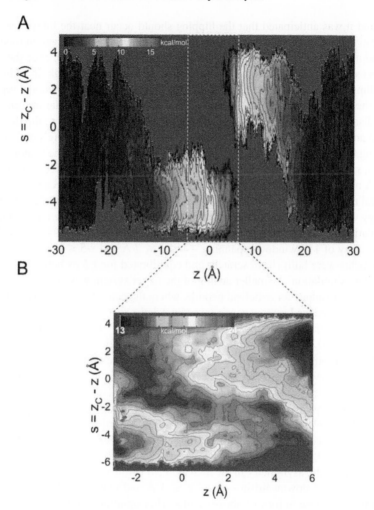

FIGURE 9 2D free energy surfaces. (A) 2D-PMF as a function of helix position (z) and relative side chain guanidine carbon–helix COM (variable $s = z_C - z$) obtained from 1D biased simulations, using Eq. (11). (B) 2D PMF in the region $-3 \leqslant z \leqslant 6$ Å obtained from a 2D biased umbrella sampling study. The merger of A and B is shown in Fig. 10A.

all orientations ($-5 \leqslant s \leqslant 3$ Å). Near the interfaces ($|z| \sim 20$ Å) the distribution narrows, with a preference for an outward-directed side chain. However, inside the core the side chain is directed downward-only in the lower core and upward-only in the upper core. At some point inside the membrane the side chain must flip i.e. it becomes thermodynamically favorable for the side chain to make contact with the other interface. According to our adiabatic energy map of Fig. 3B, there appears to be only a slight preference for one rotamer over another (down over up) and it was anticipated that the flipping should occur near the bilayer center. Instead this occurred 4 Å above center. Our MD simulations all began in rotamer *a*, and although *a–e* isomerizations were seen during equilibration, once the side chain establishes a connection to one interface, it is difficult to break. In addition, this flipping is very abrupt, as is evident from the absence of a bimodal region near the center where both up and down rotamers would be of similar free energy. The discontinuity in the free energy surface signifies a large barrier to flip the side chain and prompted us to carry out additional 2D umbrella sampling simulations, biasing both the helix position, z, and relative side chain (guanidine C atom) position, $s = z_C - z$. Because symmetrization of the PMF was observed when just a few Å near the bilayer center were omitted from unbiasing, and because only the central region is expected to be bimodal (when too far from one interface the free energy will be high for a rotamer pointing to that interface), 2D umbrella sampling simulations were carried out for helix positions within the range $-4 \leqslant z \leqslant 7$ Å. A 2D array of 110 windows spanning $-4 \leqslant z \leqslant 7$ Å and $-5 \leqslant s \leqslant 4$ Å in 1-Å increments were built (from scratch) and equilibrated for 1.5 ns before umbrella sampling simulation. A smaller analog of the larger system was generated by using 48 Leu residues for α-helical peptide, where the central residue was replaced by Arg. The reduced system consisted of 76 Å helix in a 97-Å long box to ensure translation from -5 to $+7$ Å (Dorairaj and Allen, 2007). Following equilibration for each of 110 windows, an additional $\sim$1 ns of simulation per window was then used to map out the free energy. The distributions were unbiased using Eq. (7). Figure 9B shows the resulting 2D free energy surface from this 2D biased set of simulations.

This PMF was merged with the 2D PMF from 1D biasing (Fig. 9A) by shifting to match the average free energy within a 0.5–1 Å border region (Dorairaj and Allen, 2007) leading to the complete free energy surface of Fig. 10A. We note that the merging procedure is not unique and can lead to changes of the order of 1 kcal/mol in the asymmetry in the final 2D PMF. Now we see that the side chain will be directed downward in $-13 \leqslant z \leqslant 1$ Å and upward in $1 \leqslant z \leqslant 13$ Å with flipping close to (or just above) center. This signifies that the corresponding rotameric states are actually similar in free energy, with a slight preference for the downward directed rotamer (as suggested by Fig. 3A). Off center flipping can also be attributed, in part, to the fact that the helix COM is coincident with the Arg C_α atom, but $\sim$0.8 Å above the C_β atom.

We now see why sampling was so difficult with one-dimensional simulations because the barrier to flip the side chain is 4–5 kcal/mol near the center of the membrane. Furthermore, low flipping rates due to concerted motions of the side chain and lipid head groups and the low transverse (flip-flop) lipid diffusion (Marti and Csajka, 2003) will slow down isomerizations there. While we have not sampled the dynamics of the creation of interfacial connections (to one, other or both interfaces), clearly a very slow coordinate, we attempt to extract an upper limit for the rate for flipping based on our umbrella

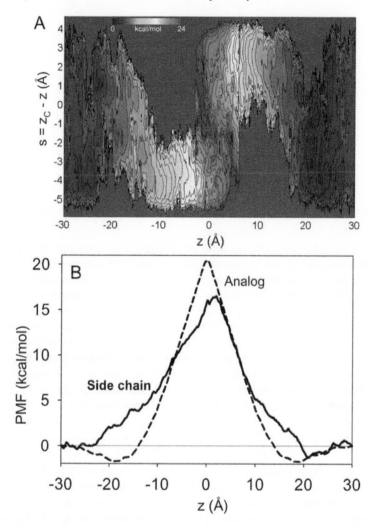

FIGURE 10   (A) Complete 2D-PMF as a function of helix position (z) and relative side chain-helix position, $s = z_C - z$, obtained from a combination of 1D- and 2D-biased simulations (adapted from (Dorairaj and Allen, 2007)). (B) Final 1D PMF for the poly-Leu + Arg TM helix (sold curve). The PMF for the methyl-guanidinium analog is shown as a dashed curve, adapted from Ref. (Li *et al.*, 2008).

sampling trajectories. The Kramer's transition rate for side chain flipping has been computed in the high friction, large activation regime (Allen *et al.*, 2003; Crouzy *et al.*, 1994)

$$k = \frac{D(s_{\text{barrier}})}{2\pi kT}\left[-W''(s_{\text{barrier}})W''(s_{\text{well}})\right]^{1/2}e^{-\Delta W_{\text{activation}}/kT}, \qquad (12)$$

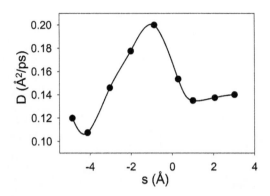

FIGURE 11   Diffusion coefficient for the relative variable s describing side chain flipping.

where $D(s_{\text{barrier}})$ is the diffusion constant measured near the top of the barrier in the s-direction, $W''(s)$ the second derivative of the PMF estimated at the barrier and wells by least squares harmonic fits, and $\Delta W_{\text{activation}}$ is the activation barrier height for the transition. To calculate the flipping rate when the Arg side chain is close to the bilayer center ($z = 0.8$ Å, corresponding to C_β at the center of the membrane), a measure of the diffusion coefficient for the side chain relative position, s, near the barrier, was obtained by analyzing 2D umbrella sampling windows close to the maximum in free energy. This can be done using an analysis of the Laplace transform of the velocity autocorrelation function based on the generalized Langevin equation for a harmonic oscillator (Crouzy *et al.*, 1994). For each simulation window the following function is evaluated

$$D(s) = \lim_{p \to 0} \frac{-\hat{C}(p; s)\langle \delta s^2 \rangle_{(i)} \langle \dot{s}^2 \rangle_{(i)}}{\hat{C}(p; s)[p\langle \delta s^2 \rangle_{(i)} + \langle \dot{s}^2 \rangle_{(i)}/p] - \langle \delta s^2 \rangle_{(i)} \langle \dot{s}^2 \rangle_{(i)}}, \qquad (13)$$

where $C(t; s)$ is the velocity autocorrelation function for the side chain relative position z, and $\hat{C}(p; s)$ is the Laplace transform of this time function into variable, p. To estimate the value of the limit as $p \to 0$, linear extrapolation from the range $15 \leqslant p \leqslant 35$ was used. The results are shown in Fig. 11. The value of the diffusion constant close to the free energy barrier in the s-direction was determined to be 0.17 Å^2/ps. We note that the s-diffusion is a maximum near this point (outward directed side chain) and drops to as low as 0.10 Å^2/ps for downward and upward directed rotamers. Using Eq. (12) the flip rate was found to be 0.012–0.13 ns^{-1}, corresponding to a flip once every 10–100 ns and explaining the need for 2D biased simulations.

Having sampled an equilibrium distribution of both helix position and side chain orientation, one may integrate over the secondary variable, s, to reveal the 1D PMF, $W(z)$,

$$e^{-W(z)/kT} = C \int ds\, e^{-W(z,s)/kT}, \qquad (14)$$

giving rise to the fairly symmetric 1D PMF of Fig. 10B (solid curve). Though there are indications in Fig. 10A of regions of high free energy outside the dashed box where the side chain will be directed away from the closest interface, these would contribute little to the final 1D PMF. Thus, as a result of careful sampling, utilizing a combination of 1D umbrella sampling with biased z-coordinate of the protein, and also with biased side chain simulations when near the membrane center, the PMF is well converged and is almost symmetric on either side of the membrane.

D. Comparison to the Side Chain Analog

We compare our results for Arg on the helix to the PMF for the side chain analog, obtained from 5 ns of umbrella sampling following 4 ns equilibration per window, shown as a dashed curve in Fig. 10B (adapted from Ref. (Li *et al.*, 2008)). The shape of this PMF is very different. Most notably, the barrier is higher by ~4 kcal/mol. Secondly, binding to the membrane interface, presumably a result of better coordination there, is evident. The helix has been seen to reduce the number of solvating head groups at the interface. But the helix also offers improved shielding from the low dielectric hydrocarbon inside the membrane, while at the same time lowering the overall dielectric properties of the side chain environment in the bulk water. The absence of the helix can therefore explain the increased barrier height in the analog PMF. We must also consider the fact that as the Arg side chain is moved into the membrane, a Leu side chain on the helix is displaced from the membrane into aqueous solution. Given the non-polarity of Leu and the free energy cost it would experience if moved from hydrocarbon to water, of ~5 kcal/mol, this would contribute to the helix PMF being higher in magnitude than the analog calculation (McCallum *et al.*, 2007). There is also a difference (around 2 kcal/mol (Vorobyov *et al.*, 2008)) between methyl-guanidinium and the actual side chain, propyl-guanidinium, due to the reduced length of the hydrophobic chain, counteracting the Leu reference free energy. The comparison of helix and analog calculations is therefore complicated by several contributing, and opposing factors.

We note that McCallum *et al.* (2007) have found a lower barrier by several kcal/mol, that may be mostly accounted for by the above factors. But there may also be some differences due to the choice of force field (McCallum *et al.* using a combination of OPLS amino acids (Jorgensen *et al.*, 1996), SPC water (Berendsen *et al.*, 1981) and united atom GROMACS lipids (Berger *et al.*, 1997)). We note that even for analog simulations caution is required when in sampling the interfacial connections close to the middle of the bilayer. Again we expect that if the analog molecule is close to the membrane center, it has a similar probability of contacting both interfaces. However, in multi-ns MD simulations, breaking and reforming of these connections with opposite interfaces is not seen. This means that on either side of the membrane center the analog molecule will experience an excess force toward the closest interface, and the PMF will climb artificially. In test simulations we have observed that connections to the opposite interface are fairly stable within a few Å of the membrane center, requiring special efforts to sample the equilibrium distribution of interfacial connections (Li *et al.*, 2008). Despite this concern about the shape and precise barrier magnitude close to the membrane center, all of these calculations are in agreement that the barrier for Arg movement across the membrane is indeed large. The membrane cannot match the bulk aqueous phase Arg hydration free energy of −62 kcal/mol (Vorobyov *et al.*, 2008).

E. Contributions to the Free Energy

It has been shown that the charged side chain is coordinated by at least one lipid head group and 4–5 water molecules, even at the center of the membrane. The side chain experiences strong attractive interactions with lipid phosphate moieties and water molecules of magnitude ∼100 and 50 kcal/mol, respectively. This provides a large driving force to pull polar groups into the core of the membrane. However, we have also seen that when the side chain is near the center of the membrane the free energy is at a maximum. This illustrates that strong attractive interactions do not demonstrate thermodynamic stability. To understand the determinants of stability we require contributions to the free energy which will be made up of a wide range of direct interactions with the protein, indirect interactions involving the membrane and electrolyte solution and, of course, entropic contributions.

One major advantage of PMF calculation is the ability to decompose the mean force. Such a decomposition has been used to reveal the roles played by membrane components as a function of depth through the bilayer (Dorairaj and Allen, 2007). We may invoke the relationship between the PMF, $W(z)$, and the mean force (Eq. (5)) to decompose the PMF into contributions, α (Roux and Karplus, 1991),

$$W_\alpha(z) = W_\alpha(z') - \int_{z'}^{z} d\zeta \, \langle F_\alpha(\zeta) \rangle, \tag{15}$$

where F_α is the instantaneous force acting directly on the helix center of mass (COM) due to component α. We choose the free energy to be zero at the bulk reference, $W(z') = 0$. The sum of all direct interactions with the protein leads to a total force on the protein, which when integrated results in the total free energy profile

$$W(z) = \sum_\alpha W_\alpha(z) = -\int_{z'}^{z} d\zeta \left\langle \sum_\alpha F_\alpha(\zeta) \right\rangle = -\int_{z'}^{z} d\zeta \langle F(\zeta) \rangle. \tag{16}$$

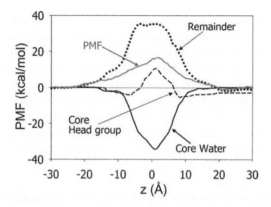

FIGURE 12 Free energy contributions from water (solid) and head groups (dashed) drawn into the bilayer core (adapted from (Dorairaj and Allen, 2007)). The gray curve is the total PMF and the dotted curve is the remainder after subtracting core group contributions.

One can also make use of the additional 2D windows for averaging the mean force due to particular interactions. A 2D histogram of mean forces was formed on the same grid as the 2D PMF, $\langle F_\alpha(z, s) \rangle$, and then the second coordinate was integrated over

$$\langle F_\alpha(z) \rangle = \frac{\int ds \, \langle F_\alpha(z, s) \rangle e^{-W(z,s)/kT}}{\int ds \, e^{-W(z,s)/kT}}. \tag{17}$$

These new mean force values can then be integrated to find a free energy contribution, via Eq. (16), which has been combined with data from 1D umbrella sampling analysis.

While any contributions to the direct interaction with the protein can, in principle, be determined, we consider here the contributions from polar (lipid head groups and water molecules) and ionic components that have entered within the core of the membrane only. It is the presence of these components inside the membrane core that results in a different physical situation to a uniform slab calculation, as usually imagined, and should explain the origins of the PMF in Fig. 10. In particular it will reveal how the perturbed membrane is helping to stabilize the charged side chain inside the membrane. The membrane core is defined as $|z| \leqslant 13$ Å and molecules are considered inside the core based on water O, lipid P atoms and ion positions. Figure 12 shows the core-located water and head group contributions to the PMF. Contributions from ions are fairly small and have not been plotted here. Each vanishes outside the membrane core due to exclusion of interactions with interfacially located groups. Core-located water and head groups stabilize the helix by as much as ∼35 and 5 kcal/mol, respectively. Notably, the core-water contribution becomes stronger and more stabilizing the deeper into the membrane core the side chain moves, being −35 kcal/mol at the bilayer center.

This is to be expected as a result of increased water penetration with side chain depth. In addition, it has been previously observed that water molecules have the ability to effectively stabilize charges away from the aqueous phase, as a result of their large dipole moments (Allen *et al.*, 2004b). However, quite surprisingly, despite the side chain remaining coordinated by 1 phosphate group at the bilayer center, the core-head groups contribute ∼10 kcal/mol less near the membrane center relative to the outer core (although they do stabilize by several kcal/mol closer to the interface). In fact, the head groups that enter deep into the core of the membrane actually cause a destabilization of the protein, with a slightly positive free energy contribution near the middle of the bilayer (though some stabilization is caused closer to the interface). How can this be understood given the strongly attractive interactions between the side chain and the lipid head groups seen? To begin, there is some weakening of side chain–head group interactions at the center relative to the interface. This can be understood because the interactions are so strong that the lipid head group must be pulled into the bilayer core, but there are large costs to do so as this places strains on the bilayer. These strains add to the total potential energy, U_{total}, which participates in the Boltzmann sum to give the average force

$$\langle F_\alpha(z) \rangle = \frac{\int d^{3N-1}\mathbf{R} F_\alpha(z) e^{-U_{total}(\mathbf{R},z)/kT}}{\int d^{3N-1}\mathbf{R} e^{-U_{total}(\mathbf{R},z)/kT}}, \tag{18}$$

i.e. configurations with stronger interaction energies are likely to correspond to larger strains and thus will be less likely in the ensemble. Thus, those bilayer strains will also contribute to the direct free energies themselves, and even if the Arg-head group interaction is fairly constant, bilayer strains lead to a destabilizing contribution to the PMF. On the other hand, we expect that water molecules can effectively stabilize the charge with less indirect cost.

The remainder curve of Fig. 12 represents everything else that contributes to the PMF: i.e. the bulk electrolyte and unperturbed lipid bilayer (without the penetrating polar groups). Therefore, if one subtracts all polar contributions from within the core, one is left with the Born dehydration free energy for a uniform slab membrane. This barrier is ∼35 kcal/mol from this analysis. This is consistent with our earlier expectations based on continuum calculations, where we indicated that the dielectric barrier for dehydration would be 10–30 kcal/mol for the dominant rotamer, when we take into account additional interactions with the membrane interfaces that give rise to a dipolar potential. Based on calculated dipolar potentials of 600–800 mV (Smondyrev and Berkowitz, 1999; Stern and Feller, 2003), this interfacial interaction would add a further 14–19 kcal/mol. Thus, a continuum dielectric calculation, plus interfacial interaction, amounts to a barrier of ∼25–50 kcal/mol, consistent with that observed in these fully atomistic free energy simulations. The fact that the barrier in the overall PMF is 17 kcal/mol is a result of lipid and water PMF contributions being insufficient to overcome this substantial electrostatic barrier presented by the membrane.

While Arg–head group interaction does experience a reduction in strength near the bilayer center, overall it is maintained in the vicinity of -100 kcal/mol (in the helix model) across the membrane, to within the fluctuations from the 1D analysis reported in Fig. 6. Yet, the mean force decomposition of Fig. 12 revealed a destabilizing contribution from this interaction, suggesting that indirect strains on the bilayer may play an important role. The hidden costs associated with indirect interactions (lipid–lipid, lipid–water, water–water, etc.) cannot be computed via a direct decomposition of the mean force (Eq. (15)), because only a sum of direct interactions can be integrated to give the overall PMF. Theoretically we may estimate the change in free energy by considering the perturbation associated with removing that indirect energy cost, U_{strain}, to the full Hamiltonian

$$W_{strain}(z) = -\left[W_{total-strain}(z) - W_{total}(z) \right]$$

$$= -\left[-kT \ln \frac{\int d^{3N-1}\mathbf{R}\, e^{+U_{strain}(\mathbf{R},z)/kT}\, e^{-U_{total}(\mathbf{R},z)/kT}}{\int d^{3N-1}\mathbf{R}\, e^{-U_{total}(\mathbf{R},z)/kT}} \right]$$

$$= kT \ln \langle e^{+U_{strain}(\mathbf{R},z)/kT} \rangle_{total}, \tag{19}$$

where the subscript (total) means that the ensemble average would be computed using simulations that include this strain cost. This indicates that, theoretically, it is possible for a contribution from an indirect interaction to be computed from our MD simulations by averaging the Boltzmann factor of the strain potential energy. This strain energy involves a large number of terms, including the dehydration of the head group pulled into the core and strains within the hydrocarbon tails. Such a Boltzmann average calculation would be difficult to converge due to the large positive strain contributions. However, for illustration only, we can consider a cumulant expansion for the random variable, U_{strain},

$$W_{strain}(z) = +kT \ln \langle e^{+U_{strain}(\mathbf{R},z)/kT} \rangle_{total}$$

$$= \langle U_{strain}(\mathbf{R}, z) \rangle_{total} + \frac{1}{2kT} \langle U_{strain}^2(\mathbf{R}, z) \rangle_{total} + \cdots, \tag{20}$$

where we observe that the strain energy on the bilayer does add to the PMF (first term in the series), despite the absence of direct iteractions with the protein. Therefore, even if the protein-head group interaction energy were to be fairly constant across the membrane, the average force resulting from strain penalties would lead to an increase in free energy when approaching the bilayer center. To converge Eq. (20) would be, however, quite difficult in practice.

To appreciate the magnitudes of the strain energies, we have carried out a short energetic analysis on one system, the side chain analog (Li *et al.*, 2008). Figure 13A shows indirect energy contributions from energy calculations using PME electrostatics on parts of the system. It reveals that the energy of the lipid molecules themselves (dotted curve) (excluding interactions with other molecules)

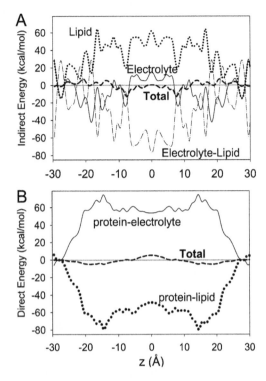

FIGURE 13 Energy analysis revealing contributions from indirect interactions and direct interactions with the methyl-guanidinium (referred to as a protein in the figure). Indirect interactions of
lipid (dotted), water + ions (electrolyte, solid) and lipid–electrolyte interaction (dash-dot) are shown
in (A), with the total plotted as a dashed curve. Direct interactions with electrolyte (solid) and lipid
(dotted) are shown in (B), with the total plotted as a dashed curve. Adapted from Ref. (Li *et al.*, 2008).

changes dramatically as the side chain analog crosses the membrane. In fact, this
indirect energy increases by up to 60 kcal/mol with respect to when the analog molecule is in the bulk aqueous solution. The contribution from electrolyte
(water and ions, solid curve) also varies, though not quite to the same extent as
the lipid value. This contribution is fairly noisy in the bulk but then reaches approximately +10 kcal/mol inside the core of the membrane. Yet, the magnitude
of these strains deep inside the membrane, compared to outside, is not evident
in the final PMF of Fig. 10. We see that the lipid–electrolyte contribution to the
energy (dash–dot curve) is balancing much of the destabilizing lipid–lipid and
water–water energy, falling to as low as around −70 kcal/mol near the center of
the bilayer. We can understand this negative energy contribution, relative to the
bulk, because lipid head groups drawn into the core of the membrane interact very
strongly with the small number of water molecules which effectively stabilize the
phosphate, choline and carbonyl moieties. The total indirect energy is plotted as

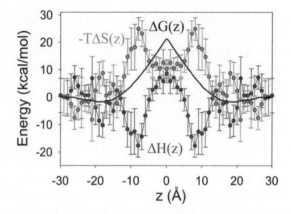

FIGURE 14 Energy (blue) and entropy (red, plotted as $-T\Delta S$) analysis for the side chain analog are shown with standard error of means as error bars. The black curve is the Arg analog PMF from Fig. 10. (For interpretation of the references to color in this figure legend, the reader is referred to the web version of this book.) Adapted from Ref. (Li *et al.*, 2008).

a dashed curve in Fig. 13A which, overall, is fairly flat. While these energies are not expected to be accurate enough to resolve small (kcal/mol order) variations, the overall cancellation of water–water, lipid–lipid with water–lipid is robust. The total indirect energy suggests a small negative contribution inside the membrane relative to bulk, but some destabilization (positive change) near the center of the membrane relative to the outer core ($z \sim \pm 8$ Å) by up to 12 kcal/mol, leading to a significant "strain force".

The direct energies are shown in Fig. 13B as a comparison. The protein-electrolyte contribution (solid curve) is very positive inside the membrane (up to +80 kcal/mol) due to dehydration of the analog molecule. The interactions with lipid molecules are very negative, also reaching nearly −80 kcal/mol, almost exactly canceling the water contributions (the two curves appear to be almost mirror images). It is remarkable how nature manages to overcome energetic gains with alternative sources of stabilization. The total direct interaction energy (dashed curve) is fairly flat as a result of this cancellation, with only a small positive energy at the bilayer center relative to the bulk of around 5 kcal/mol.

While the strain energies are large, there is a lot of cancellation of indirect terms that would lead to smaller penalties. However, there remains still significant changes in the overall free energy profile. We therefore wonder how the changes in ordering of the system due to the inclusion of a solvated charge inside the membrane core might contribute to the PMF. The usual practice for extracting the entropy is to analyze temperature variations in free energy. This would require a major undertaking given the challenges in converging the free energy at just at a single temperature. We instead sought to calculate the average energy and derive, in an approximate way, the entropy. Fig. 14 shows the total system energy for the Arg side chain analog, methyl-guanidinium, as a function of analog position (blue curve). We can see that the energy actually becomes attractive inside the membrane interface and then repulsive at the bilayer center. The variation across the membrane is up to 15 or 20 kcal/mol. We have determined that the pressure-volume work contribution to the enthalpy, $\Delta H(z)$, is negligible. If we subtract this enthalpy from the total PMF (black curve), $\Delta G(z)$, we obtain the entropic contribution, $-T\Delta S = \Delta G - \Delta H$. The result is shown as a red curve, which reaches as high as $+20$ kcal/mol relative to the bulk, and represents the dominant contribution to the free energy barrier experienced by the Arg analog. Thus, while direct interactions and strain penalties could explain significant variations in free energy within the core of the membrane, entropy is also a significant, and possibly dominant factor in the case of the side chain analog. We can understand this because many freely diffusing water molecules and head groups become immobilized as they solvate the Arg analog deep inside the hydrophobic core, leading to an entropy decrease.

F. Accuracy of Simulations

We will now establish some measure of the accuracy of the free energy results reported here by examining artifacts arising from the system setup and also limitations in the classical force fields used. The finite membrane size used in simulations introduces an artifact in the PMF (due to interactions of the protein with its periodic images changing as the side chain moves from bulk electrolyte into the membrane) (Hunenberger and McCammon, 1999; Dorairaj and Allen, 2007; Allen *et al.*, 2004b). The correction to account for the finite membrane size was obtained by MD-averaged solutions to Possion's equation, with explicit incorporation of all water and lipid contents of the hydrophobic core. This correction is small, with a maximum magnitude of ~ 0.2 kcal/mol at the membrane center (data not shown). This is similar to previous observations of small corrections for a similar size membrane (Allen *et al.*, 2006b).

In addition to the membrane being of finite size, we have chosen to simulate with constant normal pressure, fixing the area of the membrane patch. This was done to ensure that the lipid molecules maintain approximately the right area/lipid and hydrocarbon tail order (Feller and Pastor, 1999). The MD snapshots shown in Fig. 4 suggest that this could lead to an additional energy cost as many polar molecules are pulled into the bilayer core, without the bilayer patch having the capability of expanding in area as well as thickness. While the appropriate application of tension, and corresponding compressibility modulus, for a simulated membrane patch maybe problematic (Feller and Pastor, 1996), the fluctuation in membrane area and thickness are expected to be fairly large and the energetic costs of small strains should be minimal (Vorobyov *et al.*, 2008). We can also appreciate how small this cost may be based on the uncertainty in the experimental area per lipid (Nagle *et al.*, 1996). In comparison, the molecules that enter the core of the bilayer (10's of waters and an Arg side chain) correspond to only of the order of one percent of the total volume of that membrane patch. We therefore postulate that the artifact due to fixing the area of the membrane can be neglected.

We also question how accurate the empirical force field is for representing interactions for which it was not specifically parameterized. To begin, simulations with a non-polarizable force field, such as CHARMM, rely on the derivation of partial atomic charges to incorporate the overall dielectric response of molecules, despite the absence of an electronic polarizability component. For polar groups the polarizability can be accounted for, approximately, through nuclear reorientations, yet for molecules or groups not possessing a net dipole moment, this contribution is absent. The alkane dielectric constant with the CHARMM non-polarizable force field is almost indistinguishable from 1 (Vorobyov *et al.*, 2005), despite small net dipole moments, in contrast to the experimental value of $\sim$2 (Lide, 1992). In the case of bulk hydrocarbon solvents this effect would be significant. In membrane simulations, however, the charged side chain is shielded from the hydrocarbon due to membrane deformations and it is therefore important to obtain a mesure of the effect. Progress in the development of the polarizable CHARMM force fields based on the classical Drude oscillator formalism (Lamoureux and Roux, 2003; Anisimov *et al.*, 2005) (already available in CHARMM for water (Lamoureux *et al.*, 2003), alkali and halide ions (Lamoureux and Roux, 2006), hydrocarbons (Vorobyov *et al.*, 2005; Lopes *et al.*, 2007) and small neutral compounds (Noskov *et al.*, 2005; Vorobyov *et al.*, 2007)) offers an improved model for the description of electrostatic interactions capable of capturing correct dielectric response even for non-polar systems. Following these methodological advances, we have introduced polarizability to hydrocarbon lipid tails and performed multi-ns MD simulations of membrane protein systems with correct bilayer structure, and have used this Drude oscillator force-field to study the effect of the polarizability of hydrocarbon chains and channel water on ion permeation energetics (Vorobyov *et al.*, 2008). It was shown previously that this artifact can be corrected for either with approximate Poisson solutions, or with the use of a polarizable force field, by adding Drude oscillators to all lipid chain carbon atoms (Allen *et al.*, 2006b), which results in the correct dielectric properties (Vorobyov *et al.*, 2005). Energy changes upon minimization of Drude oscillators were averaged over all umbrella sampling simulation. We estimate that the maximum amplitude of this error is close to zero, with an uncertainty of $\sim$1 kcal/mol (Vorobyov *et al.*, 2008; Dorairaj and Allen, 2007). We have rerun our simulations with a polarizable membrane core to demonstrate that, indeed, the error is less than 1 kcal/mol (Vorobyov *et al.*, 2008).

Electrostatic interactions in the additive empirical force fields, such as non-polarizable CHARMM. are based on Coulombic interactions between fixed partial atomic charges, which are derived to provide a compromise between accurate description of microscopic interactions and bulk properties. The condensed phase properties such as solvation free energies, heats of vaporization and molecular volumes are primary target data in non-bonded parameter optimization of the bio-molecular force fields, whereas quantum mechanical gas phase data are used to provide a relative balance of microscopic interactions (Mackerell, 2004). However, it remains difficult to obtain accurate description for all microscopic interactions and bulk properties simultaneously within the non-polarizable framework. We therefore wish to ascertain the accuracy of the CHARMM force field for representing protein–lipid and water interactions.

First, we need to determine how accurate is the bulk solvation of the Arg side chain. In the absence of any gas phase interaction or condensed phase solvation free energy experiments for the charged (protonated) Arg side chain (though one may be estimated indirectly (Vorobyov *et al.*, 2008)), we compare the Arg-water hydration free energy value of approximately −62 kcal/mol (Vorobyov *et al.*, 2008; Deng and Roux, 2004) to the *ab initio*/implicit solvent value of −63 kcal/mol calculated by Nilsson and coworkers (Norberg *et al.*, 2005). This is also fairly consistent among popular force fields (Deng and Roux, 2004). Thus we have confidence that CHARMM Arg hydration is accurate.

Next, we need to estimate how well CHARMM represents interactions between protein, lipid and water in far from bulk aqueous environments, such as a charged side chain surrounded by head groups and water inside a membrane core. To ascertain the accuracy of these interactions we have carried out a series of *ab initio* calculations. We have computed the CHARMM interaction energy between the Arg side chain, the phospholipid head group and hydration water molecules, and compared with MP2/6-31+G(d) single-point energy calculations using counterpoise correction (CPC) (Boys and Bernardi, 1970) to account for the basis set super-position error (BSSE). This level of theory was found to be sufficient and the BSSE correction adequate by performing calculations at the MP2 and hybrid DFT PBE0 (Perdew *et al.*, 1996) methods with the aug-cc-pVDZ basis set (Dunning, 1989) for several representative complexes. Complete details have been provided elsewhere (Vorobyov *et al.*, 2008).

We extracted the Arg side chain as a methyl-guanidinium ion and the head group phosphate moiety as a dimethyl-phosphate ion. We included all first shell hydration waters for both groups to create methyl-guanidinium—dimethyl-phosphate hydrated clusters from each frame of MD trajectory. We averaged thousands of representative configurations spanning 6 ns of MD simulation with Arg at different regions of the bilayer (central core, outer core and interface) (Vorobyov *et al.*, 2008). The change in the PMF due to a change in a term in the Hamiltonian will be given by

$$\Delta W(z) = -kT \ln \langle e^{-\Delta U/kT} \rangle_{(MD)}, \qquad (21)$$

where ΔU is the difference in *ab initio* and CHARMM potential energies for each configuration sampled from MD simulation. We found, on average, a 1.6 kcal/

mol overestimation by CHARMM (being too negative). This overestimation by CHARMM alone would suggest that the barrier in the CHARMM PMF could possibly be too low by 1.6 kcal/mol. We recognize, however, that the level of accuracy in the *ab initio* calculations, due to basis set superposition errors, may be a few kcal/mol. Yet we have also shown that the overestimations inside the membrane are fairly constant, varying by less than 1 kcal/mol (Vorobyov *et al.*, 2008) telling us not only that we have a representative configurational sampling of the true ensemble for a bilayer, but that any small errors are likely to have little impact on the shape of the PMFs.

Though the current results represent our best possible effort with available force fields to study the thermodynamics of charged side chain—membrane interactions, there is room for improvement. Future studies should make use of polarizable force fields that have been parameterized to reproduce high-level gasphase QM data for water–Arg, water–phosphate and Arg–phosphate interactions as well as hydration free energies. Such efforts, in the framework of the Drude polarizable CHARMM force field, are currently underway (Alexander D. MacKerell Jr. and Benoit Roux, personal communication).

V. RECONCILING SIMULATION AND EXPERIMENT

We must explain the very different results that have emerged from these atomistic simulations and translocon-based experiments, given the significant discrepancy between 17 kcal/mol and apparent 2.5 kcal/mol free energy barriers. Translocon based measurements emerge from a complex biological system where the actual partitioning environment remains uncharacterized. In particular, the Lep protein used may remain associated with the translocon complex during detection by glycosylation, with polar faces of the H-segment shielded from lipid (Hessa *et al.*, 2005a; Plath *et al.*, 1998). This could lead to an environment for the peptide that is far more hydrophilic and possibly thinner than that of an unperturbed membrane. A further complication is associated with the undetermined protonation states of titratable residues in a membrane environment (Li *et al.*, 2008), in light of recent experimental evidence for significant pK_a reduction in nonaqueous (protein) environments (Cymes *et al.*, 2005).

Despite all of the uncertainties that exist as a result of the complex biological system used, we question the very design of the Lep protein for permitting an interpretation in terms of spatially resolved free energies. In these experiments, the (Lep) was engineered with an H-segment (19 Ala and Leu residues), flanked by Pro residues that prevent α-helix formation, and polar residues that are considered to be excluded from the membrane core (where the polar side chains and backbone would experience a dehydration barrier). We have argued that the polar residues and backbone outside the hydrophobic sequence will be unable to hold (or 'lock in') the helix across the membrane, when charged side chains are placed in the H-segment. In fact, TM helices have been observed to slide when charged side chains are attached using the glycosylation mapping technique (Killian and von Heijne, 2000). We wish to determine how strongly this hydrophobic segment is held in place and if it can withstand the large forces that a charged residue would experience if it were to be incorporated inside the core of the lipid bilayer. To answer this question we must estimate the free energy profile for displacing that helix away from its ideal transmembrane position. This free energy profile will be a function of both the cost of moving the helix (exposing its polar ends to lipid) and the free energies of interactions of the specific sequence in the H-segment with the surrounding membrane.

As we have already established that fully atomistic MD is required to study charged residues inside a membrane, calculating this PMF for the Lep protein with an attached Arg side chain would be a considerable undertaking. We therefore first estimate the PMF of the background Lep protein ($W_{\text{Lep+Leu}}(z)$, i.e., without the charged side chain) to ascertain the strength of the background helix "lock-in" potential, which can be approximated with continuum methods. We employ a Generalized Born (GB) implicit membrane (Im *et al.*, 2003a, 2003b) which adequately captures non-polar and polar contributions to the energetics of protein–membrane interactions. We have chosen a membrane that roughly mimics a DPPC bilayer. However, we have chosen to employ a uniform membrane slab of 30 Å thickness, with switched interfacial regions, which will provide an upper bound for the stiffness of the background helix. An ideal helix of 61 residues, based on the specifications provided in (Hessa *et al.*, 2005a), was placed at regular positions across the membrane such that the position of the central Leu residue ranged from $z = -15$ to 15 Å in 1-Å steps, constrained by harmonic biasing potentials of 2.5 kcal/(mol Å^2). Each of 31 simulations was carried out for 10 ns at 330 K with Langevin dynamics where all non-hydrogen atoms were assigned friction coefficients of 5 ps^{-1}. Figure 15A shows sample configurations after these simulations with the H-segment centered, or translated by ± 5 Å. Though starting configurations in each simulation were helical, they quickly unfolded at the Pro residues and termini, and several folding-unfolding events were observed over 10 ns.

The resulting PMF for the background Lep is drawn as a solid blue curve in Fig. 15B, converged to 0.06 kcal/mol. The background helix lock-in potential certainly does have a minimum at $z = 0$ and is therefore designed to hold the H-segment around the middle of the bilayer. However, the well is not infinitely steep and rises by just a few kcal/mol across the membrane core (less than 5 kcal/mol from $z = 0$ to -10 Å). To explore the robustness of the model, we also carried out calculations for a rigid TM helix with an unfolded peptide outside the H-segment. The rigid model consisted of an α-helical H segment with remaining residues stretched to be fully extended with a series of constrained minimizations. The free energy was approximated by solution to Poisson's equation, plus a non-polar

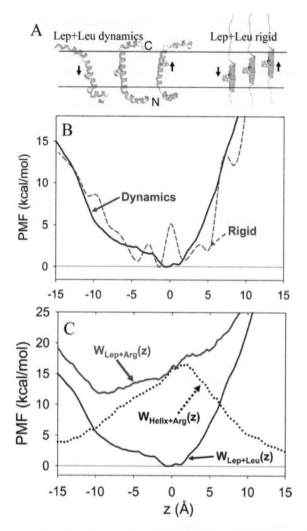

FIGURE 15 Relating to translocon experiments (adapted from (Dorairaj and Allen, 2007)). (A) Sample background helix (Lep + Leu) MD configurations (solid curve) compared to those for a rigid Lep model (dashed curve). (B) Estimates of the PMF for the background protein following 10 ns simulation in an implicit membrane (left), and for a rigid model (right). (C) Background Lep PMF (blue curve), poly-Leu + Arg PMF from Fig. 10 (dotted black curve) and total PMF (Lep + Arg) (red curve). (For interpretation of the references to color in this figure legend, the reader is referred to the web version of this book.)

contribution obtained using an approach similar to that described in (Bernèche *et al.*, 1998), using a sum over the water-accessible surface of all atoms multiplied by the surface tension of 0.033 kcal/(mol Å^2) obtained from alkane-water transfer free energies (Hermann, 1972). The resulting dashed-blue curve in Fig. 15B is remarkably similar to that for the dynamical model.

However, just as a continuum membrane would overestimate the dehydration barrier for a charged residue (by 10's of kcal/mol in the case of Arg), it also will overestimate that for polar backbone and side chains. Moreover, the conservative model membrane used was thicker (30 Å) and of lower dielectric constant (1) than even a purely saturated hydrocarbon core (~2 (Lide, 1992)), chosen to provide an upper limit to the protein's restoring force. Thus the background lock-in potential may actually be considerably weaker than these results suggest. The background helix PMF therefore provides only a small restoring force (~0.5 kcal/(mol Å) or less) for large displacements. Now, if the central residue was to be replaced by an Arg, the large force imparted by the membrane on the charged side chain (~1 kcal/(mol Å)), evident in Fig. 10, suggests that the H-segment could be drawn far from its desired transmembrane position. We can write a relationship between the two free energy profiles (with and without Arg) as

$$W_{\text{Lep+Arg}}(z) \approx W_{\text{Lep+Leu,GB}}(z) + \left\{ W_{\text{Lep+Arg}}(z) - W_{\text{Lep+Leu}}(z) \right\}_{\text{estimate}},$$
(22)

where $W_{\text{Lep+Leu,GB}}$ is the PMF for the Lep H-segment containing a central Leu residue obtained from GB calculations; a reasonable approximation for a non-polar sequence terminated by polar backbone. The term in curly brackets must then be estimated. Let us rewrite it as

$$
\begin{aligned}
&W_{\text{Lep+Arg}}(z) - W_{\text{Lep+Leu}}(z) \\
&= \left(G_{\text{Lep+Arg}}(z) - G_{\text{Lep+Arg}}(z') \right) - \left(G_{\text{Lep+Leu}}(z) - G_{\text{Lep+Leu}}(z') \right) \\
&= \left(G_{\text{Lep+Arg}}(z) - G_{\text{Lep+Leu}}(z) \right) - \left(G_{\text{Lep+Arg}}(z') - G_{\text{Lep+Leu}}(z') \right),
\end{aligned}
$$
(23)

which is the difference between the free energy to transform a central Leu residue into an Arg residue at transmembrane position z and the same mutation free energy at a second transmembrane position z'. We expect that the position within the membrane will have considerable influence on this mutation free energy as the charged side chain interacts with different regions of the lipid bilayer, but how will the contribution from the interaction of the peptide with the side chain change with position? If the peptide were instead the long helix used in our fully-atomistic MD study then this peptide contribution to the mutation free energy would be constant, but will vary a little for the finite Lep H-segment. However, for small changes in transmembrane position we can anticipate that the dominant contribution to Eq. (23) is the interaction with the membrane and this should be fairly helix-independent. Thus,

$$\left\{ W_{\text{Lep+Arg}}(z) - W_{\text{Lep+Leu}}(z) \right\}_{\text{estimate}}$$
$$= \left\{ W_{\text{Helix+Arg}}(z) - W_{\text{Helix+Leu}}(z) \right\}_{\text{MD}}$$
$$+ \left\{ \left(W_{\text{Lep+Arg}}(z) - W_{\text{Lep+Leu}}(z) \right) \right.$$
$$\left. - \left(W_{\text{Helix+Arg}}(z) - W_{\text{Helix+Leu}}(z) \right) \right\}_{\text{correction}}, \tag{24}$$

where W_{Helix} is the PMF for the long poly-Leu helix. To a first approximation the correction term in Eq. (24) maybe assumed negligible. Then, given the long helix is translationally invariant when hosting a central Leu residue, $W_{\text{Helix+Leu}}(z)$ is constant and maybe set to zero, such that

$$\left\{ W_{\text{Lep+Arg}}(z) - W_{\text{Lep+Leu}}(z) \right\}_{\text{estimate}} \approx W_{\text{Helix+Arg,MD}}(z). \tag{25}$$

This function tells us the free energy to transform a Leu into an Arg as a function of position, z, in the membrane and may now be added to the PMF for the background Lep helix, via Eq. (22), to give

$$W_{\text{Lep+Arg}}(z) \approx W_{\text{Lep+Leu,GB}}(z) + W_{\text{Helix+Arg,MD}}(z), \tag{26}$$

where $W_{\text{Helix+Arg,MD}}(z)$ is the fully atomistic PMF for a poly-Leu + Arg helix; in effect transforming a Leu into an Arg side chain. The result is a total PMF (red curve of Fig. 15B) that has a minimum at $z = -9$ Å. While the value of this total free energy cannot be compared to experiment without a reference calculation for a non-transmembrane configuration, this minimum tells us that the H-segment, hosting a central Arg residue, could slide to be accessible to the interface. Owing to the Lep sequence, this slide is toward the N-terminus. We again emphasize that the actual lock-in potential could be considerably less steep leading to even greater sliding. One can therefore expect an Arg placed at any residue in the H-segment to move to an interfacially accessible location and thus attain a similar apparent free energy. This free energy is expected to be dominated by the cost of translating the H-segment away from its perfect transmembrane position (just a few kcal/mol) and not the free energy of the Arg as a function of its depth in the membrane (up to 17 kcal/mol).

This is one possible explanation for the small variation in apparent free energies as a function of Arg position in the H-segment sequence seen experimentally (Hessa *et al.*, 2005a). The inability of Lep to sufficiently anchor a H-segment can be understood in terms of the reported free energy as a function of the location of Pro (the residue chosen to lock-in the TM helix) in the sequence, which varies by just ∼1 kcal/mol (Hessa *et al.*, 2005a), as well as by the fact that Born dehydration barriers for dipoles are less than those for monopoles. Such sliding is consistent with observations made with glycosylation mapping technique (Monne *et al.*, 1998) and does not prevent detection as transmembrane due to the placement of glycosylation sites away from the H-segment. In fact, the sequence of the engineered Lep (of which a sample was given earlier) has one glycosylation site (Asn) 9 residues (corresponding to 13.5 Å of α-helix or up to 30 Å of coil) from the H-segment on the C-terminal side and 135 residues from the H-segment on the N-terminal side (not positioned to detect whether or not the H-segment has shifted on this side).

We stress, however, that there are other possible causes for the low apparent free energies reported experimentally. Because the translocon scale exhibits a good correlation with the octanol scale (Hessa *et al.*, 2005a), one can be led to conclude that this biological scale reports the free energies of association with some fairly hydrophilic region, and not a thick hydrocarbon slab such as that in an isolated membrane. Possibilities include: sliding of H-segments toward the interface; stabilization by other polar protein residues; insertion into a perturbed membrane near the translocon; or even that the H-segments actually reside within the translocon protein when glycosylation occurs. In fact, Rapoport and colleagues have shown that the introduction of charged residues into a TM segment dramatically increases the proximity of the TM segment to the translocon, and with multiple Arg side chains near the middle of the segment only a very small percentage were stably integrated into the membrane (Heinrich *et al.*, 2000). The environment of the peptide within the translocon may be a mixture of hydrophobic and hydrophilic residues. In fact, analysis of the SecY translocon structure (van den Berg *et al.*, 2004) reveals only a thin hydrophobic belt of residues. Such a thin hydrophobic region would be all that is needed to explain small apparent free energies reported, without the protein ever residing in a membrane.

These uncertainties provide an explanation for other translocon experiments that have reported a low +0.5 kcal/mol apparent free energy to insert a single S4 segment of a voltage gated potassium channel in the ER membrane (Hessa *et al.*, 2005b). This sequence contains several Arg side chains which should be unfavorable in a transmembrane orientation (though in this case none of the charged side chains reach the bilayer center). While the free energies of individual Arg side chains on the helix are not likely to be additive (sharing membrane perturbation costs) and the presence of several non-polar residues, such as Leu, will help promote association with the membrane, it seems unlikely that the minimum free energy conformation would be transmembrane with Arg residues exposed to membrane hydrocarbon, as the non-polar residues and charged residues would be more stable if the entire segment were positioned near the interface. Alternatively, if the H-segment were to reside in a thinner hydrophobic region (inside or adjacent to the translocon), this would provide all Arg's with easy snorkeling access to polar interfaces, with small energy penalties as a function of the distance. The positioning inside the translocon or a thin slab might allow for the observation of approximate addition of small individual apparent free energies of each side chain as a function of position to give the overall apparent free energy (Hessa *et al.*, 2005b).

A. The Role of Protonation States

Another likely contributor to the lower free energies reported experimentally maybe the unknown protonation states of ionizable residues. For example, Arg has a pK_a in aqueous solution of 12–13.4 (Nozaki and Tanford, 1967; Angyal and Warburton, 1951), with corresponding deprotonation free energy change of approximately 7 kcal/mol at pH 7, and is usually assumed to be protonated, even in non-aqueous environments. However, in a low-dielectric environment one can appreciate that the relative stability of protonated and deprotonated forms of an amino acid will be shifted dramatically. This has been shown experimentally by Grosman and coworkers, who have used an ion channel, the nicotinic acetycholine receptor, to probe the protonation state of a titratable residue, position by position, by examining the rates of change between distinct ionic conductance states (Cymes et al., 2005). They have shown that the pK_a values of basic side chains drop by as much as 7 units when in a non-polar protein environment. We anticipate that large changes may also be seen in a membrane. This may lead to a balance of contributions from neutral and charged side chain deep inside the membrane, creating a complication when interpreting experiments.

We have undertaken a study to determine the free energy profile of neutral Arg, isolated or attached to a TM helix, across a lipid bilayer (Li et al., 2008), using new parameterizations for the deprotonated form of Arg determined for this purpose. Figure 16A reveals a very different situation where the membrane is essentially unperturbed as the side chain enters the core of the membrane, due to the absence of large perturbing forces. Figure 16B shows results for the neutral PMF, compared to that for the protonated case (taken from (Li et al., 2008)). The PMF was well converged, but is surprisingly challenging owing to slow equilibration and sampling of side chain hydration and hydrogen bond distributions, in the absence of the strong forces provided by the charged residue. The barrier for the neutral species is also high, being approximately 10 kcal/mol. This tells us that while deprotonation can lower the barrier to movement across a membrane, it remains large. The mechanisms are different, however, not involving membrane deformations to help overcome a large dielectric barrier for a charge, but simply involving some loss of hydration and hydrogen bonds inside the core of the membrane. Interestingly, computation of both the protonated and deprotonated Arg free energy profiles allows us to compute a pK_a shift profile across the membrane, via a thermodynamic cycle without having to carry out difficult deprotonation free energy perturbation calculations. It can be seen Fig. 16B that the pK_a will be almost unaffected throughout the membrane due to the stabilizing role of the deformations, but will drop by several units close to the bilayer center of the membrane (to between ~7 and 9 (Li et al., 2008)), consistent with results by Tieleman and coworkers (McCallum et al., 2007). In our studies we have found that the shifts are greater for the analog molecule leading to deprotonation in the bilayer center, owing to the greater energetic variations for the

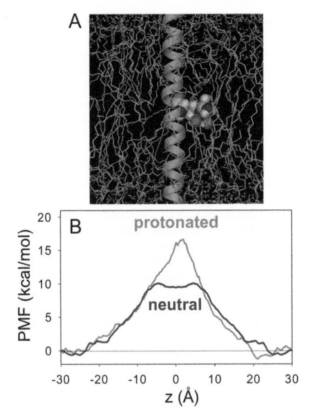

FIGURE 16   Results for deprotonated Arg side chain attached to a helix. (A) System showing no membrane perturbation for the neutral side chain (just 2 coordinating water molecules and no head group rearrangement evident). (B) PMF for neutral Arg (blue solid curve) compared to that for the charged side chain (red solid curve). Results adapted from Ref. (Li *et al.*, 2008).

simple analog molecule (Li *et al.*, 2008). This tell us that Arg, on a TM helix, will remain almost completely charged throughout the membrane at pH 7. Thus the overall barrier presented to Arg will remain like the barrier in Fig. 10 and thus deprotonation is not the main factor determining the low free energies seen experimentally.

VI. CONCLUSION AND IMPLICATIONS

Experiment has provided a wealth of insight into protein structure and function in a membrane bilayer. However, atomistic simulation, combined with statistical mechanical techniques, can assist in uncovering some hidden microscopic mechanisms. Simple model systems have been employed to allow for tractable simulations to elucidate the thermodynamics and mechanisms underlying charged side chain movement in membranes. While idealized, the use of analogs and model TM segments reduces the size and complexity of the problem and, most importantly, makes the microscopic mechanisms of interest more general and transparent.

Employing fully atomistic, explicit solvent simulations we have determined that the barrier to charged side chain movement across a membrane is large, being 17 kcal/mol for Arg on a TM helix. The shape of this free energy profile has been explained in terms of how the lipid bilayer responds to the perturbing field around a charged side chain, leading to coordination by water and lipid head groups, even at the center of the hydrophobic core of the membrane. This feature was missing in previous theoretical models, and, as a result, the shape of the free energy profile obtained with fully atomistic simulation is very different to that estimated previously with continuum approaches. Indeed the charged side chain interacts favorably with water and head groups throughout the core of the membrane, but not sufficiently to overcome the large dehydration barrier presented by the membrane and bilayer strain costs. A decomposition of the free energy revealed that, at the membrane center, only water effectively stabilizes the protein (by -35 kcal/mol). In contrast, lipid head groups drawn deep into the core present an overall destabilizing effect, explaining the shape of the PMF, which rises steeply when the side chain cannot snorkel without lipid rearrangement. We have also explored the indirect membrane strains and entropic costs of moving a protonated Arg side chain through the membrane to help explain the magnitude and shape of the barrier.

Simulation of a TM helix was found to create difficulties owing to the slow side chain reorientations close to the bilayer center. This required the use of additional 2D simulations, biasing both helix positions and side chain orientation. This approach overcame sampling difficulties and also revealed long timescales for side chain flipping events, prohibiting good sampling without additional bias. We have found that the simple side chain analog calculation also provides a large overall barrier and similar bilayer deformations. However, the free energy profile emerging from this simplified system is quite different in shape and magnitude. In a recent perspectives article (McCallum *et al.*, 2007) Tieleman and coworkers have investigated side chain analog partitioning into membranes and have also obtained a large barrier to the motion of Arg. The overall barrier is fairly consistent with that obtained here for the analog and side chain on the helix. However, the barrier is lower, by several kcal/mol, which we understand in terms of the solvation free energy of the larger propyl analog and differences in force field. While the choice of empirical force field may lead to changes in energetics, we have reported that CHARMM fairly closely reproduces energetics for those interactions. We also continue to explore methods to improve our ability to sample charged side chains deep inside the membrane core.

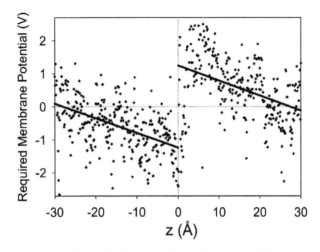

FIGURE 17 Transmembrane voltages required to overcome forces on the protein, fitted with linear regression on either side of the membrane center.

The large barriers calculated via fully atomistic MD simulation are in stark contrast to experimental values derived from biological systems that utilize the translocon in the endoplasmic reticulum to derive a thermodynamic scale for membrane protein insertion (Hessa *et al.*, 2005a, 2005b). We have discussed several possible contributing factors to those experimental results, including association with the translocon complex or perturbation of the membrane around the translocon. While these unknowns are difficult to quantify at present, we have undertaken a study to explore whether or not the very design of those experiments would allow for free energy measurement as a function of spatial position across a membrane. We have shown that the potential responsible for holding the engineered H-segment on the Lep protein in place is not infinitely steep, instead increasing by only a few kcal/mol across the membrane core. It has been estimated that the free energy for an H-segment containing a central Arg would have a minimum far from the bilayer center, corresponding to sliding of the H-segment by nearly 10 Å, allowing the side chain to snorkel to the interface. Thus, even if the peptide were isolated in the membrane, away from the translocon complex, the free energies would not tell us about the stability of Arg residues across the membrane, but instead reveal the strains on the background protein to achieve stabilization by the interface.

The PMF of Fig. 10 suggests that any conformational change of a protein requiring movement of the charged side chain while exposed to lipid would be difficult. In fact, movement from outside the membrane interface to the edge of the hydrophobic core ($|z| \sim 13$ Å) must overcome an average force of ~ 0.5 kcal/(mol Å). In the context of voltage gated ion channel gating, this would require transmembrane potentials of the order of 0.5 V, as shown in Fig. 17. The

PMF is then much steeper deep inside the core of the membrane, with typical forces of 1 kcal/(mol Å), meaning that to push the Arg to the membrane center would require ~1 V. In fact, voltages of over 2 V are required close to the membrane center. These voltages are an order of magnitude greater than physiological values. These results therefore demonstrate that lipid-exposed movement of Arg side chains must be avoided during ion channel gating.

It has been proposed that the Arg residues, contained in each K channel voltage sensing S4 segment, move across the bilayer by 15–20 Å (Jiang *et al.*, 2003; Ruta *et al.*, 2005) upon membrane depolarization, with some necessarily lipid exposed. Here, we have shown that Arg will face a large barrier and can anticipate even further destabilization if more than one Arg were to reside within the membrane core. Moreover, these Arg side chains must remain protonated to explain observed gating charges (Schoppa *et al.*, 1992). Our pK_a calculations (reported in detail elsewhere (Li *et al.*, 2008)) have shown that while deprotonation is more likely in a lipid exposed environment than bulk water, for the most part we can expect Args to remain protonated. Regardless of the protonation states of the Arg, we have shown that the free energy barrier presented to the side chains during movement through the bilayer would remain large; the dehydration of the neutral species as it enters the membrane (without the flexible bilayer deformations seen for the protonated case) leading to a considerable energy cost, with an overall free energy barrier that is comparable to the charged species.

MD simulations of an S4 segment of the KvAP channel in a bilayer (Freites *et al.*, 2005) have suggested that S4 is stable as a TM helix. The reasoning behind this was lack of large displacement over 16 ns of unbiased simulation, combined with attractive interactions with lipid head groups and water. However, we have shown that while strongly attractive interactions occur at the middle of the bilayer, this corresponds to the maximum in free energy. This clearly demonstrates that energies, and in particular side chain interaction energies alone, do not determine thermodynamic stability. However, we recognize that the system is different and that there would be other factors involved with S4 which contains several charged side chains contacting both interfaces of the membrane. The interaction of multiple Arg side chains with opposite interfaces (seemingly with no direction to slide) is likely to create a metastable state. We anticipate, however, that the minimum in free energy would correspond to a situation where the helix would reside near one interface with all lipid-exposed Arg residues hydrated.

Recent coarse grained MD studies have also shown that isolated S4 helices may spontaneously insert across lipid bilayers (Bond and Sansom, 2007). They have reported almost equal preference of the voltage sensor helix containing several Arg residues to adopt either transmembrane or interfacial bound orientation with respect to lipid bilayer, in contradiction with our theoretical calculations and fully-atomistic free energy simulations. The coarse grained force field used was developed by Marrink *et al.* (2004) for semi-quantitative simulations of lipid assemblies and was later adopted and extended by Bond and Sansom (2006) for protein simulations. This model is very simplistic; providing on average 4 to 1 mapping of non-hydrogen atoms to one interaction center and encompassing only four different main atom types. The model leads to gains of up to 3–4 orders of magnitude in the computational efficiency while providing good agreement with experimental structural and dynamic properties of lipid assemblies (Marrink *et al.*, 2004). However, it is not clear how well this force field can reproduce energetics of protein–lipid and protein–solvent interactions. In Ref. (Vorobyov *et al.*, 2008) we compare thermodynamics for Arg side chain solvation and membrane insertion to that from atomistic CHARMM simulations, revealing that the coarse grained force field, which incorporates all hydration energetics artificially into Lennard-Jones interaction terms, is almost completely missing the large energetic costs of dehydration. This results in greatly diminished free energy costs of bilayer insertion and leads us to conclude that such simplified descriptions (at least with their existing parameterizations) are unsuitable for studies of charges within membranes.

Fully atomistic MD free energy calculations therefore suggest that we must return to a more conventional picture of gating that protects charges from the lipid hydrocarbon (Hille, 2001; Ahern and Horn, 2004), or to models that require modest Arg motions across a focused electric field (Cha *et al.*, 1999; Chanda *et al.*, 2005), supported by electron paramagnetic resonance and resonance energy transfer experiments (Cuello *et al.*, 2004; Posson *et al.*, 2005). Recent simulations of the Kv1.2 Shaker (Treptow and Tarek, 2006; Jogini and Roux, 2007) and KvAP (Freites *et al.*, 2006) channels have also revealed evidence of channel-like water crevices amidst the voltage sensing segments that would effectively thin the membrane and could lead to focusing of electric fields. As a result of this water penetration, none of the four Arg side chains are exposed to lipid hydrocarbon tails (as was suggested by the original crystal structure). Focusing of the electric field then allows for smaller motions to explain the observed gating charges. We note that even with thinning the membrane to just ~15 Å (Jogini and Roux, 2007; Freites *et al.*, 2005), physiological potentials likely remain an order of magnitude too small to overcome the forces associated with the movement of lipid exposed Arg side chains. It is therefore essential that gating movements are such that Arg side chains never become exposed to lipid hydrocarbon tails, but instead remain protected within the confines of water crevices or protein.

While the full ion channel structure may undergo a more complicated mechanism of charge movement in a complex environment, idealized model systems have helped us to understand the underlying mechanisms of charged protein side chain insertion into, and movement across biological membranes. These results also have implications for understanding a wide range of biological phenomena including the actions of membrane proteins, antibiotic and viral peptides, toxins, hormone receptors , pharmacological agents and biomembrane nanodevices.

Acknowledgements
We thank Benoit Roux, Tom Rapoport, Nir Ben Tal, Peter Tieleman, Olaf Andersen, and Roger Koeppe for interesting discussions. Some simulations were carried out on Pittsburgh Supercomputing facilities, allocation MCB050002, and this work was supported by NSF Career award MCB-0546768.

References

Ahern, C.A., Horn, R. (2004). Stirring up controversy with a voltage sensor paddle. *Trends Neurosci.* **27**, 303–307.

Allen, T.W., Andersen, O.S., Roux, B. (2003). The structure of gramicidin A in a lipid bilayer environment determined using molecular dynamics simulations and solid-state NMR data. *J. Am. Chem. Soc.* **125**, 9868–9877.

Allen, T.W., Andersen, O.S., Roux, B. (2004a). On the importance of flexibility in studies of ion permeation. *J. Gen. Physiol.* **124**, 679–690.

Allen, T.W., Andersen, O.S., Roux, B. (2004b). Energetics of ion conduction through the gramicidin channel. *Proc. Natl. Acad. Sci.* **101**, 117–122.

Allen, T.W., Andersen, O.S., Roux, B. (2006a). Molecular dynamics potential of mean force calculations as a tool for understanding ion permeation and selectivity in narrow channels. *Biophys. Chem.* **124**, 251–267.

Allen, T.W., Andersen, O.S., Roux, B. (2006b). Ion permeation through a narrow channel: Using gramicidin to ascertain all-atom molecular dynamics potential of mean force methodology and biomolecular force fields. *Biophys. J.* **90**, 3447–3468.

Andersen, O.S., Koeppe, R.E. (1992). Molecular determinants of channel function. *Physiol. Rev.* **72**, S89–S158.

Angyal, S.J., Warburton, W.K. (1951). The basic strengths of methylated guanidines. *J. Chem. Soc.* **1951**, 2492–2494.

Anisimov, V.M., Lamoureux, G., Vorobyov, I.V., Huang, N., Roux, B., MacKerell, A.D. (2005). Determination of electrostatic parameters for a polarizable force field based on the classical drude oscillator. *J. Chem. Theor. Comput.* **1**, 153–168.

Armstrong, C.M., Bezanilla, F. (1973). Currents related to movement of the gating particles of the sodium channels. *Nature* **242**, 459–461.

Ashkin, A. (2000). History of optical trapping and manipulation of small-neutral particle, atoms, and molecules. *IEEE J. Sel. Top. Quant. Electr.* **6**, 841–856.

Beglov, D., Roux, B. (1994). Finite representation of an infinite bulk system: Solvent boundary potential for computer simulations. *J. Chem. Phys.* **100**, 9050–9063.

Berendsen, H.J.C., Postma, J.P.M., van Gunsteren, W.F., Hermans, J. (1981). Interaction models for water in relation to proteins hydration. In: "Intermolecular Forces". Reidel, Dordrecht, pp. 331–342.

Berger, O., Edholm, O., Jahnig, F. (1997). Molecular dynamics simulations of a fluid bilayer of dipalmitoylphosphatidylcholine at full hydration, constant pressure, and constant temperature. *Biophys. J.* **72**, 2002–2013.

Bernèche, S., Nina, M., Roux, B. (1998). Molecular dynamics simulation of melittin in a dimyristoylphosphatidylcholine bilayer membrane. *Biophys. J.* **75**, 1603–1618.

Binning, G., Rohrer, H., Geber, C., Weibel, E. (1982). Surface studies by scanning tunneling microscopy. *Phys. Rev. Lett.* **49**, 57–61.

Block, S.M. (1992). Making light work with optical tweezers. *Nature* **360**, 493–495.

Bond, P.J., Sansom, M.S.P. (2006). Insertion and assembly of membrane proteins via simulation. *J. Am. Chem. Soc.* **128**, 2697–2704.

Bond, P.J., Sansom, M.S.P. (2007). Bilayer deformation by the Kv channel voltage sensor domain revealed by self-assembly simulations. *Proc. Natl. Acad. Sci. USA* **104**, 2631–2636.

Born, M. (1920). Volumen und hydratationswarme der ionen. *Z. Phys.* **1**, 45.

Boys, S.F., Bernardi, F. (1970). The calculation of small molecular interactions by the differences of separate total energies. Some procedures with reduced errors. *Mol. Phys.* **19**, 553–566.

Brooks, B.R., Bruccoleri, R.E., Olafson, B.D., States, D.J., Swaminathan, S., Karplus, M. (1983). CHARMM: A program for macromolecular energy minimization and dynamics calculations. *J. Comput. Chem.* **4**, 187–217.

Brown, K.L., Hancock, R.E. (2006). Cationic host defense (antimicrobial) peptides. *Curr. Opin. Immunol.* **18**, 24–30.

Cha, A., Snyder, G.E., Selvin, P.R., Bezanilla, F. (1999). Atomic scale movement of the voltage-sensing region in a potassium channel measured via spectroscopy. *Nature* **402**, 809–813.

Chanda, B., Asamoah, O.K., Blunck, R., Roux, B., Bezanilla, F. (2005). Gating charge displacement in voltage-gated ion channels involves limited transmembrane movement. *Nature* **436**, 852–856.

Chen, B., Siepmann, J.I. (2006). Microscopic structure and solvation in dry and wet octanol. *J. Phys. Chem. B* **110**, 3555–3563.

Chiu, S.W., Clark, M.M., Jakobsson, E., Subramaniam, S., Scott, H.L. (1999). Application of a combined Monte Carlo and molecular dynamics method to the simulation of a dipalmitoyl phosphatidylcholine lipid bilayer. *J. Comput. Chem.* **20**, 1153–1164.

Cohn, E.J., Edsall, J.T. (1943). "Proteins, Amino Acids, and Peptides as Ions and Dipolar Ions". Reinhold, New York.

Crouzy, S., Woolf, T.B., Roux, B. (1994). A molecular dynamics study of gating in dioxolane-linked gramicidin a channels. *Biophys. J.* **67**, 1370–1386.

Cuello, L.G., Romero, J.G., Cortes, D.M., Perozo, E. (1998). pH-dependent gating in the streptomyces lividans K^+ channel. *Biochemistry* **37**, 3229–3236.

Cuello, L.G., Cortes, D.M., Perozo, E. (2004). Molecular architecture of the KvAP voltage-dependent K^+ channel in a lipid bilayer. *Science* **306**, 491–495.

Cymes, G.D., Ni, Y., Grosman, C. (2005). Probing ion-channel pores one proton at a time. *Nature* **438**, 975–980.

Darden, T., York, D., Pedersen, L. (1993). Particle mesh Ewald: A $n \log(n)$ method for Ewald sums in large systems. *J. Chem. Phys.* **98**, 10089–10092.

Davis, J.H., Clare, D.M., Hodges, R.S., Bloom, M. (1983). Interaction of a synthetic amphiphilic polypeptide and lipids in a bilayer structure. *Biochemistry* **22**, 5298–5305.

Deng, Y.Q., Roux, B. (2004). Hydration of amino acid side chains: Nonpolar and electrostatic contributions calculated from staged molecular dynamics free energy simulations with explicit water. *J. Phys. Chem. B* **108**, 16567–16576.

Dorairaj, S., Allen, T.W. (2007). On the thermodynamic stability of a charged arginine side chain in a transmembrane helix. *Proc. Natl. Acad. Sci. USA* **104**, 4943–4948.

Doyle, D.A., Cabral, J.M., Pfuetzner, R.A., Kuo, A., Gulbis, J.M., Cohen, S.L., Chait, B.T., MacKinnon, R. (1998). The structure of the potassium channel: Molecular basis of K^+ conduction and selectivity. *Science* **280**, 69–77.

Dunbrack, R.L., Karplus, M. (1993). Backbone-dependent rotamer library for proteins. Application to side-chain prediction. *J. Mol. Biol.* **230**, 543–574.

Dunning, T.H. (1989). Gaussian basis sets for use in correlated molecular calculations. 1. The atoms boron through neon and hydrogen. *J. Chem. Phys.* **90**, 1007–1023.

Elinder, F., Arhem, P., Larsson, H.P. (2001). Localization of the extracellular end of the voltage sensor S4 in a potassium channel. *Biophys. J.* **80**, 1802–1809.

Engelman, D.M., Steitz, T.A., Goldman, A. (1986). Identifying nonpolar transmembrane helices in amino acid sequences of membrane proteins. *Annu. Rev. Biophys. Biophys. Chem.* **15**, 321–353.

Feller, S.E., MacKerel, A.D. (2000). An improved empirical potential energy function for molecular simulations of phospholipids. *J. Phys. Chem. B* **104**, 7510–7515.

Feller, S.E., Pastor, R.W. (1996). On simulating lipid bilayers with an applied surface tension: Periodic boundary conditions and undulations. *Biophys. J.* **71**, 1350–1355.

Feller, S.E., Pastor, R.W. (1999). Constant surface tension simulations of lipid bilayers: The sensitivity of surface areas and compressibilities. *J. Chem. Phys.* **111**, 1281–1287.

Feller, S.E., Zhang, Y.H., Pastor, R.W., Brooks, B.R. (1995). Constant pressure molecular dynamics simulation—the Langevin piston method. *J. Chem. Phys.* **103**, 4613–4621.

Feller, S.E., Venable, R.M., Pastor, R.W. (1997). Computer simulation of a DPPC phospholipid bilayer. Structural changes as a function of molecular surface area. *Langmuir* **13**, 6555–6561.

Freites, J.A., Tobias, D.J., von Heijne, G., White, S.H. (2005). Interface connections of a transmembrane voltage sensor. *Proc. Natl. Acad. Sci. USA* **102**, 15059–15064.

Freites, J.A., Tobias, D.J., White, S.H. (2006). A voltage-sensor water pore. *Biophys. J.* **91**, L90–L92.

Gandhavadi, M., Allende, D., Vidal, A., Simon, S.A., McIntosh, T.J. (2002). Structure, composition, and peptide binding properties of detergent soluble bilayers and detergent resistant rafts. *Biophys. J.* **82**, 1469–1482.

Gennis, R.B. (1989). "Biomembranes: Molecular Structure and Functions". Springer-Verlag, New York, NY.

Grabe, M., Lecar, H., Jan, Y.N., Jan, L.Y. (2004). A quantitative assessment of models for voltage-dependent gating of ion channels. *Proc. Natl. Acad. Sci. USA* **101**, 17640–17645.

Han, X., Bushweller, J.H., Cafiso, D.S., Tamm, L.K. (2001). Membrane structure and fusion-triggering conformational change of the fusion domain from influenza hemagglutinin. *Nat. Struct. Biol.* **8**, 715–720.

Heinrich, S.U., Mothes, W., Brunner, J., Rapoport, T.A. (2000). The sec61p complex mediates the integration of a membrane protein by allowing lipid partitioning of the transmembrane domain. *Cell* **102**, 233–244.

Hermann, R.B. (1972). Theory of hydrophobic bonding. II. The correlation of hydrocarbon solubility in water with solvent cavity surface area. *J. Phys. Chem.* **76**, 2754–2759.

Hessa, T., Kim, H., Bihlmaier, K., Lundin, C., Boekel, J., Andersson, H., Nilsson, I., White, S.H., von Heijne, G. (2005a). Recognition of transmembrane helices by the endoplasmic reticulum translocon. *Nature* **433**, 377–381.

Hessa, T., White, S.H., von Heijne, G. (2005b). Membrane insertion of a potassium-channel voltage sensor. *Science* **307**, 1427.

Hille, B.H. (2001). "Ionic Channels of Excitable Membranes". Sinauer, Sunderland, MA.

Honig, B.L., Hubbell, W.L. (1984). Stability of "Salt Bridges" in membrane proteins. *Proc. Natl. Acad. Sci. USA* **81**, 5412–5416.

Hovmoller, S., Zhou, T., Ohlson, T. (2002). Conformations of amino acids in proteins. *Acta Crystallogr. D* **58**, 768–776.

Huang, H.C., Briggs, J.M. (2002). The association between a negatively charged ligand and the electronegative binding pocket of its receptor. *Biopolymers* **63**, 247–260.

Hunenberger, P.H., McCammon, J.A. (1999). Effect of artificial periodicity in simulations of biomolecules under Ewald boundary conditions: A continuum electrostatics study. *Biophys. Chem.* **78**, 69–88.

Im, W., Feig, M., Brooks, C.L. (2003a). An implicit membrane generalized born theory for the study of structure, stability, and interactions of membrane proteins. *Biophys. J.* **85**, 2900–2918.

Im, W., Lee, M.S., Brooks, C.L. (2003b). Generalized born model with a simple smoothing function. *J. Comput. Chem.* **24**, 1691–1702.

Jiang, Y.X., Ruta, V., Chen, J.Y., Lee, A., MacKinnon, R. (2003). The principle of gating charge movement in a voltage-dependent K^+ channel. *Nature* **423**, 42–48.

Jogini, V., Roux, B. (2007). Dynamics of the Kv1.2 voltage-gated K^+ channel in a membrane environment. *Biophys. J.* **93**, 3070–3082.

Johansson, A.C.V., Lindahl, E. (2006). Amino-acid solvation structure in transmembrane helices from molecular dynamics simulations. *Biophys. J.* **91**, 4450–4463.

Johansson, A., Smith, G.A., Metcalfe, J.C. (1981). The effect of bilayer thickness on the activity of ($Na^+ + K^+$)-atpase. *Biochem. Biophys. Acta* **641**, 416–421.

Jorgensen, W.L., Chandrasekhar, J., Madura, J.D., Impey, R.W., Klein, M.L. (1983). Comparison of simple potential functions for simulating liquid water. *J. Chem. Phys.* **79**, 926–935.

Jorgensen, W.L., Maxwell, D.S., Tirado-Rives, J. (1996). Development and testing of the OPLS all-atom force-field on conformational energetics and properties of organic liquids. *J. Am. Chem. Soc.* **118**, 11225–11236.

Jost, P.C., Hayes-Griffith, O., Capaldi, R.A., Vanderkooi, G. (1973). Evidence for boundary lipid in membranes. *Proc. Natl. Acad. Sci.* **70**, 480–484.

Kalderon, D., Richardson, W.D., Markham, A.F., Smith, A.E. (1984). Sequence requirements for nuclear location of sv40 large-t antigen. *Nature* **311**, 33–38.

Karplus, M. (2002). Molecular dynamics simulations of biomolecules. *Acc. Chem. Res.* **35**, 321–323.

Kessel, A., Ben-Tal, N. (2002). Free energy determinants of peptide association with lipid bilayers. *Curr. Top. Membr.* **52**, 205–253.

Killian, J.A. (2003). Synthetic peptides as models for intrinsic membrane proteins. *FEBS Lett.* **555**, 134–138.

Killian, J.A., von Heijne, G. (2000). How proteins adapt to a membrane–water interface. *Trends Biochem. Sci.* **25**, 429–434.

Killian, J.A., Salemink, I., de Planque, M.R.R., Lindblom, G., Koeppe II, R.E., Greathouse, D.V. (1996). Induction of nonbilayer structures in diacylphosphatidylcholine model membranes by transmembrane alpha-helical peptides: Importance of hydrophobic mismatch and proposed role of tryptophans. *Biochemistry* **35**, 1037–1045.

Kuhn, L.A., Leigh, J.S. (1985). A statistical technique for predicting membrane protein structure. *Biochim. Biophys. Acta* **828**, 351–361.

Kumar, S., Bouzida, D., Swendsen, R.H., Kollman, P.A., Rosenberg, J.M. (1992). The weighted histogram analysis method for free-energy calculations on biomolecules. I. The method. *J. Comput. Chem.* **13**, 1011–1021.

Kyte, J., Doolittle, R.F. (1982). A simple method for displaying the hydropathic character of a protein. *J. Mol. Biol.* **157**, 105–132.

Lamoureux, G., Roux, B. (2003). Modeling induced polarization with classical Drude oscillators: Theory and molecular dynamics simulation algorithm. *J. Chem. Phys.* **119**, 3025–3039.

Lamoureux, G., Roux, B. (2006). Absolute hydration free energy scale for alkali and halide ions established from simulations with a polarizable force field. *J. Phys. Chem. B* **110**, 3308–3322.

Lamoureux, G., MacKerell Jr., A.D., Benoit, R. (2003). A simple polarizable model of water based on classical drude oscillators. *J. Chem. Phys.* **119**, 5185–5197.

Lehtonen, J.Y.A., Kinnnunen, P.K.J. (1995). Poly(ethylene glycol)-induced and temperature-dependent phase-separation in fluid binary phospholipid membranes. *Biophys. J.* **68**, 525–535.

Li, L.B., Vorobyov, I., Allen, T.W. (2008). Potential of mean force and pK_a profile calculation for a lipid membrane-exposed Arg side chain. *J. Phys. Chem. B*, submitted for publication.

Lide, D.R. (Ed.) (1992). "CRC Handbook of Chemistry and Physics". CRC Press, Boston.

Liu, R.N.A.H., Lewis, R.S., Hodges, F., McElhaney, R.N. (2004). Effect of variations in the structure of a polyleucine-based-helical transmembrane peptide on its interaction with phosphatidylethanolamine bilayers. *Biophys. J.* **87**, 2470–2482.

Long, S.B., Campbell, E.B., MacKinnon, R. (2005). Crystal structure of a mammalian voltage-dependent *shaker* family K$^+$ channel. *Science* **309**, 897.

Lopes, P.E.M., Lamoureux, G., Roux, B., MacKerell, A.D. (2007). Polarizable empirical force field for aromatic compounds based on the classical drude oscillator. *J. Phys. Chem. B* **111**, 2873–2885.

Lundbaek, J.A., Bim, P., Girshman, J., Hansen, A.J., Andersen, O.S. (1996). Membrane stiffness and channel function. *Biochemistry* **35**, 3825–3830.

Lundbaek, J.A., Maer, A.M., Andersen, O.S. (1997). Lipid bilayer electrostatic energy, curvature stress and assembly of gramicidin channels. *Biochemistry* **36**, 5695–5701.

Luscombe, N.M., Laskowski, R.A., Thornton, J.M. (2001). Amino acid–base interactions: A three-dimensional analysis of protein–DNA interactions at an atomic level. *Nucleic Acids Res.* **29**, 2860–2874.

MacKerell Jr., A.D. (2004). Empirical force fields for biological macromolecules: Overview and issues. *J. Comput. Chem.* **25**, 1584–1604.

MacKerell Jr., A.D., Bashford, D., Bellot, M., Dunbrack, R.L., Evanseck, J.D., Field, M.J., Fischer, S., Gao, J., Guo, H., Ha, S., Joseph-McCarthy, D., Kuchnir, L., Kuczera, K., Lau, F.T.K., Mattos, C., Michnick, S., Ngo, T., Nguyen, D.T., Prodhom, B., Reiher-III, W.E., Roux, B., Schlenkrich, B., Smith, J., Stote, R., Straub, J., Watanabe, M., Wiorkiewicz-Kuczera, J., Karplus, M. (1998). All-atom empirical potential for molecular modeling and dynamics studies of proteins. *J. Phys. Chem. B* **102**, 3586–3616.

Marrink, S.J., deVries, A.H., Mark, A.E. (2004). Coarse grained model for semiquantitative lipid simulations. *J. Phys. Chem. B* **108**, 750–760.

Marti, J., Csajka, F.S. (2003). Flip-flop dynamics in a model lipid bilayer membrane. *Europhys. Lett.* **61**, 409–414.

Mathies, R.A., Lin, S.W., Ames, J.B., Pollard, W.T. (1991). From femtoseconds to biology: Mechanism of bacteriorhodopsin's light-driven proton pump. *Annu. Rev. Biophys. Biophys. Chem.* **20**, 491–518.

McCallum, J., Bennet, W.F.D., Tieleman, D.P. (2007). Partitioning of amino acid sidechains into lipid bilayers: Results from computer simulations and comparison to experiment. *J. Gen. Physiol.* **129**, 371–377.

Monne, M., Nilsson, I., Johansson, M., Elmhed, N., von Heijne, G. (1998). Positively and negatively charged residues have different effects on the position in the membrane of a model transmembrane helix. *J. Mol. Biol.* **284**, 1177–1183.

Mori, H. (1965). Transport, collective motion, and Brownian motion. *Prog. Theor. Phys.* **33**, 423–455.

Mouritsen, O.G., Bloom, M. (1984). Mattress model of lipid–protein interactions in membranes. *Biophys. J.* **46**, 141–153.

Nagle, J.F. (1993). Area/lipid of bilayers from NMR. *Biophys. J.* **64**, 1476–1481.

Nagle, J.F., Tristram-Nagle, S. (2000). Structure of lipid bilayers. *Biochim. Biophys. Acta* **1469**, 159–195.

Nagle, J.R., Zhang, R., Tristram-Nagle, S., Sun, W., Petrache, H., Suter, R.M. (1996). X-ray structure determination of l, phase DPPC bilayers. *Biophys. J.* **70**, 1419–1431.

Nina, M., Beglov, D., Roux, B. (1997). Atomic radii for continuum electrostatics calculations based on molecular dynamics free energy simulations. *J. Phys. Chem.* **101**, 5239–5248.

Norberg, J., Foloppe, N., Nilsson, L. (2005). Intrinsic relative stabilities of the neutral tautomers of arginine side-chain models. *J. Chem. Theory Comput.* **1**, 986–993.

Noskov, S.Yu., Lamoureux, G., Roux, B. (2005). Molecular dynamics study of hydration in ethanol-water mixtures using a polarizable force field. *J. Phys. Chem. B* **109**, 6705–6713.

Nozaki, Y., Tanford, C. (1967). Examination of titration behavior. *Methods Enzymol.* **11**, 715–734.

Ozdirekcan, D.T.S., Rijkers, R.M.J., Liskamp, S., Killian, J.A. (2005). Influence of flanking residues on tilt and rotation angles of transmembrane peptides in lipid bilayers. A solid state ^{2}H NMR study. *Biochemistry* **44**, 1004–1012.

Pastor, R.W., Venable, R.M., Feller, S.E. (2002). *Acc. Chem. Res.* **35**, 438.

Pauling, L., Corey, R.B. (1951). The pleated sheet, a new layer configuration of polypeptide chains. *Proc. Natl. Acad. Sci.* **37**, 251–256.

Pauling, R.B., Corey, L., Branson, H.R. (1951). The structure of proteins: Two hydrogen-bonded helical configurations of the polypeptide chain. *Proc. Natl. Acad. Sci.* **37**, 205–211.

Perdew, K., Burke, J.P., Ernzerhof, M. (1996). Generalized gradient approximation made simple. *Phys. Rev. Lett.* **77**, 3865–3868.

Plath, K., Mothes, W., Wilkinson, B.M., Stirling, C.J., Rapoport, T.A. (1998). Signal sequence recognition in posttranslational protein transport across the yeast ER membrane. *Cell* **94**, 795–807.

Posson, D.J., Ge, P., Miller, C., Bezanilla, F., Selvin, P.R. (2005). Small vertical movement of a K^+ channel voltage sensor measured with luminescence energy transfer. *Nature* **436**, 848–851.

Radzicka, A., Wolfenden, R. (1988). Comparing the polarities of the amino acids: Side-chain distribution coefficients between the vapor phase, cyclohexane, 1-octanol, and neutral aqueous solution. *Biochemistry* **27**, 1664–1677.

Rao, J.K.M., Argos, P. (1986). A conformational preference parameter to predict helices in integral membrane proteins. *Biochem. Biophys. Acta* **859**, 197–214.

Roux, B., Karplus, M. (1991). Ion transport in a gramicidin-like channel: Structure and thermodynamics. *Biophys. J.* **59**, 961–981.

Ruta, V., Chen, J., MacKinnon, R. (2005). Calibrated measurement of gating-charge arginine displacement in the KvAP voltage-dependent K^+ channel. *Cell* **123**, 463–475.

Ryckaert, J.P., Ciccotti, G., Berendsen, H.J.C. (1977). Numerical integration of the cartesian equation of motions of a system with constraints: Molecular dynamics of *n*-alkanes. *J. Comput. Chem.* **23**, 327–341.

Sangster, J. (Ed.) (1997). "Octanol–Water Partitioning Coefficients: Fundamentals and Physical Chemistry". Wiley, Chichester, UK.

Schoppa, N.E., McCormack, K., Tanouye, M.A., Sigworth, F.J. (1992). The size of gating charge in wild-type and mutant shaker potassium channels. *Science* **255**, 1712–1715.

Scotto, A.W., Zakim, D. (1985). Reconstitution of membrane proteins: Catalysis by cholesterol of insertion of integral membrane proteins into preformed lipid bilayers. *Biochemistry* **25**, 1555–1561.

Seeman, P. (1997). The membrane actions of anesthetics and tranquilizers. *Pharmacol. Rev.* **24**, 583–655.

Segrest, J.P., DeLoof, H., Dohlman, J.G., Brouillette, C.G., Anantharamaiah, G.M. (1990). Amphipathic helix motif: Classes and properties. *Protein Sci.* **8**, 103–117.

Smondyrev, A.M., Berkowitz, M.L. (1999). Structure of dipalmitoylphosphatidylcholine/cholesterol bilayer at low and high cholesterol concentrations: Molecular dynamics simulation. *Biophys. J.* **77**, 2075–2089.

Sperotto, M.M., Mouritsen, O.G. (1993). Lipid enrichment and selectivity of integral membrane proteins in two-component lipid bilayers. *Eur. Biophys. J.* **22**, 323–328.

Stern, H.A., Feller, S.E. (2003). Calculation of the dielectric permittivity profile for a nonuniform system: Application to a lipid bilayer simulation. *J. Chem. Phys.* **118**, 3401–3412.

Strandberg, E., Killian, J.A. (2003). Snorkeling of lysine side chains in transmembrane helices: How easy can it get? *FEBS Lett.* **544**, 69–73.

Tieleman, D.P., Marrink, S.J., Berendsen, H.J.C. (1997). A computer perspective of membranes: Molecular dynamics studies of lipid bilayers systems. *Biochim. Biophys. Acta Rev. Biomembranes* **1331**, 235–270.

Tobias, D.J., Tu, K., Klein, M.L. (1997). Atomic-scale molecular dynamics simulations of lipid membranes. *Curr. Opin. Colloid Interface Sci.* **2**, 15–27.

Torrie, G.M., Valleau, J.P. (1977). Nonphysical sampling distributions in Monte Carlo free-energy estimation: Umbrella sampling. *J. Comput. Phys.* **23**, 187–199.

Treptow, W., Tarek, M. (2006). Environment of the gating charges in the Kv1.2 shaker potassium channel. *Biophys. J. Biophys. Lett.* **106**, 64–66.

van den Berg Jr., B., Clemons, W.M., Collinson, I., Modis, Y., Hartmann, E., Harrison, S.C., Rapoport, T.A. (2004). X-ray structure of a protein-conducting channel. *Nature* **427**, 36–44.

Vorobyov, I.V., Anisimov, V.M., MacKerell, A.D. (2005). Polarizable empirical force field for alkanes based on the classical drude oscillator model. *J. Phys. Chem. B* **109**, 18988–18999.

Vorobyov, I., Anisimov, V.M., Greene, S., Venable, R.M., Moser, A., Pastor, R.W., MacKerell, A.D. (2007). Additive and classical drude polarizable force fields for linear and cyclic ethers. *J. Chem. Theor. Comput.* **3**, 1120–1133.

Vorobyov, I., Li, L.B., Allen, T.W. (2008). Assessing atomistic and coarse-grained force fields for protein–lipid interactions: The formidable challenge of an ionizable side chain in a membrane. *J. Phys. Chem. B*, submitted for publication.

Weiss, T.M., van der Wel, P.C.A., Killian, J.A., Koeppe II, R.E., Huang, H.W. (2003). Hydrophobic mismatch between helices and lipid bilayers. *Biophys. J.* **84**, 379–385.

White, S.H. (2003). Translocons, thermodynamics, and the folding of membrane proteins. *FEBS Lett.* **555**, 116–121.

White, S.H., Wiener, M.C. (1996). The liquid crystallographic structure of fluid lipid bilayer membranes. In: Merz, K.M., Roux, B. (Eds.), "Biological Membranes. A Molecular Perspective from Computation and Experiment". Birkhauser, Boston, pp. 127–144.

White, S.H., von Heijne, G. (2005). Transmembrane helices before, during, and after insertion. *Curr. Opin. Struct. Biol.* **15**, 378–386.

White, S.H., Wimley, W.C. (1998). Hydrophobic interactions of peptides with membrane interfaces. *Biochim. Biophys. Acta* **1376**, 339–352.

Wimley, W.C., White, S.H. (1992). Partitioning of tryptophan side-chain analogs between water and cyclohexane. *Biochemistry* **31**, 12813–12818.

Wimley, W.C., White, S.H. (1993). Membrane partitioning: Distinguishing bilayer effects from the hydrophobic effect. *Biochemistry* **32**, 6307–6312.

Wimley, W.C., White, S.H. (1996). Experimentally determined hydrophobicity scale for proteins at membrane interfaces. *Nat. Struct. Biol.* **3**, 842–848.

Wimley, W.C., Creamer, T.P., White, S.H. (1996). Solvation energies of amino acid side chains and backbone in a family of host-guest pentapeptides. *Biochemistry* **35**, 5109.

Wolfenden, R. (2007). Experimental measures of amino acid hydrophobicity and the topology of transmembrane and globular proteins. *J. Gen. Physiol.* **129**, 357–362.

Wolfenden, R., Andersson, L., Cullis, P.M., Southgate, C.C.B. (1981). Affinities for amino acid side chains for water. *Biochemistry* **20**, 849–855.

Woolf, T.B. (1998). Molecular dynamics simulations of individual alpha-helices of bacteriorhodopsin, in dimyristoylphosphatidylcholine. II. Interaction energy analysis. *Biophys. J.* **74**, 115–131.

Woolf, T., Roux, B. (1996). Structure, energetics and dynamics of lipid–protein interactions: A molecular dynamics study of gramicidin a channel in DMPC bilayer. *Proteins: Struct. Funct. Gen.* **24**, 92–114.

Yang, A.-S., Honig, B. (1992). Electrostatic effects on protein stability. *Curr. Opin. Struct. Biol.* **2**, 40–45.

Yin, D.X., MacKerell, A.D. (1998). Combined ab initio empirical approach for optimization of Lennard-Jones parameters. *J. Comput. Chem.* **19**, 334–348.

Zwanzig, R.W. (2001). "Nonequilibrium Statistical Mechanics". Oxford University Press, New York and Oxford.

Index

Printed and bound by CPI Group (UK) Ltd, Croydon, CR0 4YY

03/10/2024

01040414-0008